Topics in Medicinal Chemistry

Volume 41

Topics in Medicinal Chemistry (TMC) covers all relevant aspects of medicinal chemistry research, e.g. pathobiochemistry of diseases, identification and validation of (emerging) drug targets, structural biology, drugability of targets, drug design approaches, chemogenomics, synthetic chemistry including combinatorial methods, bioorganic chemistry, natural compounds, high-throughput screening, pharmacological in vitro and in vivo investigations, drug-receptor interactions on the molecular level, structure-activity relationships, drug absorption, distribution, metabolism, elimination, toxicology and pharmacogenomics. Drug research requires interdisciplinary team-work at the interface between chemistry, biology and medicine. To fulfil this need, TMC is intended for researchers and experts working in academia and in the pharmaceutical industry, and also for graduates that look for a carefully selected collection of high quality review articles on their respective field of expertise.

Medicinal chemistry is both science and art. The science of medicinal chemistry offers mankind one of its best hopes for improving the quality of life. The art of medicinal chemistry continues to challenge its practitioners with the need for both intuition and experience to discover new drugs. Hence sharing the experience of drug research is uniquely beneficial to the field of medicinal chemistry.

All chapters from Topics in Medicinal Chemistry are published OnlineFirst with an individual DOI. In references, Topics in Medicinal Chemistry is abbreviated as Top Med Chem and cited as a journal.

Vittoria Colotta • Claudiu T. Supuran
Editors

Purinergic Receptors and their Modulators

With contributions by

L. Antonioli · J. K. Awalt · E. Barresi · D. Bassani · L. Benvenuti · M. Buccioni · S. Calenda · C. Capasso · D. Catarzi · E. Cescon · C. Chang · F. Cherchi · V. Colotta · E. Coppi · V. D'Antongiovanni · T. Da Ros · F. Da Settimo · D. Dal Ben · I. Dettori · C. Di Salvo · S. Federico · M. Fornai · B. Francucci · C. Giacomelli · I. Grieco · G. Haskó · K. A. Jacobson · M. Jörg · C. Lambertucci · C. Martini · G. Marucci · L. T. May · S. Moro · A. Nocentini · P. Oliva · C. Payne · F. Pedata · C. Pellegrini · M. Persico · F. Prencipe · A. M. Pugliese · G. Spalluto · A. Spinaci · C. T. Supuran · R. R. Suresh · S. Taliani · M. L. Trincavelli · J. D. A. Tyndall · F. Varano · M. Venturini · A. J. Vernall · E. Vigiani · R. Volpini

Springer

Editors
Vittoria Colotta
Department of Neurofarba, Pharmaceutical and Nutraceutical Section
University of Florence
Florence, Italy

Claudiu T. Supuran
Department of Neurofarba, Pharmaceutical and Nutraceutical Section
University of Florence
Florence, Italy

ISSN 1862-2461 ISSN 1862-247X (electronic)
Topics in Medicinal Chemistry
ISBN 978-3-031-39724-0 ISBN 978-3-031-39725-7 (eBook)
https://doi.org/10.1007/978-3-031-39725-7

This Springer imprint is published by the registered company Springer Nature Switzerland AG
The registered company address is: Gewerbestrasse 11, 6330 Cham, Switzerland

Paper in this product is recyclable.

Preface

Purinergic receptors, termed P1 and P2, are activated, respectively, by adenosine and different nucleotides, such as ATP and ADP. Purinergic signaling plays a key role in human physiological and pathological processes and many selective P1 and P2 ligands have been developed as therapeutics for a variety of diseases, such as thrombosis, cerebral ischemia, dry eye, neurodegenerative diseases, and cancer. Several P1 and P2 receptor modulators entered clinical trials and some have been approved for marketing. In this volume of the book series Topics in Medicinal Chemistry, we propose a series of reviews that provide updated information on purinergic receptors, their modulators, and related therapeutic applications and also give suggestions for possible future direction of research in this area.

The first chapter, by Catarzi et al., is a detailed overview of adenosine and its receptors that helps the reader to enter the intriguing world of this purine nucleoside. Adenosine is an autacoid that locally regulates several tissue functions and also acts as a neuromodulator. It exerts its effects by activating G protein-coupled receptors, termed P1 receptors, and classified into A_1, A_{2A}, A_{2B}, and A_3 subtypes. Under metabolically unfavorable conditions such as tissue hypoxia or stress, adenosine counteracts tissue damage, but its excessive and chronic production can be harmful, leading to fibrosis, inflammation, and other pathological situations. This evidence made soon consider adenosine receptors (ARs) as potential drug targets and stimulated medicinal chemistry studies to produce a large number of AR ligands. The chapter summarizes the most relevant series of selective agonists and antagonists of the four AR subtypes, with a particular emphasis on the compounds employed as pharmacological tools, and as radioligands. A section reviews the effects and usefulness of AR ligands in different therapeutic areas (central nervous system, cardiovascular, inflammatory, and immune system, pain). Particular attention has been paid to binding kinetics and biased signaling of ARs that should be considered in the design of new AR ligands as more effective and safer drugs.

The second chapter, by Federico et al., focuses on the twenty-year history of one of the most fruitful and deeply investigated series of AR antagonists, the pyrazolo [4,3-*e*][1,2,4]triazolo[1,5-*c*]pyrimidines (PTP). This series furnished many potent and selective antagonists for the A_{2A} and A_3 subtypes some of which have been used as molecular probes for the study of structure and function of the A_{2A} and A_3 ARs. Both these AR subtypes have aroused great interest in their role in Parkinson's disease and immune-oncology (A_{2A}) and in cancer, glaucoma, and inflammation (A_3). Structure-activity relationships at all ARs are discussed, pointing out the structural requirements to bind selectively the A_{2A} or the A_{3A} AR subtypes. PTP derivatives have also been taken as lead to the design of new molecular probes and multifunctional molecules. Several PTP-fluorophore conjugates were developed obtaining fluorescent antagonists toward A_{2A} and/or A_3 ARs, whose applications range from binding assays to fluorescence microscopy and receptor detection. PTP-based molecules were also described as dual ligands blocking both A_{2A}AR and histone deacetylase as new potential drugs in cancer immunotherapy.

The third chapter, by Varano et al., addresses particular potential therapeutic applications of AR ligands, i.e. treatment of wound healing and fibrosis. Agonists of the A_{2A} AR, such as CGS21680 and sonedenoson (MRE-0094), accelerated wound closure and healing in animal models by a mechanism that depends on tissue plasminogen activator and angiogenesis promotion. Excessive activation of ARs can lead to matrix overproduction, scarring, and fibrosis of the skin, and organs such as the liver, heart, lungs, and kidneys. Thus, AR antagonists could represent novel therapeutic agents for the treatment of fibrotic diseases. A_{2A} AR antagonists proved to be potentially useful against dermal and hepatic fibrosis, and the A_{2B} AR blockade potentially leads to pulmonary fibrosis control.

The high therapeutic potential demonstrated by A_{2A}AR antagonists has prompted medicinal chemists to design and develop several series of compounds belonging to different chemical classes. The fourth chapter, by Volpini et al., reviews these series, from the first alkyl xanthine compounds to the most potent and selective heterocyclic derivatives unrelated to xanthines. Structure–activity relationship studies are described for the most relevant classes. An overview of the A_{2A}AR antagonists evaluated in clinical trials is provided and one section is dedicated to the representation of published crystallographic structures and molecular modeling studies. Furthermore, the use of A_{2A} AR antagonists as neuroprotective agents and their potential application in the anti-cancer immune therapy is discussed.

In contrast to the A_{2A} AR, the A_{2B} AR has been less investigated due to the scarce availability of truly selective agonists and antagonists of this AR subtype. The fifth chapter, by Cherchi et al, is an overview of recent knowledge acquired on the role of the A_{2B} AR subtype in brain ischemia. The results of different research are critically discussed pointing out the diverse and contradictory roles played by this receptor after ischemia, being harmful or protective, depending on the timing and its cellular localization. The role of the A_{2B}AR in demyelinating conditions, occurring after ischemic injury, has also been reviewed highlighting that the A_{2B}AR antagonists improve remyelination and, consequently, white matter integrity in stroke and demyelinating pathologies.

The A_3AR receptor was the last of the four ARs to be discovered, it was identified by cloning and first described as an orphan receptor. The sixth chapter, by Jacobson et al., focuses on this AR subtype, dealing with the medicinal chemistry of its selective agonists, antagonists, and allosteric modulators, from the early discoveries to the most recent advances. Radiolabeled ligands developed as receptor probes are described, as well as prodrugs of some derivatives and their efficacy. A section reports on the grown number of clinical trials conducted on A_3AR ligands which proved to be therapeutic candidate molecules for a variety of chronic (e.g., psoriasis, liver disease, pain, cancer, glaucoma) and acute (stroke, cardiac ischemia) conditions.

Bifunctional compounds designed to target ARs and comprising bivalent, bitopic, and dual-acting ligands are reviewed in Chap. 7, by Vernall et al. The beginning of the chapter provides a clear description of the bifunctional ligands. In the last years, the interest in multi-target ligands has progressively increased due to their potential benefits for the treatment of complex diseases such as cancer and CNS disorders. The development of bivalent ligands is based on the widely accepted evidence that G protein-coupled receptors, including ARs, can exist as homodimers or heterodimers. Bivalent ligands targeting the heterodimers A_1AR-β_2R, A_1AR-D_1R, $A_{2A}AR$-D_2R, A_3AR-A_1AR are reported. Dual-acting ligands behaving as A_1AR-$A_{2A}AR$ antagonists, $A_{2A}AR$ antagonists-MAO-B inhibitors, and A_{2A}-D_1 antagonists, all developed as potential anti-PD agents, have been inserted. A section is devoted to discussing the fundamental role of computational methods and structural biology in guiding the rational design of bifunctional ligands.

Chapter 8, by Taliani et al., deals with allosteric modulators of ARs, the first GPCRs known to be allosterically regulated. The chapter begins with a detailed description of the main mechanisms of GPCRs' allosteric modulation, the methodological aspect of allosteric ligand characterization, and the advantages and disadvantages of allosteric ligands compared to the orthosteric ones. This section can help the reader to get into the complex world of allosterism. The second part starts with a description of the mechanism of AR allostery and continues with the medicinal chemistry of allosteric modulators directed toward all four AR subtypes, Structure–activity relationships and the therapeutic potential of each class have been discussed, pointing out the most recent advances in the development of AR modulators.

The discovery and development of AR ligands has been improved during the last decades thanks to advances in techniques and methods which belong to the great family of "Computer-Aided Drug Design." Chapter 9, by Moro et al., deals with the application of main computational design methods in the field of ARs. Particular attention is devoted to molecular docking and molecular dynamics as the most used structure-based methods in the adenosine field, thanks to the availability of X-ray crystal structures and Cryo-EM structures of all ARs, except the A_3AR. Each computational technique is critically described pointing out its merits and drawbacks. The last paragraph deals with machine learning and artificial intelligence in drug discovery, starting from a general description of the process to its application in the AR field.

A receptor of the P2 receptor family is the topic of the tenth chapter, by Antonioli et al. P2 purine receptors are classified into P2X and P2Y families, each including different receptor subtypes. The P2X receptors are ligand-gated ion channel receptors and have been classified as $P2X_1$ to $P2X_7$. This chapter deals with the $P2X_4$

subtype, an ATP-gated cation channel that allows Na^+, K^+, and Ca^{2+} ion fluxes, and is widely expressed in cells of the immune system. The structure and molecular biology of this receptor, and its role on the functions of immune cells and microglia is reviewed. The potential of the $P2X_4$ receptor as a drug target to treat disorders, such as neuropathic pain, neurodegenerative disease, or metabolic syndrome, is discussed.

Two chapters of the book, Chaps. 11 and 12, from Supuran's group deal with CD73 ligands in drug design campaigns for identifying antitumor agents (human enzymes) and as potential antibacterials (case in which CD73 of bacterial origin are considered). These enzymes are ectonucleases that hydrolyze ADP and less efficiently other nucleotides too, through a zinc ion-dependent mechanism. Their 3D structures are known and in the case of the mammalian enzymes this led to the design of potent inhibitors, some of which are in clinical trials as antitumor agents. The bacterial enzymes were less investigated as drug targets but might be interesting for the development of antibacterials with novel mechanisms of action. All these aspects are dealt with in the last two chapters.

Overall, the present book presents the state of the art in this fascinating field of purinergic receptors.

Florence, Italy
July 2023

Vittoria Colotta
Claudiu T. Supuran

Contents

Top Med Chem (2023) 41: 1–46
https://doi.org/10.1007/7355_2023_158

Published online: 8 March 2023

Once Upon a Time Adenosine and Its Receptors: Historical Survey and Perspectives as Potential Targets for Therapy in Human Diseases

Daniela Catarzi, Flavia Varano, Sara Calenda, Erica Vigiani, and Vittoria Colotta

Contents

Abstract Endogenous autacoid adenosine plays important physiological roles in various tissues and organs by acting through A_1, A_{2A}, A_{2B}, and A_3 receptors. Dysregulated extracellular adenosine concentrations can cause many disorders such as inflammation, neurodegeneration, pain, and cancer, thus suggesting adenosine receptors (ARs) as interesting therapeutic targets. Medicinal chemistry studies have generated a large number of AR ligands, which have been pharmacologically characterized and optimized. Improvement in AR structural biology stimulated new computer-aided drug design approaches for drug discovery. Some AR ligands have reached clinical trials, and a limited number of them are on the market. To achieve these results, great progress has been made since the first steps in the intriguing world of adenosine and its receptors. New knowledge has been achieved, many

D. Catarzi (✉), F. Varano, S. Calenda, E. Vigiani, and V. Colotta
Dipartimento di Neuroscienze, Psicologia, Area del Farmaco e Salute del Bambino, Sezione di Farmaceutica e Nutraceutica, Università degli Studi di Firenze, Sesto Fiorentino (Firenze), Italy
e-mail: daniela.catarzi@unifi.it

issues have been solved, and new important aspects, previously ignored or underestimated, have been highlighted. In particular, but not only, taking into account important factors such as binding kinetics and biased signaling of ARs can lead to a rational design of more effective and safer drugs. Multitarget-directed ligands that combine, in the same molecule, the ability to interact with AR(s) and other target(s) are getting much attention as innovative therapeutic agents for treating multifactorial disorders such as neurodegenerative diseases and cancer.

Keywords Adenosine receptor ligands, Adenosine receptors, Cancer, Inflammation, Neurodegeneration, Pain

1 Introduction

Our scientific journey into the magical world of adenosine and its receptors began more than 30 years ago when the knowledge on this topic was still in its infancy. One of our first readings focused on a book chapter entitled "Purine Receptors," written by Michael William, where the beginning of a story was told [1]. Then, we let ourselves be carried away by this boundless world ready to be amazed. Once upon a time, there was an intriguing new biological "actress" named adenosine, which made its appearance on the scene as early as the late 1920s and proved to have strong effects on humans. It influenced heart rate, not increasing, but reducing it, and producing profound hypotension. An unexpected effect, if we want to see adenosine as an intriguing lady, but certainly an effect that aroused a lot of attention from insiders. However, interest in adenosine and its potential as an antihypertensive agent rapidly declined when its short half-life was evidenced. It was not until the 1950s that adenosine was again evaluated for its physiological roles. And the story goes on. Adenosine takes care of maintaining homeostasis and for this reason, it has been identified as one of the main regulators of local tissue functions and defined as a guardian angel [2]. However, our angel adenosine, though prone to its regulatory role, also knows how to arm itself to defend tissues and organs, thus becoming functional in all situations in which the organism suffers an attack or an insult. The concentration of adenosine increases under metabolically unfavorable conditions such as tissue hypoxia or stress to counteract tissue damage [3, 4]. As the guardian angel of territory that it rules and controls to great effect, adenosine has a personal army made up of four G protein-coupled adenosine receptors (ARs), termed A_1, A_{2A}, A_{2B}, and A_3, which can be considered as its warrior angels. After having played with adenosine and its receptors for a while, we leave the world of supernatural creatures and return to the scientific and rigorous sphere in which, however, adenosine continues to play a guardian angel role. In the last 50 years, a myriad of new evidence on adenosine has been discovered that allows us to draw an increasingly accurate profile. Much new information on its pathophysiological roles and effects mediated by activation of its receptors is available. Worth mentioning is the progress that has been made in the design of new molecules capable of interacting with its receptors and producing useful effects to counteract old and new pathologies. This is an extraordinary fairy tale. However, although adenosine plays the role of homeostatic regulator, there are conditions in which excessive and chronic production of adenosine can be harmful, leading to fibrosis, inflammation, and other pathological situations [5]. It should be remembered that the results obtained to date in this field are due to the ingenuity, perspicacity, and tenacity of many researchers as well as to their ability to inspire new talents. Successes and failures have followed each other over the years, as well as moments of great ferment alternating with periods of stagnation in research. To honor the progress and successes achieved, we can trace the journey taken by those who, full of curiosity and with a pioneering spirit, sometimes overcoming the disbelief and skepticism of their colleagues, started this adventure almost 250 years ago.

2 Brief History of Adenosine and Its Receptors

The history of adenosine is related to that of all other nucleosides and more directly to that of natural purine derivatives. Purine chemistry began in 1776, with Karl Scheele, a Swedish pharmacist, who isolated uric acid from a bladder stone [6]. From then on, many scientists contributed to clarifying the structure and role of uric acid, until Emil Fischer, who prepared and utilized it to synthesize all naturally occurring purines [7, 8]. Later, he solved the structure of caffeine (1,3,7-trimethylxanthine) and related compounds and denoted them as purines [9]. At the same time, Albert Kossel isolated adenine and the other nucleic acid bases [10, 11]. Many efforts were made until the mid-1920s to determine the structure of nucleic acids. However, the interest was moved away from adenosine as a xenobiotic molecule with physiological actions, rather than just a component of nucleic acids. Some progress still has to be made, such as the determination of the sugar part of nucleosides and nucleotides. At the beginning of the twentieth century, the structure of AMP was proposed and D-ribose was identified as the previously unknown sugar portion [12]. ATP was isolated in 1929 independently by two research groups, but many doubts remain about the partnership of its discovery [13]. Based on growing evidence that ATP was important in cellular energy metabolism, in the early 1930s, beef heart muscle extracts containing AMP and adenosine were considered to have medicinal value. The biological effects of adenosine were described in the late 1920s when Drury and Szent-Gyorgyi described the cardiovascular effects of intravenous administration of extracts of cardiac tissues and other organs in several mammalian species [14]. They observed profound changes in cardiac rhythm and a slowing of heart rate to complete cardiac arrest. Furthermore, a potent hypotensive action was observed secondary to dilation of coronary blood flow. Intrigued, they proceeded to the isolation of the biologically active substance which they identified as an adenine derivative, i.e., adenosine-5′-monophosphate (5′-AMP). Moreover, administration

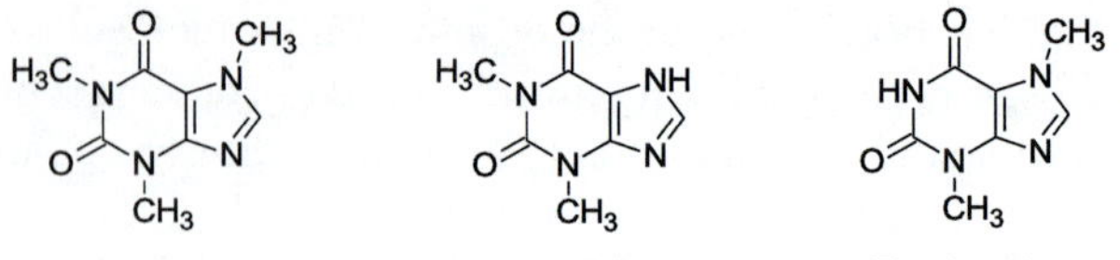

Fig. 1 Methylxanthine derivatives as adenosine receptor antagonists

of a mixture of 5′-AMP and adenosine caused qualitative and quantitative effects comparable to those produced by heart extracts.

The potential of adenosine as an antihypertensive agent was attractive. IG Farben preparation, named Lacarnol, was studied, confirming that purine nucleosides and nucleotides produced profound vasodilatory effects on coronaries [15] and changes in blood pressure [16]. In these studies, it emerged that adenosine was more potent than its nucleotides in determining vasodilation or inducing hypotension. Its short half-life prevents side effects mediated by A_1AR activation, such as cardiac block. The Locarnol extract was also evaluated in humans with heart disease, but the clinical trials failed and were stopped due to the lability of the natural nucleoside [17].

The study of Drury and Szent-Gyorgyi [14] can be considered as the starting point for both investigations on the physiological roles of adenosine and clarification of the biology of its receptors. However, 30 years passed before Berne and Gerlach, through independent experiments, proved that adenosine, produced by the degradation of ATP, caused hypoxic coronary vasodilation [18, 19]. However, their hypothesis has been the subject of discussion and controversy within the scientific community for several years. From then on, much research on the metabolism and physiological roles of adenosine was conducted. In the 1970s, the role of adenosine as an extracellular signaling molecule was proposed [20]. It was clarified that adenosine existed both in normally oxygenated and hypoxic myocardium [21, 22] and later its roles as a marker of hypoxia were confirmed leading to evidence of its potential effects on tissue protection [23]. At the same time, adenosine was postulated to exert its effect by its acting on specific receptors that stimulated or inhibited adenylate cyclase (AC). In particular, the induction of large accumulations of cyclic AMP in brain slices by adenosine was demonstrated [24, 25]. Moreover, the methylxanthines caffeine and theophylline (1,3-dimethylxanthine) (Fig. 1) showed to antagonize the effects of adenosine. Thus, the existence of ARs linked to cAMP production was hypothesized, and the natural methylxanthines, including theobromine, were accepted as antagonists at these receptors [4]. This study opens the door to a new vision of the nucleoside adenosine as a modulator of brain functions [24].

Other evidence had sparked interest in adenosine as a physiological regulator. One of the most incisive was the observation that the effects of exogenous adenosine can be enhanced by the adenosine uptake inhibitor, dipyridamole, which was introduced to the market in the 1950s [26–28].

In these years of great ferment, the pathways that lead to the formation and degradation of purine were also identified [29–31]. The confirmation that the enzyme responsible for the degradation of adenosine, that is, adenosine deaminase, was localized in limited areas of the brain, strengthened the hypothesis of the role of the modulator of the nucleoside itself [32].

Until now, the physiological role of nucleosides and purine nucleotides has been increasingly emerging, often leading to heated debates. Burnstock developed his idea on the new concept of purinergic signaling that challenged the skepticism of many colleagues [20]. In 1978, he proposed the first classification of purinergic receptors based on their sensitivity to adenine nucleosides and nucleotides. In particular, the P1 receptors were the most sensitive to adenosine and AMP, modified AC activity leading to altered cAMP levels, and were competitively antagonized by methylxanthines. In contrast, P2 receptors that had high sensitivity to ATP and ADP were not antagonized by methylxanthines and caused an increase in prostaglandin production [33].

Contemporarily, progress was being made on the delineation of the adenosinergic system. In the late 1970s, the existence of specific receptors for adenosine was increasingly accepted and two AR subtypes were described, termed A_1 and A_2, based on their opposite effect on AC activity [34]. After the hypothesis of Sattin and Rall [24], Van Calker and co-workers confirm the existence of two different ARs [35]. In particular, the authors demonstrated the inhibitory effect of adenosine on cAMP production, while the Sattin and Rall experiments had previously shown its stimulatory action. On this basis, the receptors were classified as the inhibitory receptor A_1 and the stimulatory A_2 receptor. Almost simultaneously, Londos and co-workers pointed out that adenosine-like compounds exhibited different potency at the two receptors, which were classified in R_i and R_a based on the receptor-mediated effect on AC activity [36]. However, in the same period, effectors other than AC were identified, such as ion channels and phospholipases [37–39].

The discovery of high and low affinity receptors that activate AC led to further subdivision of the A_2 subtype [40]. The high affinity receptor was termed A_{2A}AR, and the low affinity subtype was termed A_{2B} AR.

The effects of adenosine on cAMP formation, especially in brain tissue, have been extensively studied. In the late 1970s, two groups independently demonstrated that in a specific region of the brain, rich in dopamine, AC was activated by low concentrations of adenosine analogs. This was in contrast to the results obtained in cortical brain slices where a high level of adenosine analogs was necessary to activate AC through the A_2 receptor [41, 42]. Daly and co-workers confirmed the existence of the two A_{2A} and A_{2B} receptors based on binding studies on caffeine and related methylxanthines [43]. During the molecular cloning experiments of the adenosine A_1, A_{2A}, and A_{2B} receptors, a new functional receptor, different from those already known, was identified and classified as the A_3 receptor. The existence of A_3AR was hypothesized in the mid-1980s [44] but it still took a few years before Ali and co-workers unquestionably characterized the A_3 subtype as the fourth AR [45]. The receptor finally cloned [46] was different from the putative A_3AR first isolated by Ribeiro [44].

In the following years, all the AR known today were cloned from several species, including humans. A report by the International Union of Basic and Clinical

Pharmacology on the nomenclature and classification of ARs was published at the beginning of the 2000s [47] with an update in 2011 [48].

ARs belong to the family of G protein-coupled receptors. The A_{2A} and A_{2B} receptors were defined based on their ability to stimulate AC. Thus, they preferably interact with members of the G_s family of G proteins. As a consequence, an increase in cAMP levels was determined, leading to the activation of a pool of mediators, depending on the signaling triggered by cAMP in specific cells. In contrast, the inhibitory A_1 and A_3 ARs were supposed to interact with $G_{i/o}$ proteins [35, 40]. However, other G protein interactions have also been described [47]. Each AR is characterized by specific secondary signal transductors which can vary in receptor tissue localization, thus influencing specific physiological and/or pathological effects. Different and new signal transduction mechanisms associated with the different ARs have been identified and clarified over the years [49]. This peculiarity opens new frontiers for future drug development.

The detection and tissue localization of ARs has been facilitated and made possible by the development of specific radioligands for the different ARs. The first attempt dates back to 1978 when [^{3}H]-adenosine was used for binding studies in fat cells [50]. The experiments failed due to concurring factors, the first among all the rapid metabolism of the nucleoside. Two years later, several radioligand binding studies on A_1AR were conducted independently, leading to confirmation of the proposed classification. Labeled adenosine-like compounds and xanthine derivatives such as [^{3}H]-CADO ([^{3}H]-2-chloroadenosine) and [^{3}H]-DPCPX ([^{3}H]-8-cyclopentyl-1,3-dipropylxanthine), respectively (Fig. 2), were used to bind the A_1 receptor [51–53].

Radioligand binding studies on the A_{2A} receptor were much more difficult as the first experiments failed due to high non-specific binding. The first positive results appeared in 1986 when Bruns eliminated the undesirable A_1 component of a labeled non-selective A_{2A} agonist, using an A_1-selective ligand [40]. Greater success was achieved over time as new selective A_{2A}AR agonists and antagonists, such as [^{3}H]-CGS21680 ([^{3}H]-3-(4-(2-((6-amino-9-((2R,3R,4S,5S)-5-(ethylcarbamoyl)-3,4-dihydroxytetrahydrofuran-2-yl)-9H-purin-2-yl)amino)ethyl)phenyl)propanoic acid) and [^{3}H]-Scheme 58261 ([^{3}H]-7-(2-phenylethyl)-5-amino-2-(2-furyl)-pyrazolo-[4,3-e]-1,2,4-triazolo[1,5-c]pyrimidine), respectively (Fig. 2), were developed [54–56].

One of the first radiolabeled agonists for the A_3AR was [^{125}I]-AB-MECA (Fig. 2), endowed with high specific radioactivity (>140 Ci/mmol) as a radioiodinated compound [57]. This peculiarity allowed the detection of receptors expressed in tissues at low levels. Successively, [^{3}H]-MRE 3008-F20 ([^{3}H]-(5-N-(4-methoxyphenylcarbamoyl)amino-8-propyl-2-(2-furyl)pyrazolo[4,3-e]-1,2,4-triazolo[1,5- c]pyrimidine) (Fig. 2), a high affinity radioligand antagonist for human A_3AR, was reported in 2000. Both radiolabeled ligands represented a useful tool for further characterization of the A_3AR subtype [58].

Until now, no selective tritiated-A_{2B}AR agonists had been reported. [^{3}H]-NECA ([^{3}H]-5′-N-ethylcarboxamidoadenosine) is currently the only useful radioligand to determine the affinity of ligands at this AR subtype. A tritium-labeled form of a

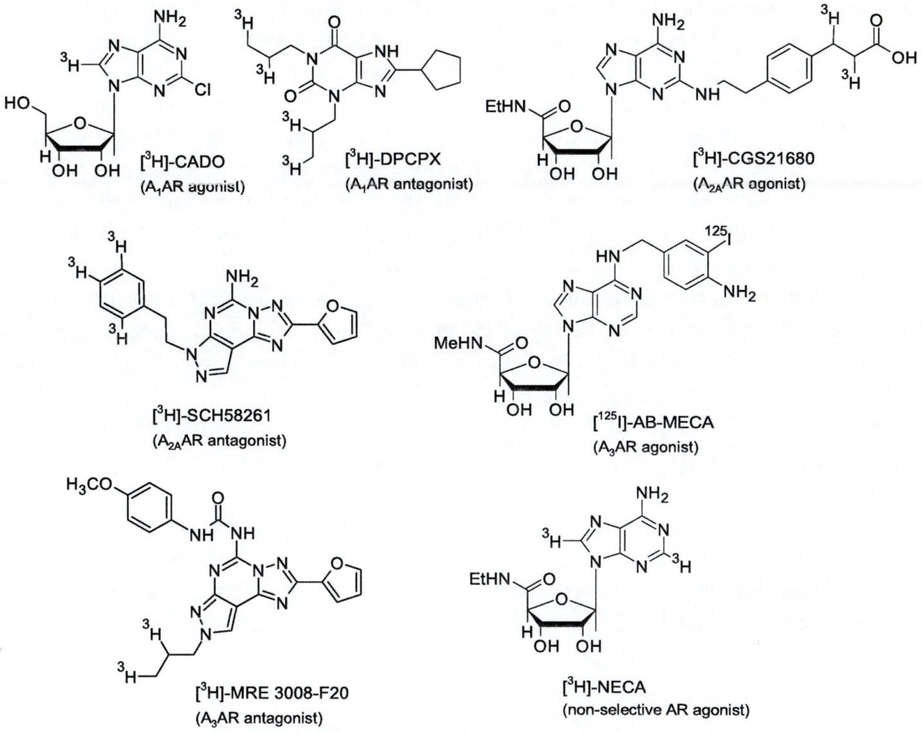

Fig. 2 Some of the earliest AR radioligands used for in vitro studies

non-adenosine-like A_{2B}AR partial agonist, belonging to the 3,5-dicyanopyridine series, was synthesized and evaluated as a radioligand binding probe. The study failed probably due to the high non-specific binding and moderate affinity of the ligand at the A_{2B} receptor. However, the results obtained allowed us to hypothesize that nucleoside and non-nucleoside A_{2B}AR agonists bind to different conformations of the receptor [59].

Many efforts have been devoted to the development of highly selective radioligands for each AR subtype, both agonists and antagonists. The most representative examples have recently been reported in some reviews which are recommended for further information on chemical probes for ARs [60, 61].

The availability of highly selective AR radioligands helped not only the tissue localization of the different ARs but also the clarification of their functional coupling and regulation. In addition to radioligands useful for the in vitro characterization of receptors, also specific chemical probes for ARs were developed, which are useful for drug discovery, diagnostics, and for application to structural biology studies [60]. Currently, the application of the methodology for covalent probes represents an important tool for stabilizing target proteins to obtain X-ray crystal structure [48, 61, 62]. All of these advances have been made possible by the intense development of a large number of AR agonists and antagonists, both by pharmaceutical industries and academic research groups. Although the first studies date back to the 1950s [28], the following years represented a period of great ferment for the identification of new pharmacological tools. Among the first compounds studied, the adenosine-like derivatives emerge mainly as agonists and the xanthine derivatives are among antagonists [63]. Many non-xanthine compounds have also been designed to antagonize the ARs. Bicyclic antagonists with a purine core were developed by replacing the ribose of adenosine with smaller substituents. In fact, in the late 1980s, it was evidenced that the elimination of the ribose moiety from the adenosine structure led to AR antagonists [64].

This experimental observation consolidated the idea that the presence of the ribose moiety was essential for agonist activity. William and co-workers reported a series of new 1,2,4-triazolo[1,5-c]quinazolines that paved the way for many derivatives with a tricyclic structure [65].

Since then, hundreds of new AR ligands endowed with different degrees of affinity and selectivity have been synthesized and extensively reviewed [4, 63, 66–77].

More recent achievements are related to the clinical development of AR ligands and also their marketing [77, 78]. A great deal of progress has been made since 1985, when adenosine was approved as a drug to treat supraventricular tachycardia (Adenocard®) [79] and, later, to act as a vasodilatory agent in myocardial perfusion imaging (Adenoscan®) [80].

3 The State of the Art

The historical survey carried out to synthesize the most important milestones highlights how difficult it was to reach the current level of knowledge on adenosine and its receptors. To date, we know that adenosine is a ubiquitous extracellular signaling molecule that produces its effects through the activation of G protein-coupled ARs that are divided into four subtypes, termed A_1, A_{2A}, A_{2B}, and A_3 [49]. The widespread distribution of its receptors indicates that adenosine plays a fundamental role in the regulation of many important functions in tissues and organs. Under physiological conditions, extracellular adenosine reaches a concentration of 30 to 200 nM, which is high enough to activate both the A_1 and A_{2A} ARs. On the contrary, under stress conditions such as hypoxia or high metabolism, the level of extracellular adenosine can grow to very high micromolar values, thus activating also low affinity A_{2B} and A_3 subtypes. Thus, adenosine can be considered as a homeostatic regulator or, more correctly, a “retaliatory metabolite,” the role it has earned in the past [81].

3.1 *Adenosine Production and Metabolism*

At the extracellular level, adenosine is generated from nucleotide (i.e., ATP, ADP, and AMP) dephosphorylation, mainly through the activation of ectonucleotidases CD39 and CD73 [49, 82]. Adenosine precursors are released by different cells under stressful conditions and then transformed into the protective mediator adenosine. Under physiological conditions, adenosine can also be generated intracellularly or transported within the cell from the extracellular space. In particular, hydrolysis of S-adenosylhomocysteine or AMP through cytosolic S-adenosyl-homocysteine hydrolase and endo-5′-nucleotidase, respectively, leads to the formation of adenosine [83]. Otherwise, because of different types of nucleoside transporters, adenosine can cross the cell membrane with different mechanisms and increase its intracellular concentration. Obviously, in stressful conditions such as hypoxia, the direction of adenosine flux can be reversed [84]. In extracellular spaces, adenosine is transformed into inosine by adenosine deaminase (ADA) or phosphorylated to AMP by adenosine kinase (AK). Adenosine that has not been taken up or deaminated interacts with its ARs.

3.2 *Distribution of Adenosine Receptors and Their Main Physiological Effects*

The wide distribution of ARs throughout the body highlights the important physiological roles, both in the central nervous system (CNS) and in the periphery, attributed to adenosine [85]. Each receptor is characterized by a peculiar distribution at the tissue and organ level and, sometimes, by multiple secondary signal transductors that can mediate a plethora of physiological effects. Additionally, ARs show different degrees of affinity for adenosine, ranging from the highest, typical of the A_1AR ($K_i = 70$ nM), passing through the intermediate of the A_{2A} receptor (150 nM) to the lowest of the A_{2B} and A_3 ARs (5,100 and 6,500 nM, respectively). The affinity of adenosine for its receptors is modulated by ecto-adenosine deaminase (ecto-ADA). In particular, ecto-ADA interacts with specific binding sites on the ARs leading to increased receptor affinity and signaling [86, 87].

The A_1AR is highly represented in different areas of the CNS (brain cortex, hippocampus, cerebellum, spinal cord, and glial cells) [83], a distribution that justifies the wide range of physiological effects mediated by it such as anticonvulsant, anxiolytic, pain reduction, sleep/wakefulness cycle control, modulation of neurotransmitter release [88, 89]. This subtype is also widely distributed in the cardiovascular system with the major expression in the heart, where it exerts negative inotropic, chronotropic, and dromotropic effects. In adipose tissue and pancreas, A_1AR modulates glucose and lipid metabolism by inducing inhibition of lipolysis and secretion of insulin, respectively [90]. At the kidney level, where it is located in cells of the juxtaglomerular apparatus, collecting ducts of the papilla and inner medulla, the A_1 subtype reduces renal blood flow, glomerular filtration rate, and renin release [91, 92].

It is also distributed in the respiratory system (airway epithelium and smooth muscles), where it produces bronchoconstriction, and in the immune system (neutrophils, eosinophils, macrophages, and monocytes) where its main effects are pro-inflammatory [93, 94].

Furthermore, the A_{2A}AR subtype is widely distributed in the body, but with the highest expression in the central nervous (striatum, olfactory tubercle) and immune systems. Lower levels of this subtype were found in other CNS districts (cerebral cortex, hippocampus) and the heart. In the peripheral immune system, the A_{2A}AR, located in leukocytes, mediates numerous anti-inflammatory effects. This receptor subtype was identified as an important mediator of adenosine inhibition of platelet aggregation [95].

A_{2B}AR is less expressed in the brain than in the periphery, where it is highly expressed in the bowel, bladder, lung, and also in different types of cells such as fibroblasts, platelets, endothelial and immune cells. This subtype is gaining a prominent role as a modulator of inflammation and immune responses in many diseases including cancer, diabetes, and other vascular disorders. Moreover, its upregulated expression in selected stressful conditions (hypoxia, inflammation) underlines its pathophysiological relevance [5].

The A_3AR subtype has a low concentration in the brain, where it can be found in restricted areas, including the cortex, hippocampus, thalamus, and hypothalamus, as well as motor nerve terminals. It is also expressed in selected cells and tissues such as microglia and astrocytes, through which it mediates the neuroinflammatory response directly associated with a reduction of neuropathic pain [96].

Controlling of inflammation was also evidenced at the peripheral level where the A_3 receptor is located in inflammatory cells such as lymphocytes, eosinophils, neutrophils, monocytes, macrophages, and mast cells, which are involved in the A_3AR anti-inflammatory effects [97].

Overexpression of A_3AR in different cancer cells and tissues is also reported, thus attributing an important antitumoral role to this receptor subtype [98]. At the periphery, A_3AR is also distributed in other districts, including enteric neurons, lung parenchyma, bronchi, and, not less important in the heart (coronary and carotid arteries) where it mediates cardioprotective effects [99].

3.3 Signal Transduction of Adenosine Receptors

Specific signal transduction systems and a great variety of switchable-associated mediators are connected with each AR, which mediates physiological responses also depending on the type of cell or tissue involved. The A_1 and A_3 ARs are coupled to heterotrimeric G proteins belonging to the G_i/G_o family, while the A_{2A} and A_{2B} receptors predominantly signal through G_s proteins. G_i/G_o proteins inhibit AC, thereby lowering cAMP levels (A_1 and A_3 ARs), and increase both inositol

1,4,5-trisphosphate (IP3) and intracellular Ca^{2+} levels by activating phospholipase C (PLC) via Gβγ subunits (A_3AR). In contrast, activation of G_s proteins leads to a cAMP level increase by stimulation of AC activity. Nevertheless, increasing evidence suggests that ARs can couple to multiple G proteins [49, 100, 101].

For example, A_3 and A_{2B} receptors are also coupled to G_q proteins, whose stimulation activates PLC and produces an increase in calcium levels [48, 49]. A_{2A}AR is also reported to couple to AC via G_{olf} in the striatum [102]. All ARs are involved in the intracellular phosphorylation cascade of the mitogen-activated protein kinase (MAPK) family, giving them a role in cell growth, survival, death, and differentiation [49, 103].

More recently, it has been evidenced that ARs can also signal through G protein-independent effectors [49]. In particular, one of these G protein-independent pathways, with β-arrestins as signal transducers, was demonstrated for the A_{2B} and A_3 ARs, with limited evidence for the A_1 and A_{2A} subtypes [104].

The identification of G protein-independent pathways other than those coupled to G proteins and associated with a specific AR, allowed to open a new frontier for the development of drugs with fewer side effects. In fact, each signal pathway, associated with a specific receptor subtype, can produce different effects depending on the cell type or tissue. Therefore, the availability of AR agonists, defined as biased, able to distinguish between the different signaling pathways and to choose one over the others, could allow obtaining targeted therapeutic actions. In particular, avoiding pathways associated with undesirable effects could lead to increased drug safety. A first step forward seems to have been made. Wall and co-workers highlighted the importance of the concept of biased agonism. An A_1AR-selective agonist, which was reported to activate specific G proteins, was demonstrated to be a potent analgesic devoid of side effects at cardiovascular, respiratory, and CNS levels [105].

3.4 Adenosine Receptor Structures

The four receptor subtypes were purified and successfully cloned from mammalian and non-mammalian species, particularly rats, mice, and humans [47]. Considerable species differences were observed in some cases, which are more pronounced for the A_3 subtype [106, 107]. Based on sequence similarity and G protein-coupling specificity, A_1 and A_3 receptors share 49% sequence while A_{2A} and A_{2B} receptors are almost 59% identical. The most conserved region may be found in the extracellular loops of the receptor where the homology sequence reaches 71%.

All ARs are characterized by a common structure consisting of seven transmembrane α-helices (7TM, helices 1–7), an extracellular amino-terminus (N-terminus), a cytosolic carboxy-terminus (C-terminus), three extracellular loops (EL1-3), and three intracellular loops (IL1-3) of different lengths and functions, depending on the AR subtype [47, 108]. These domains confer specific characteristics that are important for receptor–ligand interactions. The N-terminus has one or more glycosylation sites, and the C-terminus contains phosphorylation and palmitoylation regions, which are involved in both receptor desensitization and internalization processes.

The complex structures of A_1 and A_{2A} ARs were determined using X-ray crystallography, NMR techniques, or, more recently, cryo-electron microscopy [78].

The first investigations that led to the elucidation of the A_{2A}AR structure through X-ray crystallography [109] helped the other AR architecture to be solved. Other studies include crystal structure determinations of complexes with antagonists [109, 110] and agonists [111, 112], and for a ternary complex of A_{2A}AR with an agonist and a mimetic engineered G protein [113].

Regarding A_1AR, many molecular modeling and mutagenesis studies have been conducted before crystal structures of the A_1AR were proposed [114, 115]. The first simulations were performed computationally using homology models of the rhodopsin or A_{2A}AR structure [116].

In 2017 Glukhova et al. proposed a crystal structure of the A_1AR bound to a covalent xanthine antagonist [117]. The computational and mutagenesis analysis of the A_1AR identified peculiar differences from the previously solved A_{2A}AR structure. In particular, a distinct conformation of the second extracellular loop and a wider extracellular cavity endowed with a secondary binding pocket for orthosteric and allosteric ligands were highlighted.

Crystallization of a receptor, often necessary for structural biology studies, is still a very challenging task. In fact, referring to ARs, their low expression in native tissue, and their instability once extracted from the membrane, can represent important limitations.

More recently, cryogenic electron microscopy (cryo-EM) has attracted wide attention as an innovative approach for receptor structure elucidation without the use of crystallization. By using this innovative technique, many detailed structures of A_{2A}AR have been reported over the years. In fact, the A_{2A} has become one of the most investigated GPCR structures [118, 119]. In addition, structures of the A_1 and A_{2A} ARs, coupled with the suitable G protein or with its engineered parent, have been determined. These are the only examples of GPCR-G protein complex structures reported in the literature till now. In particular, the adenosine-bound A_1AR coupled with the G_i protein [120] and the NECA-bound A_{2A}AR coupled with an engineered G_s protein [121] were studied. The complex structures provided important insights into the active conformations of the A_1 and A_{2A} ARs and the G proteins coupled to them. However, the dynamic mechanism of the specific interactions remains unclear.

Studies performed to clarify the AR structure using advanced techniques (X-ray crystallography, cryo-EM, and NMR) have generated several structures for ARs. Their availability has stimulated different new structure-based and computer-aided drug design (CADD) approaches for the development of new AR ligands [122]. To date, the Protein Data Bank (PDB) [123] has recorded approximately 64 structures of ARs under different functional states. However, structural studies on A_{2B} and A_3 AR have not yet led to the desired results.

Furthermore, there is a major limitation regarding the structural studies of all AR subtypes. In fact, there are no experimental techniques available that can evaluate the flexibility of the receptors, which plays a critical role in the regulation of their biological functions. Moreover, only static images are available in PDB structures that cannot capture and describe the dynamics of the ligand–receptor interaction. However, molecular dynamics (MD) simulations [124] seem to be useful for this purpose, providing models of the ligand-binding/dissociation processes and new insights into the functional mechanisms of ARs. The Gaussian accelerated molecular dynamics (GaMD) method has been used to clarify the mechanism of A_1 and A_{2A} AR–G protein interactions [125]. A review of simulation studies has recently been reported which emphasizes the high potential of these types of techniques for the rational design of drugs targeting ARs [126].

3.5 *Medicinal Chemistry of Adenosine Receptor*

3.5.1 Adenosine Receptor Ligands, in Short

ARs are widely distributed throughout the body and their activation modulates a wide variety of effects in different tissues and organs. Thus, the regulation of their activity could have therapeutic potential, but at the same time, open the possibility of side effects occurring. For these reasons, the development of selective agonists and antagonists is particularly intriguing. However, modulation of AR activity can be realized not only by using orthosteric ligands but also by allosteric modulators. Allosteric ligands recognize a distinct binding site, conformationally linked to that of orthosteric ligands [127]. The therapeutic utility of AR allosteric modulators, in particular positive allosteric modulators (PAMs), has been established in various preclinical animal models. In this regard, a reading of recently published in-depth reviews on AR allosteric enhancers also focused on selected ARs [4, 68, 128, 129] is recommended.

Over the years, different classes of ligands have been evaluated at ARs because of the availability of potent and selective radioligands which emerged as essential to support the drug discovery process. In fact, the first screening of newly synthesized compounds consists of a rapid radioligand binding assay. Together with functional assays, useful for evaluating the pharmacological profile of ligands, binding studies have helped define the recurring structural features that can affect the affinity or efficacy of compounds for a specific AR [4]. This was not an easy challenge in the case of AR antagonists, as a result of both the large number and variety of their scaffolds that made it difficult to identify common structural requirements. Although antagonists are characterized by large structural variability, the agonist profile has been associated for a long time with an adenosine-like structure.

3.5.2 Emerging Adenosine Receptor Ligands as Tool Compounds

When the biological relevance of adenosine emerged, the availability of specific tool compounds for AR studies was an important goal to be achieved. Subsequently, the discovery of specific AR radioligands, and also fluorescent and covalent ligands, all useful to facilitate receptor characterization, structural biology studies, drug discovery and diagnostics, emerged [60, 61].

The AR agonists and antagonists collected in the following can be considered as some of the main players that have contributed greatly to tool compounds to deepen the knowledge of the world of adenosine and its receptors. AR affinities of the reported compounds can be found in the references cited.

Agonists

Adenosine itself has been used for therapeutic and imaging purposes for several years. Although the rapid metabolism of adenosine in the bloodstream has allowed its use as a diagnostic tool or agent in emergency therapy, nevertheless it has also represented a disadvantage for other potential applications. In the past decades, the medicinal chemistry approach to obtain AR agonists was to perform selected structural modification at N6- and 2-positions of the adenosine nucleus, and/or at the 3′-, 4′- or 5′-positions of the ribose moiety. Efforts to synthesize more stable analogs began in the 1960s, producing some compounds including 2-chloroadenosine (CADO) and NECA [130] (Fig. 3). Suitable substitutions of the adenosine core led to AR agonists with varying degrees of subtype selectivity. Much effort has been made to obtain clear and complete structure-activity relationships (SARs) of adenosine-like derivatives [4, 63, 66–68, 70–74] which have helped the discovery of more suitable compounds to define the pharmacology of AR subtypes.

Although the adenine nucleus can be decorated with more flexibility to yield potent and selective AR agonists, modifications at the ribose moiety require particular attention. In fact, structural alterations at this level can produce a modulation not only of AR affinity and selectivity, but also of the efficacy of the target compound. The rationale for these modifications was first to obtain compounds endowed with subtype selectivity by changing the electronic and steric features of the molecules. A challenge was to improve the physicochemical properties of compounds, with the improper pharmacokinetic profile often being a drawback in promoting a drug candidate to a therapeutic agent.

Some AR agonists were obtained through substitution at the adenosine N6-position. An example is the N6-cyclopentyl substituted derivative CPA (N6-cyclopentyladenosine) which was shown to selectively bind to the human A_1AR. CADO is a metabolically stable analog of adenosine that behaves as a non-selective AR agonist. It has been used to define the AR pharmacological profile in a particular cell/tissue. Tritiated CADO was used as a radioligand probe for AR studies [131].

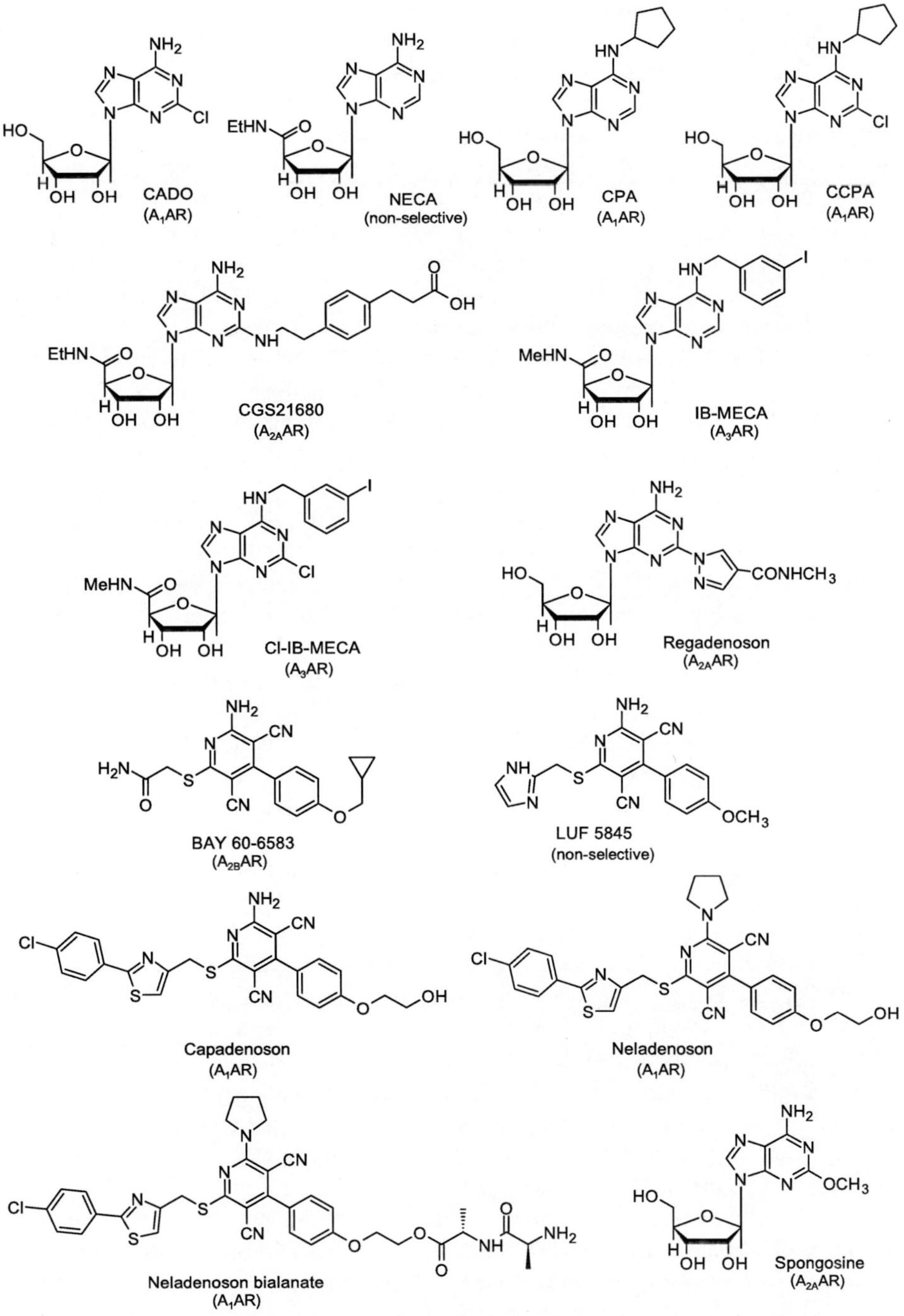

Fig. 3 Adenosine receptor agonists as tool compounds

2-Chloro-N6-cyclopentyladenosine (CCPA) displayed slightly higher human A_1AR affinity than the parent compound CPA. Among the first adenosine-like compounds that appeared, the 5′-N-ethylcarboxamido-derivative NECA behaved as a broad-spectrum AR agonist, with low affinity for the A_{2B} subtype [40]. However, it has been one of the most widely used AR agonists, especially as a useful tool for studying the A_{2B} subtype, which lacks potent and selective agonists with high efficacy [132]. Substitution with the small 5′-N-alkyluronamide group on the ribose moiety of adenosine as in NECA, in general, provided increased potency at all ARs. This structural feature is present in the first selective A_{2A} receptor agonist CGS21680 (2-[4-(2-carboxyethyl)phenylethylamino]-5′-N-ethylcarboxamidoadenosine), which was shown to be potent at $A_{2A}AR$ in rat and mouse, but with lower selectivity in humans. This 2-(2-phenylethyl)amino modification was particularly effective in improving affinity at the A_{2A} subtype. CGS21680 was tested in a Phase 2A clinical trial as a potential antihypertensive agent. However, administration caused a large fall in blood pressure and consequent increase in cardiac output. The introduction of a 3-iodophenyl moiety at the N6 position in the NECA structure led to the A_3AR agonist IB-MECA (Piclodenoson, 1-deoxy-1-[6-[[(3-iodophenyl)methyl]amino]-9H-purin-9-yl]-N-methyl-β-D-ribofuranuronamide), whose selectivity versus the A_1 and A_{2A} subtypes was very low. Its 2-chloro analog Cl-IB-MECA (namodenoson, 1-[2-chloro-6-[[(3-iodophenyl)methyl]amino]-9H-purin-9-yl]-1-deoxy-N-methyl-β-D-ribofuranuronamide) emerged as a very potent and highly selective A_3AR agonist. Both IB-MECA and its parent compound Cl-IB-MECA have been in clinical development. Regadenoson (CVT3146, Lexiscan®)(1-[6-amino-9-[(2R,3R,4S,5R)-3,4-dihydroxy-5-(hydroxymethyl)oxolan-2-yl]purin-2-yl]-N-methylpyrazole-4-carboxamide), though not a highly selective $A_{2A}AR$ agonist in humans, entered clinical trials for many diseases (transplantation and lung inflammation). Many other new and old AR agonists have been and are being clinically evaluated for human use [77, 78]. In fact, from the first results, over the years many AR agonists, endowed with high affinity and selectivity, have been reported that were modified differently at the ribose moiety and substituted at the N6 and C2 adenine positions [4, 63, 66–68, 70–74, 77]. The positive results obtained are an expression of all the efforts made by the scientific community to obtain drug candidates with high efficacy and a suitable pharmacokinetic profile.

Although the agonist profile has long been associated with an adenosine-like structure, more recent studies reported some non-nucleoside compounds belonging to the 6-aminopyridine-3,5-dicarbonitrile series that showed different degrees of efficacies at the different ARs (Fig. 3).

Some of these ligands, although lacking the ribose portion, have significant affinity and efficacy for human ARs. For example, BAY60-6583 (2-[[6-amino-3,5-dicyano-4-[4-(cyclopropylmethoxy)phenyl]-2-pyridinyl]thio]-acetamide) [75] and LUF 5845 (2-amino-4-(4-methoxy-phenyl)-6-(1H-imidazol-2-ylmethylsulfanyl)-pyridine-3,5-dicarbonitrile) [75, 133] are among the prototypical 6-aminopyridine-3,5-dicarbonitriles reported in the literature.

Although BAY60-6583 is a potent and selective human $A_{2B}AR$ agonist discovered by Bayer Healthcare, LUF 5845 and its analogs displayed in general nanomolar affinity for all ARs, and in some cases even considerable efficacy. BAY 60-6583 behaves as an A_{2B} partial agonist [134], but showed to act as an antagonist at the other ARs [106].

In fact, the pharmacological profile of this compound can change depending on the tissue, receptor expression level, and in situ adenosine concentration. Amino-3,5-dicyanopyridine compounds have common features, that is two cyano groups at C3 and C5 and a thiomethyl group with an imidazole-2-yl moiety at the 6-position. The 4-phenyl moiety can be substituted with a suitable group (H, OH, OCH_3, OCH_2-cC_3H_5) in the *para-* or *meta-*position. Modification of LUF series compounds at the level of thiomethyl-side-chain moiety led to the development of a potent and selective partial A_1AR agonist which was developed as a radioligand for positron emission tomography (PET) imaging in the brain [77, 135].

BAY 60-6583 was introduced as a selective non-nucleoside $A_{2B}AR$ agonist with high affinity/potency, depending on the type and condition of the assay [134, 136, 137]. It was used for several pharmacological studies, thus allowing its beneficial effects in obesity [138] and insulin resistance to emerge [139, 140], but also cardiovascular disorders [141, 142]. The exhaustive review of Dal Ben and colleagues [75] describes developments in this field from 2000 to recent times.

The lucky structural modifications of dicyanopyridine derivatives led to the development of compounds with a substituted thiazole group linked to the scaffold at the 2/6 position through a thiomethyl function. Capadenoson (2-amino-6-[[2-(4-chlorophenyl)-1,3-thiazol-4-yl]methylsulfanyl]-4-[4-(2-hydroxyethoxy)phenyl]pyridine-3,5-dicarbonitrile) is currently the most representative compound of this series, as it represents a reference A_1AR partial agonist, endowed with sub-nanomolar activity at the human A_1AR [143]. Capadenoson has been shown to be useful in different stressful and diseased conditions [144–148] showing a favorable pharmacokinetic profile [146]. Moreover, the advantages of the use of a partial A_1AR agonist are primarily based on the lower risk of producing severe side effects compared to full agonists [146, 149]. Structural modifications at the exocyclic amine group of Capadenoson led to the development of other dicyanopyridine derivatives whose lead compound was the A_1AR agonist Neladenoson (2-(((2-(4-chlorophenyl)-1,3-thiazol-4-ylmethyl)sulfanyl)-4-(4-(2-hydroxyethoxy)phenyl)-6-(pyrrolidin-l-yl)pyridine-3,5-dicarbonitrile). It is a selective A_1AR partial agonist [150, 151] with a pharmacological profile similar to that of Capadenoson, but endowed with lower central undesirable effects. Its prodrug, in the form of its bialanate derivative (alanine–alanine ester), Neladenoson bialanate (BAY 1067197), was developed that led to an improvement in the pharmacokinetic profile. Analogs of Capadenoson were designed and synthesized, focusing attention on their dissociation kinetic properties at the A_1AR [152]. Some of these derivatives showed a residence time longer than that of Capadenoson.

Recently, a thorough review of current and historical AR agonists in preclinical and clinical development has appeared [77].

Antagonists

In general, antagonists are nitrogen-containing heterocycles endowed with a planar π-electron rich structure, generally decorated with hydrophobic groups that modulate not only AR affinity but even selectivity. Finally, they lack ribose moiety (Fig. 4). The prototypical AR antagonists are caffeine and theophylline, natural psychostimulant xanthines, endowed with micromolar affinity and low selectivity for ARs (Fig. 1) [153]. Taking these latter as lead compounds, many other differently substituted xanthine derivatives were designed and synthesized yielding the definition of detailed SARs which have been confirmed over the following years [4, 154].

The introduction of suitable substituents at the 1-, 3-, and 8-positions of the xanthine scaffold shifts the affinity toward a particular receptor subtype. A very large number of selective AR antagonists have been developed as derived from natural xanthines. A_1AR selective compounds typically feature a bulky cycloalkyl substituent at 8-position such as 1,3-dipropyl-8-cyclopentylxanthine (DPCPX) [18] which is, however, only moderately selective for this subtype. Alkylation of the nitrogen atom at N1 and N3 positions with N-propyl groups increases the affinity at all AR subtypes, even if it was demonstrated later that A_{2B}AR prefers N3-unsubstituted compounds. The presence of a styryl moiety at the 8-position of the xanthine scaffold improves the selectivity for the A_{2A} subtype as demonstrated by (E)-8-(3,4-dimethoxystyryl)-1,3-diethyl-7-methylxanthine (Istradefylline, KW-6002) which is the first A_{2A}AR antagonist approved as a drug [155]. It is highly selective versus A_3 and A_{2B} ARs but not the A_1 subtype. An associated problem is its E/Z isomerization and also light-induced dimerization [156].

During the past decade, potent and selective A_{2B}AR antagonists belonging to the xanthine series, such as PSB-603 (8-(4-(4-(4-chlorophenyl)piperazine-1-sulfonyl) phenyl)-1-propylxanthine), have been developed, that displays a K_i value of 0.553 nM at the human A_{2B} receptor and >15,000-fold selectivity over other ARs. A tritiated form of the compound was prepared as a new radioligand and characterized in kinetic, saturation, and competition studies [157].

Further modifications at position-8 have resulted in many other highly selective AR antagonists of interest. MRS 1754 (8-[4-[((4-cyanophenyl)carbamoylmethyl) oxy]phenyl]-1,3-di(n-propyl)xanthine) is a selective antagonist for A_{2B}AR with a very low affinity for the A_1 and A_3 receptors of both humans and rats, which has been prepared as a radioligand [132, 158].

More recently, many xanthine-based AR antagonists emerged, showing also promising efficacy in vitro/in vivo animal models of Parkinson's disease (PD) [159], diuresis [160], and asthma [161, 162]. In addition, starting from the adenine structure, various non-xanthine cores were taken into consideration to develop new AR antagonists. In fact, the decoration of bicyclic and tricyclic ring systems led to the identification of highly potent and selective AR antagonists with improved affinity and selectivity with respect to xanthine derivatives (Fig. 4).

9-Chloro-2-(furan-2-yl)-[1,2,4]triazolo[1,5-c]quinazolin-5-amine (CGS15943) was one of the most potent tricyclic non-xanthine AR antagonists reported in the literature as non-selective AR ligand [65]. Despite this, it has been considered for a

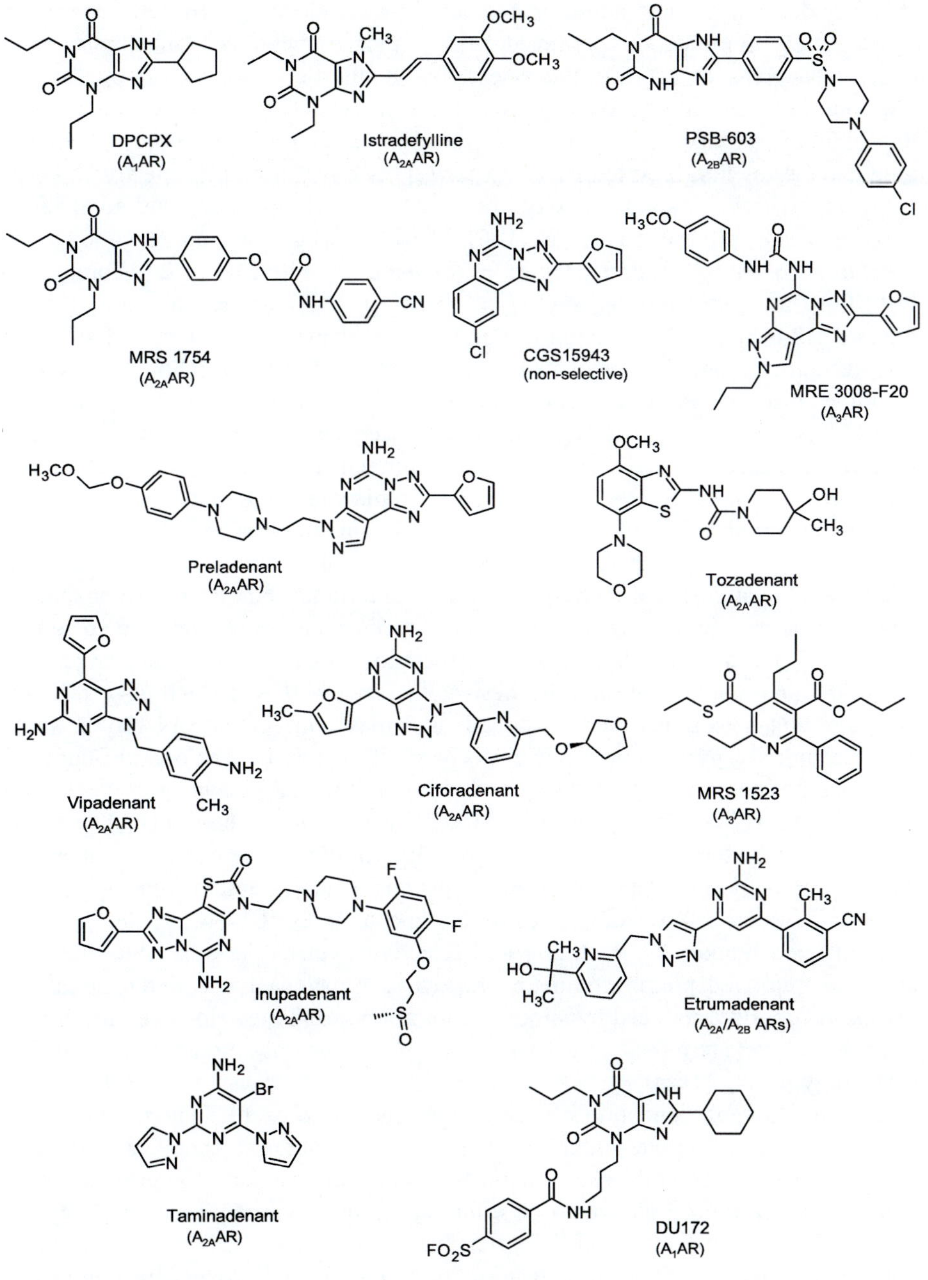

Fig. 4 Adenosine receptor antagonists as tool compounds

long time as the lead compound for the new generation of tricyclic AR antagonists of which the most representative are reported in Refs. [4, 66–69, 71–73, 76]. CGS15943 is substituted at position 2 with a furan-2-yl group, which is considered an important feature for A_{2A} activity and selectivity [163]. Successively, its structural modification produced many other tricyclic AR antagonists with different degrees of affinity and selectivity at all four ARs [4, 66–69, 71–73, 76]. One of these emerging compounds is MRE 3008-F20 which is derived from the structure-activity optimization of pyrazolo[4,3-e]-1,2,4-triazolo[1,5-c]pyrimidine derivatives first reported by Gatta et al. [164]. It is a highly selective hA_3AR antagonist with comparable activity in the cAMP functional assay and good selectivity versus the other AR subtypes [165]. The same drug design approach led to the identification of Preladenant (Scheme 420814, 2-(2-furanyl)-7-[2-[4-[4-(2-methoxyethoxy)phenyl]-1-piperazinyl]ethyl]7H-pyrazolo[4,3-e][1,2,4] triazolo[1,5-c]pyrimidine-5-amine), which is one of the most potent and selective A_{2A} receptor antagonists evaluated in a clinical trial for PD. Though it is well tolerated no significant beneficial effects were obtained following its therapeutic use.

The discovery of $A_{2A}AR$ antagonists with acceptable selectivity, physicochemical, and pharmacological properties has continued to be an intriguing challenge [166]. Hoffmann-La Roche identified a benzothiazole-based potent and selective $A_{2A}AR$ antagonist, Tozadenant (4-hydroxy-N-(4-methoxy-7-morpholinobenzo[d] thiazol-2-yl)-4-methylpiperidine-1-carboxamide), which reached phase 3 clinical trials for idiopathic PD but the process was interrupted based on an unexpected emerging safety. Earlier SAR studies on bicyclic core compounds led to highly potent $hA_{2A}AR$ antagonists of the next generation to CGS15943. Some pyrazolo-, -pyrrolo-, and imidazo-pyrimidine derivatives endowed with high A_{2A} receptor affinities and selectivity were obtained as Vipadenant precursors [167]. Some selected imidazo-pyrimidines showed to be orally active in a haloperidol-induced hypolocomotion mouse model assay. In addition, low oral bioavailability and poor in vivo stability were observed in rat models. By replacing the imidazole nucleus with a triazole, a series of 7-aryl triazolo[4,5-d]pyrimidine compounds was obtained. Optimization of the latter yielded Vipadenant (3-[(4-amino-3-methylphenyl) methyl]-7-(furan-2-yl)triazolo[4,5-d]pyrimidin-5-amine, BIIB014, V2006) which showed good human $A_{2A}R$ affinity and selectivity versus the other AR subtypes, and also improved pharmacokinetic properties. Vipadenant was demonstrated to reverse haloperidol-induced hypolocomotion in mice and rats. However, its clinical trials for PD were terminated in 2010 due to safety concerns emerging in preclinical toxicology testing [168].

Investigation of benzothiazole derivatives as Tozadenant analogs led to the discovery of a new potent and selective $A_{2A}AR$ antagonist endowed with high aqueous medium solubility and stability versus microsomal enzymes. Selected compounds have been shown to be effective in an in vitro PD model [169]. To address the chemical structural liabilities that probably have led to concerns about Vipadenant toxicity, further modification led to the lucky discovery by Vernalis of Ciforadenant (7-(5-methylfuran-2-yl)-3-[[6-[[(3S)-oxolan-3-yl]oxymethyl]pyridin-2-yl]methyl]triazolo[4,5-d]pyrimidin-5-amine, CPI-444, V81444), which is currently under clinical evaluation for cancer therapy [78].

Further structural modification of Tozadenant yielded other $A_{2A}AR$ antagonists showing high affinity and selectivity for this subtype. The radiolabeling of some fluorinated compounds with ^{18}F and in vitro autoradiography revealed that these compounds represent potential candidates for the visualization of the A_{2A} receptor [170].

In addition to the bicyclic scaffold, various monocyclic compounds were also reported including suitably substituted pyridines, pyrimidines, and thiazoles [4, 66–69, 71–73, 76] (Fig. 4). For example, MRS 1523 (6-ethyl-5-[(ethylthio)carbonyl]-2-phenyl-4-propyl-3-pyridinecarboxylic acid propyl ester) is an A_3 AR antagonist at both human and rat. The compound is only moderately potent but very selective in humans, and, more importantly, it was the first derivative to possess sub-micromolar affinity at the rat A_3ARsubtype [171]. Furthermore, MRS 1523 was useful in clarifying the involvement of the A_3AR in the inhibition of melanoma cell growth. In fact, it has been shown to counteract, both in vitro and in vivo, the modulation in the expression of key signaling proteins, demonstrating that the A_3AR response leads to inhibition of tumor growth [172]. MRS 1523 can exert an antihyperalgesic effect through the blockage of the N-type Ca^{2+} channel in isolated neurons of the rat dorsal root ganglion (DRG). This effect is in common with its ancestors, 1,4-dihydropyridines, known as calcium channel inhibitors, from which it derives by oxidation [173].

Saini et al. reported an in-depth review of AR antagonists covering literature data from 2013 to 2021 [76]. Jacobson et al. proposed a historical overview of the chemical scaffolds typical of AR ligands, highlighting those compounds, respectively, derived from xanthine, adenine, and prototypical monocyclic scaffolds [174].

The virtual screening of chemical libraries using computational approaches has contributed to the discovery of novel AR antagonists. Molecular docking studies at the four ARs, improved by the availability of the X-ray crystal structure of the A_1 and A_{2A} subtypes, have also allowed rationalizing the SARs of a large number of chemical classes of AR antagonists and highlight similarities and differences between the AR binding sites [117, 175–181].

Many compounds that act at the ARs were evaluated that showed their potential as therapeutics. Thus, numerous clinical trials have been conducted, leading to the identification of many agents capable of modulating adenosinergic signaling. Although most of these trials have not been pursued, in some cases the preliminary results were encouraging [77, 78, 182–189]. Currently, in addition to adenosine, only Regadenoson [190] (Fig. 3) and Istradefylline [156] (Fig. 4), agonist and antagonist, respectively, of the A_{2A}AR, are approved for use in humans. In parallel to preclinical and clinical trials, many studies are in progress to investigate new promising compounds as modulators of the adenosine system. The dichotomy of the adenosine system and the consequent controversies that still exist about the pathophysiological roles of ARs make the goal more challenging and stimulating [2, 5].

3.6 Pathophysiological Roles of Adenosine Receptors and Emerging Therapeutic Intervention

Generally, both acute and physiological increases in adenosine production are associated with beneficial effects such as vasodilation and decreased inflammation [2]. However, in important pathological states, adenosine levels can increase due to chronic overproduction, leading to harmful effects such as inflammation and organ damage [5]. Under these conditions, adenosine takes off its guardian angel clothes and becomes a bad angel. Due to the dichotomy of adenosine behavior, the adenosine system can be considered in all respects as an important and strategic target for counteracting old and new unmet diseases. As a consequence, agonists and antagonists of the same AR subtype have been investigated for the treatment of similar pathologies. This approach is not paradoxical considering that the design of an effective therapy may depend on critical factors such as the state of disease progression, the dosage, the time of drug application, and the delivery method. Thus, AR agonists and antagonists, due to their potential, have been evaluated as drug candidates with many therapeutic applications [4, 76, 77, 182]. Since ARs are ubiquitous and involved in major physiological processes, biologically active compounds that interact with the adenosine system at different levels are still of interest.

3.6.1 Central Nervous System Disorders

Studies on adenosine and its receptors in the CNS are instrumental in understanding the pathogenic mechanism in many neurodegenerative disorders such as PD, Alzheimer's (AD), Huntington's disease, and ischemia. Furthermore, much evidence has highlighted that adenosine is also involved in pain [88, 89, 191]. In general, A_1, A_{2A}, and A_3 ARs contribute to neuroprotection but are also involved in neurodegenerative processes. A_{2A}AR is a promising therapeutic target for PD due to its interaction with the dopamine D_2 receptor. The existence of hetero-receptor complexes (ARs-DRs), and the beneficial effect in PD of reciprocal antagonistic interaction [58] were demonstrated. Many A_{2A}AR antagonists are under development and Istradefylline (Fig. 4), endowed with a safety profile and approved in combination with Levodopa for PD therapy, is commercially available in Japan. Furthermore, it was approved in 2019 even by the US Food and Drug Administration (FDA) for therapeutic use in PD [156]. Its approval process in Europe is still in progress [155, 156, 192].

The clinical evaluation of A_{2A}AR antagonists, performed to test their efficacy in PD, also included Preladenant (Fig. 4) whose Phase III trial did not support evidence on its efficacy as monotherapy [193]. It has been repurposed to treat advanced solid tumors (also in combination with pembrolizumab), but the data did not support the study endpoints. Positive results after the clinical trial of another A_{2A}AR antagonist, Tozadenant (Fig. 4), suggested its more promising profile with respect to

Preladenant [194]. However, the evaluation was discontinued and stopped due to serious side effects.

$A_{2A}AR$ is also a therapeutic target for the development of AD and HD drugs. While in HD both A_{2A} receptor agonists and antagonists have shown beneficial effects, leaving the question open [195], in AD selective $A_{2A}AR$ antagonists improve memory and can reduce brain β-amyloid (Aβ) levels [196]. It is worth noting that caffeine (Fig. 1), the most widely used drug, since contained in many consumer beverages, has been reported to reduce the risk of developing AD, leading to decreased brain Aβ formation, and prevent amyloid load and cognitive decline [197]. Furthermore, caffeine intake was associated with protective effects on cognitive impairment in both animals and humans [198, 199]. The health benefit of both coffee and tea consumption is related to folk medicine and experimental evidence [200]. However, caffeine is currently used as a CNS stimulant and in combination with analgesic [200] and has also been evaluated in clinical trials.

The therapeutic use of AR ligands in cerebral ischemia was widely reported [201]. Although the roles of A_3AR are controversial [202], the inhibitory effect of A_1AR activation on excitotoxic glutamate release has been reported to lead to neuroprotection. Unfortunately, the use of selective A_1AR agonists is hampered by peripheral side effects [68]. Several studies reported the neuroprotective mediated-actions of A_3AR. In fact, as with A_1AR, during ischemia, CA1 hippocampal A_3 receptors affect neurotransmission leading to neuroprotection. However, the A_3 subtype, depending on the situation, can also induce damage [202]. Some experimental evidence suggested that A_1AR partial agonists could be used effectively as therapeutics in cerebral ischemia. Two CPA derivatives were reported to have in vitro neuroprotective effects after the application of oxygen-glucose deprivation (OGD) [203].

3.6.2 Pain

Many preclinical and clinical studies revealed that A_1 and, to a lesser extent, $A_{2A}AR$ agonists were effective in counteracting pain. However, their therapeutic application has been hampered by adverse cardiovascular and motor side effects [89]. A_1AR agonists have been reported to inhibit nociceptive input into the dorsal spinal cord [204] and to reduce pain in the periphery [2].

$A_{2A}AR$ agonists, though they exhibit some pronociceptive effects in the periphery, may be useful for inflammatory and neuropathic pain. Activation of the A_{2A} receptor increases the release of interleukin-10 (IL-10) which is a potent anti-inflammatory cytokine and leads to the suppression of pain. The $A_{2A}AR$ agonist Spongosine (2-methoxyadenosine, Fig. 3) isolated from a tropical marine sponge (Caribbean sponge Tectitethya crypta) demonstrated a diverse bioactivity profile that included anti-inflammatory activity and analgesic and vasodilation properties. It showed significant pain relief in an animal without producing typical side effects of AR agonists. Therefore, Spongosine was evaluated in a clinical trial for diabetic neuropathic pain, but the study was interrupted independently of the results obtained [205].

It is important to note that the role of A_1 and A_{2A} ARs in pain is still being debated because both agonists and antagonists have been reported to counteract pain of various origins [89, 206–211]. Although the A_1 subtype has generally been reported to be a neuroprotective receptor [212, 213], some data suggest that prolonged stimulation of A_1AR can lead to neurodegeneration. This result is due to the neuroprotective effects exerted by caffeine and other mixed A_1/A_{2A} AR antagonists, whose action has been attributed to the blockade of both A_1 and A_{2A} ARs [206, 214]. The potential of A_{2A} receptor antagonists in treating pain was suggested [215] and successively confirmed [216, 218]. Clinical studies carried out to evaluate A_1AR agonists for the treatment of chronic pain failed, although they showed efficacy in animal models [217].

Recurring evidence underlines that A_3AR plays a role in pain [219–221]. In particular, A_3AR agonists exerted remarkable and robust antinociceptive properties in acute and chronic pain while being devoid of side effects. This observation is supported by their good safety profile evidenced in non-pain clinical trials [222].

Some studies suggested that the use of A_1AR partial agonists could be an alternative strategy to exploit the therapeutic potential of A_1AR activation to treat pain. Adenosine-like compounds have been reported to show an antihyperalgesic effect comparable to that exerted by a full A_1AR agonist as a reference compound. However, the reduction in pain was lower due to an improper pharmacokinetic profile [223].

3.6.3 Inflammation and Immune Diseases

Experimental evidence supports the use of antagonists and agonists of all ARs in the control of inflammatory and immune system responses [97, 224]. Since ARs are widely expressed on the surface of several immune cell families, the mediator adenosine can modulate their multiple functions, leading to maintenance of tissue integrity. However, excessive adenosine signaling can produce immune system suppression that promotes cancer development and sepsis progression [224]. This prolonged immune response together with dysregulation of immunomodulatory adenosine signaling can contribute to the development of autoimmune disorders.

The natural alkaloid theophylline (Fig. 1) is currently used for the treatment of asthma. However, more potent and safer therapeutic agents have limited its use [225].

An overview of the roles of adenosine in modulation of the immune system with careful attention to potential adenosine-based therapies for autoimmune disorders was recently reported [226]. The review focused on the A_3AR which can represent a very promising therapeutic target due to its overexpression in inflammatory cells. Another aspect that makes this AR subtype particularly attractive is that it is mainly activated in diseased tissues due to its up-regulation in autoimmune diseases [227].

Currently, IB-MECA (Fig. 3), a moderately selective A_3AR agonist, is in a Phase III clinical trial for patients affected by psoriasis due to its demonstrated efficacy and safety profile [78, 228]. It has also been clinically evaluated for the treatment of rheumatoid arthritis, showing a significant anti-rheumatic effect. Its analog Cl-IB-MECA (Fig. 3) is clinically evaluated for autoimmune inflammatory diseases [49, 78].

Even A_{2A} and A_{2B} ARs are involved in inflammation and both agonists and antagonists have been proposed as therapeutic agents [182]. A_{2A}AR agonists, showing good outcomes in an animal model of asthma and chronic obstructive pulmonary disease (COPD), have been evaluated in affected patients. Unfortunately, the clinical studies failed [229]. The FDA-approved synthetic A_{2A}AR agonist Regadenoson (Fig. 3) is currently clinically used as a diagnostic tool for myocardial perfusion imaging [230]. The approval for human treatment allows its evaluation in various clinical trials [77]. Regadenoson is currently being evaluated clinically in lung transplantation due to its efficacy to attenuate ischemia-reperfusion injury at this level [231].

Selective A_{2A} agonists have been administered by inhalation to overcome severe cardiocirculatory effects in clinical trials to treat asthma. However, the trials failed due to their low efficacy [49].

3.6.4 Cancer

Adenosine levels increase in hypoxia that occurs in the solid tumor environment, thus affecting cancer growth. However, most ARs may have opposing roles in modulating the cell cycle or immunity, this depends on different factors [227, 232]. Interestingly, A_3AR is overexpressed in several cancer cells and tissues and is therefore likely to have an important antitumor role [233]. Adenosine, through the A_3AR subtype, activates the protection of tissues. In line with this assumption, many A_3AR agonists showed their efficacy in different animal models of cancer [4, 98, 172, 234]. Two of these, i.e., IB-MECA and Cl-IB-MECA (Fig. 3), are currently in clinical trials for different cancer. In particular, Cl-IB-MECA, which shows both high safety and efficacy profiles, is in phase II trial for hepatocellular carcinoma [235]. The A_3AR agonist IB-MECA was shown to be effective in an animal model of different types of metastatic carcinomas after oral administration [49]. Its synergistic action with cyclophosphamide in the murine lung metastatic melanoma model was demonstrated [97].

The high expression of the A_3AR in many types of cancer cells, together with the growing experimental evidence accumulated over the years, makes it a promising target for drug development and also an effective tumor biomarker [184].

Other innovative tools can be the A_{2A}AR antagonists as immune checkpoint blockers in oncology. Their effectiveness in counteracting cancer can be attributed to the ability to potentiate the immune system response [4, 184, 232, 236] and is supported by their safety profiles evidenced in clinical trials for PD. Several pharmaceutical companies have started clinical evaluations of A_{2A}AR antagonists for oncology immunotherapy. As an example, the A_{2A}AR antagonist Preladenant, whose phase 3 trial for PD failed, has been evaluated for the treatment of advanced solid tumors [78].

Several A_{2A}AR antagonists have been evaluated, but their clinical study has failed, and some others are still in clinical trials in oncology. Recently, iTeos Therapeutics produced a report on the Phase 1 trial of the A_{2A} receptor antagonist Inupadenant (EOS-850, 7-amino-10-[2-[4-[2,4-difluoro-5-[2-[(S)-methylsulfinyl] ethoxy]phenyl]piperazin-1-yl]ethyl]-4-(furan-2-yl)-12-thia-3,5,6,8,10-pentazatricyclo[7.3.0.0]dodeca-1(9),2,4,7-tetraen-11-one) (Fig. 4) indicating its immunomodulatory and antitumor activity when administered in monotherapy [237]. It is reported to produce T-lymphocytes activation and proliferation and stimulation of a T-cell-mediated immune response against tumor cells. Inupadenant was generally well tolerated with the initial evidence of clinical benefit, consistent with previously reported safety data [238]. It is also orally active and does not penetrate brain. The company will continue to evaluate Inupadenant in the ongoing Phase 1b/2a trial in combination with a monoclonal antibody (Pembrolizumab) and chemotherapy, initially focusing on patients with prostate cancer, resistant melanoma, and breast cancer.

Even A_{2B}AR is highly expressed in tumor cells, representing a potential target for cancer therapy. In fact, some A_{2B}AR antagonists are in clinical trials as antitumor agents [97, 98]. However, some evidence demonstrated that the roles of A_{2B}AR in cancer are complex and sometimes controversial. Synergistic action of A_{2A}/A_{2B} AR antagonists was taken into consideration for the treatment of cancer. In fact, they are under clinical evaluation for a polypharmacological treatment in combination with a checkpoint inhibitor [236]. Concomitant AR blockade may improve the efficacy of the chemo/immunotherapy protocol. In particular, small-molecule dual A_{2A}/A_{2B} receptor antagonist, Etrumadenant (AB928, 3-[2-amino-6-[1-[[6-(2-hydroxypropan-2-yl)pyridin-2-yl]methyl]triazol-4-yl]pyrimidin-4-yl]-2-methylbenzonitrile) (Fig. 4), when tested in Phase 1 clinical trials showed good safety and pharmacokinetic profiles after oral administration [239]. More importantly, a significant decrease in tumor growth rate was observed when administered alone or better in combination with chemotherapy.

The ongoing clinical trials include the evaluation of two potent and selective A_{2A} receptor antagonists, Taminadenant (NIR178, PBF-509, 5-bromo-2,6-di (1H-pyrazol-1-yl)pyrimidin-4-amine) and Ciforadenant (Fig. 4). Taminadenant is a potent, no-furan, pyrimidine-based, and orally active antagonist developed by Novartis and Palobiofarma. It was endowed with a potential antineoplastic activity that could reactivate the antitumor immune response. Taminadenant has been introduced in clinical trials for many diseases such as PD and various types of cancer, in combination with a checkpoint inhibitor. In association with Spartalizumab (anti-PD-1 Ab), it showed preliminary antitumor activity in a Phase 1b study in non-small cell lung carcinoma (NSCLC) [240]. The results of the Phase 2 study in patients with microsatellite stable (MSS) colorectal cancer (CRC) showed the efficacy of the therapeutic combination that was well tolerated, leading to clinical benefit [241].

Ciforadenant is an orally active A_{2A}AR antagonist created by Vernalis. It was designed to address the chemical structural liabilities that may have led to concerns about toxicity of Vipadenant [168]. Ciforadenant blocks A_{2A}AR expressed on the surface of immune cells (T-lymphocytes, natural killer (NK) cells, macrophages), thus abrogating adenosine-induced immunosuppression in the tumor microenvironment. This may stimulate antitumor immune responses leading to tumor regression. The preclinical evaluation of Ciforadenant in combination with other checkpoint inhibitors in solid tumors showed the potentiality of this therapeutic approach. Tumors were eliminated in up to 90% of treated mice. Thus, these results suggested the evaluation of Ciforadenant as a therapeutic agent in patients with solid tumors [242]. Ciforadenant was shown to be also active in other preclinical tumor models, both as monotherapy and in combination with anti-PD-L1 [243]. Different clinical studies on Ciforadenant started (safety and tolerability, against refractory multiple myeloma, solid tumors, and advanced cancers), most of which were completed but without posted results. A report from a Phase 1/1b clinical trial demonstrated that Ciforadenant was active alone and in combination with Atezolizumab (anti_PD-L1) in patients with advanced refractory renal cell cancer [244].

3.6.5 Cardiovascular and Related Diseases

Each AR has been taken into consideration as a target for the development of ligand-based therapies against a wide range of cardiovascular and related diseases.

Adenosine itself is currently used for myocardial perfusion imaging for the diagnosis of coronary disease. Through the activation of A_{2A} receptors, it causes coronary vasodilation with consequent variations in blood flow. However, fast adenosine metabolism prevents severe side effects such as cardiac block mediated by the A_1 subtype. In contrast, acute bronchospasm is a recognized side effect related to activation of "off-target" A_{2B}AR. Therefore, in recent times, Regadenoson [190, 191] (Fig. 3) has replaced adenosine for coronary stress imaging. It has been approved for use in humans both in the USA and in the European Union. Regadenoson is a short half-life selective A_{2A} agonist that can be the best choice for patients with known bronchospasms [245]. In addition, the antiarrhythmic action of adenosine is exploited for the treatment of supraventricular tachycardia [246].

The scarcely selective A_3 agonist IB-MECA (Fig. 3) was shown to be cardioprotective in cardiomyocyte progenitor cells [247]. This effect was ascribed to activation of both A_3 and A_{2A} ARs. Cl-IB-MECA also exerted protection against cardiotoxicity by reducing the inflammatory response and apoptosis [248].

Several A_1AR ligands are currently in preclinical studies or are being developed for therapeutic applications. The non-nucleoside A_1AR partial agonist, Capadenoson [249] (Fig. 3), was evaluated in a clinical trial in patients with angina [144], but this study was discontinued. In another phase II clinical study for the treatment of persistent or permanent atrial fibrillation, it did not show to effectively modulate heart rate [146]. However, an analog of Capadenoson, i.e., Neladenoson in the form of its prodrug (Neladenoson bialanate) [150] (Fig. 3), was shown to be safe for the treatment of chronic heart failure without cardiac or neurological side effects [151] but the clinical trial failed [78]. In fact, a lack of restorative effects on cardiac structure and function, and on exercise capacity, was reported, as well as a decrease in renal function [250, 251]. Cardiovascular activity of Capadenoson [10] must be attributed not only to activation of the A_1AR but also the A_{2B} subtype, which promotes cardioprotection and modulates cardiac fibrosis in the diseased heart [148, 252]. A_{2B}AR activation is required for A_1AR-mediated cardioprotection [253]. The greater effectiveness and safety of the Neladenoson bialanate prodrug with respect to Capadenoson is probably due to its higher A_1AR selectivity leading to fewer undesirable side effects [151]. Interestingly, A_1AR partial agonists have also been shown to be effective in reducing blood glucose levels by improving the insulin sensitivity [90, 254].

A_{2A}AR agonists can be considered potential therapeutics in cardiovascular-related diseases. In fact, some of them, in addition to being cardioprotective as a consequence of their anti-inflammatory effects, modulate platelet aggregation [255–257] and cholesterol homeostasis [258] leading to the potential development of new cardiovascular therapies [259].

4 Perspectives

Despite extensive research efforts from the pharmaceutical industry and academia, only a limited number of drugs targeting ARs have reached the market [78]. This is mainly due to the complexity of signaling, as well as the ubiquitous distribution of ARs in both healthy and diseased tissues. In fact, as a consequence, adenosine activates the protection of tissues and cells against injury, but it also stimulates various adverse effects, especially in the cardiovascular and respiratory systems [5].

Although the countless limitations still do not allow us to clarify the numerous dark aspects of adenosine and its receptors, it is undeniable to recognize the great progress that all researchers working in this field have contributed to making. Curiosity continues to fuel research on this topic, and a glance at the current publications gives a measure of the ever-new interest.

Over the years, many important issues have been solved and, at the same time, many important new aspects, up to now ignored or underestimated, have emerged as essential for the development of new drugs. What follows is just a brief mention of the critical issues that have emerged and the new frontiers to be set as objectives for the years to come.

The importance of ligand-binding kinetics in the drug discovery process has been underestimated. Although the affinity of a ligand is important for predicting its therapeutic potential, all kinetic characteristics can highly influence the possibility that a biologically active compound could be developed as a drug. More attention should be paid to this concept to reduce the failures of future clinical trials. The availability of the residence time of orthosteric ligands may allow the choice of a candidate compound with the appropriate residence times for a specific therapeutic use [78]. Furthermore, kinetic studies can also shed light on the mechanism of AR allosteric modulators. These latter potentially represent a novel therapeutic strategy for some pathological conditions as an alternative to orthosteric ligands that often lack sufficient on-target selectivity thus showing adverse effects [128, 182, 260]. The allosteric modulator can stabilize active and/or inactive receptor conformation(s), thus influencing the kinetics of a ligand bound *within* the orthosteric site, in addition to its affinity and/or efficacy. The ability of an AR allosteric ligand to modulate adenosine signaling predominantly in damaged tissues and organs, thus allowing spatiotemporal control, results in a reduction in adverse effects [127, 261]. Allosteric modulators can selectively elicit a physiological response only where and when the orthosteric ligand is released.

Modulation of kinetic characteristics has been useful in generating compounds with an "infinite" residence time since their bond to the target persists over time. These are covalent ligands, i.e., compounds that irreversibly bind to the receptor because they are endowed with a reactive moiety versus specific amino acid residues of the binding site. Covalent ligands as well as many other kinds of AR probes gave a contribution to clarifying the structural biology of ARs that suffer from numerous limitations [62]. For example, the structure of human A_1AR was elucidated by X-ray crystallography with the covalent antagonist, DU172 (4-[3-(8-cyclohexyl-2,6-dioxo-1-propyl-7H-purin-3-yl)propylcarbamoyl]benzenesulfonyl fluoride) [117] (Fig. 4).

Over the past decades, a diverse array of molecular probes that can be used to investigate receptor structure and function have proven invaluable in AR research. However, though advanced techniques have generated a great number of structures for GPCRs including A_1 and A_{2A} ARs, structure elucidation of A_{2B} and A_3 receptors has not yet been successful. These will be the challenges that experts in this field will face in the coming years. Moreover, the lack of experimental techniques capable of evaluating dynamics opens the question of characterization of the thermodynamic and kinetic properties of ligand binding to ARs [262].

An additional focus has to be placed on the particular phenomenon of some AR ligands, biased agonism, also known as functional selectivity. Exploiting the ability of a receptor to differentially activate therapeutic signaling pathways while avoiding others that could lead to side effects appears to be an intriguing strategy. Biased AR agonists, which can preferentially signal through a G protein-independent pathway, can be considered innovative smart drugs potentially endowed with fewer side effects. This new aspect of AR signaling should be taken into account for the design of new drugs, to reduce the possibility of clinical trial failure [104].

Although different AR ligands have been studied at the preclinical level to fight cancer, a few molecules have progressed to clinical trials. The A_3AR agonist Cl-IB-MECA is under clinical evaluation for the treatment of advanced hepatocellular carcinoma treatment. However, the A_3AR appears to have a dual nature, as evidence of pro-tumoral and antitumoral activity suggests that other mechanisms are involved in the anticancer properties of A_3AR ligands [227, 232]. Therefore, an in-depth investigation must be performed taking into account also that extrapolation from preclinical animal studies to humans is hazardous due to the different expression and functions of A_3AR in the various animal species [106].

Several A_{2A}AR antagonists are being evaluated. In fact, A_{2A}AR has recently received renewed attention for its roles in immuno-oncology [263]. Although early clinical studies have shown modest benefits with A_{2A}AR blockade, considerable work is needed to identify new polypharmacological strategies in which checkpoint inhibition is associated with other immune-based therapies [264]. In addition to A_{2A}AR, the A_{2B} subtype was also proposed as a novel target for the (immuno)-therapy of cancer. In fact, the discovery of new molecules capable of antagonizing A_{2B}AR or both A_{2B} and A_{2A} subtypes opened a new perspective in pharmacotherapy. However, only a few A_{2B}AR antagonists have reached clinical trials [78]. Currently, the most likely scenario is that AR modulation could serve as an additional therapy to other anticancer strategies (chemo-, immune- and radio-therapy) to combat a wide variety of cancers.

A polypharmacological approach has a higher probability of success than a one-target therapy to treat not only cancer but also many other diseases with high social impact [265]. The multifactorial character of these diseases suggests exploiting the simultaneous modulation of multiple targets involved in the disorder that must be fought.

Thus, a new goal may be the discovery of multitarget-directed ligands (MTDLs) (two or more birds with one stone) to provide a strong synergistic effect against a specific pathology. MTDLs can be designed by combining two (or more) different pharmacophoric features by using diverse strategies leading, in general, to bitopic, bivalent, and dual-acting ligands [266, 267]. The advantages of a ligand obtained with the merging approach over ligands designed by pharmacophoric linkage or fusion (reduction of molecular weight, topological polar surface area, and clogP) could result in improved "drug-like" properties. The MTDL approach resulted from the observation that compounds modulating a single target might not be optimal drug candidates for the treatment of complex diseases [268]. Thus, to overcome this issue, a single molecule with dual activity could be a more effective drug because it is endowed with higher efficacy, better pharmacokinetic profiles, and also compliance with the patient. This strategy is expected to have many advantages compared to the coadministration of different drugs. Over the years, some progress has been made. Interesting ligands targeting one or more ARs, or one AR and other GPCRs (such as β-adrenergic, dopaminergic, histaminergic), or an enzyme (monoamine oxidase (MAO), carbonic anhydrases (CAs)) have been obtained and studied. Some of the more recent and/or interesting reports are suggested [269–274]. Although few general reviews on the peculiarities of these types of ligands are available [266, 267], a detailed overview of progress in the field of MTDLs targeting ARs is needed.

5 Concluding Remarks

In conclusion, the rapid progress of studies on adenosine and its receptors has had a strong impact on the drug discovery process. Much evidence indicated that ARs are involved in the physiological regulation of several homeostatic processes and, as a consequence, in the etiology of different pathologies. Moreover, understanding the molecular biology of each AR has provided a solid basis for considering them as targets for adenosine-based pharmacotherapies. Previously and newly reported discoveries allowed the identification of increasingly active and selective AR ligands endowed with a safety profile. Thus, the therapeutic potential of these AR ligands was evidenced and the frontiers for their admission to clinical trial programs were opened.

Perspectives for compounds endowed with adequate kinetic properties or that interfere with adenosine signaling are very encouraging. Emerging as potential innovative therapeutic agents, they are likely to reach the role of drugs for clinical use.

Attention should be paid to $A_{2A}AR$ antagonists whose innovative mechanism of action in oncology has been validated. In fact, the development of $A_{2A}AR$ antagonists as anticancer agents seems to rapidly progress, and some of them are already being evaluated in a clinical trial. This attractive therapeutic use is highly competitive and is supported by the safety profile of these compounds that was evidenced in clinical trials for PD. Similar consideration can be given to A_3AR agonists that have emerged not only as promising anticancer agents but also as successful drugs to counteract pain. Therefore, they are or are being tested in clinical trials for neuropathic pain [275]. As in the case of $A_{2A}AR$ antagonists, their good safety profile was evidenced in non-pain clinical studies, thus supporting their potential role as therapeutics in humans.

Last but not least, it is important to underline the well-known efficacy of A_1AR partial agonists in cardiovascular diseases, such as chronic heart failure, atrial fibrillation, and angina, as documented by the results obtained in clinical trials. However, the anti-inflammatory effect that makes them effective in cardiovascular-related diseases could work in synergy to develop new cardiovascular therapies.

From a future perspective, polypharmacology seems to represent an innovative therapeutic strategy [265]. In particular, small "drug-like" molecules targeting two different receptor binding sites simultaneously could be the best choice. The development of MTDLs has become a promising approach in drug design and discovery to overcome various problems related to the administered drugs and/or therapeutic protocols. In fact, it remains one of the major challenges in drug development and opens new avenues for rational design of the next generation of more effective but less toxic therapeutic agents.

The search for novel AR-targeted MTDLs with adequate balanced activities could represent a new strategy to combat multifactorial-dependent diseases in which adenosine is involved. The goal of developing AR ligands as drugs accepted for clinical use in humans appears to be getting closer and closer. This is what every researcher working in this field continues to do with the tenacity to try to achieve with a contribution, be it small or large.

Compliance with Ethical Standards Conflict of Interest

The author declares that they have no conflict of interest.
Funding Original research of our team is funded by the Italian Ministry for University and Research (MIUR, PRIN 2017MT3993_004 project).

Ethical Approval This chapter does not contain any studies with human participants or animals performed by the authors.

References

1. Williams RA (1989) Purine receptors. In: Williams M, Glennon RA, Timmermans PBMWM (eds) Receptor pharmacology and function. Weiner M, Dekker M, New York, pp 503–525
2. Borea PA, Gessi S, Merighi S, Varani K (2016) Adenosine as a multi-signalling guardian angel in human diseases: when, where and how does it exert its protective effects? Trends Pharmacol Sci 37:419–434
3. Cunha RA (2001) Adenosine as a neuromodulator and as a homeostatic regulator in the nervous system: different roles, different sources and different receptors. Neurochem Int 38: 107–125
4. Borea PA, Varani K, Gessi S, Merighi S, Vincenzi F (2019) The adenosine receptors. In: Di Giovanni G (ed) The receptors, vol 34. Humana Press, pp 1–593
5. Borea PA, Gessi S, Merighi S, Vincenzi F, Varani K (2017) Pathological overproduction: the bad side of adenosine. Br J Pharmacol 174:1945–1960
6. Scheele KW (1776) Analysis of the calculus vesicae. Essay IX. In: The chemical essays of Charles-William Scheele. Translated from the transactions of the Academy of Sciences at Stockholm, with additions (1786). J. Murray, London, pp 199–214
7. Fischer E (1890) Synthesen in der Zuckergruppe. Ber Deut Chem Ges 23:2114–2141
8. Fischer E (1899) Synthesen in der Puringruppe. Ber Deut Chem Ges 32:435–504
9. Fischer E (1907) Untersuchungen in Der Puringruppe: (1882–1906). Springer Verlag, Berlin
10. Kossel A (1885) Ueber eine neue Base aus dem Tierkorper. Ber Deut Chem Ges 18:79–81
11. Kossel A (1888) Ueber eine neue Base aus dem Pflanzenreich. Ber Deut Chem Ges 21:2164–2167
12. Levene PA, Jacobs WA (1911) Ueber die Inosinsaure. Vierte Mitteilung Ber Deut Chem Ges 44:746–752
13. Maruyama K (1991) The discovery of adenosine triphosphate and the establishment of its structure. J Hist Biol 24:145–154
14. Drury AN, Szent-Gyorgyi A (1929) The physiological activity of adenine compounds with especial reference to their action on the mammalian heart. J Physiol 68:213–237
15. Bennett DW, Drury AN (1931) Further observations relating to the physiological activity of adenine compounds. J Physiol 72:28–320
16. Gillespie JH (1934) The biological significance of the linkages in adenosine triphosphoric acid. J Physiol 80:345–359
17. Honey RM, Ritchie WT, Thomson WAR (1930) The action of adenosine upon the human heart. QJM os-23(92):485–489
18. Berne RM (1963) Cardiac nucleotides in hypoxia: possible role in regulation of coronary blood flow. Am J Physiol 204:317–322

19. Gerlach E, Deuticke B, Dreisbach RH (1963) Die Nucleotid-Abbauim Herzmuskel bei Sauerstoffmangel und seine mogliche Bedeutung fur die Coronardurchblutung. Naturwissenschaft
20. Burnstock G (1972) Purinergic nerves. Pharmacol Rev 24:509–581
21. Rubio R, Berne RM (1969) Release of adenosine by the normal myocardium in dogs and its relationship to the regulation of coronary resistance. Circ Res 25:407–415
22. Olsson RA (1970) Changes in content of purine nucleoside in canine myocardium during coronary occlusion. Circ Res 26:301–306
23. Fox AC, Reed GE, Glassman E, Kaltman AJ, Silk BB (1974) Release of adenosine from human hearts during angina induced by rapid atrial pacing. J Clin Invest 53:1447–1457
24. Sattin A, Rall TW (1970) The effect of adenosine and adenine nucleotides on the cyclic adenosine 3′,5′-phosphate content of Guinea pig cerebral cortex slices. Mol Pharmacol 6:13–23
25. Shimizu H, Daly J (1970) Formation of cyclic adenosine 3′,5′-monophosphate from adenosine in brain slices. Biochim Biophys Acta 222(2):465–473
26. Bretschneider HJ, Frank A, Bernard U, Kochsiek K, Scheler F (1959) Effect of a pyrimido-pyrimidine derivative on the oxygen supply of the heart muscle. Arzneimittelforschung 9:49–59
27. Pfleger K, Seifen E, Schöndorf H (1969) Inosine potentiation of the effect of adenosine on the heart. Biochem Pharmacol 18:43–51
28. Dietmann K, Schaumann W, Thiel M (1981) Die Geschichte der Adenosin-Derivate. Ein Abenteuer der Arzniemittelforschung. Getbooks GmbH, Bad Camberg, Germany
29. Rubio R, Berne RM (1980) Location of purine and pyrimidine nucleoside phosphorylases in heart, kidney and liver. Am J Physiol 239:H721–H730
30. Hershfield MS, Kredich NM (1978) S-adenosylhomocysteine hydrolase is an adenosine binding protein: a target for adenosine toxicity. Science 202:757–760
31. Schrader J, Nees S, Gerlach E (1981) Role of S-adenosylhomocysteine hydrolase in adenosine metabolism in mammalian heart. Biochem J 196:65–70
32. Nagy JI, La Bella LA, Buss M, Daddona PE (1984) Immunohistochemistry of adenosine deaminase: implications for adenosine neurotransmission. Science 224(4645):166–168
33. Burnstock G (1978) A basis for distinguishing two types of purinergic receptor. In: Straub RW, Boils L (eds) Cell membrane receptors for drugs and hormones: a multidisciplinary approach. Raven Press, New York, pp 107–118
34. Fredholm BB (1982) Adenosine receptors. Med Biol 60(6):289–293
35. Van Calker D, Müller M, Hamprecht B (1979) Adenosine regulates via two different types of receptors, the accumulation of cyclic AMP in cultured brain cells. J Neurochem 33:999–1005
36. Londos C, Wolff J (1977) Two distinct adenosine-sensitive sites on adenylate cyclase. Proc Natl Acad Sci U S A 74:5482–5486
37. Schrader J, Baumann G, Gerlach E (1977) Adenosine as inhibitor of myocardial effects of catecholamines. Pflugers Arch Eur J Physiol 372:29–35
38. Böhm M, Brückner R, Neumann J, Schmitz W, Scholz H, Starbatty J (1986) Role of guanine nucleotide-binding protein in the regulation by adenosine of cardiac potassium conductance and force of contraction. Evaluation with pertussis toxin. Naunyn Schmiedebergs Arch Pharmacol 332:403–405
39. Dolphin AC, Forda SR, Scott RH (1986) Calcium-dependent currents in cultured rat dorsal root ganglion neurones are inhibited by an adenosine analogue. J Physiol 373:47–61
40. Bruns RF, Lu GH, Pugsley TA (1986) Characterization of the A2 adenosine receptor labeled by [^{3}H]NECA in rat striatal membranes. Mol Pharmacol 29:331–346
41. Fredholm BB (1977) Activation of adenylate cyclase from rat striatum and tuberculum olfactorium by adenosine. Med Biol 55:262–267
42. Premont J, Perez M, Bockaert J (1977) Adenosine-sensitive adenylate cyclase in rat striatal homogenates and its relationship to dopamine- and Ca^{2+}-sensitive adenylate cyclases. Mol Pharmacol 13:662–670

43. Daly JW, Butts-Lamb P, Padgett W (1983) Subclasses of adenosine receptors in the central nervous system: interaction with caffeine and related methylxanthines. Cell Mol Neurobiol 3: 69–80
44. Ribeiro JA, Sebastião AM (1986) Adenosine receptors and calcium: basis for proposing a third (A3) adenosine receptor. Prog Neurobiol 26:179–209
45. Ali H, Cunha-Melo JR, Saul WF, Beaven MA (1990) Activation of phospholipase C via adenosine receptors provides synergisti signals for secretion in antigen-stimulated RBL-2H3 cells. Evidence for a novel adenosine receptor. J Biol Chem 265:745–753
46. Zhou QY, Li C, Olah ME, Johnson RA, Stiles GL, Civelli O (1992) Molecular cloning and characterization of an adenosine receptor: the A_3 adenosine receptor. Proc Natl Acad Sci U S A 89:7432–7436
47. Fredholm BB, Ijzerman AP, Jacobson KA, Klotz KN, Linden J (2001) International Union of Pharmacology. XXV. Nomenclature and classification of adenosine receptors. Pharmacol Rev 53:527–552
48. Fredholm BB, IJzerman AP, Jacobson KA, Linden J, Müller CE (2011) International union of basic and clinical pharmacology. LXXXI. Nomenclature and classification of adenosine receptors – an update. Pharmacol Rev 63:1–34
49. Borea PA, Gessi S, Merighi S, Vincenzi F, Varani K (2018) Pharmacology of adenosine receptors: the state of the art. Physiol Rev 98:1591–1625
50. Malbon CC, Hert RC, Fain JN (1978) Characterization of [^{3}H]adenosine binding to fat cell membranes. J Biol Chem 253:3114–3122
51. Schwabe U, Trost T (1980) Characterization of adenosine receptors in rat brain by (–)[^{3}H]N^6-phenylisopropyladenosine. Naunyn Schmiedebergs Arch Pharmacol 313:179–187
52. Williams M, Risley EA (1980) Biochemical characterization of putative central purinergic receptors by using 2-chloro[^{3}H]adenosine, a stable analog of adenosine. Proc Natl Acad Sci U S A 77:6892–6896
53. Bruns RF, Fergus JH, Badger EW, Bristol JA, Santay LA, Hartman JD, Hays SJ, Huang CC (1987) Binding of the A_1-selective adenosine antagonist 8-cyclopentyl-1,3-dipropylxanthine to rat membranes. Naunyn Schmied Arch Pharmacol 335:59–63
54. Jarvis MF, Williams M (1989) Direct autoradiographic localization of adenosine A_2 receptors in the rat brain using the A_2-selective agonist, [^{3}H]CGS 21680. Eur J Pharmacol 168:243–246
55. Zocchi C, Ongini E, Conti A, Monopoli A, Negretti A, Baraldi PG, Dionisotti S (1996) The non-xanthine heterocyclic compound SCH 58261 is a new potent and selective A2a adenosine receptor antagonist. J Pharmacol Exp Ther 276(2):398–404
56. Poucher SM, Keddie JR, Singh P, Stoggall SM, Caulkett PW, Jones G, Coll MG (1995) The in vitro pharmacology of ZM 241385, a potent, non-xanthine A_2a selective adenosine receptor antagonist. Br J Pharmacol 115:1096–1102
57. Olah ME, Gallo-Rodriguez C, Jacobson KA, Stiles GL (1994) ^{125}I-4-aminobenzyl-5X-N methylcarboxamidoadenosine, a high affinity radioligand for the rat A_3 adenosine receptor. Mol Pharmacol 45:978–982
58. Baraldi PG, Cacciari B, Romagnoli R, Varani K, Merighi S, Gessi S, Borea PA, Leung E, Hickey SL, Spalluto G (2000) Synthesis and preliminary biological evaluation of [^{3}H]-MRE 3008-F20: the first high affinity radioligand antagonist for the human A_3 adenosine receptors. Bioorg Med Chem Lett 10:209–211
59. Hinz S, Alnouri WM, Pleiss U, Müller CE (2018) Tritium-labeled agonists as tools for studying adenosine A_{2B} receptors. Purinergic Signal 14:223–233
60. Federico S, Lassiani L, Spalluto G (2019) Chemical probes for the adenosine receptors. Pharmaceuticals 12:168
61. Yang X, Heitman LH, IJzerman AP, van der Es D (2021) Molecular probes for the human adenosine receptors. Purinergic Signal 17(1):85–108
62. Weichert D, Gmeiner P (2015) Covalent molecular probes for class A G protein-coupled receptors: advances and applications. ACS Chem Biol 10:1376–1386

63. Klotz KN (2000) Adenosine receptors and their ligands. Naunyn Schmiedebergs Arch Pharmacol 362(4–5):382–391
64. Ukena D, Padgett WL, Hong O, Daly JW, Daly DT, Olsson RA (1987) N6-substituted 9-methyladenines: a new class of adenosine receptor antagonists. FEBS Lett 215(2):203–208
65. Francis JE, Cash WD, Psychoyos S, Ghai G, Wenk P, Friedmann RC, Atkins C, Warren V, Furness P (1988) Structure-activity profile of a series of novel triazoloquinazoline adenosine antagonists. J Med Chem 31(5):1014–1020
66. Kiesman WF, Elzein E, Zablocki J (2009) A_1 adenosine receptor antagonists, agonists, and allosteric enhancers. In: Wilson C, Mustafa S (eds) Adenosine receptors in health and disease. Handbook of experimental pharmacology, vol 193. Springer, Berlin, pp 25–58
67. Cristalli G, Müller CE, Volpini R (2009) Recent developments in adenosine A_{2A} receptor ligands. In: Wilson CN, Mustafa SJ (eds) Adenosine receptors in health and disease. Handbook of experimental pharmacology, vol 193. Springer, Berlin, pp 59–98
68. Deb PK, Deka S, Borah P, Abed SN, Klotz K-N (2019) Medicinal chemistry and therapeutic potential of agonists, antagonists and allosteric modulators of A_1 adenosine receptor: current status and perspectives. Curr Pharm Des 25:2697–2715
69. Preti D, Baraldi PG, Moorman AR, Borea PA, Varani K (2015) History and perspectives of A_{2A} adenosine receptor antagonists as potential therapeutic agents. Med Res Rev 35:790–848
70. Guerrero A (2018) A_{2A} adenosine receptor agonists and their potential therapeutic applications. An update. Curr Med Chem 25:3597–3612
71. Kalla RV, Zablocki J, Tabrizi MA, Baraldi PG (2009) Recent developments in A2B adenosine receptor ligands. In: Wilson C, Mustafa S (eds) Adenosine receptors in health and disease. Handbook of experimental pharmacology, vol 193. Springer, Berlin, pp 99–122
72. Jacobson KA, Klutz A, Tosh DK, Ivanov AA, Preti D, Baraldi PG (2009) Medicinal chemistry of the A_3 adenosine receptor: agonists, antagonists, and receptor engineering. In: Wilson C, Mustafa S (eds) Adenosine receptors in health and disease. Handbook of experimental pharmacology, vol 193. Springer, Berlin, pp 123–159
73. Müller CE, Jacobson KA (2011) Recent developments in adenosine receptor ligands and their potential as novel drugs. Biochim Biophys Acta 1808:1290–1308
74. Gao Z-G, Jacobson KA (2011) Emerging adenosine receptor agonists-an update. Expert Opin Emerg Drugs 16(4):597–602
75. Dal Ben D, Lambertucci C, Buccioni M, Navia AM, Marucci G, Spinaci A, Volpini R (2019) Non-nucleoside agonists of the adenosine receptors: an overview. Pharmaceuticals 12:150
76. Saini A, Patel R, Gaba S, Singh G, Gupta GD, Monga V (2022) Adenosine receptor antagonists: recent advances and therapeutic perspective. Eur J Med Chem 227:113907
77. Jacobson KA, Tosh DK, Jain S, Gao Z-G (2019) Historical and current adenosine receptor agonists in preclinical and clinical development. Front Cell Neurosci 13:124
78. IJzerman AP, Jacobson KA, Müller CE, Cronstein BN, Cunha RA (2022) International Union of Basic and Clinical Pharmacology. CXII: adenosine receptors: a further update. Pharmacol Rev 74:340–372
79. Di Marco JP, Sellers TD, Berne RM, West GA, Belardinelli L (1985) Diagnostic and therapeutic use of adenosine in patients with supraventricular tachyarrhythmias. J Am Coll Cardiol 6:417–425
80. Verani M (1991) Pharmacological stress with adenosine for myocardial perfusion imaging. Semin Nucl Med 21:266–272
81. Newby AC (1984) Adenosine and the concept of "retaliatory metabolites". Trends Biochem Sci 9:42–44
82. Zimmermann H (2000) Extracellular metabolism of ATP and other nucleotides. Naunyn Schmiedebergs Arch Pharmacol 362:299–309
83. Chen JF, Eltzschig HK, Fredholm BB (2013) Adenosine receptors as drug targets – what are the challenges? Nat Rev Drug Discov 12:265–286

84. Deussen A, Stappert M, Schäfer S, Kelm M (1999) Quantification of extracellular and intracellular adenosine production: understanding the transmembranous concentration gradient. Circulation 99:2041–2047
85. Peleli M, Fredholm BB, Sobrevia L, Carlström M (2017) Pharmacological targeting of adenosine receptor signaling. Mol Aspects Med 55:4–8
86. Gracia E, Farré D, Cortés A, Ferrer-Costa C, Orozco M, Mallol J, Lluís C, Canela EI, McCormick PJ, Franco R, Fanelli F, Casadó V (2013) The catalytic site structural gate of adenosine deaminase allosterically modulates ligand binding to adenosine receptors. FASEB J 27:1048–1061
87. Pacheco R, Martinez-Navio JM, Lejeune M, Climent N, Oliva H, Gatell JM, Gallart T, Mallol J, Lluis C, Franco R (2005) CD26, adenosine deaminase, and adenosine receptors mediate costimulatory signals in the immunological synapse. Proc Natl Acad Sci U S A 102: 9583–9588
88. Liu Y-J, Chen J, Li X, Zhou X, Hu Y-M, Chu S-F, Peng Y, Chen N-H (2019) Research progress on adenosine in central nervous system diseases. CNS Neurosci Ther 25:899–910
89. Sawynok J (2016) Adenosine receptor targets for pain. Neuroscience 338:1–18
90. Dhalla AK, Chisholm JW, Reaven GM, Belardinelli L (2009) A_1 adenosine receptor: role in diabetes and obesity. Handb Exp Pharmacol 193:271–295
91. Varani K, Vincenzi F, Merighi S, Gessi S, Borea PA (2017) Biochemical and pharmacological role of A_1 adenosine receptors and their modulation as novel therapeutic strategy. Adv Exp Med Biol 1051:193–232
92. Rabadi MM, Lee HT (2015) Adenosine receptors and renal ischaemia reperfusion injury. Acta Physiol 213:222–231
93. Hua X, Erikson CJ, Chason KD, Rosebrock CN, Deshpande DA, Penn RB, Tilley SL (2007) Involvement of A_1 adenosine receptors and neural pathways in adenosine-induced bronchoconstriction in mice. Am J Physiol Lung Cell Mol Physiol 293:L25–L32
94. Wilson CN, Nadeem A, Spina D, Brown R, Page CP, Mustafa SJ (2009) Adenosine receptors and asthma. Handb Exp Pharmacol 193:329–362
95. Wolska N, Rozalski M (2019) Blood platelet adenosine receptors as potential targets for anti-platelet therapy. Int J Mol Sci 20(21):5475
96. Janes K, Esposito E, Doyle T, Cuzzocrea S, Tosh DK, Jacobson KA, Salvemini D (2014) A_3 adenosine receptor agonist prevents the development of paclitaxel-induced neuropathic pain by modulating spinal glial-restricted redox-dependent signaling pathways. Pain 155:2560–2567
97. Jacobson KA, Merighi S, Varani K, Borea PA, Baraldi S, Aghazadeh Tabrizi M, Romagnoli R, Baraldi PA, Ciancetta A, Tosh DK, Gao Z-G, Gessi S (2018) A_3 adenosine receptors as modulators of inflammation: from medicinal chemistry to therapy. Med Res Rev 38(4):1031–1072
98. Marucci G, Santinelli C, Buccioni M, Martí Navia A, Lambertucci C, Zhurina A, Yli-Harjab O, Volpini R, Kandhavelu M (2018) Anticancer activity study of A3 adenosine receptor agonists. Life Sci 205:155–163
99. Hussain A, Gharanei AM, Nagra AS, Maddock HL (2014) Caspase inhibition via A_3 adenosine receptors: a new cardioprotective mechanism against myocardial infarction. Cardiovasc Drugs Ther 28:19–32
100. Gao Z-G, Inoue A, Jacobson KA (2018) On the G protein-coupling selectivity of the native A_{2B} adenosine receptor. Biochem Pharmacol 151:201–213
101. Cordeaux Y, IJzerman AP, Hill SJ (2004) Coupling of the human A_1 adenosine receptor to different heterotrimeric G proteins: evidence for agonist-specific G protein activation. Br J Pharmacol 143:705–714
102. Kull B, Svenningsson P, Fredholm BB (2000) Adenosine A2A receptors are colocalized with and activate golf in rat striatum. Mol Pharmacol 58(4):771–777

103. Schulte G, Fredholm BB (2000) Human adenosine A(1), A(2A), A(2B), and A(3) receptors expressed in Chinese hamster ovary cells all mediate the phosphorylation of extracellular-regulated kinase 1/2. Mol Pharmacol 58:477–482
104. McNeill SM, Jo-Anne B, White PJ, May Lauren T (2021) Biased agonism at adenosine receptors. Cell Signal 82:109954
105. Wall MJ, Hill E, Huckstepp R, Barkan K, Deganutti G, Leuenberger M, Preti B, Winfield I, Wei H, Imlach W et al (2020) A biased adenosine A_1R agonist elicits analgesia without cardiorespiratory depression. bioRxiv. 2020.04.04.023945
106. Alnouri MW, Jepards S, Casari A, Schiedel AC, Hinz S, Müller CE (2015) Selectivity is species-dependent: characterization of standard agonists and antagonists at human, rat, and mouse adenosine receptors. Purinergic Signal 11:389–407
107. Du L, Gao ZG, Paoletta S, Wan TC, Gizewski ET, Barbour S, van Veldhoven JPD, IJzerman AP, Jacobson KA, Auchampach JA (2018) Species differences and mechanism of action of A_3 adenosine receptor allosteric modulators. Purinergic Signal 14:59–71
108. Fredholm BB, Arslan G, Halldner L et al (2000) Structure and function of adenosine receptors and their genes. Naunyn Schmiedebergs Arch Pharmacol 362:364–374
109. Jaakola V-P, Griffith MT, Hanson MA, Cherezov V, Chien EYT, Lane JR, Ijzerman AP, Stevens RC (2008) The 2.6 angstrom crystal structure of a human A_{2A} adenosine receptor bound to an antagonist. Science 322:1211–1217
110. Doré AS, Robertson N, Errey JC, Ng I, Hollenstein K, Tehan B, Hurrell E, Bennett K, Congreve M, Magnani F, Tate CG, Weir M, Marshall FH (2011) Structure of the adenosine A(2A) receptor in complex with ZM241385 and the xanthines XAC and caffeine. Structure 19:1283–1293
111. Lebon G, Warne T, Edwards PC, Bennett K, Langmead CJ, Leslie AGW, Tate CG (2011) Agonist-bound adenosine A_{2A} receptor structures reveal common features of GPCR activation. Nature 474:521–525
112. Xu F, Wu H, Katritch V, Han GW, Jacobson KA, Gao Z-G, Cherezov V, Stevens RC (2011) Structure of an agonist-bound human A_{2A} adenosine receptor. Science 332:322–327
113. Carpenter B, Nehmé R, Warne T, Leslie AG, Tate CG (2016) Structure of the adenosine A (2A) receptor bound to an engineered G protein. Nature 536:104–107
114. Barrington WW, Jacobson KA, Stiles GL (1989) Demonstration of distinct agonist and antagonist conformations of the A_1 adenosine receptor. J Biol Chem 264:13157–13164
115. Olah ME, Jacobson KA, Stiles GL (1994) Role of the second extracellular loop of adenosine receptors in agonist and antagonist binding: analysis of chimeric A_1/A_3 adenosine receptors. J Biol Chem 269:24692–24698
116. Tosh DK, Paoletta S, Deflorian F et al (2012) Structural sweet spot for A_1 adenosine receptor activation by truncated (N)-methanocarba nucleosides: receptor docking and potent anticonvulsant activity. J Med Chem 55:8075–8090
117. Glukhova A, Thal DM, Nguyen AT, Vecchio EA, Jörg M, Scammells PJ, May LT, Sexton PM, Christopoulos A (2017) Structure of the adenosine A_1 receptor reveals the basis for subtype selectivity. Cell 168:867–877
118. Cheng Y (2018) Single-particle cryo-EM-how did it get here and where will it go. Science 361: 876–880
119. Ceska T, Chung CW, Cooke R, Phillips C, Williams PA (2019) Cryo-EM in drug discovery. Biochem Soc Trans 47:281–293
120. Draper-Joyce CJ, Khoshouei M, Thal DM, Liang YL, Nguyen ATN, Furness SGB, Venugopal H, Baltos JA, Plitzko JM, Danev R (2018) Structure of the adenosine-bound human adenosine A_1 receptor-Gi complex. Nature 558:559–563
121. García-Nafría J, Lee Y, Bai X, Carpente B, Tate CG (2018) Cryo-EM structure of the adenosine A_{2A} receptor coupled to an engineered heterotrimeric G protein. Elife 7:e35946
122. Wang J, Bhattarai A, Ahmad WI, Farnan TS, John KP, Miao Y (2020) Computer-aided GPCR drug discovery. In: Jastrzebska B, Park PSH (eds) GPCRs. Academic Press, Cambridge, pp 283–293

123. Sussman JL, Lin D, Jiang J, Manning NO, Prilusky J, Ritter O, Abola EE (1998) Protein Data Bank (PDB): database of three-dimensional structural information of biological macromolecules. Acta Crystallogr Sect D Biol Crystallogr 54:1078–1084
124. Karplus M, Mc Cammon JA (2002) Molecular dynamics simulations of biomolecules. Nat Struct Mol Biol 9:646–652
125. Wang J, Miao Y (2019) Mechanistic insights into specific G protein interactions with adenosine receptors. J Phys Chem B 123(30):6462–6473
126. Wang J, Bhattarai A, Do HN, Akhter S, Miao Y (2022) Molecular simulations and drug discovery of adenosine receptors. Molecules 27:2054
127. May LT, Leach K, Sexton PM, Christopoulos A (2007) Allosteric modulation of G protein-coupled receptors. Annu Rev Pharmacol Toxicol 47:1–51
128. Vecchio EA, Baltos J-A, Nguyen ATN, Christopoulos A, White PJ, May LT (2018) New paradigms in adenosine receptor pharmacology: allostery, oligomerization and biased agonism. Br J Pharmacol 175(21):4036–4046
129. Korkutata M, Agrawal L, Lazarus M (2022) Allosteric modulation of adenosine A_{2A} receptors as a new therapeutic avenue. Int J Mol Sci 23(4):2101
130. Poulsen S-A, Quinn RJ (1998) Adenosine receptors: new opportunities for future drugs. Bioorg Med Chem 6(6):619–641
131. Wu PH, Phillis JW (1982) Adenosine receptor in rat brain membranes: characterization of high affinity binding of [^{3}H]-2-chloroadenosine. Int J Biochem 14:399–404
132. Müller CE, Baqi Y, Hinz S, Namasivayam V (2018) Medicinal chemistry of A_{2B} adenosine receptors. In: Borea PA, Varani K, Gessi S, Merighi S, Vincenzi F (eds) The adenosine receptors. The receptors, vol 34. Springer, Cham, pp 137–168
133. Beukers MW, Chang LCW, von Frijtag Drabbe Kunzel JK, Mulder-Krieger T, Spanjersberg RF, Brussee J, IJzermann AP (2004) New, non-adenosine, high-potency agonists for the human adenosine A_{2B} receptor with an improved selectivity profile compared to the reference agonist N-ethylcarboxamidoadenosine. J Med Chem 47:3707–3709
134. Hinz S, Lacher SK, Seibt BF, Müller CE (2014) BAY60-6583 acts as a partial agonist at adenosine A_{2B} receptors. J Pharmacol Exp Ther 349:427–436
135. Guo M, Gao ZG, Tyler R, Stodden T, Li Y, Ramsey J, Zhao WJ, Wang GJ, Wiers CE, Fowler JS, Rice KC, Jacobson KA, Kim SW, Volkow ND (2018) Preclinical evaluation of the first adenosine A_1 receptor partial agonist radioligand for positron emission tomography imaging. J Med Chem 61:9966–9975
136. Bertarelli DC, Diekmann M, Hayallah AM, Rüsing D, Iqbal J, Preiss B, Verspohl EJ, Müller CE (2006) Characterization of human and rodent native and recombinant adenosine A (2B) receptors by radioligand binding studies. Purinergic Signal 2:559–571
137. Betti M, Catarzi D, Varano F, Falsini M, Varani K, Vincenzi F, Dal Ben D, Lambertucci C, Colotta V (2018) The aminopyridine-3,5-dicarbonitrile core for the design of new non-nucleoside-like agonists of the human adenosine A_{2B} receptor. Eur J Med Chem 150:127–139
138. Johnston-Cox H, Koupenova M, Yang D, Corkey B, Gokce N, Farb MG, LeBrasseur N, Ravid K (2012) The A_{2B} adenosine receptor modulates glucose homeostasis and obesity. PloS One 7: e40584
139. Johnston-Cox H, Eisenstein AS, Koupenova M, Carroll S, Ravid K (2014) The macrophage A2B adenosine receptor regulates tissue insulin sensitivity. PloS One 9:e98775
140. Csoka B, Koscso B, Toro G, Kokai E, Virag L, Nemeth ZH, Pacher P, Bai P, Hasko G (2014) A_{2B} adenosine receptors prevent insulin resistance by inhibiting adipose tissue inflammation via maintaining alternative macrophage activation. Diabetes 63:850–866
141. Tian Y, Piras BA, Kron IL, French BA, Yang Z (2015) Adenosine 2B receptor activation reduces myocardial reperfusion injury by promoting anti-inflammatory macrophages differentiation via PI3K/Akt pathway. Oxid Med Cell Longev 2015:585297
142. Ni Y, Liang D, Tian Y, Kron IL, French BA, Yang Z (2018) Infarct-sparing effect of adenosine A_{2B} receptor agonist is primarily due to its action on splenic leukocytes via a PI3K/Akt/IL-10 pathway. J Surg Res 232:442–449

143. Bott-Flugel L, Bernshausen A, Schneider H, Luppa P, Zimmermann K, Albrecht-Kupper B, Kast R, Laugwitz KL, Ehmke H, Knorr A, Seyfarth M (2011) Selective attenuation of norepinephrine release and stress-induced heart rate increase by partial adenosine A_1 agonism. PloS One 6(3):e18048
144. Tendera M, Gaszewska-Zurek E, Parma Z, Ponikowski P, Jankowska E, Kawecka-Jaszcz K, Czarnecka D, Krzeminska-Pakula M, Bednarkiewicz Z, Sosnowski M, Kilama MO, Agrawal R (2012) The new oral adenosine A_1 receptor agonist capadenoson in male patients with stable angina. Clin Res Cardiol 101:585–591
145. Sabbah HN, Gupta RC, Kohli S, Wang RS, Zhang K, Zimmermann K, Diedrichs N, Albrecht-Kupper BE (2013) Chronic therapy with a partial adenosine A_1-receptor agonist improves left ventricular function and remodeling in dogs with advanced heart failure. Circ Heart Fail 6: 563–571
146. Albrecht-Kupper BE, Leineweber K, Nell PG (2012) Partial adenosine A_1 receptor agonists for cardiovascular therapies. Purinergic Signal 8:91–99
147. Bailey IR, Laughlin B, Moore LA, Bogren LK, Barati Z (2017) Drew KL optimization of thermolytic response to A_1 adenosine receptor agonists in rats. J Pharm Exp Ther 362:424–430
148. Baltos JA, Vecchio EA, Harris MA, Qin CX, Ritchie RH, Christopoulos A, White PJ, May LT (2017) Capadenoson, a clinically trialed partial adenosine A1 receptor agonist, can stimulate adenosine A_{2B} receptor biased agonism. Biochem Pharmacol 135:79–89
149. Rueda P, Merlin J, Chimenti S, Feletou M, Paysant J, White PJ, Christopoulos A, Sexton PM, Summers RJ, Charman WN, May LT, Langmead CJ (2021) Pharmacological insights into safety and efficacy determinants for the development of adenosine receptor biased agonists in the treatment of heart failure. Front Pharmacol 12:628060
150. Meibom D, Albrecht-Kupper B, Diedrich N, Hubsch W, Kast R, Kramer T, Krenz U, Lerchen HG, Mittendorf J, Nell PG, Süssmeier F, Vakalopoulos A, Zimmermann K (2017) Neladenoson bialanate hydrochloride: a prodrug of a partial adenosine A_1 receptor agonist for the chronic treatment of heart diseases. Chem Med Chem 12:728–737
151. Voors AA, Dungen HD, Senni M, Nodari S, Agostoni P, Ponikowski P, Bax JJ, Butler J, Kim RJ, Dorhout B, Dinh W, Gheorghiade M (2017) Safety and tolerability of neladenoson bialanate, a novel oral partial adenosine A_1 receptor agonist, in patients with chronic heart failure. J Clin Pharmacol 57:440–451
152. Louvel J, Guo D, Soethoudt M, Mocking TA, Lenselink EB, Mulder-Krieger T, Heitman LH, IJzerman AP (2015) Structure-kinetics relationships of Capadenoson derivatives as adenosine A_1 receptor agonists. Eur J Med Chem 101:681–691
153. Fredholm BB (1985) On the mechanism of action of theophylline and caffeine. Acta Med Scand 217:149–153
154. Müller CE, Jacobson KA (2011) Xanthines as adenosine receptor antagonists. Handb Exp Pharmacol 200:151–199
155. Takahashi M, Fujita M, Asai N, Saki M, Mori A (2018) Safety and effectiveness of istradefylline in patients with Parkinson's disease: interim analysis of a postmarketing surveillance study in Japan. Expert Opin Pharmacother 19:1635–1642
156. Chen J-F, Cunha RA (2020) The belated US FDA approval of the adenosine A_{2A} receptor antagonist istradefylline for treatment of Parkinson's disease. Purinergic Signal 16(2):167–174
157. Borrmann T, Hinz S, Bertarelli DCG, Li W, Florin NC, Scheiff AB, Müller CE (2009) 1-Alkyl-8-(piperazine-1-sulfonyl)phenylxanthines: development and characterization of adenosine A_{2B} receptor antagonists and a new radioligand with subnanomolar affinity and subtype specificity. J Med Chem 52(13):3994–4006
158. Ji X, Kim YC, Ahern DG, Linden J, Jacobson KA (2001) [^{3}H]MRS 1754, a selective antagonist radioligand for A_{2B} adenosine receptors. Biochem Pharmacol 61:657–663
159. Rohilla S, Bansal R, Kachler S, Klotz KN (2019) Synthesis, biological evaluation and molecular modelling studies of 1, 3, 7, 8-tetrasubstituted xanthines as potent and selective A_{2A} AR ligands with in vivo efficacy against animal model of Parkinson's disease. Bioorg Chem 87:601–612

160. Koul S, Ramdas V, Barawkar DA, Waman YB, Prasad N, Madadi SK, Shejul YD, Bonagiri R, Basu S, Menon S, Reddy SB (2017) Design and synthesis of novel, potent and selective hypoxanthine analogs as adenosine A_1 receptor antagonists and their biological evaluation. Bioorg Med Chem 25:1963–1975
161. Basu S, Barawkar DA, Ramdas V, Waman Y, Patel M, Panmand A, Kumar S, Thorat S, Bonagiri R, Jadhav D, Mukhopadhyay P (2017) A_{2B} adenosine receptor antagonists: design, synthesis and biological evaluation of novel xanthine derivatives. Eur J Med Chem 127:986–996
162. Basu S, Barawkar DA, Ramdas V, Patel M, Waman Y, Panmand A, Kumar S, Thorat S, Naykodi M, Goswami A, Reddy BS (2017) Design and synthesis of novel xanthine derivatives as potent and selective A_{2B} adenosine receptor antagonists for the treatment of chronic inflammatory airway diseases. Eur J Med Chem 134:218–229
163. de Lera RM, Lim Y-H, Zeng J (2014) Adenosine A_{2A} receptor as a drug discovery target. J Med Chem 57:3623–3650
164. Gatta F, Del Giudice MR, Borioni A, Borea PA, Dionisotti S, Ongini E (1993) Synthesis of imidazo[1,2-c]pyrazolo[4,3-e]pyrimidines, pyrazolo[4,3-e]1,2,4-triazolo[1,5-c]pyrimidines and 1,2,4- triazolo[5,1-i]purines: new potent adenosine A2 receptor antagonists. Eur J Med Chem 28:569–576
165. Baraldi PG, Tabrizi MA, Fruttarolo F, Bovero A, Avitabile B, Preti D, Romagnoli R, Merighi S, Gessi S, Varani K, Borea PA (2003) Recent developments in the field of A_3 adenosine receptor antagonists. Drug Dev Res 58:315–329
166. Le Witt PA, Aradi SD, Hauser RA, Rascol O (2020) The challenge of developing adenosine A_{2A} antagonists for Parkinson disease: istradefylline, preladenant, and tozadenant. Parkinsonism Relat Disord 80(Suppl 1):S54–S63
167. Gillespie RJ, Cliffe IA, Dawson C, Dourish CT, Gaur S, Jordan AM, Knight AR, Lerpiniere J, Misra A, Pratt RM, Roffey J, Stratton GC, Upton R, Weiss SM, Williamson DS (2008) Antagonists of the human adenosine A_{2A} receptor. Part 3: design and synthesis of pyrazolo [3,4-d]pyrimidines, pyrrolo[2,3-d]-pyrimidines and 6-arylpurines. Bioorg Med Chem Lett 18(9):2924–2929
168. Gillespie RJ, Bamford SJ, Botting R, Come M, Denny S, Gaur S, Griffin M, Jordan AM, Knight AR, Lerpiniere J, Leonardi S, Lightowler S, Mc Ateer S, Merrett A, Misra A, Padfield A, Reece M, Saadi M, Selwood DL, Stratton GC, Surry D, Todd R, Tong X, Ruston V, Upton R, Weiss SM (2009) Antagonist of the human A_{2A} adenosine receptor. 4. Design, synthesis and preclinical evaluations of 7-aryltriazolo[4,5-d]pyrimidines. J Med Chem 52(1):33–47
169. Basu S, Barawkar DA, Thorat S, Shejul YD, Patel M, Naykodi M, Jain V, Salve Y, Prasad V, Chaudhary S, Ghosh I (2017) Design, synthesis of novel, potent, selective, orally bioavailable adenosine A_{2A} receptor antagonists and their biological evaluation. J Med Chem 60:681–694
170. Renk DR, Skraban M, Bier D, Schulze A, Wabbals E, Wedekind F, Neumaier F, Neumaier B, Holschbach M (2021) Design, synthesis and biological evaluation of Tozadenant analogues as adenosine A_{2A} receptor ligands. Eur J Med Chem 15(214):113214
171. Li A-N, Moro S, Forsyth N, Melman N, Ji X-D, Jacobson KA (1999) Synthesis, CoMFA analysis, and receptor docking of 3,5-diacyl-2,4-dialkylpyridine derivatives as selective A_3 adenosine receptor antagonists. J Med Chem 42:706–721
172. Madi L, Bar-Yehuda S, Barer F, Ardon E, Ochaion A, Fishman P (2003) A3 adenosine receptor activation in melanoma cells: association between receptor fate and tumor growth inhibition. J Biol Chem 278(43):42121–42130
173. Moro S, Gao ZG, Jacobson KA, Spalluto G (2006) Progress in pursuit of therapeutic adenosine receptor antagonists. Med Res Rev 26:131–159
174. Jacobson KA, IJzerman AP, Müller CE (2021) Medicinal chemistry of P2 and adenosine receptors: common scaffolds adapted for multiple targets. Biochem Pharmacol 187:114311

175. Cheng RKY, Segala E, Robertson N, Deflorian F, Doré AS, Errey JC, Fiez-Vandal C, Marshall FH, Cooke RM (2017) Structures of human A_1 and A_{2A} adenosine receptors xanthines reveal determinants of selectivity. Structure 25:1275–1285
176. Jespers W, Schiedel AC, Heitman LH, Cooke R, Gutiérrez-de-Terán H (2018) Structural mapping of adenosine receptor mutations: ligand binding and signaling mechanisms. Trends Pharmacol Sci 39:75–89
177. Carpenter B, Lebon G (2017) Human adenosine A_{2A} receptor: molecular mechanism of ligand binding and activation. Front Pharmacol 8:898
178. Rucktooa P, Cheng RKY, Segala E, Geng T, Errey JC, Brown GA, Cooke RM, Marshall FH, Doré AS (2018) Towards high throughput GPCR crystallography: in Meso soaking of adenosine A_{2A} receptor crystals. Sci Rep 8:41
179. Costanzi S, Ivanov AA, Tikhonova IG, Jacobson KA (2007) Structure and function of G protein-coupled receptors studied using sequence analysis, molecular modeling and receptor engineering: adenosine receptors. Front Drug Des Discov 3:63–79
180. Dal Ben D, Lambertucci C, Marucci G, Volpini R, Cristalli G (2010) Adenosine receptor modeling: what does the A_{2A} crystal structure tell us? Curr Top Med Chem 10:993–1018
181. Ciancetta A, Jacobson KA (2017) Structural probing and molecular modeling of the A_3 adenosine receptor: a focus on agonist binding. Molecules 22:449
182. Effendi WI, Nagano T, Kobayashi K, Nishimura Y (2020) Focusing on adenosine receptors as a potential targeted therapy in human diseases. Cell 9:785
183. Sousa JB, Fresco P, Diniz C, Goncalves J (2018) Adenosine receptor ligands on cancer therapy. A review of patent literature. Recent Pat Anticancer Drug Discov 13:40–69
184. Merighi S, Battistello E, Giacomelli L, Varani K, Vincenzi F, Borea PA (2019) Targeting A_3 and A_{2A} adenosine receptors in the fight against cancer. Exp Opin Ther Targets 23:669–678
185. Yu F, Zhu C, Xie Q, Wang Y (2020) Adenosine A_{2A} receptor antagonists for cancer immunotherapy. J Med Chem 63:12196–12212
186. Guglielmi P, Carradori S, Campestre C, Poce G (2019) Novel therapies for glaucoma: a patent review (2013-2019). Exp Opin Ther Patents 29:769–780
187. Zheng J, Zhang X, Zhen X (2019) Development of adenosine A_{2A} receptor antagonists for the treatment of Parkinson's disease: a recent update and challenge. ACS Chem Nerosci 10:783–791
188. Borah P, Deka S, Mailavaram RP, Deb PK (2019) P1 receptor agonists/antagonists in clinical trials – potential drug candidates of the future. Curr Pharm Des 25:2792–2807
189. Chandrasekaran B, Samarneh S, Jaber AMY, Kassab G, Agrawal N (2019) Therapeutic potentials of A_{2B} adenosine receptor ligands: current status and perspectives. Curr Pharm Des 25:2741–2771
190. Iskandrian AE, Bateman TM, Belardinelli L, Blackburn B, Cerqueira MD, Hendel RC, Lieu H, Mahmarian JJ, Olmsted A, Underwood SR, Vitola J, Wang W (2007) Adenosine versus regadenoson comparative evaluation in myocardial perfusion imaging: results of the ADVANCE phase 3 multicenter international trial. J Nucl Cardiol 14(5):645–658
191. Stone TW, Ceruti S, Abbracchio MP (2009) Adenosine receptors and neurological disease: neuroprotection and neurodegeneration. In: Wilson CN, Mustafa SJ (eds) Adenosine receptors in health and disease, vol 193. Springer, Berlin, pp 535–587
192. Jenner P, Mori A, Aradi SD, Hauser RA (2021) Istradefylline – a first generation adenosine A_{2A} antagonist for the treatment of Parkinson's disease. Expert Rev Neurother 21:317–333
193. Stocchi F, Rascol O, Hauser RA, Huyck S, Tzontcheva A, Capece R, Ho TW, Sklar P, Lines C, Michelson D, Hewitt DJ (2017) Randomized trial of preladenant, given as monotherapy, in patients with early Parkinson disease. Neurology 88:2198–2206
194. Hauser RA, Olanow CW, Kieburtz KD, Pourcher E, Docu-Axelerad A, Lew M, Kozyolkin O, Neale A, Resburg C, Meya U, Kenney C, Bandak S (2014) Tozadenant (SYN115) in patients with Parkinson's disease who have motor fluctuations on levodopa: a phase 2b, double-blind, randomised trial. Lancet Neurol 13:767–776

195. Popoli P, Blum D, Domenici M, Burnouf S, Chern YA (2008) Critical evaluation of adenosine A_{2A} receptors as potentially "Druggable" targets in Huntington's disease. Curr Pharm Des 14: 1500–1511
196. Faivre E, Coelho JE, Zornbach K, Malik E, Baqi Y, Schneider M, Cellai L, Carvalho K, Sebda S, Figeac M, Eddarkaoui S, Caillierez R, Chern Y, Heneka M, Sergeant N, Müller CE, Halle A, Buée L, Lopes LV, Blum D (2018) Beneficial effect of a selective adenosine A_{2A} receptor antagonist in the APPswe/PS1dE9 mouse model of Alzheimer's disease. Front Mol Neurosci 11:235
197. Eskelinen MH, Kivipelto M (2010) Caffeine as a protective factor in dementia and Alzheimer's disease. J Alzheimers Dis 20:S167–S174
198. Santos C, Lunet N, Azevedo A, de Mendonça A, Ritchie K, Barros H (2010) Caffeine intake is associated with a lower risk of cognitive decline: a cohort study from Portugal. J Alzheimers Dis 20(Suppl 1):S175–S185
199. Chen JQA, Scheltens P, Groot C, Ossenkoppele R (2020) Associations between caffeine consumption, cognitive decline, and dementia: a systematic review. J Alzheimers Dis 78(4): 1519–1546
200. van Dam RM, Hu FB, Willett WC (2020) Coffee, caffeine, and health. N Engl J Med 383:369–378
201. Melani A, Pugliese AM, Pedata F (2014) Chapter thirteen-adenosine receptors in cerebral ischemia. Int Rev Neurobiol 119:309–348
202. Pedata F, Pugliese AM, Coppi E, Popoli P, Morelli M, Schwarzschild MA, Melani A (2007) Adenosine in the central nervous system: effects on neurotransmission and neuroprotection. Immun Endoc Metab Agents Med Chem 7:304–321
203. Martire A, Lambertucci C, Pepponi R, Ferrante A, Benati N, Buccioni M, Dal Ben D, Marucci G, Klotz K-N, Volpini R, Popoli P (2019) Neuroprotective potential of adenosine A_1 receptor partial agonists in experimental models of cerebral ischemia. J Neurochem 149: 211–230
204. Dickenson AH, Suzuki R, Reeve AJ (2000) Adenosine as a potential analgesic target in inflammatory and neuropathic pains. CNS Drugs 13:77–85
205. Knezevic NN, Cicmil N, Knezevic I, Candido KD (2015) Discontinued neuropathic pain therapy between 2009-2015. Expert Opin Investig Drugs 24:1631–1646
206. Rivera-Oliver M, Díaz-Ríos M (2014) Using caffeine and other adenosine receptor antagonists and agonists as therapeutic tools against neurodegenerative diseases: a review. Life Sci 101:1–9
207. Bilkei-Gorzo A, Abo-Salem OM, Hayallah AM, Michel K, Müller CE, Zimmer A (2008) Adenosine receptor subtype-selective antagonists in inflammation and hyperalgesia. Naunyn Schmiedebergs Arch Pharmacol 377:65–76
208. Pechlivanov DM, Georgiev VP (2005) Effects of single and long-term theophylline treatment on the threshold of mechanical nociception: contribution of adenosine A_1 and α2-adrenoceptors. Methods Find Exp Clin Pharmacol 27:659–664
209. Gong Q-J, Li Y-Y, Xin W-J, Wei X-H, Cui Y, Wang J, Liu Y, Liu C-C, Li Y-Y (2010) Differential effects of adenosine A_1 receptor on pain-related behavior in normal and nerve-injured rats. Brain Res 1361:23–30
210. Katz NK, Ryals JM, Wright DE (2015) Central or peripheral delivery of an adenosine A1 receptor agonist improves mechanical allodynia in mouse model of painful diabetic neuropathy. Neuroscience 285:312–323
211. Ferre S, Diamond I, Goldberg SR, Yao L, Hourani SMO, Huang ZL, Urade Y, Kitchen I (2007) Adenosine A_{2A} receptor in ventral striatum, hypothalamus and nociceptive circuitry. Implication for drug addiction, sleep and pain. Prog Neurobiol 83:332–347
212. Cunha RA (2005) Neuroprotection by adenosine in the brain: from A_1 receptor activation to A_{2A} receptor blockade. Purinergic Signal 1:111–134
213. Stockwell J, Jakova E, Cayabyab FS (2017) Adenosine A_1 and A_{2A} receptors in the brain: current research and their role in neurodegeneration. Molecules 22(4):676

214. Betti M, Catarzi D, Varano F, Falsini M, Varani K, Vincenzi F, Pasquini S, Di Cesare ML, Ghelardini C, Lucarini E, Dal Ben D, Spinaci A, Bartolucci G, Menicatti M, Colotta V (2019) Modifications on the Amino-3,5-dicyanopyridine core to obtain multifaceted adenosine receptor ligands with antineuropathic activity. J Med Chem 62:6894–6912
215. Ledent C, Vaugeois JM, Shiffmann SN, Pedrazzini T, El Yacoubini M, Vanderhaeghen JJ, Costentin J, Heath JK, Vassert G, Parmentier M (1997) Aggressiveness, hypoalgesia and high blood pressure in mice lacking the adenosine A_{2A} receptor. Nature 388:647–678
216. Falsini M, Catarzi D, Varano F, Ceni C, Dal Ben D, Marucci G, Buccioni M, Volpini R, Di Cesare ML, Lucarini E, Ghelardini C, Bartolucci G, Menicatti M, Colotta V (2019) Antioxidant-conjugated 1,2,4-triazolo[4,3-a]pyrazin-3-one derivatives: highly potent and selective human A_{2A} adenosine receptor antagonists possessing protective efficacy in neuropathic. J Med Chem 62:8511–8531
217. Imlach WL, Bhola RF, May LT, Christopoulos A, Christie MJ (2015) A positive allosteric modulator of the A_1 AR selectively inhibits primary afferent synaptic transmission in a neuropathic pain model. Mol Pharmacol 88:460–468
218. Varano F, Catarzi D, Vincenzi F, Betti M, Falsini M, Ravani A, Borea PA, Colotta V, Varani K (2016) Design, synthesis, and pharmacological characterization of 2-(2-furanyl)thiazolo [5,4-d]pyrimidine-5,7-diamine derivatives: new highly potent A_{2A} adenosine receptor inverse agonists with antinociceptive activity. J Med Chem 59:10564–10576
219. Salvemini D, Jacobson KA (2017) Highly selective A_3 adenosine receptor agonists relieve chronic neuropathic pain. Expert Opin Ther Patents 27:967
220. Janes K, Symons-Liguori AM, Jacobson KA, Salvemini D (2016) Identification of A_3 adenosine receptor agonists as novel non-narcotic analgesics. Br J Pharmacol 173:1253–1267
221. Terayama R, Tabata M, Maruhama K, Iida S (2018) A_3 adenosine receptor agonist attenuates neuropathic pain by suppressing activation of microglia and convergence of nociceptive inputs in the spinal dorsal horn. Exp Brain Res 236:3203–3213
222. Van Troostenburg AR, Clark EV, Carey WDH, Warrington SJ, Kerns WD, Cohn I, Silverman MH, Bar-Yehuda S, Fong K-LL, Fishman P (2004) Tolerability, pharmacokinetics and concentration-dependent hemodynamic effects of oral CF101, an A_3 adenosine receptor agonist, in healthy young men. Int J Clin Pharmacol Ther 42:534–542
223. Schaddelee MP, Collins SD, DeJongh J, de Boer AG, Ijzerman AP, Danhof M (2005) Pharmacokinetic/pharmacodynamic modelling of the anti-hyperalgesic and anti-nociceptive effect of adenosine A_1 receptor partial agonists in neuropathic pain. Eur J Pharmacol 514:131–140
224. Antonioli L, Fornai M, Blandizzi C, Pacher P, Haskó G (2019) Adenosine signaling and the immune system: when a lot could be too much. Immunol Lett 205:9–15
225. Journey JD, Bentley TP (2020) Theophylline toxicity. In: StatPearls, StatPearls Publishing LLC, Treasure Island. https://www.ncbi.nlm.nih.gov/books/NBK532962/
226. Magni G, Ceruti S (2020) Adenosine signaling in autoimmune disorders. Pharmaceuticals 13:260
227. Gessi S, Merighi S, Varani K, Leung E, Mac Lennan S, Borea PA (2008) The A_3 adenosine receptor: an enigmatic player in cell biology. Pharmacol Ther 117:123–140
228. David M, Akerman L, Ziv M, Kadurina M, Gospodinov D, Pavlotsky F, Yankova R, Kouzeva V, Ramon M, Silverman MH, Fishman P (2012) Treatment of plaque-type psoriasis with oral CF101: data from an exploratory randomized phase 2 clinical trial. J Eur Acad Dermatol Venereol 26:361–367
229. Mantell S, Jones R, Trevethick M (2010) Design and application of locally delivered agonists of the adenosine A_{2A} receptor. Expert Rev Clin Pharmacol 3:55–72
230. Palani G, Ananthasubramaniam K (2013) Regadenoson: review of its established role in myocardial perfusion imaging and emerging applications. Cardiol Rev 21:42–48
231. Sharma AK, La Par DJ, Stone ML, Zhao Y, Mehta CK, Kron IL, Laubach VE (2016) NOX2 activation of natural killer T cells is blocked by the adenosine A_{2A} receptor to inhibit lung ischemia-reperfusion injury. Am J Respir Crit Care Med 193:988–999

232. Gessi S, Merighi S, Sacchetto V, Simioni C, Borea PA (2011) Adenosine receptors and cancer. Biochim Biophys Acta 1808:1400–1412
233. Ren T, Tian T, Feng X, Ye S, Wang H, Wu W, Qiu Y, Yu C, He Y, Zeng J, Cen J, Zhou Y (2015) An adenosine A_3 receptor agonist inhibits DSS-induced colitis in mice through modulation of the NF-κB signaling pathway. Sci Rep 5:9047
234. Ohana G, Bar-Yehuda S, Arich A, Madi L, Dreznick Z, Rath-Wolfson L, Silberman D, Slosman G, Fishman P (2003) Inhibition of primary colon carcinoma growth and liver metastasis by the A_3 adenosine receptor agonist CF101. Br J Cancer 89:1552–1558
235. Stemmer SM, Manojlovic NS, Vasile Marinca M, Petrov P, Cherciu N, Ganea D, Ciuleanu T-E, Puscas IA, Shaalan Beg M, Purcell WT, Croitoru A-E, Nedyalkova Ilieva R, Natošević S, Lavinir Nita A, Kalev DN, Harpaz Z, Farbstein M, Silverman MH, Fishman P, Llovet JM (2019) A phase II, randomized, double-blind, placebo-controlled trial evaluating efficacy and safety of namodenoson (CF102), an A_3 adenosine receptor agonist (A_3AR), as a second-line treatment in patients with child-Pugh B (CPB) advanced hepatocellular carcinoma (HCC). J Clin Oncol 37:2503–2503
236. Congreve M, Brown GA, Borodovsky A, Lamb ML (2018) Targeting adenosine A_{2A} receptor antagonism for treatment of cancer. Expert Opin Drug Discovery 13:997–1003
237. Buisseret L, Rottey S, De Bono JS, Migeotte A, Delafontaine B, Manickavasagar T, Martinoli C, Wald N, Rossetti M, Gangolli EA, Wiegert E, Ni MI, Lager JJ, Machiels J-PH (2021) Phase 1 trial of the adenosine A_{2A} receptor antagonist inupadenant (EOS-850): update on tolerability, and antitumor activity potentially associated with the expression of the A_{2A} receptor within the tumor. J Clin Oncol 39(15):2562–2562
238. Houthuys E, Brouwer M, Nyawouame F, Pirson R, Marillier R, Deregnaucourt T, Marchante J, Swiercz J, Moulin C, Bol V, Driessens G, Detheux M, Quéva C, Crosignani S, Gomes B (2017) A novel adenosine A_{2A} receptor antagonist optimized for high potency in adenosine-rich tumor microenvironment boost antitumor immunity. Cancer Res 77(13):1683
239. Seitz L, Jin L, Leleti M, Ashok D, Jeffrey J, Rieger A, Tiessen RG, Arold G, Tan JBL, Powers JP, Walters MJ, Karakunnel J (2019) Safety, tolerability, and pharmacology of AB928, a novel dual adenosine receptor antagonist, in a randomized, phase 1 study in healthy volunteers. Invest New Drugs 37:711–721
240. Chiappori AA, Creelan B, Tanvetyanon T, Gray JE, Haura EB, Thapa R, Barlow L, Chen Z, Chen DT, Beg AA, Boyle TA, Castro J, Morgan L, Morris E, Aregay M, FeK H, Manenti L, Antonia S (2022) Phase I study of taminadenant (PBF509/NIR178), an A_{2A} receptor antagonist, with or without spartalizumab (PDR001), in patients with advanced non-small cell lung cancer. Clin Cancer Res 28(11):2313–2320
241. Saavedra Santa Gadea O, Greil R, de Jonge MJA, Tan D, Jerusalem G, Damian S, Grell P, Wainberg ZA, Wolf J, Joerger M, Carlino MS, Kasper S, Yap TA, Otero J, Yang X, Nesbitt V, Kim J, Ho Lee L, Choudhury S, Leal TA (2021) Phase (Ph) II study of taminadenant (NIR178) + spartalizumab (PDR001) in patients (pts) with microsatellite stable (MSS) colorectal cancer (CRC). Ann Oncol 32(5):S553
242. Willingham SB, Ho Po Y, Hotson A, Hill C, Piccione EC, Hsieh J, Liu L, Buggy JJ, McCaffery I, Miller RA (2018) A_{2A}R antagonism with CPI-444 induces antitumor responses and augments efficacy to anti-PD-(L)1 and anti-CTLA-4 in preclinical models. Cancer Immunol Res 6(10):1136–1149
243. Leone RD, Sun I-M, Oh M-H, Sun I-H, Wen J, Englert J, Powell JD (2018) Inhibition of the adenosine A2a receptor modulates expression of T cell coinhibitory receptors and improves effector function for enhanced checkpoint blockade and ACT in murine cancer models. Cancer Immunol Immunother 67(8):1271–1284
244. Fong HA, Powderly JD, Sznol M, Heist RS, Choueiri TK, George S, Hughes BGM, Hellmann MD, Shepard DR, Rini BI, Kummar S, Weise AM, Riese MJ, Markman B, Emens LA, Mahadevan D, Luke JJ, Laport G, Brody JD, Hernandez-Aya L, Bonomi P, Goldman JW, Berim L, Renouf DJ, Goodwin RA, Munneke B, Ho PY, Hsieh J, McCaffery I, Kwei L,

Willingham SB, Miller RA (2020) Adenosine A_{2A} receptor blockade as an immunotherapy for treatment-refractory renal cell cancer. Cancer Discov 10(1):40–53
245. Patel JJ, Alzahrani T (2021) Myocardial perfusion scan. In: StatPearls, StatPearls Publishing LLC, Treasure Island. https://www.ncbi.nlm.nih.gov/books/NBK539772/
246. Singh S, McKintosh R (2022) Adenosine. In: StatPearls. StatPearls Publishing LLC, Treasure Island. https://www.ncbi.nlm.nih.gov/books/NBK519049/
247. Tian Y, Marshall M, French BA, Linden J, Yang Z (2015) The infarct-sparing effect of IB-MECA against myocardial ischemia/reperfusion injury in mice is mediated by sequential activation of adenosine A_3 and A_{2A} receptors. Basic Res Cardiol 110:16
248. Galal A, El-Bakly WM, Al Haleem ENA, El-Demerdash E (2016) Selective A_3 adenosine receptor agonist protects against doxorubicin-induced cardiotoxicity. Cancer Chemother Pharmacol 77:309–322
249. Rosentreter U, Krämer T, Shimada M, Hübsch W, Diedrichs N, Krahn T, Henninger K, Stasch J-P, Wischnat R (2003) Substituted 2-thio-3,5-dicyano-4-phenyl-6-aminopyridines and the use of the same. WO 2003053441 A1
250. Voors AA, Bax JJ, Hernandez AF, Wirtz AB, Pap AF, Ferreira AC, Senni M, van der Laan M, Butler J (2019) Safety and efficacy of the partial adenosine A_1 receptor agonist neladenoson bialanate in patients with chronic heart failure with reduced ejection fraction: a phase IIb, randomized, double-blind, placebo-controlled trial. Eur J Heart Fail 21:1426–1433
251. Shah S.J, Voors AA, McMurray JJV, Kitzman DW, Viethen T, Bomfim Wirtz A, Huang E, Pap AF, Solomon SD (2012) Effect of neladenoson bialanate on exercise capacity among patients with heart failure with preserved ejection fraction: a randomized clinical trial. JAMA 321:2101–2112
252. Lasley RD (2018) Adenosine receptor-mediated cardioprotection-current limitations and future directions. Front Pharmacol 9:310
253. Zhan E, McIntosh VJ, Lasley RD (2011) Adenosine A_{2A} and A_{2B} receptors are both required for adenosine A_1 receptor-mediated cardioprotection. Am J Physiol Heart Circ Physiol 301: H1183–H1189
254. Dhalla AK, Santikul M, Smith M, Wong M-Y, Shryock JC, Belardinelli L (2007) Antilipolytic activity of a novel partial A_1 adenosine receptor agonist devoid of cardiovascular effects: comparison with nicotinic acid. J Pharmacol Exp Ther 321(1):327–333
255. Fuentes E, Fuentes M, Caballero J, Palomo I, Hinz S, El-Tayeb A, Müller CE (2018) Adenosine A_{2A} receptor agonists with potent antiplatelet activity. Platelets 29:292–300
256. Wolska N, Kassassir NH, Luzak B, Watala C, Rozalski M (2020) Adenosine receptor agonists increase the inhibition of platelet function by P2Y12 antagonists in a cAMP and calcium-dependent manner. Pharmaceuticals 13:177
257. Johnston-Cox HA, Ravid K (2011) Adenosine and blood platelets. Purinergic Signal 7:357–365
258. Voloshyna I, Littlefield MJ, Kaplan L, Rieger JM, Figler R, Reiss AB (2013) Adenosine A_{2A} receptor agonists regulate cholesterol homeostasis in mouse bone marrow derived macrophages (BMDM). FASEB J 27
259. Reiss AB, Grossfeld D, Kasselman LJ, Renna HA, Vernice NA, Drewes W, Konig J, Carsons SE, De Leon J (2019) Adenosine and the cardiovascular system. Am J Cardiovasc Drugs 19: 449–464
260. Gregory KJ, Bridges TM, Gogliotti RG, Stauffer SR, Noetzel MJ, Jones CK, Craig WL, Conn PJ, Niswender CM (2019) In vitro to in vivo translation of allosteric modulator concentration-effect relationships: implications for drug discovery. ACS Pharmacol Transl Sci 2:442–452
261. Lane JR, May LT, Parton RG, Sexton PM, Christopoulos A (2017) A kinetic view of GPCR allostery and biased agonism. Nat Chem Biol 13:929–937
262. Deganutti G, Moro S (2017) Estimation of kinetic and thermodynamic ligand-binding parameters using computational strategies. Future Med Chem 9(5)
263. Zhang J, Yan W, Duan W, Wüthrich K, Cheng J (2020) Tumor immunotherapy using A_{2A} adenosine receptor antagonists. Pharmaceuticals 13(9):237

264. Augustin RC, Leone RD, Naing A, Fong L, Bao R, Luke JJ (2022) Next steps for clinical translation of adenosine pathway inhibition in cancer immunotherapy. J Immunother Cancer 10(2):e004089
265. Anighoro A, Bajorath J, Rastelli G (2014) Polypharmacology: challenges and opportunities in drug discovery. J Med Chem 57(19):7874–7887
266. Ma H, Huang B, Zhang Y (2020) Recent advances in multitarget-directed ligands targeting G-protein-coupled receptors. Drug Discov Today 25(9):1682–1692
267. Zhou J, Jiang X, He S, Jiang H, Feng F, Liu W, Wei Q, Sun H (2019) Rational design of multitarget-directed ligands: strategies and emerging paradigms. J Med Chem 62:8881–8914
268. Morphy R, Rankovic Z (2005) Designed multiple ligands. An emerging drug discovery paradigm. J Med Chem 48(21):6523–6543
269. Wang M, Hou S, Wei Y, Li D, Lin J (2021) Discovery of novel dual adenosine A_1/A_{2A} receptor antagonists using deep learning, pharmacophore modeling and molecular docking. PLoS Comput Biol 17(3):e1008821
270. Barlow N, Baker SP, Scammells PJ (2013) Effect of linker length and composition on heterobivalent ligand-mediated receptor cross-talk between the A_1 adenosine and beta2 adrenergic receptors. ChemMedChem 8(12):2036–2046
271. Pulido D, Casadó-Anguera V, Gómez-Autet M, Llopart N, Moreno E, Casajuana-Martin N, Ferré S, Pardo L, Casadó V, Royo M (2022) Heterobivalent ligand for the adenosine A_{2A}-dopamine D2 receptor heteromer. J Med Chem 65(1):616–632
272. Hagenow S, Affini A, Pioli EY, Hinz S, Zhao Y, Porras G, Namasivayam V, Müller CE, Lin J-S, Bezard E, Holger S (2021) Adenosine $A_{2A}R/A_1R$ antagonists enabling additional H_3R antagonism for the treatment of Parkinson's disease. J Med Chem 64(12):8246–8262
273. Brunschweiger A, Koch P, Schlenk M, Rafehi M, Radjainia H, Küppers P, Hinz S, Pineda F, Wiese M, Hockemeyer J, Heer J, Denonne F, Müller CE (2016) 8-Substituted 1,3-dimethyltetrahydropyrazino [2,1-f]purinediones: water-soluble adenosine receptor antagonists and monoamine oxidase B inhibitors. Bioorg Med Chem 24(21):5462–5480
274. Ceni C, Catarzi D, Varano F, Dal Ben D, Marucci G, Buccioni M, Volpini R, Angeli A, Nocentini A, Gratteri P, Supuran CT, Colotta V (2020) Discovery of first-in-class multi-target adenosine A_{2A} receptor antagonists-carbonic anhydrase IX and XII inhibitors. 8-Amino-6-aryl-2-phenyl-1,2,4-triazolo [4,3-a]pyrazin-3-one derivatives as new potential antitumor agents. Eur J Med Chem 201:112478
275. Durante M, Squillace S, Lauro F, Giancotti LA, Coppi E, Cherchi F, Di Cesare ML, Ghelardini C, Kolar G, Wahlman C, Opejin A, Xiao C, Reitman ML, Tosh DK, Hawiger D, Jacobson KA, Salvemini D (2021) Adenosine A_3 agonists reverse neuropathic pain via T cell-mediated production of IL-10. J Clin Invest 131:e139299

Top Med Chem (2023) 41: 47–88
https://doi.org/10.1007/7355_2023_163

Published online: 19 August 2023

Adenosine Receptor Ligands, Probes, and Functional Conjugates: A 20-Year History of Pyrazolo[4,3-*e*][1,2,4]Triazolo[1,5-*c*]Pyrimidines (PTP)

Filippo Prencipe, Tatiana Da Ros, Eleonora Cescon, Ilenia Grieco, Margherita Persico, Giampiero Spalluto, and Stephanie Federico

Contents

Abstract Adenosine receptors are G protein-coupled receptors (GPCRs) involved in several processes in the human body. Their involvement in pathologies like coronary vasodilation, neurodegenerative diseases, inflammation, cancer, and pain make them interesting therapeutic targets. In particular, antagonism at the A_{2A} and A_3 adenosine receptor subtypes is of particular interest due to the central role in Parkinson's disease and immunoncology of the first, and in cancer, glaucoma, and inflammation of the second. Pyrazolo[4,3-*e*][1,2,4]triazolo[1,5-*c*]pyrimidine nucleus is one of the most studied scaffolds to develop A_{2A} and A_3 adenosine receptor antagonists. In fact, a well distinct structure–activity relationship exists for these

F. Prencipe, T. Da Ros, E. Cescon, I. Grieco, M. Persico, G. Spalluto, and S. Federico (✉)
Department of Chemical and Pharmaceutical Sciences, University of Trieste, Trieste, Italy
e-mail: sfederico@units.it

receptors. As a consequence, pyrazolo[4,3-*e*][1,2,4]triazolo[1,5-*c*]pyrimidine-based antagonists have also represented an optimal starting point to design probes and multifunction molecules that potently and selectively target A_{2A} and A_3 adenosine receptors. In this review, we reported the structure–activity relationship of pyrazolo [4,3-*e*][1,2,4]triazolo[1,5-*c*]pyrimidines at all the adenosine receptors along with all functionalization performed till now with their applications to study the structure and function of A_3, but especially A_{2A} adenosine receptors. Thus, pyrazolo[4,3-*e*][1,2,4] triazolo[1,5-*c*]pyrimidines and adenosine receptors are simply examples that would serve as a model that could be applied to every GPCR when a robust structure–activity relationship of a ligand class is known.

Keywords Adenosine receptors, Binding, Carbon nanotubes, Dendrimers, Fluorescent ligands, GPCR, Preladenant, Probes, Pyrazolo[4,3-*e*][1,2,4]triazolo [1,5-*c*]pyrimidines, Radioligands, Radiotracers, SCH442416, SCH58261

1 Introduction

Adenosine is an endogenous nucleoside widely distributed in mammalian organisms where it modulates several important physiological functions [1]. It derives from the degradation of intracellular adenosine monophosphate (AMP) and it is released from cells by a concentration-dependent, bi-directional nucleoside transporter, and it may then exert its extracellular effects. The adenosine produced from extracellular AMP by ecto-5′-nucleotidases may also contribute to rise extracellular adenosine levels [2]. Adenosine is removed from the extracellular space through the action of specialized transport proteins [3], while, intracellularly, it may be degraded by adenosine deaminase (ADA) to inosine, which is then transformed into uric acid.

Adenosine possesses various properties similar to neurotransmitters, in particular: (1) it acts on specific receptors; (2) it can be antagonized by specific compounds; (3) enzymes producing adenosine are present in synapses; (4) its effect could be terminated by an efficient reuptake and a metabolic pathway. Nevertheless, adenosine is usually considered a neuromodulator because there is no evidence that it is stored in or released by specific purinergic nerves [4]. Adenosine's role appears to be a homeostatic regulation under normal, physiological conditions, and a protective role in emergency situations [5]. Furthermore, adenosine could indirectly influence the action of neurotransmitters and other neuromodulators, behaving as a modulator of modulators. It could be considered as a fine-tuner and, in this way, it contributes to a very sophisticated interplay between its own receptors and the receptors for other neurotransmitters or neuromodulators [6].

Adenosine action occurs through the stimulation of the purinergic receptors that were classified as P1 receptors, while the receptors activated by phosphorylated nucleotides were classified as P2 receptors [7]. The P1 family could be subdivided into four receptor subtypes named A_1, A_{2A}, A_{2B}, and A_3 adenosine receptors (ARs), respectively, which have been cloned and pharmacologically characterized. All these receptor subtypes are G protein-coupled receptors (GPCRs) and can be distinguished according to their preferred mechanism of signal transduction. A_1 and A_3 ARs interact with pertussis toxin-sensitive G proteins of the G_i and G_o family, while the canonical signaling mechanism of A_{2A} and A_{2B} ARs is the stimulation of adenylyl cyclase via G_s proteins. The consequence of this activation is a reduction (A_1 AR, A_3 AR) or an increase (A_{2A} AR, A_{2B} AR) of the levels of cyclic adenosine monophosphate (cAMP) as second messenger. In addition to the coupling to adenylyl cyclase, all four subtypes may positively couple to phospholipase C via different G protein subunits [1, 8].

It has also been demonstrated that adenosine receptors could activate signal pathways that are not related to G proteins, in particular the β-arrestin one, inducing responses different from the G protein signals [8].

Considering the large distribution of adenosine in the organism, it is evident that adenosine receptors could be a significant target for several pathologies and in fact, in the last decades, different classes of potent and selective derivatives, both agonists and antagonists, have been synthesized with the aim to characterize and study the physio-pathological role of these receptors and their possible role in several disorders [9, 10].

1.1 A_1 AR

Several studies have demonstrated the role of A_1 ARs in respiratory diseases. As an example, the A_1 AR antagonism blocks allergic responses to house dust mites in allergic rabbit model of asthma [11]. In contrast, in mice, A_1 AR also mediates anti-inflammatory effects via macrophages and reduces polymorphonuclear infiltration by inhibiting the release of chemotactic cytokines [12, 13]. Moreover, it has been demonstrated that A_1 AR agonists decreased inflammation, edema, and neutrophil chemotaxis [14] and these receptors promoted the recruitment of leukocytes to infected lungs and attenuated their injury [15]. In inflammation processes, there are evidences that A_1 AR activation could induce both proinflammatory and anti-inflammatory responses [16]. Low concentration of adenosine induces neutrophil chemotaxis and adherence to endothelium via A_1 AR [17], in contrast, A_1 AR resulted to be a potent anti-inflammatory mediator in various kidney, heart, lung, and brain injury models [18].

In the cardiovascular system, A_1 AR mediated negative chronotropic and dromotropic effects through inactivation of the inwardly rectifying K^+ current, inhibition of the inward Ca^{2+} current, and activation of nitric oxide synthase. Furthermore, A_1 AR activation showed both antiarrhythmic and proarrhythmic effects [19] and various A_1 AR agonists have been evaluated as antiarrhythmic agents [20–23].

In the central nervous system (CNS), A_1 ARs contribute to neuroprotection but they are also involved in neurodegeneration. In fact, their activation under hypoxia leads to the inhibition of presynaptic Ca^{2+} influx-related release of neurotransmitters, in particular glutamate, to generate neuroprotection [24, 25]. Nevertheless, it was assumed that prolonged activation of A_1 AR increased A_{2A} AR levels, which generate neurodegeneration in ischemic stroke [26]. It has also been observed that the use of A_1 AR agonists amplifies nociceptive thresholds in spinal glial in neuropathic pain [27], it is also able to localize epileptic foci [28], and it decreases tremors in Parkinson's disease [29].

Moreover, A_1 ARs play a significant role in some metabolic diseases [30], in fact their agonists have potential use as antilipolytic agents, and, they are under development for the treatment of type 2 diabetes mellitus [31].

1.2 A_{2A} AR

A_{2A} ARs act as the most dominant anti-inflammatory effectors of extracellular adenosine [32], in fact the activation of these receptors inhibits neutrophil adhesion to endothelial cells, and formation of reactive oxygen species [16]. In the respiratory system, their activation affects multiple aspects of inflammatory process, in particular in asthmatic airways suppressing inflammation by reduced neutrophil adherence to the endothelium, and, moreover, there is the inhibition of histamine and tryptase release. In fact, in animal models of asthma and chronic obstructive pulmonary diseases (COPD), A_{2A} AR agonists have resulted in good outcomes [33, 34]. In the cardiovascular system, A_{2A} ARs seem to protect against ischemia and increase the occurrence of cardiac arrhythmias [35]. In the CNS, A_{2A} ARs are in general associated to neurodegeneration, due to their excitatory effects [36]. Nevertheless several ligands, both agonists and antagonists, are under investigation for the treatment of several CNS disorders. In particular, agonists are under investigations for sleep disorders [36], schizophrenia [37], pain [38, 39], and drug addiction [40], while antagonists for the treatment of stroke [41], anxiety/depression [42], Alzheimer [43], Huntington [44], Parkinson [45–47], schizophrenia [37], and epilepsy [48]. At the present, the most appealing use of A_{2A} AR antagonists is in cancer immunotherapy. In fact, it has been demonstrated that adenosine is capable to create a protection in the tumor microenvironment for both solid and hematologic tumors [49]. Adenosine acts both on A_{2A} and A_{2B} ARs inducing an anti-inflammatory phenotype in T cells, macrophages, and other cells, and antagonists have a beneficial effect when combined with immunotherapy [50].

1.3 A_{2B} AR

A_{2B} ARs are expressed on most inflammatory cells and possess both proinflammatory and anti-inflammatory effects [51]. In general, they generate anti-inflammatory responses by coupling with G_s protein and proinflammatory effects by coupling with G_q protein [52]. For these reasons, the use of agonists or antagonists for the treatment of inflammatory diseases is still controversial [53]. As mentioned, A_{2B} AR binds both with G_s and G_q proteins to mediate airway reactivity, inflammation, and asthma [33]. Activation of A_{2B} ARs in mast cells induces release of inflammatory mediators with bronchoconstriction as a consequence [54]. In fact, A_{2B} AR antagonists prove to be effective in the treatment of asthma in mouse model [55]. A_{2B} AR agonists reduce pulmonary edema and improve histologic lung injury, diminishing lung inflammation [56, 57].

The studies performed to clarify the role of A_{2B} AR in the cardiovascular system are conflicting. It seems that A_{2B} AR plays an important role in ischemia preconditioning [58] but, at the same time, other studies reported that A_{2B} AR is not involved [59]. A_{2B} ARs have also been proposed to inhibit smooth muscle cell-associated hypertension [60], and they may contribute to the pathogenesis of atherosclerosis by preventing atherosclerosis plaque formation [61].

It is also known that A_{2B} ARs are involved in homeostasis and lipid metabolism, insulin secretion and resistance, β-cell survivor and kidney protection, and the use of agonists, in fact, inhibits diabetes and protects pancreas [62].

A_{2B} ARs are also highly expressed in tumor cells and promote tumor cell proliferation; in fact, their antagonists seem to reduce cell growth in tumors and metastasis, permitting to consider A_{2B} AR as a target for novel or combined therapies in cancer [63, 64].

1.4 A_3 AR

The dual effect of A_3 ARs in the inflammation process involves many cells that can have overlapping and opposite function [16]. A_3 ARs produce a proinflammatory effect by promoting the release of histamine and other allergic mediators [65, 66].

Nevertheless, A_3 ARs inhibit lipopolysaccharide (LPS)-stimulated release of tumor necrosis factor-α (TNF-α) and nitric oxide (NO), by increasing neutrophil chemotaxis and, in contrast, mediate the inhibition of oxidative burst in human neutrophil and promyelocytic HL60 cells [67, 68].

Clinical trials demonstrated that A_3 AR allosteric modulators induce an anti-inflammatory effect in animal models of arthritis [69], and the use of A_3 AR agonists could be useful for treatment of patients with moderate to severe plaque psoriasis [70]. In contrast, the use of A_3 AR antagonists inhibits shrinkage of human non-pigmented ciliary epithelial cells and reduces mouse intraocular pressure in an animal model of glaucoma [71, 72].

A_3 ARs also play a role in asthma, chronic obstructive pulmonary disease (COPD), lung fibrosis, and pulmonary inflammation. Nebulized agonists directly induced lung mast cell degranulation, while having no effect in adenosine receptor knock-out (KO) mice [73].

Taking into account that A_3 ARs are not present in human lung mast cells and that they inhibit only degranulation of eosinophils, they could be useful only in eosinophil-dependent allergic disorders. On these bases, we could assert that A_3 ARs are involved in both proinflammatory and anti-inflammatory responses depending on the cell type involved [74].

Activation of the A_3 ARs either prior to ischemia or during reperfusion is useful in cardioprotection: [75] administration of agonists before ischemia or during reperfusion reduces infarct size or inhibits myocardial ischemia-reperfusion injury [76, 77]. In contrast, other experimental observations clearly indicate that low-level expression in the heart provides effective protection against ischemic injury without adverse effects, while higher levels lead to the development of a dilated cardiomyopathy [78].

In the last years, it has been demonstrated that selective A_3 AR antagonists could be useful for the treatment of atherosclerosis, and hypercholesterolemia, and could ameliorate diabetic kidney complications [79]. Nevertheless, the role of cardioprotection by A_3 ARs is not yet completely clarified [80].

The expression of A_3 ARs in the brain is low but several studies have reported a neuroprotective function [81], as in ischemia. In this case, it is possibly mediated by several pathways including microtubule-associated protein 2 (MAP-2), enhancement of the expression of glial fibrillary acidic protein, depression of nitric oxide synthase (NOS), stimulation of glial C-C motif chemokine ligand 2 (CCL2) synthesis, and delay of irreversible synaptic failure [82, 83]. Very recently an interesting application of A_3 AR agonists for the treatment of neuropathic pain has also been reported [84].

In contrast, the use of antagonists seems to increase the stability of I_{GABA} in different epileptic tissues, giving new opportunities in the treatment of epilepsy [85].

A_3 ARs seem to be involved in the protection in inflammatory gastrointestinal diseases, in fact the treatment with agonists reduces oxidative damage and inflammatory mediators in the colon [86]. Instead, the use of A_3 AR antagonists produces protection in renal failure [87].

A_3 ARs play an important role in cancer, in fact primary and metastatic solid tumors show high expression of this receptor subtype [88], and it is well known that A_3 ARs are involved in the regulation of the cell cycle. In the event of tumor inhibition, the antiproliferative effect occurs mainly via A_3 ARs-induced cell cycle arrest in the G_0/G_1 phase and decreases in telomeric signaling in these cells. It has been demonstrated that agonists protect the retinal ganglion by increasing survival and they mediate a tonic proliferative effect in colon tumor cells [89–93].

A_3 AR agonists inhibited tumor cell growth of HCT-116 human and CT-26 murine colon carcinoma cell and B16-F10 melanoma cells, and in in vivo studies have shown that in PC-3 prostate carcinoma cells and in N1S1 rat hepatocellular carcinoma cells induced apoptosis and tumor growth inhibition [94–96].

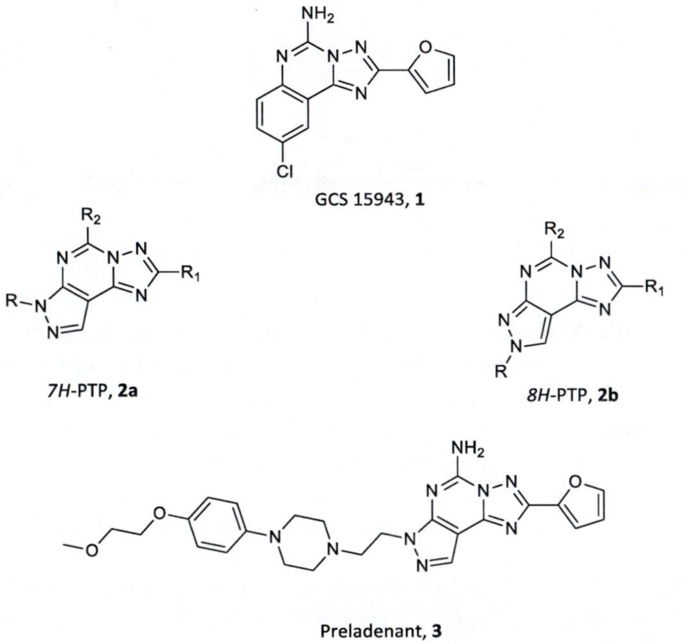

Fig. 1 5-Amino-9-chloro-2-(2-furyl)1,2,4-triazolo[1,5-*c*]quinazoline and pyrazolo[4,3-*e*][1,2,4]triazolo[1,5-*c*]pyrimidines

2 Pyrazolo-Triazolo-Pyrimidines (PTP) as AR Antagonists

PTP tricyclic scaffold represents the pharmacophoric core of several series of non-xanthine heterocyclic compounds displaying antagonistic activity at A_{2A} and A_3 ARs.

The initial studies aimed at the discovery of potent and selective adenosine receptor antagonists led to the identification of several classes of non-xanthine heterocyclic compounds possessing antagonistic properties either at A_1 or A_{2A} adenosine receptors such as triazoloquinazolines [97], triazoloquinoxalines [98], and imidazoquinoline [99]. Among these structurally diverse chemical entities, compound 5-amino-9-chloro-2-(2-furyl)1,2,4-triazolo[1,5-*c*]quinazoline derivative GCS 15943 (**1**, Fig. 1) was reported to exert high potency as A_{2A} receptor antagonist while showing low selectivity at the A_{2A} AR vs A_1 AR [97].

On the basis of these findings and with the aim to enhance potency and A_{2A} selectivity, in 1993, Gatta et al. explored new chemical entities reporting on the first series of pyrazolo[4,3-*e*][1,2,4]triazolo[1,5-*c*]pyrimidine derivatives as A_{2A} adenosine receptor antagonists (PTP **2 a-b**) (Fig. 1) [100]. The pyrazolo[4,3-*e*][1,2,4]triazolo[1,5-*c*]pyrimidine derivatives described in the paper exhibited nanomolar affinity at the A_{2A} AR, representing the first step for the development of a new class of potent and selective AR antagonists.

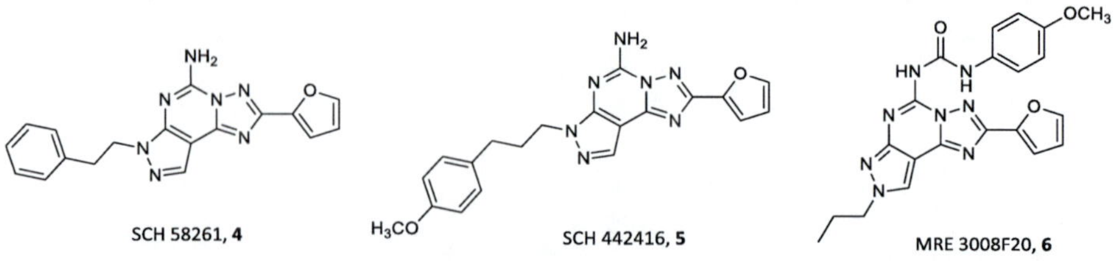

Fig. 2 Structures of AR antagonists as tools for functional characterization

Following this preliminary work, great efforts by medicinal chemists and pharmacologists have been made in order to produce more potent and selective A_{2A} and A_3 AR antagonists. Beginning in 1994, several series of PTP have been reported, some with good selectivity and potency [101–113]. Given the vastness of the subject, the synthetic approaches employed for the synthesis and the related optimization studies of pyrazolo-triazolo-pyrimidine derivatives are behind the scope of this review but a detailed dissertation can be found elsewhere [114].

Among the most advanced PTP derivatives, preladenant (Scheme 420814/MK-3814, **3**, Fig. 1), an A_{2A} AR antagonist arising from an optimization study of the PTP tricyclic scaffold, has reached the phase III clinical trial for the treatment of Parkinson's disease (PD), but failed for lack of efficacy (Fig. 1) [109, 115–117].

The discovery of potent and selective AR antagonists over the years has provided researchers important tools (Fig. 2) for the characterization of the physiological/pathological role of the ARs and their distribution in mammalian organisms. One of such compounds is Scheme 58261 (**4**), a PTP derivative synthesized by Baraldi et al. [101, 102]. Scheme 58261 has been shown to have an affinity in the low nM range (K_i value of 1–2 nM) for A_{2A} ARs expressed by a variety of tissues and cell types, and a good A_{2A} vs A_1 ARs selectivity (about 50- to 100-fold in rat and bovine brain, respectively), whereas it has no or little affinity for A_{2B} and A_3 ARs [118]. Its tritiated form has been used to specifically label A_{2A} ARs, in particular regions of the rat brain such as caudate-putamen, nucleus accumbens, and tuberculum olfactorium [119]. Scheme 58261 (**4**) antagonizes in a competitive manner the effects of rabbit platelet aggregation and porcine coronary artery relaxation induced by the A_{2A} AR selective agonist CGS 21680 2-[4-(2-carboxyethyl)-phenethyl-amino]-5′-N-ethylcarboxamidoadenosine [118]. In rat, after peritoneal administration, Scheme 58261 (**4**) has demonstrated to counteract the effect of A_{2A} AR agonists, enhancing locomotor activity [120], increasing waking behavior [121], and slightly increasing both blood pressure and heart rate [122]. In rat models mimicking CNS disorders, selective blockade of A_{2A} AR by Scheme 58261 (**4**) potentiates the antiparkinsonian effects of L-dopa, indicating a therapeutic potential of this class of compounds in PD treatment [123, 124].

A_{2A} AR antagonists have been considered attractive tools to improve the treatment of neurological disorders by preventing neurons damage and Scheme 442416 (**5**), described by Baraldi et al. [104], has been shown to increase the glutamate uptake activity in retinal Muller cells under increased hydrostatic pressure. This molecule, deriving from the same family of PTP compounds, increased expression levels of glutamine synthase and glutamate aspartate transporter thus acting as neuroprotective agent, and it has been considered a potential candidate for the treatment of glaucoma by accelerating the clearance of extracellular glutamate [125]. Moreover, it displayed binding with different affinity to particular populations of postsynaptic A_{2A} ARs in the rat striatum, suggesting the existence of at least two functionally and pharmacologically different populations of striatal postsynaptic A_{2A} ARs [126].

Instead, the pharmacological characterization of the human A_3 AR has been achieved by using MRE 3008F20 (**6**), a highly selective and high-affinity human A_3 AR antagonist [105, 127]. In fact, it has been shown to display high selectivity for human A_3 AR (K_i = 0.85 nM) compared with its affinity to human A_1 (K_i = 1,100 nM), A_{2A} (K_i = 140 nM), and A_{2B} (K_i = 2,100 nM) ARs [128]. It has been employed to investigate the presence of the A_3 AR on human colon cancer cells, and to evaluate its functional effect on colon cancer cell biology. The adenosine-mediated stimulatory effect, which results in proliferative effect, was reduced by MRE 3008F20, suggesting a potential role of the A_3 AR in human colon cancer cell lines such as Caco2, DLDI, and HT29 (Fig. 2) [93].

2.1 A_{2A} AR Antagonists

The first series of PTP derivatives as AR antagonists has been described by Gatta and coworkers [100]. As depicted in Fig. 3, in this preliminary study, the synthesis of five different PTP scaffolds has been described: 7-substituted-3,7-dihydro-2*H*-pyrazolo[4,3-*e*][1,2,4]triazolo[1,5-*c*]pyrimidin-2-one (**7**), 2,7-disubstituted-7*H*-pyrazolo[4,3-*e*][1,2,4]triazolo[1,5-*c*]pyrimidine (**8**), 5-amino-2,7-disubstituted-7*H*-pyrazolo[4,3-*e*][1,2,4]triazolo[1,5-*c*]pyrimidine (**9**), and 5-amino-2,8-disubstituted-8*H*-pyrazolo[4,3-*e*][1,2,4]triazolo[1,5-*c*]pyrimidine (**10**). Among the synthesized compounds, the most promising A_{2A} AR antagonists were found in the 5-amino-pyrazolo[4,3-*e*][1,2,4]triazolo[1,5-*c*]pyrimidines series (**9**,**10**). Specifically, compound 8-FB-PTP (**11**) showed the highest affinity at the A_{2A} AR (hA_{2A} K_i = 1.2 nM), whereas 7-FB-PTP (**12**) displayed a slight decrease in affinity at the A_{2A} AR (hA_{2A} K_i = 12 nM) but a better selectivity against the A_1 AR than compound **11** (**12**, A_1/A_{2A} = 15.8 vs **11**, A_1/A_{2A} = 2.8) [100, 102]. Both compounds bear a free amino group at the N^5 position, a 4-fluorobenzyl moiety at the 7 or 8 positions, and a 2-furyl ring at the 2 position. Free amino and furyl groups at the N^5 and 2 positions, respectively, were essential for high affinity at both A_1 and A_{2A} ARs (Fig. 3). Several substitutions on the pyrazole ring, at the 7 and 8 positions, with alkyl or aralkyl moieties were investigated indicating that substitution at the 8 position, in general, resulted in compounds with good affinity for adenosine receptors but with low potential in discriminating different AR subtypes. Substitution in 7 leads to improved selectivity at the A_{2A} AR as demonstrated by comparing compound **4** with its corresponding 8-β-phenylethyl derivative, rA_1/rA_{2A} = 52.6 and 3.4, respectively.

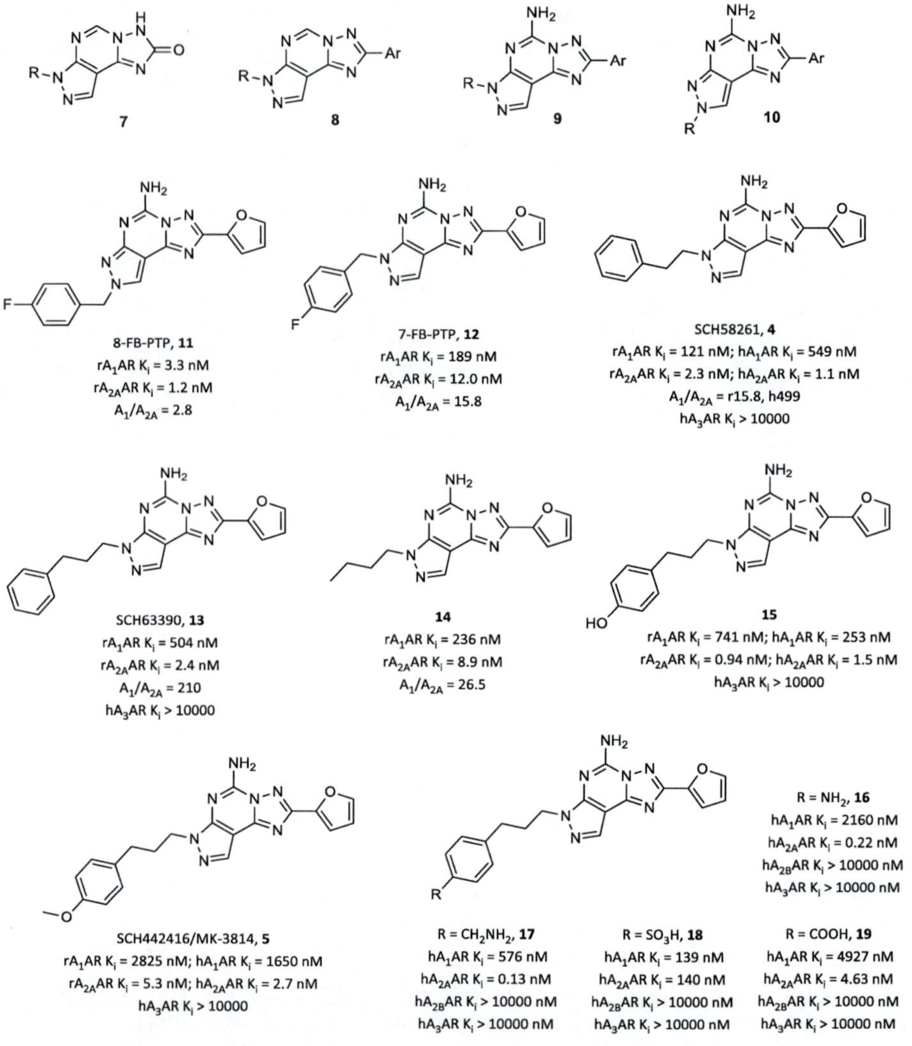

Fig. 3 PTP derivatives as A_{2A} AR antagonists: general structures **7–10** and compounds **11, 12, 4,** and **13–19**

Moreover, among 7 substituted derivatives, compounds bearing an arylalkyl moiety (e.g., 7-(3-phenyl)propyl, compound **13**, SCH63390) were found to be superior in terms of selectivity at the A_{2A} AR than compounds bearing an alkyl moiety (e.g., 7-*n*-butyl, compound **14**) [101, 102]. In an in vivo cardiovascular model, SCH58261 (**4**) retained antagonist activity at the A_{2A} AR [102].

Compounds **11–14** and compound **4** are highly lipophilic, therefore exhibiting poor water solubility, which is a common drawback for A_{2A} AR antagonists. So, modifications on the aryl group of the substituent in 7 have been investigated by introducing several polar moieties in order to improve the hydrophilic character of

PTP derivatives while retaining high affinity and selectivity at the hA_{2A} AR, which seems related to the phenyl ring [104, 107]. Moreover, the propyl chain at the 7 position, used to introduce the aryl moiety, was confirmed as best length for the linker. The proposed compounds displayed high affinity for hA_{2A} AR with different degrees of selectivity over the hA_1 AR. On the contrary, all compounds were found to be inactive at the hA_{2B} and hA_3 ARs. Hence, the introduction of a hydroxyl function at the *para* position of the phenyl ring of Scheme 63390 (**13**) resulted in both higher affinity and selectivity at the rA_{2A} AR (compound **15**). In the same way, the *para*-methoxy analog, Scheme 442416 **5**, was found to be a potent and selective A_{2A} AR antagonist, both at the human and rat receptors (rA_{2A} $K_i = 5.3$ nM, $rA_1/rA_{2A} = 487$; hA_{2A} $K_i = 2.7$ nM, $hA_1/hA_{2A} = 611$) [104]. In addition, different substituents on the phenyl ring, like basic or acidic functions which allow the preparation of soluble salts or prodrugs, were evaluated. The *para*-amino (**16**) and the *para*-methanamino (**17**) derivatives showed high affinity and selectivity at the hA_{2A} AR (compound **16**, hA_{2A} $K_i = 0.22$ nM, $hA_1/hA_{2A} = 9{,}818$; compound **17**, hA_{2A} $K_i = 0.13$ nM, $hA_1/hA_{2A} = 4{,}430$) but no improvements were obtained in terms of water solubility. The introduction of a sulfonic acid group (compound **18**) produced a marked increase in water solubility but at the expense of both the affinity and the selectivity at the hA_{2A} AR, which were significantly reduced (**18**, hA_{2A} $K_i = 140$ nM, $hA_1/hA_{2A} = 1$). An attempt to retain the water solubility of sulfonic derivatives while restoring affinity and selectivity at the hA_{2A} AR was made by introducing a weaker acidic function such as carboxylic acid. Compound **19** displayed an acceptable affinity and selectivity at the hA_{2A} AR (**19**, hA_{2A} $K_i = 4.63$ nM, $hA_1/hA_{2A} = 1{,}064$) but the water solubility was lower than compound **18** [107].

Unfortunately, the strategy of producing water-soluble compounds by introducing a polar group on the phenyl ring connected to the PTP scaffold through the alkyl chain proved to be unsuccessful; therefore, a different approach was explored. Modifications were directed to the 7-arylalkyl side chain where the distance between the alkyl chain at the 7 position and the phenyl ring was increased by introducing a polar piperazine ring as spacer, with the introduction of various substituents on the phenyl ring, especially at the *para* position. The arylpiperazine derivatives displayed both high A_{2A} binding affinity in the low nanomolar range and high selectivity over A_1 receptor (Fig. 4) [109]. Additionally, the majority of derivatives demonstrated considerable oral activity in the rat haloperidol-induced catalepsy model, being considered as potential therapeutic agents for PD.

Unsubstituted phenylpiperazine **20** showed potent oral activity 1 h after dose administration but the activity was lost after 4 h, due to metabolic conversion to derivative **21,** which resulted inactive when tested at an oral dose of 1 mg/kg even though its high affinity at the A_{2A} AR ($hA_{2A}K_i = 1.3$ nM). The in vivo metabolism issue was tackled by substituting the phenyl ring with moieties, like fluorophenyl unit, that are not prone to metabolism. One of such compounds, the 2,4-difluoro analog (Scheme 412348, **22**) displayed potent oral activity but its water solubility was poor. Compounds with improved solubility were prepared by introducing ether-linked substituents on the phenyl ring leading to derivatives endowed with

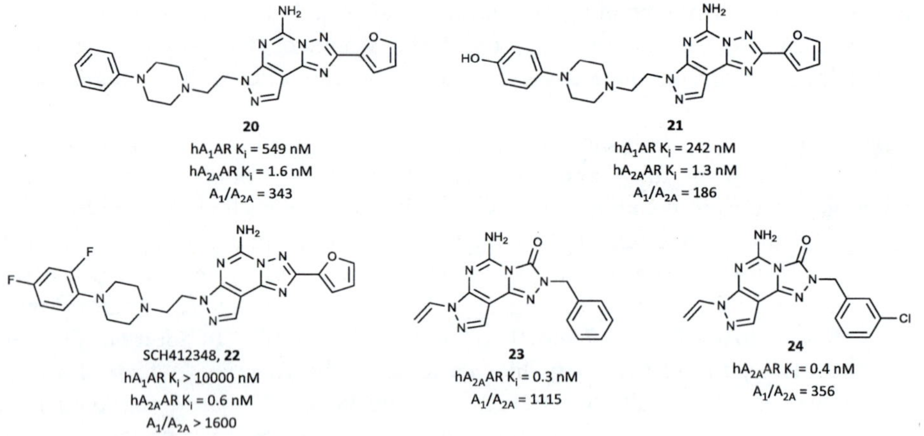

Fig. 4 PTP derivatives as A_{2A} AR antagonists: compounds **20–24**

subnanomolar affinity at the hA_{2A} AR and prolonged oral anti-cataleptic activity. The methoxy derivative Scheme 420814/MK-3814 (**3**), for example, exhibited high affinity and selectivity profile (hA_{2A} K_i = 1.1 nM, rA_{2A} K_i = 2.5 nM; hA_1/hA_{2A} = 1,340), robust oral anti-cataleptic activity in the rat model, and good solubility in water as crystalline free base (0.2 μM at native pH of 5.1 and 2 mM in 0.01 N HCl). For these reasons, it was subjected to extensive profiling and additionally further evaluated in clinical trials under the name of preladenant [115–117]. Although several high-resolution crystal structures of the A_{2A} AR were obtained, no structure in complex with preladenant has been reported, thus its exact binding mode and interactions within the receptor binding site are still unknown. Claff et al. have presented the first high-resolution crystal structure of A_{2A}-PSB1-bRIL, a thermostabilized A_{2A} AR mutant harboring only a single point mutation, in complex with a fluorescent preladenant conjugate providing insights into the interaction of the preladenant scaffold with the orthosteric binding site of the receptor as discussed in the fluorescent conjugate section of this review (Fig. 12) [129].

In general, modifications regarding the 2 position of the PTP scaffold have been constrained by the observation that the presence of the furanyl group was a key requirement for an effective binding of the ARs and its replacement with phenyl or substituted phenyl or as long as with other heterocycles rings led to a marked loss of affinity at all the AR subtypes. In addition, in most cases within these series of PTP derivatives, substitution of the pyrazole ring took place at the 7 rather than at the 8 position and hence, the binding mode of these compounds might not be the same as for their N^8 analogs [108].

Interestingly, in an additional effort to improve solubility with respect to preladenant (**3**), the pyrazolo[4,3-*e*]-1,2,4-triazolo[4,3-*c*]pyrimidin-3-one scaffold has been investigated, resulting in a novel class of potent and selective A_{2A} AR antagonists (Fig. 4). Introduction of a vinyl group at the 7 position and furan replacement at the 2 position with a benzyl or 3-chlorobenzyl moieties (compounds

23 and **24**, respectively) afforded potent and selective A_{2A} AR antagonists (**23**, hA_{2A} $K_i = 0.3$ nM, $hA_1/hA_{2A} = 1{,}115$; **24**, hA_{2A} $K_i = 0.4$ nM, $hA_1/hA_{2A} = 356$). However, these compounds were inactive in the rat catalepsy assay due to poor pharmacokinetic properties and exposure in rat brain compared to preladenant **3** [111].

2.2 A_3 AR Antagonists

The A_3 AR has been cloned from different species such as rat, human, sheep, mouse, and rabbit [130–133, 134], and significant differences in sequence homology for A_3 receptors have been observed among species [131]. In particular, rat A_3 AR is significantly different from human (73.8% of identical sequence) and behaves anomalously in ligand binding assays, where different affinities values were obtained at rat and human A_3 AR, respectively, when comparing the same antagonist ligands [106, 137]. Moreover, tissue distribution of A_3 AR is very different in rat as compared to man [135].

The starting point for the development of potent and selective A_3 AR antagonists can be dated back in 1996 when Jacobson and coworkers reported on a series of triazoloquinazoline derivatives among which compound **1**, GCS 15493, displayed a certain affinity for the human A_3 AR (hA_3 $Ki = 14$ nM). Acylation of the free amino group at position 5 of GCS 15493 (**1**) with a phenylacetyl group led to MRS 1220 (**25**), a potent but not highly selective A_3 AR antagonist (rA_3 $Ki = 0.65$ nM, $A_1/A_3 = 469$, $A_{2A}/A_3 = 80$) [136]. In the same year, Baraldi and coworkers described a series of N^6-(substituted-phenylcarbamoyl)adenosine-5′-uronamide derivatives, such as compound **26**, acting as potent A_3 AR agonists (Fig. 5) [137]. The pivotal step was the idea to functionalize the amino group at the 5 position of the PTP scaffold, known to confer antagonism activity at the A_{2A} ARs, with the arylcarbamoyl moieties of potent derivatives within the adenosine-5′-uronamide A_3 agonists series [105]. Interestingly the researchers described the exclusive preparation of the 8-substituted derivatives, since the coupling reaction with isocyanate, leading to the formation of the carbamoyl moiety, failed in the case of the 7-substituted PTP derivatives. The synthesized compounds displayed high affinity at the hA_3 AR with the level of selectivity being influenced by the presence of both arylureido and (ar)alkyl substituents at the 5 and 8 positions, respectively (Fig. 6). This behavior is in contrast if compared with the corresponding N^5-unsubstituted derivatives which were also assayed at the various AR subtypes, showing high affinity for the A_{2A} AR as long with low selectivity vs the A_1 AR and poor activity at hA_3 AR [105, 138]. Specifically, the aforementioned difference could be clearly seen by comparing N^5-unsubstituted derivative **27**, which was a poorly selective A_{2A} AR antagonist (hA_{2A} $K_i = 0.34$ nM, $A_1/A_{2A} = 2.9$), with compound **28** where the amino group functionalization with the *para*-methoxycarbamoyl moiety afforded a potent and selective A_3 AR antagonist (hA_3 $K_i = 0.20$ nM, $A_1/A_3 = 5{,}485$, $A_{2A}/A_3 = 6{,}950$) [138]. The following optimization efforts were focused on

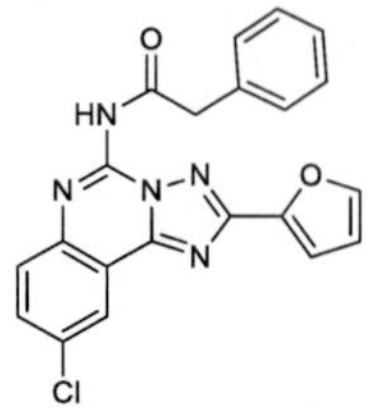

MRS 1220, **25**

$rA_1AR\ K_i = 305\ nM$

$rA_{2A}AR\ K_i = 52\ nM$

$rA_3AR\ K_i = 0.65\ nM$

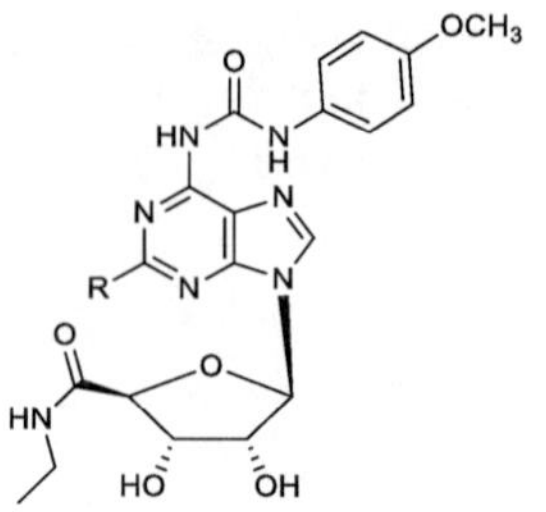

A_3 AR agonist, **26**

$rA_1AR\ K_i = 33\ nM$

$rA_{2A}AR\ K_i = 3363\ nM$

$rA_3AR\ K_i = 6.6\ nM$

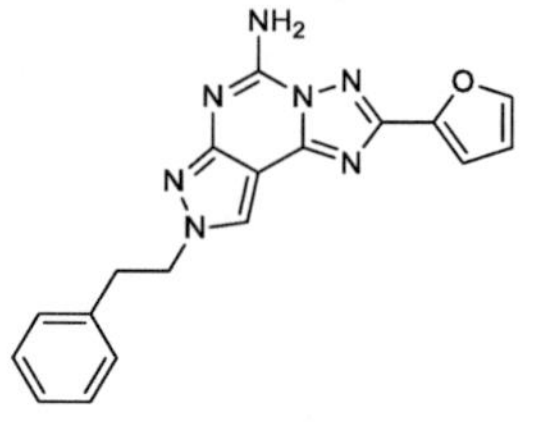

27

$rA_1AR\ K_i = 2.16\ nM$; $hA_1AR\ K_i = 1\ nM$

$rA_{2A}AR\ K_i = 0.70\ nM$; $hA_{2A}AR\ K_i = 0.34\ nM$

$hA_{2B}AR\ K_i = 5.1\ nM$

$hA_3AR\ K_i = 280\ nM$

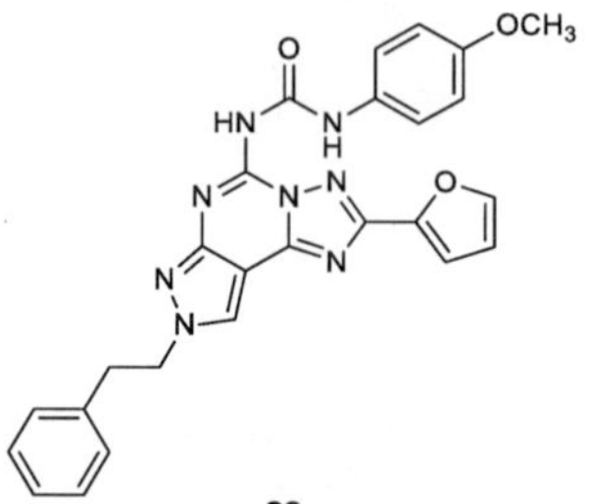

28

$rA_1AR\ K_i = 1282\ nM$; $hA_1AR\ K_i = 201\ nM$

$rA_{2A}AR\ K_i = 1398\ nM$; $hA_{2A}AR\ K_i = 120\ nM$

$hA_{2B}AR\ K_i = 5.1\ nM$

$hA_3AR\ K_i = 0.98\ nM$

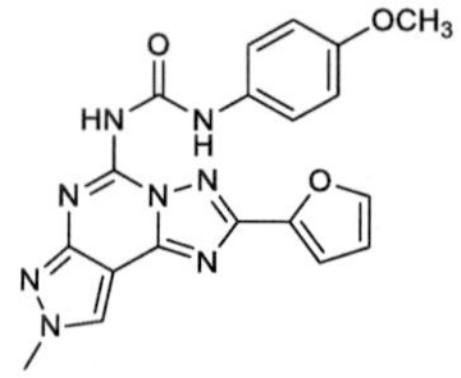

29

$hA_1AR\ K_i = 1097\ nM$

$hA_{2A}AR\ K_i = 1390\ nM$

$hA_{2B}AR\ K_i = 261\ nM$

$hA_3AR\ K_i = 0.20\ nM$

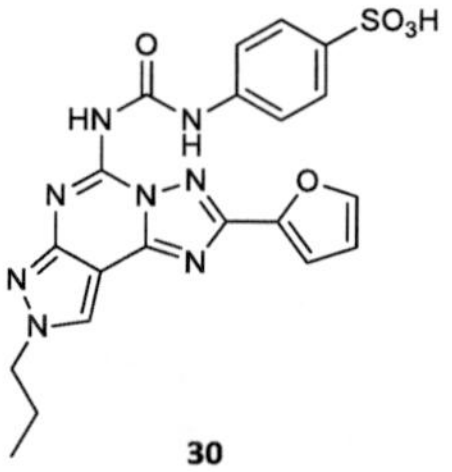

30

$hA_1AR\ K_i > 10000\ nM$

$hA_{2A}AR\ K_i = 305\ nM$

$hA_{2B}AR\ K_i > 10000\ nM$

$hA_3AR\ K_i = 30\ nM$

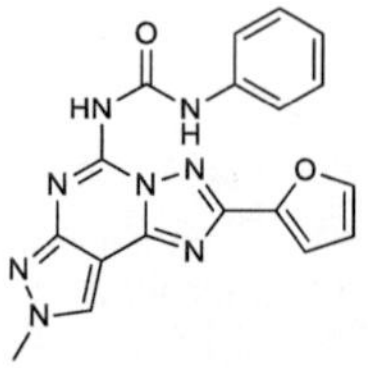

31

$hA_1AR\ K_i = 594\ nM$

$hA_{2A}AR\ K_i = 381\ nM$

$hA_{2B}AR\ K_i = 222\ nM$

$hA_3AR\ K_i = 0.16\ nM$

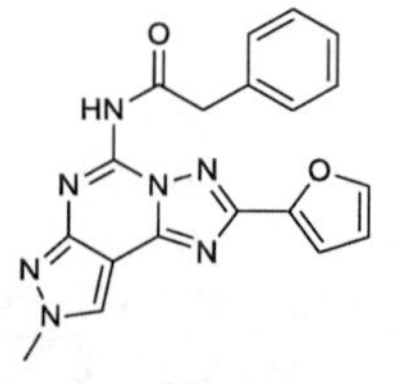

32

$hA_1AR\ K_i = 702\ nM$

$hA_{2A}AR\ K_i = 423\ nM$

$hA_{2B}AR\ K_i = 165\ nM$

$hA_3AR\ K_i = 0.81\ nM$

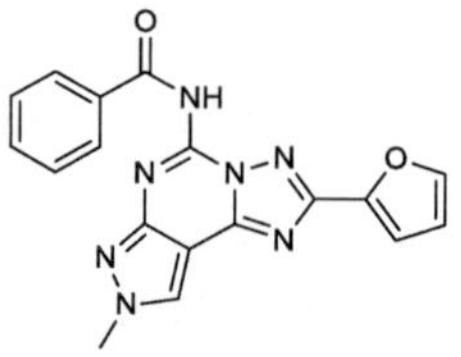

33

$hA_1AR\ K_i = 2030\ nM$

$hA_{2A}AR\ K_i = 879\ nM$

$hA_{2B}AR\ K_i > 30000\ nM$

$hA_3AR\ K_i = 15.7\ nM$

Fig. 5 PTP derivatives as A_3 AR antagonists: compounds **25–33**

modifications of the chain in 8 while maintaining either the 4-methoxyphenylcarbamoyl or the 3-chlorophenylcarbamoyl moieties at the N^5 position. Several alkyl and arylalkyl moieties were evaluated in order to identify the best substitution pattern in terms of steric and lipophilic features for the specific recognition of the A_3 AR subtype. From the obtained binding results it was quite evident that at the 8 position in the pyrazole ring the presence of a small alkyl chain,

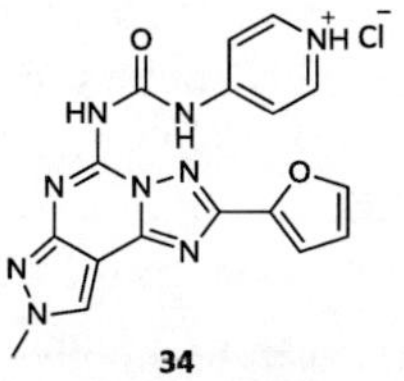

34
hA_1AR K_i = 355 nM
$hA_{2A}AR$ K_i = 110 nM
$hA_{2B}AR$ K_i > 1000 nM
hA_3AR K_i = 0.014 nM

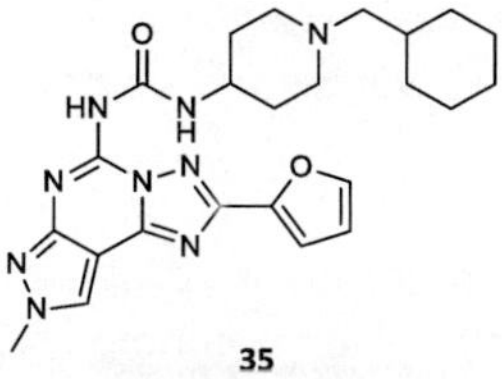

35
hA_1AR K_i = 3408 nM
$hA_{2A}AR$ K_i > 5000 nM
$hA_{2B}AR$ K_i > 5000 nM
hA_3AR K_i = 9.7 nM

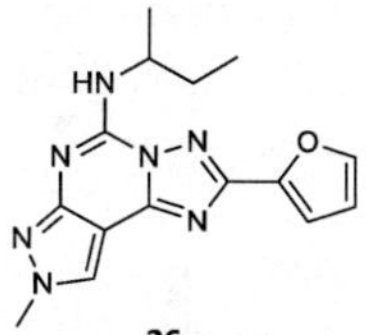

36
hA_1AR K_i = 131 nM
$hA_{2A}AR$ K_i = 248 nM
$hA_{2B}AR$ K_i = 3860 nM
hA_3AR K_i = 5.23 nM

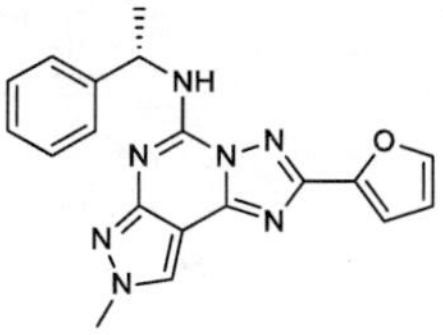

37
hA_1AR K_i = 338 nM
$hA_{2A}AR$ K_i = 55.2 nM
$hA_{2B}AR$ K_i > 5290 nM
hA_3AR K_i = 0.30 nM

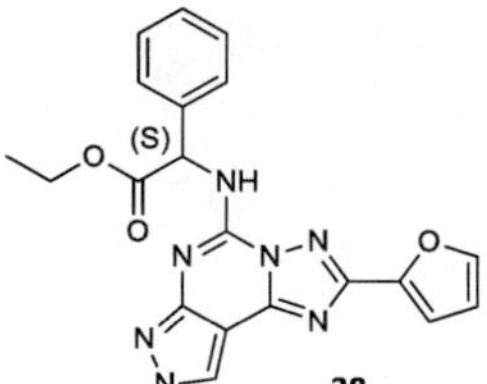

38
hA_1AR K_i = 1820 nM
$hA_{2A}AR$ K_i = 1300 nM
$hA_{2B}AR$ K_i > 10000 nM
hA_3AR K_i = 50.6 nM

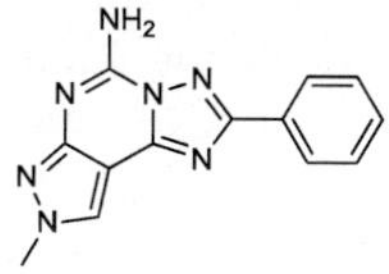

39
hA_1AR K_i = 339 nM
$hA_{2A}AR$ K_i = 121 nM
$hA_{2B}AR$ K_i > 10000 nM
hA_3AR K_i = 75 nM

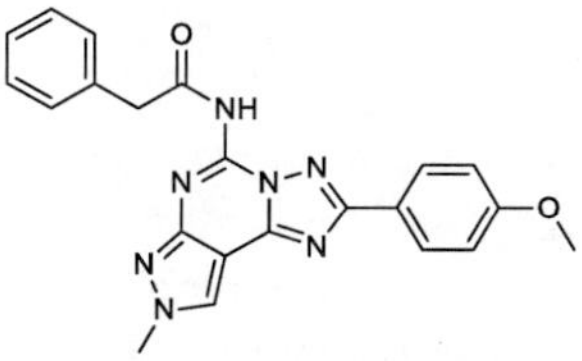

40
hA_1AR K_i > 30000 nM
$hA_{2A}AR$ K_i > 100000 nM
$hA_{2B}AR$ K_i > 10000 nM
hA_3AR K_i = 0.241 nM

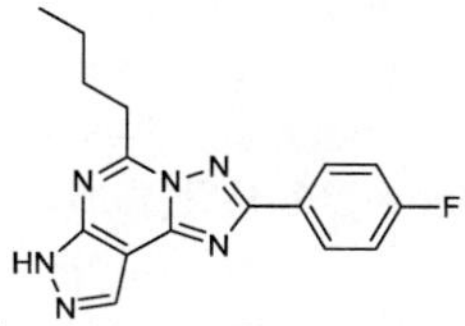

41
hA_1AR IC_{50} = 610 nM
$hA_{2A}AR$ IC_{50} > 10000 nM
$hA_{2B}AR$ IC_{50} > 9400 nM
hA_3AR IC_{50} = 1.9 nM

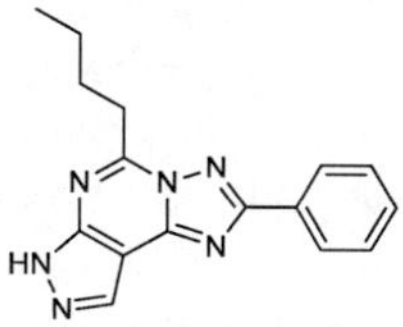

42
hA_1AR IC_{50} = 27 nM
$hA_{2A}AR$ IC_{50} = 190 nM
$hA_{2B}AR$ IC_{50} = 2600 nM
hA_3AR IC_{50} = 2.1 nM

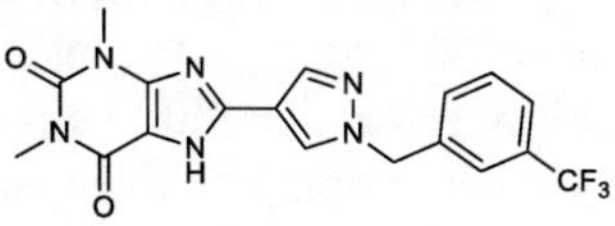

CVT-6975, 43
hA_1AR K_i = 990 nM
$hA_{2A}AR$ K_i = 690 nM
$hA_{2B}AR$ K_i = 1 nM
hA_3AR K_i = 1000 nM

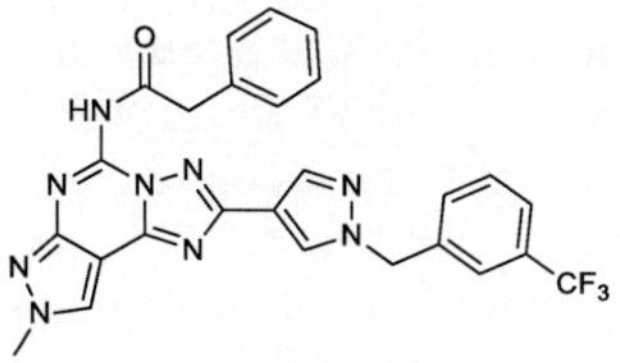

44
hA_1AR K_i > 100000 nM
$hA_{2A}AR$ K_i > 100000 nM
$hA_{2B}AR$ K_i > 10000 nM
hA_3AR K_i = 11 nM

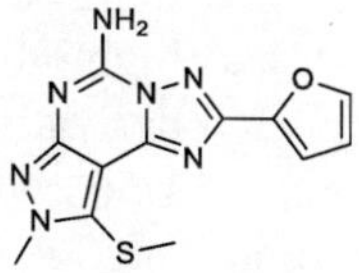

45
hA_1AR K_i = 8.4 nM
$hA_{2A}AR$ K_i = 1.2 nM
$hA_{2B}AR$ K_i = 10.3 nM
hA_3AR K_i = 35 nM

Fig. 6 PTP derivatives as A_3 AR antagonists: compounds **34–42**, **44–45**, and compound **43** as A_{2B} AR antagonists such as methyl, endowed the newly synthesized PTP derivatives with high affinity at the A_3 AR and good selectivity versus the other AR subtypes (i.e., compound **29**, hA_3AR $K_i = 0.20$ nM; $A_1/A_3 = 5{,}485$, $A_{2A}/A_3 = 6{,}950$) [138]. Next, a further investigation on the effect of diverse substituents (e.g., chloro, fluoro, bromo, methyl, nitro, sulfonic acid, trifluoromethyl) at the various positions on the phenyl ring of the N^5-phenylcarbamoyl group combined with the presence of a small alkyl group at N^8 position has been carried out. Interestingly, the synthesized compounds displayed affinity values at the A_3 AR in the subnanomolar range (0.16–3.7 nM). In particular, the lower affinity value obtained for the *para*-sulfonic acid analog (compound **30**, hA_3 $K_i = 25$ nM) was explained by the presence of a hydrophobic environment in the A_3 AR region surrounding the *para* position. Moreover, molecular modeling studies highlighted the presence of a steric control at the *para* and *meta* position which explained why bulky and polar substituents were not well tolerated at these positions [139]. Thus, the most potent and selective compound of the series was the unsubstituted derivative **31** displaying an A_3 K_i value of 0.16 nM, being more than 2000-fold selective for A_3 over A_1 and A_{2A} ARs. When the phenylureido moiety was replaced with a phenylacetyl moiety, a typical feature of MRS 1220 (**25**), the affinity at the hA_3 AR was maintained, whereas a slight drop in selectivity was noted (compound **31** vs compound **32**, Fig. 5). On the contrary, the shortening of the amide chain by deletion of a methylene group (compound **33**) produced a drop of affinity at the A_3 AR, while retention of selectivity was observed (**33**, hA_3 $K_i = 15.7$ nM; $hA_1/hA_3 = 129$, $hA_{2A}/hA_3 = 56$) [140].

Although good results have been achieved in terms of potency and selectivity at the A_3 AR, the major problem with PTP derivatives was still the typical low water solubility which hampered their use as pharmacological and diagnostic tools. Based on the molecular modeling study suggesting the presence of steric control around the *para* position, an attempt to improve water solubility was made by bioisosteric replacement of the phenyl ring of the phenylureido moiety at the N^5 position with a pyridine ring, which in turn allowed the preparation of pyridinium hydrochloride salts. Moreover, the position and the basicity of the pyridine nitrogen, as long as the replacement of pyridine moiety with different heteroaryl groups, were evaluated (Fig. 6). The best substituent profile was found to be the 4-pyridinium hydrochloride salt, as highlighted by compound **34** which, when compared with the unsubstituted phenyl derivative **31**, displayed an increased water solubility (**31** $R_m = 4.06$ vs **34** $R_m = 1.66$, where $R_m = \log(1/R_f - 1)$; maximum concentration in water of compound **34**: 15 mM) as well as enhanced potency and selectivity at the hA_3 AR (**34**, hA_3 $K_i = 0.014$ nM; $hA_1/hA_3 = 25{,}357$, $hA_{2A}/hA_3 = 7{,}857$) [141, 142].

To evaluate the pharmacological potential of this series of PTP derivatives, aqueous formulations at physiological pH, suitable for intravenous administration, were prepared. Despite the increased water solubility with respect to that of compound **31**, at physiological pH the solubility of **34** dropped to less than 0.1 mg/ml. Thus, a new series of derivatives, in which the pyridine-4-yl moiety was replaced by a 1-(substituted)-piperidin-4-yl ring, was designed with the idea of evaluating the higher basicity of piperidine and its potential to form stable, water-soluble salts. The best binding profile was shown by compound **35** (hA_3 $K_i = 9.7$ nM; $A_1/A_3 = 351$,

$A_{2A}/A_3 > 515$, Fig. 6) which remarkably exhibited a solubility of 8 mg/ml at physiological pH, allowing the preparation of stable aqueous solutions suitable for intravenous administration [112].

A more recent study reported on an in-depth evaluation of several polyfunctionalized chains such as alkylamino, arylalkylamino, heterocycles moieties replacing the aryl carbamoyl or arylacetyl moieties at the N^5 position, providing useful information about the extracellular environment surrounding the N^5 position of the PTP nucleus. These specific modifications proved to be successful as demonstrated by the newly synthesized compounds displaying affinity for the hA_3 AR in the nanomolar and subnanomolar range with different degrees of selectivity over the other AR subtypes. The introduction at the 5 position of small branched alkylamines and several benzylamines produced good results in terms of affinity but at the expenses of selectivity over the other AR subtypes which was poor (e.g., compound **36,** hA_3 $K_i = 131$ nM; $A_1/A_3 = 25$, $A_{2A}/A_3 = 47.4$). The best results in terms of potency and selectivity were achieved through the synthesis of compound **37** bearing an (S) α-phenylethyl chain at the N^5 position (**37**, hA_3 $K_i = 0.3$ nM; $A_1/A_3 = 1{,}127$, $A_{2A}/A_3 = 184$) [143]. Analogously, Federico et al. explored a related modification at the 5 position by introducing several aminoesters (e.g., compound **38**) obtaining novel PTP derivatives endowed with good affinity at the hA_3 AR. Furthermore, the aminoester moiety represents an optimal linker allowing further functionalization of the carboxylic group in order to obtain more complex conjugated derivatives for diagnostic/theranostic purposes [144]. Although compound **38** was not as potent at the hA_3 AR as previously reported antagonists, it was chosen for molecular modeling simulations to evaluate the possible interactions of these PTP derivatives within the hA_3 AR binding site. As highlighted by the docking pose of compound **38** at the hA_3 AR, which is the pose observed at the receptor binding site among four different binding modes isolated from the docking simulations, and the related energy fingerprint (Fig. 7), the ligand formed hydrogen bonds (HBs) with residue N250, established electrostatic interactions with N250 and L90 and several hydrophobic contacts especially with F168, L246, and I268. Van der Waals interaction with F168 is repulsive but the hydrophobic terms tended to overcome it. These data indicated that this particular binding mode could represent a plausible solution for the binding of these PTP derivatives at the hA_3 AR binding site [144].

Taking into account the modifications at the 2 position, the indication that N^8 substituted PTP derivatives might display a different binding mode compared to N^7 substituted derivatives (showing high selectivity at the A_{2A} AR), and considering that several tricyclic hA_3 AR antagonists are functionalized with a phenyl ring at the position equivalent to that of furan in PTP derivatives [145], Cheong and coworkers proposed a new structure–activity relationship evaluation by designing and synthesizing a series of PTP derivatives bearing a (*para*-substituted)-phenyl moiety at the 2 position maintaining either methyl or phenyl-ethyl groups at 8 position and a free amino, phenylacetamide or (bis-)benzamide at the 5 position. The binding assay results clearly indicated that the free amino group at 5 position poorly discriminates among the AR subtypes (compound **39**). On the contrary, when a phenylacetamido group was introduced at the 5 position, the resulting derivatives retained good

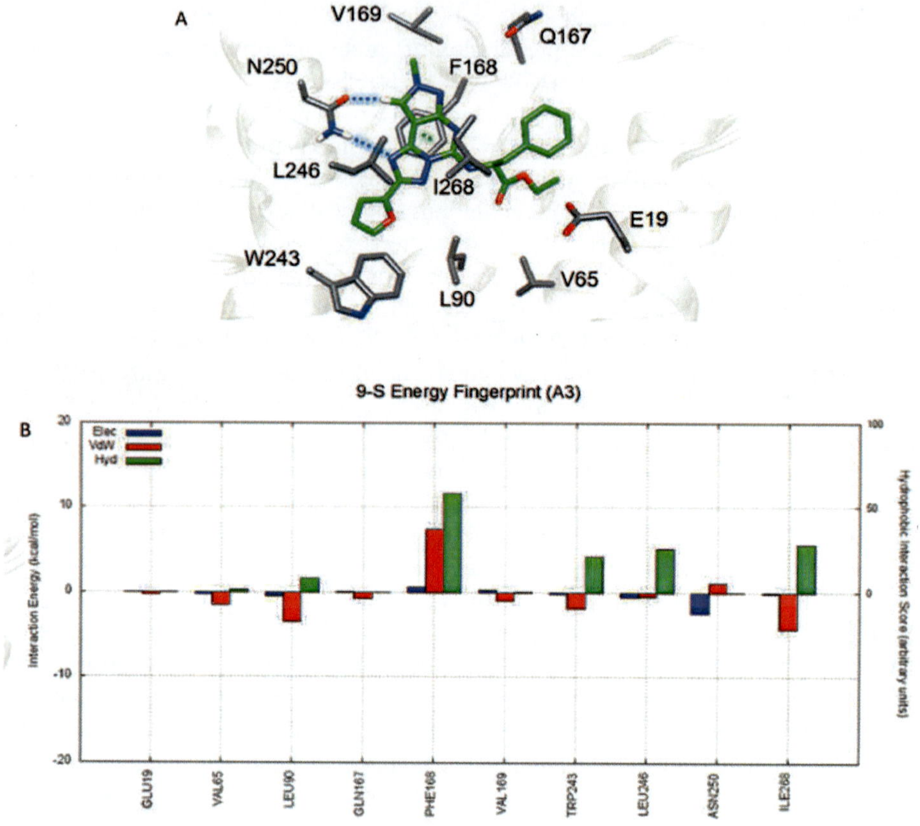

Fig. 7 Per-residue energy histograms (panel **a**) and docking pose (panel **b**) of compound **38** at the hA_3 AR. Reproduced from [144] with permission from the Royal Society of Chemistry

affinity towards the hA_3 AR subtype and a notable increase in hA_3 AR selectivity over other AR subtypes was observed in comparison to the 2-furyl PTP derivatives (e.g., compound **40**, hA_3 $K_i = 0.241$ nM; $A_1/A_3 > 124{,}000$, $A_{2A}/A_3 = 415{,}000$ vs compound **32**, hA_3 $K_i = 0.81$ nM, $A_1/A_3 = 877$, $A_{2A}/A_3 = 522$ or vs compound **31**, hA_3 $K_i = 0.16$ nM; $A_1/A_3 = 3{,}713$, $A_{2A}/A_3 = 2{,}381$ [110, 144].

Moreover, Okamura et al. prepared different fused 1,2,4-triazolo[1,5-*c*]pyrimidine derivatives, including 5-*n*-butyl-pyrazolo[4,3-*e*]-1,2,4-triazolo[1,5-*c*]pyrimidines (Fig. 6), confirming that both the type and the position of the substituent on the phenyl ring at the 2 position affected the potency and the selectivity at the hA_3 AR over the other AR subtypes. A higher hA_3 AR selectivity over the hA_{2A} AR was observed when the phenyl ring was substituted at the *para* position rather than at the *meta* or *ortho* positions as well as in comparison with unsubstituted phenyl derivatives (**41** (4-F-Ph), hA_3 $IC_{50} = 1.9$ nM; $A_1/A_3 = 321$, $A_{2A}/A_3 = 5{,}263$ vs compound **42** (Ph), hA_3 $IC_{50} = 2.1$ nM; $A_1/A_3 = 12.9$, $A_{2A}/A_3 = 90.5$) [72].

Interestingly, in a further study aimed at the identification of new non-xanthine hA_{2B} AR antagonists, Federico et al. reported on a novel series of PTP derivatives in which the furyl group at the 2 position was replaced with a 1-(3-trifluoromethyl-benzyl)1*H*-pyrazol-4-yl moiety, a substituent known to confer high affinity for the hA_{2B} AR when introduced at the 8 position of xanthine derivatives such as CVT-6975 (**43**), a potent hA_{2B} AR antagonist. Surprisingly, the synthesized hybrids were inactive at the A_{2B} AR while behaving as potent and selective hA_3 AR antagonists. Among the synthesized derivatives, the best results were obtained with compound **44**, which was not as potent at the hA_3 AR as its corresponding 2-furyl derivative **32**, but demonstrated a higher selectivity at the hA_3 AR over the other receptor subtypes, thus indicating that replacement of the furyl group at the 2 PTP position with longer chain and more sterically hindering group such as 1-(3-trifluoromethyl-benzyl)1*H*-pyrazol-4-yl moiety is also well tolerated, opening new possibilities in the development of even more potent and selective PTP derivatives as A_3 AR antagonists [113].

Finally, Baraldi et al. investigated also the 9 position of the PTP scaffold by evaluating the effects produced upon the introduction of several substituents with different steric and hydrophilic/lipophilic properties, such as alkylamino, arylamino, alkylthio, or N-methylpiperazino moieties while maintaining the key features for antagonism, that are the methyl group at the 8 position and a free amino or phenylureido as long with arylacetamido moieties at the 5 position of the PTP scaffold for A_{2A} AR antagonism and A_3 AR antagonism, respectively. In general, the introduction of a substituent at the 9 position resulted in a complete loss of selectivity even though the receptor affinity was maintained. Among the different substituents screened, the methylthio group at the 9 position was the best tolerated, as for compound **45**, which displayed nanomolar affinity at the four AR subtypes (**45**, hA_1 K_i = 8.4 nM; hA_{2A} K_i = 1.2 nM; hA_{2B} K_i = 10.3 nM; hA_3 K_i = 35 nM) [108].

2.3 A_{2B} and A_1 ARs Antagonists

Whilst the identification of several potent and selective antagonists has allowed to outline a clear SAR profile for the PTP related compounds at the A_{2A} and A_3 AR subtypes, the same level of understanding regarding the structural requirements has not yet been achieved to obtain highly potent and selective PTP ligands for of the A_{2B} and A_1 AR subtypes. Many efforts have been spent and some typical features could be identified among some potent, but unselective, PTP A_{2B} antagonists reported so far. In general, substitution at the 8 position, rather than at the 7 position of the pyrazole ring, seems to be favorable for the interaction with the A_{2B} AR (Fig. 8).

Specifically, Baraldi et al. reported that N^5 unsubstituted PTP derivatives with long alkyl chain or a phenylethyl moiety at the 8 position displayed interesting affinity at the A_{2B} AR in the nanomolar range but with low selectivity over the other

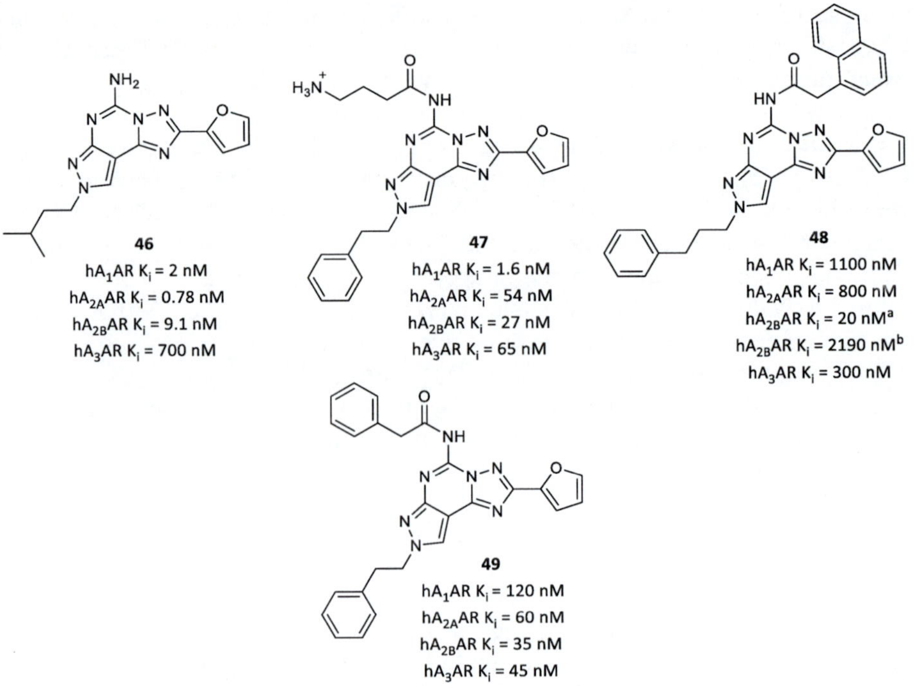

Fig. 8 PTP derivatives as unselective AR antagonists: compounds **45–49**.[a] Displacement of specific [^{3}H]-DPCPX binding at human A_{2B} AR expressed in HEK-293 cells; [b] Displacement of specific [^{3}H]-ZM241385 binding at human A_{2B} AR expressed in chinese hamster ovary (CHO cells)

AR subtypes (e.g., compound **27**, hA_{2B} K_i = 5.1 nM; compound **46**, hA_{2B} K_i = 9.1 nM) [138]. The introduction of a γ-aminobutyramido at the 5 position (compound **47**) resulted in a slight decrease of affinity at the A_{2B} AR, while an enhanced selectivity was observed over the A_{2A} AR subtype [146].

Further optimization studies of the substituent at the N^5 position led to the observation that a combination of branched or arylalkyl chain at the 8 position together with arylacetyl moieties at the N^5 position promotes the binding with the A_{2B} AR. In particular, the presence of a bulky α-naphthylacetyl moiety at the N^5 position, such as in compound **48**, produced a significant increase in selectivity at the A_{2B} AR over the other receptor subtypes resulting in the first potent and selective PTP A_{2B} AR antagonist [147]. Nevertheless, further evaluation of the binding properties of compound **48** produced controversial results. In fact, in a subsequent study, the potency of several A_{2A} AR antagonists was determined through displacement of specific binding of [^{3}H]-ZM241385, a dual A_{2A}/A_{2B} antagonist, at human A_{2B} ARs expressed in CHO cell. Surprisingly, in this specific assay, compound **48** displayed a K_i value at the hA_{2B} receptor of 2,190 nM, which is in evident contrast with the K_i value of 20 nM previously reported. These results have casted doubts about the effectiveness of compound **48** as an A_{2B} AR antagonist [148].

As discussed in the previous section, a hybridization approach has been employed to identify new non-xanthine hA_{2B} AR antagonists by introducing at the 2 position of the PTP scaffold a 1-(3-trifluoromethyl-benzyl)1*H*-pyrazol-4-yl moiety, which, when present at the 8 position of xanthine derivatives such as CVT-6975 (**43**), confers high affinity for the hA_{2B} AR. Unfortunately, this approach proved to be unsuccessful, being the newly synthesized compound inactive at the hA_{2B} AR, but surprisingly this modification led to some potent and selective hA_3 antagonists, such as compound **44** [113].

A similar behavior could be observed for the A_1 AR too. In general, 8-substituted PTP analogs displayed a better binding affinity at this receptor subtype compared to the 7-substituted PTP derivatives while showing no selectivity at the A_{2B} AR over the A_{2A} AR in particular, and generally over the other AR subtypes. This could be clearly seen by comparing compounds **12** and **4** with compounds **11** and **27**, which bear a 4-fluoro benzyl and a phenylethyl moiety at the 7 and 8 position, respectively (**12**, rA_1 $K_i = 189$ nM; $rA_1/rA_{2A} = 15.8$ and **4**, hA_1 $K_i = 549$ nM; $hA_1/hA_{2A} = 499$ vs **11**, rA_1 $K_i = 3.3$ nM; $rA_1/rA_{2A} = 2.8$ and **27**, hA_1 $K_i = 1$ nM; $hA_1/hA_{2A} = 3.22$) [102, 105, 137]. Moreover, the introduction of a substituent at the amino group at the 5 position resulted in a reduced affinity at the A_1 AR. In particular, compound **49** displayed an affinity about 120-fold lower compared to its N^5 unsubstituted analog, compound **27** (**49**, hA_1 $K_i = 120$ nM vs **27**, hA_1 $K_i = 1$ nM) (Fig. 8) [138, 147].

3 Probes and Functional Conjugates Targeting Adenosine Receptors Bearing a PTP Scaffold

3.1 Covalent Derivatives

Irreversible ligands represent an useful tool to study receptor functions and structure, such as physiological functioning, receptor reserve determination, and ligand binding site mapping [149]. In addition, covalent ligands are effective at stabilizing GPCRs to obtain X-ray crystal structures [150, 151]. At the A_{2A} AR, reaction of SCH58261 (**4**) with fluorosulfonic acid affords the A_{2A} AR irreversible antagonist 5-amino-7-[2-(4-fluorosulfonyl)phenylethyl]-2-(2-furyl)-pryazolo[4,3-*e*]-1,2,4-triazolo[1,5-*c*]pyrimidine (FSPTP, **50**, Fig. 9), which has been used in coronary conductance experiments on guinea pig hearts, showing that there is a large receptor reserve, which justifies the high agonist-mediated coronary vasodilation, that evidently is not consistent with agonist affinity values [152]. Another useful method to study protein targets is photoaffinity labeling (PAL). Regarding GPCR, a photoaffinity probe is useful both to localize receptor and to have information about receptor binding sites [153]. In particular, identification of the cross-link site could be obtained after UV-mediated reaction with a solubilized protein followed by digestion and LC-MS/MS analysis. Compounds **51** and **52** (Fig. 9) were developed by replacing the fluorine atom in 8-FB-PTP (**11**) and 7-FB-PTP (**12**) with

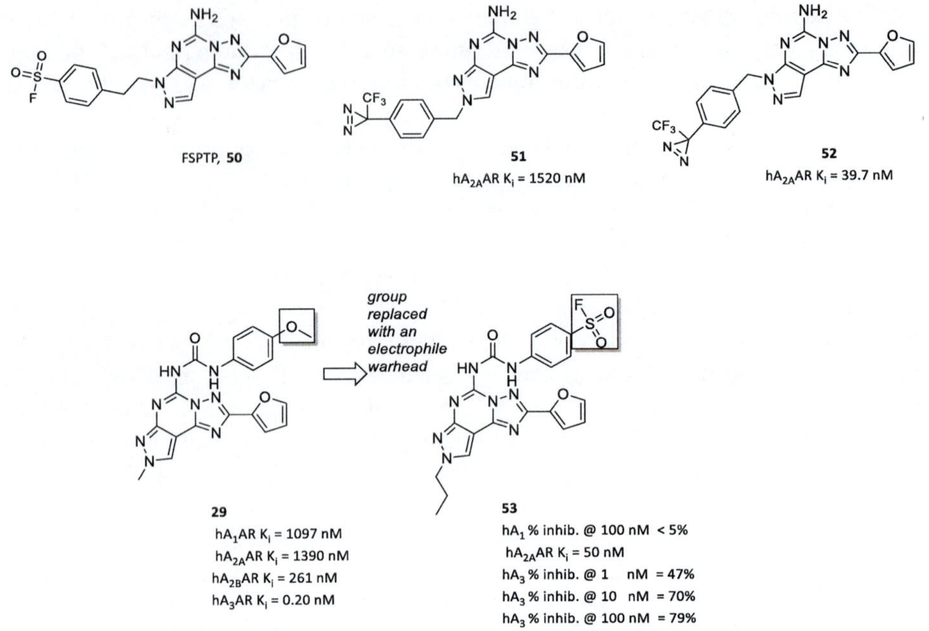

Fig. 9 PTP irreversible compounds

the photophore trifluoromethylphenyl diazirine (TPD) group. Fluorescent polarization binding assay has been performed on solubilized hA_{2A} AR showing K_i values of 39.7 and 1,520 nM for compounds **51** and **52**, respectively. Thus, PAL of the receptor was performed using compound **52** and Tyr271$^{7.36}$ has been identified as the most likely cross-linked amino acid [154].

A PTP scaffold has also been used to develop irreversible antagonists for the A_3 AR [106]. In particular, the design of covalent derivatives was based on the selective A_3 AR antagonist, which bear a short linear alkyl chain at the N^8 position combined with a 4-methoxyphenylureido moiety at the 5 position. The replacement of the methoxy group at the *para* position of the arylurea was proposed introducing two different electrophile groups: fluorosulfonyl and bis-(β-chloroethyl)amino groups. The first can react with any kind of nucleophiles, thus with both amines, alcohols or thiols, while nitrogen mustard is less reactive than fluorosulfonyl moiety and, consequently, it reacts preferably with amines. Among the series, only fluorosulfonyl compounds exhibited an irreversible mechanism of action towards A_3 AR, with the best compound bearing a propyl moiety at the N^8 position (**53**, Fig. 9). A_3 ARs were preincubated with the compounds, washed to remove the free antagonist and then submitted to classical radioligand binding. Docking studies suggested that in the proximity of the transmembrane (TM) binding cleft there are at least two residues possibly implicated in the irreversible binding: Ser247 and Cys251. Even remaining more potent at the A_3 AR (47% of inhibition at 1 nM and 70% of inhibition at 10 nM) compound **53** exhibited also a good affinity against A_{2A} AR (K_i A_{2A} AR = 50 nM) [106].

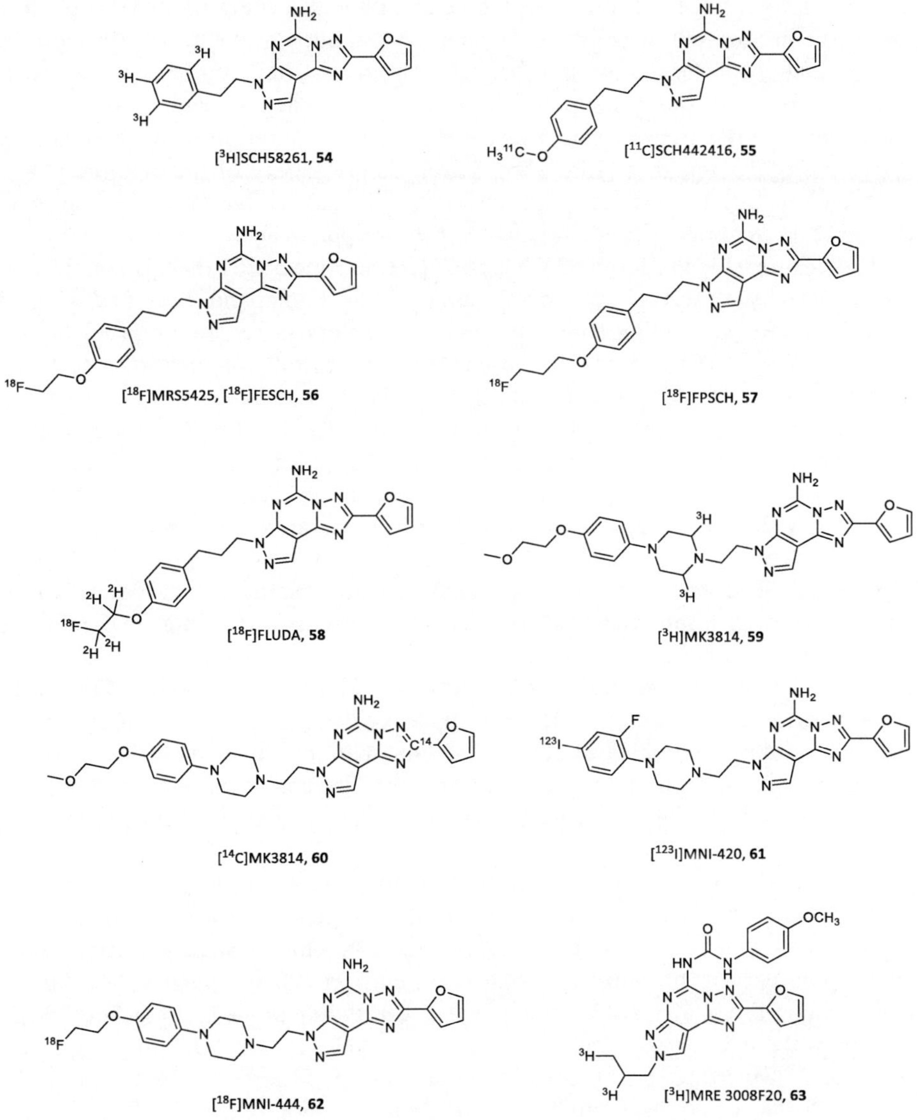

Fig. 10 PTP radioligands and radiotracers for the A_{2A} AR (**54–62**) and A_3AR (**63**)

3.2 *Radiolabeled Derivatives*

PTP derivatives have been used to develop radiolabeled compounds principally targeting the A_{2A} AR subtype. Different radioactive atoms have been inserted as ^{3}H (tritium), ^{11}C, ^{14}C, and ^{18}F (Fig. 10). Radioligands are developed both to study target receptor and for drug development, both in vitro and in vivo. Radioligands are used in various stages of drug development: simple screening by in vitro binding assays; demonstration of target engagement, distribution at the target tissue/organ; and evaluation of the relationship between plasma concentration and target occupancy in vivo. Last points are particularly important to determine the dose range in the clinic. In vivo studies were feasible thanks to positron emission tomography (PET) molecular imaging [155, 156]. Given the versatility of this approach, it is therefore obvious that a variety of probes containing radioactive atoms with different half-lives are required, depending on the specific application.

The first developed tritium PTP has been [^{3}H]Scheme 58261 (**54**). Due to its A_{2A} AR selectivity observed in binding assays, this compound becomes a good radioligand for autoradiographic studies on receptors in rat, mouse, and human brains [119, 157–161]. In addition, it has been successfully applied in scintillation proximity assay (SPA)-based ligand screening [162]. [^{3}H]Scheme 58261 (**54**) was found to label A_{2A} receptors with high affinity (hA_{2A} $K_D = 2.3$ nM) and with a rapid, saturable, and reversible specific binding of 70% [158]. Another PTP A_{2A} AR antagonist explored to obtain radioligands useful for PET imaging is Scheme 442416 (**5**). In this case, it was labeled with a ^{11}C atom, [^{11}C]Scheme 442416 (**55**). Because A_{2A} AR antagonists are studied for the treatment of Parkinson's disease, a radioligand for this receptor should be able to permeate blood-brain barrier along with high affinity and selectivity, low nonspecific binding, and minimal metabolism. Another key factor is that half-life of these PET radiotracers should be minimal, thus radiosynthesis needs to be easy and fast. [^{11}C]Scheme 442416 (**55**) is easily obtained by direct alkylation of hydroxy derivative with [^{11}C]CH_3I under alkaline conditions. Reaction and purification require a total of 40 min with a radiochemical yield of 29% [163]. Studies on rats and primates identify [^{11}C] Scheme 442416 (**55**) as a promising tracer for the in vivo PET imaging of A_{2A} AR [164].

Because A_{2A} AR density has been reportedly increased in the caudate–putamen in PD patients, A_{2A} AR-mediated PET imaging can help for early identification of this disease [165]. With respect to ^{11}C, ^{18}F isotope has higher specific activity and a longer physical half-life (109.8 vs 20.4 min), thus providing longer biodistribution and scanning times, and, not least, it allows a distribution of the radiotracer to image centers without cyclotron facilities [166]. For these reasons, Bhattacharjee and coworkers developed a ^{18}F-containing analog of Scheme 442416 (**5**), [^{18}F] MRS5425 (**56**) where the methoxy group at *para* position of the N^7-phenyl propyl moiety has been replaced by the 2-[^{18}F]fluoroethoxy group (Fig. 10). [^{18}F]MRS5425 ([^{18}F]FESCH, **56**) radioactivity in rat striatum, after reaching the maximum, decreases slower than with [^{11}C]Scheme 442416 (**55**), demonstrating to be a good probe [167]. Also a 3-[^{18}F]fluoropropoxy derivative, [^{18}F]FPSCH (**57**), has been developed in order to evaluate the effect of a longer fluoroalkyl chain, but the authors concluded that these compounds performed equally well in rats [166]. In a study aimed to highlight changes in the availability of A_{2A}/D_2 heterodimeric receptors in a mouse model of Parkinson's disease, [^{18}F]MRS5425 (**56**) led to inconclusive data, at least in part due to the brain penetration of a non-negligible fraction of a single radiometabolite [168]. To overcome this problem, a deuterated isotopologue of [^{18}F]

MRS5425 (**56**), [^{18}F]FLUDA (**58**), with enhanced metabolic stability has been developed. As expected, no brain-penetrant radiometabolites have been detected [169, 170].

Obviously, after its discovery, preladenant (**3**) became the focus of the research on A_{2A} AR antagonists for the treatment of PD. In particular, Merck, the proprietary of this molecule following its acquisition of Schering-Plough Corp, developed three isotopically labeled forms of this molecule. [^{3}H]MK3814 (**59**) and [^{14}C]MK3814 (**60**) were used for a preliminary and more definitive (e.g., in human) absorption, distribution, metabolism, and excretion data (ADME) evaluation of the compound, respectively. In this work also the non-radioactive deuterated compound [^{2}H]MK 3814 (e.g., fully deuterated ethyl chain between pyrazole and piperazine) was prepared as an internal standard to be used in a liquid chromatography mass spectrometry-based bioanalytical method [171]. Preladenant (**3**) represents also a starting point to develop other new radiotracers. In particular, in order to obtain a single photon emission computed tomography (SPECT) radiotracer, a longer-lived isotope, ^{123}I, has been used in analog [^{123}I]MNI-420 (**61**). In fact, a ^{123}I-based radiotracer could be dispensed from a central pharmacy rather than generated on-site [172]. During evaluation on humans, [^{123}I]MNI-420 (**61**) demonstrated to possess the same characteristics than currently used SPECT radiotracers for other human neuroreceptors imaging (Fig. 10) [173]. In 2014, Barret et al. reported [^{18}F] MNI-444 (**62**), obtained by reaction between the tosyl derivative 2-(4-(4-(2-(5-amino-2-(furan-2-yl)-7H-pyrazolo[4,3-*e*][1,2,4]triazolo[1,5-*c*] pyrimidin-7-yl)ethyl)piperazin-1-yl)phenoxy)ethyl-4-methylbenzenesulfonate and $^{18}F^{-}$. [^{18}F]MNI-444 (**62**) has been developed to investigate receptor occupancy and plasma levels of preladenant (**3**) and tozadenant, a benzothiazole-based A_{2A} AR antagonists, in nonhuman primates (NHP) [174]. In these studies, compound **62** showed a better specific-to-nonspecific ratio than other A_{2A} AR PET radiotracers available at that time, e.g. [^{11}C]TMSX and [^{11}C]Scheme 442416 (**55**), successively confirmed also in humans brain [174, 175].

Concerning A_1 and A_{2B} ARs both agonists and antagonists radioligands are reported in literature, but none bearing a PTP scaffold [176]. Instead, on A_3 AR, MRE 3008F20 (**6**) has been successfully developed as tritiated derivative [^{3}H]MRE 3008F20 (**63**) (Fig. 10) [128, 177]. [^{3}H]MRE 3008F20 (**63**) has been used to detect the presence of A_3 AR on plasma membrane of human lymphocytes, correlating the up-regulation of A_3 receptors functionally coupled with adenylyl cyclase in activated T cells with a potential biological role for adenosine-mediated responses in the immune system [178]. Gessi et al. used this radioligand also to investigate A_3 AR density in colon carcinomas, finding that it was higher than in normal mucosa of the same patient. Interestingly, overexpression of A_3 AR in tumors was reflected in peripheral blood cells, indicating that this receptor could be a diagnostic marker for colon cancer [179].

3.3 *Conjugated Derivatives*

3.3.1 Fluorescent Derivatives

Fluorescent ligands are of great interest because they allow real-time monitoring of the localization and function of the receptors in unmodified human cells, both in health and disease, at endogenous expression levels and this knowledge is fundamental to develop new therapies [180].

In the last 10 years, PTPs were conjugated to different fluorophores obtaining fluorescent antagonists towards A_{2A} and/or A_3 ARs. Applications of these compounds range from binding assay to fluorescence microscopy and receptor detection.

Taking inspiration from Scheme 442416 (**5**), Jacobson and coworkers designed the Alexa Fluor 488 derivative MRS5346 (**64**), which showed a K_i of 111 nM on A_{2A} AR (Fig. 11). The problem in developing fluorescent ligands is that fluorophore conjugation should not interfere with the ligand recognition in the orthosteric binding site of the receptor. For compound **64**, docking studies suggested that the linker and the fluorophore protruded towards the extracellular region of the A_{2A} AR. In addition, MRS5346 (**64**) was successfully employed as ligand in a fluorescent polarization (FP) binding assay which represents an alternative to the classical radioligand-based binding assay [181]. Alexa Fluor 488 has a lifetime appropriated to be used in FP (4.5 ns); shorter lifetimes like in TAMRA (**65**) and BODIPY650/665 derivatives (MRS5418, **66**) were not suitable for this application, but these fluorescent ligands could be used for other purposes like fluorescence microscopy [182]. The same group reported also other two Scheme 442416-fluorescent analogs, compounds **67–68**, MRS7416 (**69**), and MRS7396 (**70**) bearing Alexa Fluor 647, Alexa Fluor 488, and BODIPY630/650-X as fluorophore moiety, respectively (Fig. 11). MRS7416 (**69**) and MRS7396 (**70**) display highest affinity at the A_{2A} AR, thus has been subjected to docking studies using the high-resolution A_{2A} AR X-ray structure (PDB ID: 4EIY). Molecular modeling analysis revealed that these compounds behave as bitopic ligands, where PTP resides in the orthosteric binding site of A_{2A} AR, while fluorophores allocate in different adjacent clefts [183]. Subsequent work demonstrated that MRS7416 (**69**) and MRS7396 (**70**) bitopic ligands are allosteric antagonists; authors studied several different fluorescent ligands, demonstrating that they can behave both as allosteric or competitive antagonists, depending on the nature of the linker and fluorophore [184].

An approach similar to that proposed by Jacobson's laboratory has been used by Kellam and coworkers on preladenant (**3**), where the binding pose at the hA_{2A} AR revealed that the chain on the phenyl ring is oriented towards the more solvent-exposed part of the receptor binding cleft, thus representing a good anchoring point for a fluorophore moiety. These data have been confirmed obtaining different A_{2A} AR fluorescent probes (**71–76**), based on the preladenant structure that maintained a good affinity towards the receptor (Fig. 11) [180]. Compounds **71–76** were applied in a bioluminescence resonance energy transfer (BRET) binding ligand assay,

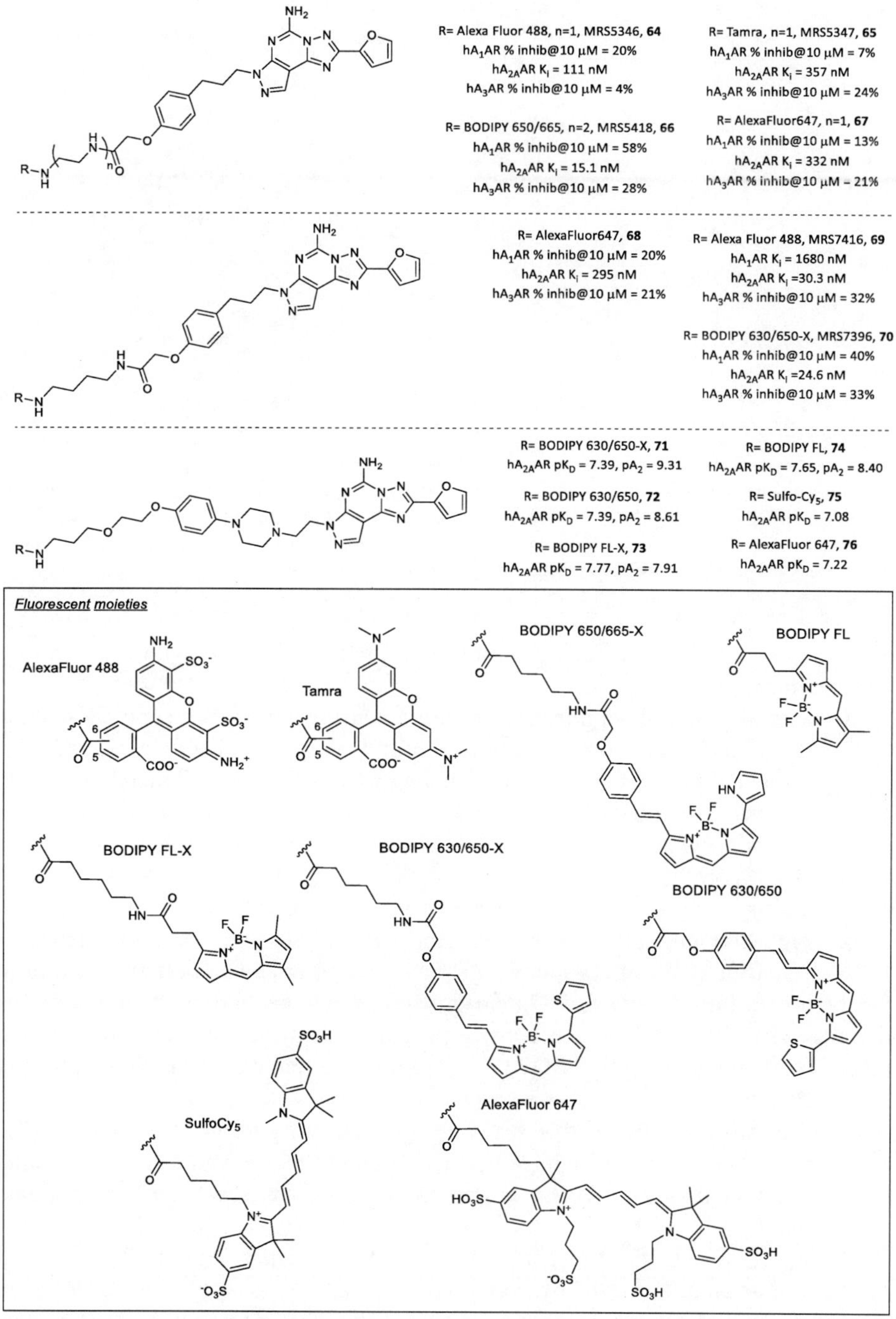

Fig. 11 PTP fluorescent A_{2A} AR antagonists

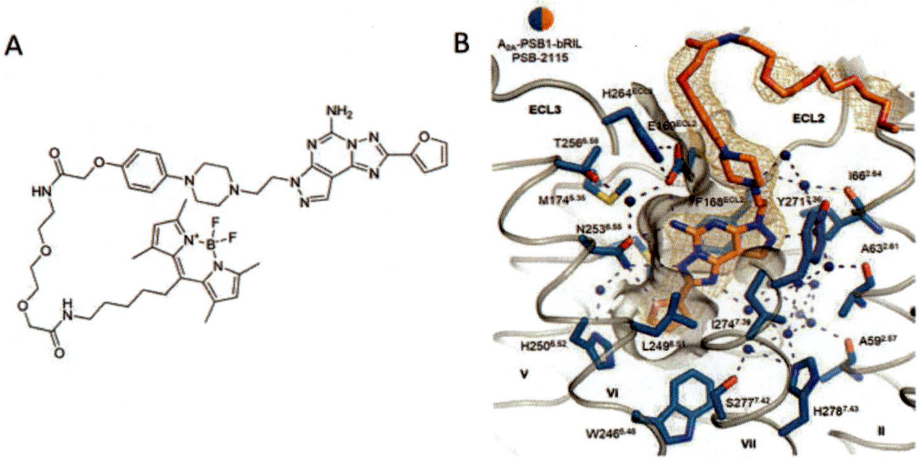

Fig. 12 (**a**) Structure of the A_{2A} AR fluorescent ligand PSB-2115 (**77**); (**b**) Ligand binding pocket of A_{2A}-PSB1-bRIL-PSB-2115 (77) (PDB:7PYR) taken from Angew. Chem. Int. Ed. 2022, 61, e202115545 International Edition: doi.org/10.1002/anie.202115545 [129]. The $2F_o$-F_c electron density of PSB-2115 is shown in orange mesh (contoured at 1.0 σ)

exhibiting a clear saturable component of specific binding along with low levels of nonspecific binding.

Because one possible application of fluorescent ligands is in receptor visualization, compounds **71** and **74** were evaluated for this purpose in confocal microscopy. Images indicated a successful specific binding to A_{2A} AR at the cell membranes with undetectable uptake into the cytosol. Finally, due to its optimal characteristics, which comprise a good water solubility, Alexa Fluor 647 derivative **76** has been further validated as fluorescent ligand in BRET binding assay of known hA_{2A} AR antagonist, resulting in a promising pharmacological tool [180]. Recently, an X-ray crystallographic structure of the new preladenant-derived fluorescent A_{2A} AR antagonist PSB-2115 (BODIPY as fluorophore, **77**) has been reported by Claff et al. Unfortunately, no electron density could be observed for the linker and the fluorophore, but the pose of PTP scaffold in the binding pocket is identical to that reported for the precursor, demonstrating that BODIPY does not interfere with PTP binding to A_{2A} AR [129].

As shown in Fig. 12, in this structure, preladenant portion of PSB-2115 (**77**) shows direct interactions to helices V, VI, VII and to the second extracellular loop (ECL2) of A_{2A} AR. The typical key hydrogen bond networks between 5-amino group and $N253^{6.55}$/$E169^{ECL2}$ and between furan oxygen and $N253^{6.55}$ can be observed, along with the tricyclic π–π stacking interaction with $F168^{ECL2}$. Other important interactions are the hydrophobic contacts with $L249^{6.51}$ and $I274^{7.39}$. As predicted by previous computational studies, the phenylpiperazinylethyl moiety attached to the N^7 in the PTP scaffold effectively extends towards the extracellular surface of the A_{2A} AR, stabilized by π–π stacking to $H264^{ECL3}$ [129].

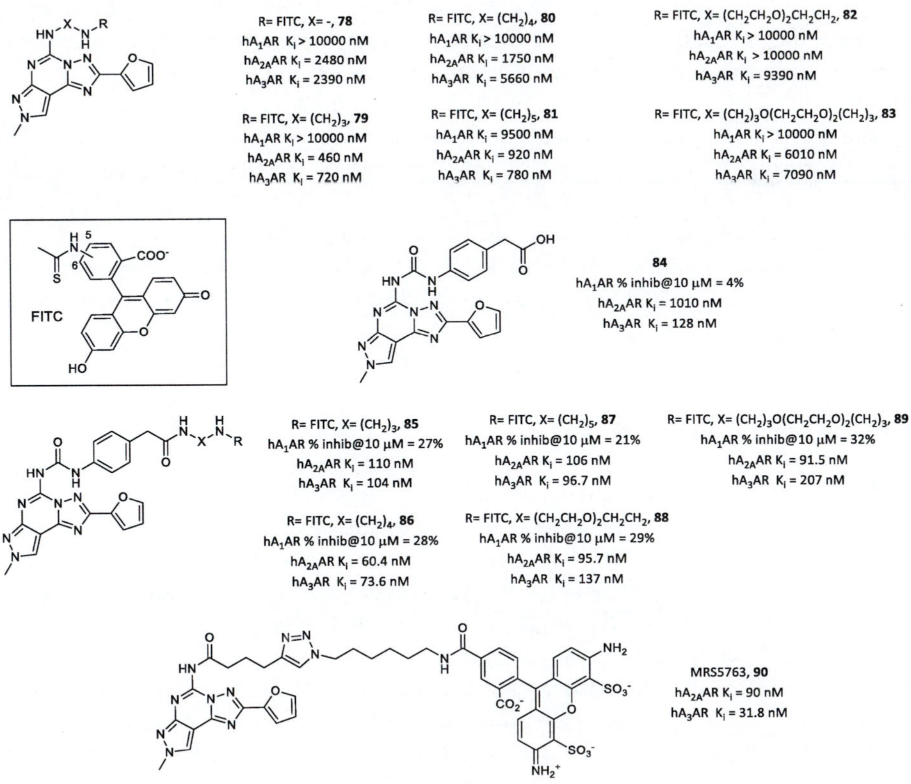

Fig. 13 PTP fluorescent A_3 AR antagonists

Finally, a commercially available preladenant-based fluorescent ligand, SCH-red, has been used by Ciruela and colleagues to demonstrate the existence of native D_2R-A_{2A} AR oligomers in time-resolved fluorescence resonance energy transfer (TR-FRET) experiments, in striatal rat membranes both in normal conditions and in a Parkinson's disease model. Studies demonstrated a decreased D_2R-A_{2A} AR oligomer formation in the 6-OHDA-lesioned striatal membranes [185].

Due to the well-known SAR of PTP on the A_3 AR, different attempts to obtain potent and selective A_3 AR fluorescent antagonists have been conducted by inserting the linker, and thus the fluorophore (e.g., fluorescein isothiocyanate, FITC), at the 5 position of the 8-methyl-PTP scaffold [186–188]. In a first attempt, simple alkyldiamino moieties were directly introduced at the 5 position as linkers, leading to the low affinity and mostly dual A_{2A} AR and A_3 AR ligands **78–83** (Fig. 13). Computational studies tentatively ascribed the lack of selectivity towards A_3 AR to the presence of low significant hydrophobic interactions in the binding cleft [186, 187]. For this reason, authors extended the pharmacophoric portion of PTP, using as starting compound not the simple 5-amino-8-methyl-PTP but compound **84,** where the benzylamido moiety confers high affinity and selectivity at the A_3 AR and the carboxylic acid at the *para* position of the phenyl moiety represents the anchoring point for the linker [188]. A firm increase in affinity was obtained in compounds

85–89 (Fig. 13), reaching a K_i value around 100 nM towards A_3 AR independently from the linker length. Surprisingly, a rise in affinity was observed also against A_{2A} AR, thus defining these compounds as dual fluorescent antagonists. Finally, a series of alkyne derivatives of PTP was synthetized with the aim to use the classical click chemistry approach to introduce the fluorophore (e.g., as azide derivative). The formed triazole is an aromatic ring that, when present at the 5 position, is known to give favorable interactions at the A_3 AR subtype. Alkyne derivative with a length of 6 carbon atoms has been used to make the conjugate with an azido derivative of Alexa Fluor 488, leading to MRS5763 (**90**) that is the best A_3 AR fluorescent ligand bearing a PTP scaffold, exhibiting a K_i of 31.8 nM and a 2.8-fold selectivity against A_{2A} AR.

3.3.2 Other Conjugates

PTPs were used also to design other kind of conjugates with different purposes, such as biotin conjugate, that, thanks to its binding to streptavidin, could be used to set up a variety of experiments both for screening, receptor detection or imaging purposes (Fig. 14). Specifically, compound **91** was developed on the Scheme 442416 (**5**) structure, thus as tool of the A_{2A} AR. In the same work, Kumar et al. reported dendrimers functionalized again with two different amino derivatives of Scheme 442416 (**92,93**). In particular, polyamidoamine (PAMAM) G3.5 dendrimer 15 (containing a butane core), that possesses peripheral carboxylic acids, has been used [182]. ARs are present on the cell surface as multi homomeric or heteromeric (with other GPCRs) entities, thus the interest in creation of multivalent confined structures able to interact with them [189]. Dendrimers **92** and **93** (Fig. 14) were highly potent and selective towards A_{2A} AR, with practically no difference between two linkers in terms of affinity at the target receptor [182].

With the same objective to develop multivalent structures, an A_3 AR antagonist has been conjugated with magnetic carbon nanotubes. Because A_3 AR has been found to be overexpressed in different cancer cell lines, the magnetic property of these Fe-filled carbon nanotubes could be useful to perform cancer magnetic cell sorting and thermal therapy. The final nanostructures Fe-filled-CNTs-HEG-PTP (**94**) and Fe-filled-CNTs-TEG-PTP (**95**), differing for the used linker, have shown to compete with A_3 AR radioligand in a concentration-dependent manner, but structure **94** failed to selectively target hA_3 AR overexpressing cells. The problem could be ascribed to electrostatic interactions between the unreacted free amino groups on the carbon nanotubes and anionic residue on the cell membranes [190].

PTP were also implied to develop multitarget ligands, in particular a dual ligand targeting A_{2A} AR and Histone Deacetylase (HDAC) as a possible therapeutic strategy in cancer immunotherapy. Structural features of A_{2A} AR antagonists and HDAC inhibitors were combined thanks to a structure-based drug design leading to the discovery of IHCH-3064 (**96**), a PTP derivative exhibiting a K_i value towards A_{2A} AR of 2.2 nM and an IC_{50} value towards HDAC of 80.2 nM. IHCH-3064 (**96**) showed good antiproliferative activity against tumor cell lines in vitro and, when administered intraperitoneally in mice, inhibited tumor growth [191].

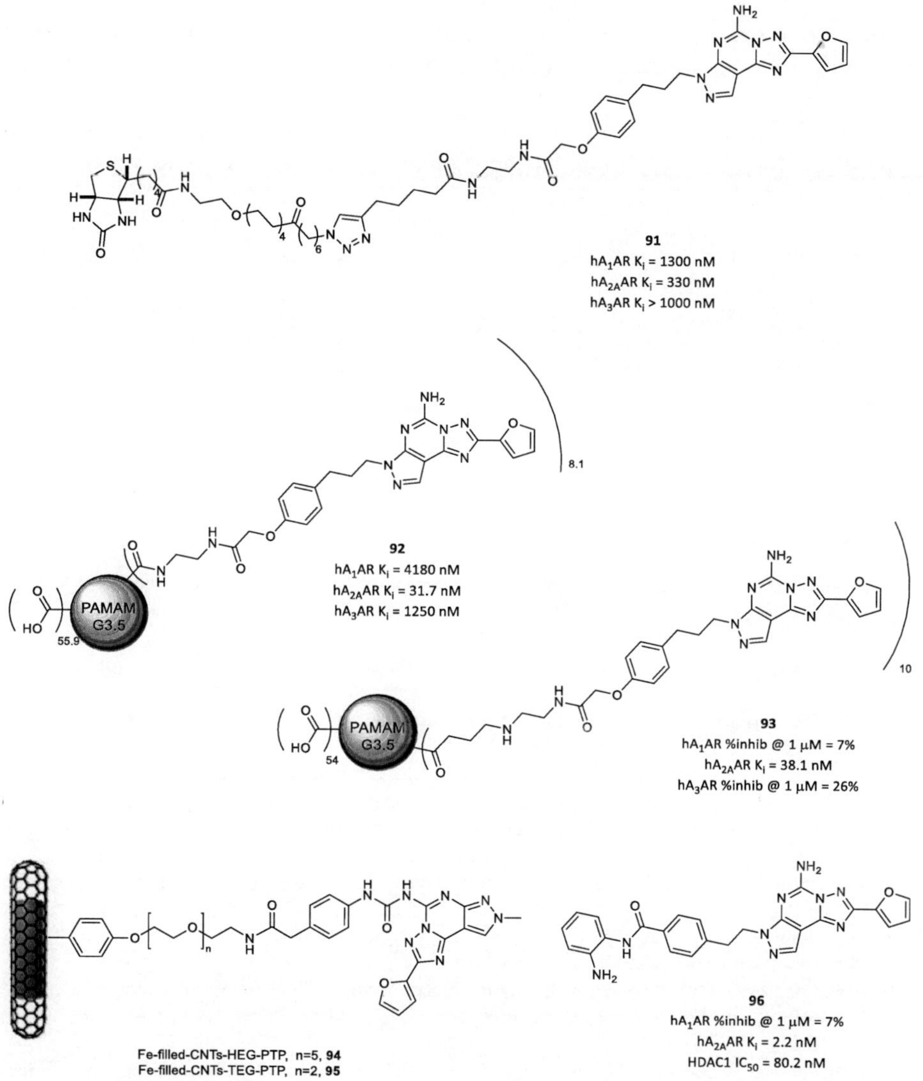

Fig. 14 Other PTP conjugates

4 Conclusions

ARs, and in particular A_{2A} ARs, are among the most studied GPCRs, particularly from a structural point of view. For this reason, they represent a model for other GPCRs and several pharmacological tools and probes have been developed over the years. Among them, PTP-based antagonists were extensively investigated both at the A_{2A} and A_3 ARs. In fact, in addition to a very complete and distinct SAR profile to both receptors, PTP antagonists have been used as skeleton to develop more complex structures bearing specific functions. These range from simple covalent interaction or radiolabeling to conjugation to linkers in order to obtain multipurpose molecules such as fluorescent, multitarget, and/or multivalent ligands. Thus, PTP represent the perfect example on how it is possible to play with a known GPCR class of ligands in order to increase knowledge on the GPCR itself, crucial to discover new pathophysiological implications of the protein and, consequently, to develop new valuable therapies.

Funding The authors have no conflicts of interest to declare that are relevant to the content of this chapter.

Ethical Approval Ethical approval is not applicable for this article.

Informed Consent There are no human subjects in this work and informed consent is not applicable.

References

1. Borea PA, Gessi S, Merighi S et al (2018) Pharmacology of adenosine receptors: the state of the art. Physiol Rev 98:1591–1625
2. Pearson JD (1985) Ectonucleotidases. Measurement of activities and use of inhibitors. In: Paton DM (ed) Methods in pharmacology. Methods used in adenosine research, vol 6. Plenum Press, New York, pp 83–107
3. Ontyd J, Schrader J (1984) Measurement of adenosine, inosine, and hypoxanthine in human plasma. J Chromatogr 307:404–409
4. Kostopoulos GK (1988) Adenosine: a molecule for synaptic homeostasis? Evolution of current concepts on the physiological and pathophysiological roles of adenosine in the brain. In: Avoli M, Reader TA, Dykes RW, Gloor P (eds) Neurotransmitters and cortical function: from molecules to mind. Plenum Press, New York, pp 415–435
5. Dragunow M, Faull RLM (1988) Neuroprotective effects of adenosine. Trends Pharmacol Sci 9:193–194
6. Sebastiao AM, Ribeiro JA (2000) Fine-tuning neuromodulation by adenosine. Trends Pharmacol Sci 21:341–346
7. Burnstock G (1978) A basis for distinguishing two types of purinergic receptor. In: Bolis L, Straub RW (eds) Cell membrane receptors for drugs and hormones. Raven Press, New York, pp 107–118
8. Ijzerman AP, Jacobson KA, Muller CE et al (2022) International union of basic and clinical pharmacology. CXII: a further update. Pharmacol Rev 74:340–372
9. Jacobson KA, Gao ZG (2006) Adenosine receptor as therapeutic target. Nat Rev Drug Discov 5:247–264
10. Effendi WI, Nagano T, Kobayashi K et al (2020) Focusing on adenosine receptors as a potential targeted therapy in human diseases. Cell 9:785
11. Obiefuna PCM, Batra VK, Nadeem A et al (2005) A novel A_1 adenosine receptor antagonist, L-97-1 [3-[2-(4-aminophenyl)-ethyl]-8-benzyl-7-{2-ethyl-(2-hydroxy-ethyl)-amino-ethyl}-1-propyl-3,7-dihydro-purine-2,6-dione], reduces allergic responses to house dust mite in an allergic rabbit model of asthma. J Pharmacol Exp Ther 315:329–336
12. Chunn JL, Young HW, Banerjee SK et al (2001) Adenosine-dependent airway inflammation and hyper responsiveness in partially adenosine deaminase-deficient mice. J Immunol 167: 4676–4685

13. Ngamsri KC, Wagner R, Vollmer I et al (2010) Adenosine receptor A_1 regulates polymorphonuclear cell tracking and microvascular permeability in lipopolysaccharide-induced lung injury. J Immunol 185:4374–4384
14. Fernandez LG, Sharma AK, Lapar DJ et al (2013) Adenosine A1 receptor activation attenuates lung ischemia-reperfusion injury. J Thorac Cardiovasc Surg 145:1654–1659
15. Affner F, Woods PS, Davis IC (2014) Activation of A_1-adenosine receptors promotes leukocyte recruitment to the lung and attenuates acute lung injury in mice infected with influenza A/WSN/33 (H1N1) virus. J Virol 88:10214–10227
16. Blackburn MR, Vance CO, Morschl E, Wilson CN (2009) Adenosine receptors and inflammation. In: Wilson CN, Mustafa SJ (eds) Adenosine receptors in health and disease. Springer, Berlin, pp 215–269
17. Cronstein BN, Levin RI, Philips M et al (1992) Neutrophil adherence to endothelium is enhanced via adenosine A1 receptors and inhibited via adenosine A_2 receptors. J Immunol 148:2201–2206
18. Gessi S, Merighi S, Fazzi D et al (2011) Adenosine receptor targeting in health and disease. Expert Opin Investig Drugs 20:1591–1609
19. Headrick JP, Peart JN, Reichelt ME et al (2011) Adenosine and its receptors in the heart: regulation, retaliation and adaptation. Biochim Biophys Acta Biomembr 1808:1413–1428
20. Jacobson KA, Tosh DK, Jain S et al (2019) Historical and current adenosine receptor agonists in preclinical and clinical development. Front Cell Neurosci 13:124
21. Ellenbogen KA, O'Neill G, Prystowsky EN et al (2005) Trial to evaluate the management of paroxysmal supraventricular tachycardia during an electrophysiology study with tecadenoson. Circulation 111:3202–3208
22. Mason PK, DiMarco JP (2009) New pharmacological agents for arrhythmias. Circ Arrhythm Electrophysiol 2:588–597
23. Szentmiklosi A, Galajda Z, Cseppento A et al (2015) The Janus face of adenosine: antiarrhythmic and proarrhythmic actions. Curr Pharm Des 21:965–976
24. Wu LG, Saggau P (1997) Presynaptic inhibition of elicited neurotransmitter release. Trends Neurosci 20:204–212
25. Stone TW, Ceruti S, Abbracchio MP (2009) Adenosine receptors and neurological disease: neuroprotection and neurodegeneration. In: Wilson CN, Mustafa SJ (eds) Adenosine receptors in health and disease, vol 193. Springer, Berlin, pp 535–587
26. Borea PA, Gessi S, Merighi S et al (2016) Adenosine as a multi-signalling Guardian angel in human diseases: when, where and how does it exert its protective effects? Trends Pharmacol Sci 37:419–434
27. Luongo L, Petrelli R, Gatta L et al (2012) 5'-Chloro-5′-deoxy-(±)-ENBA, a potent and selective adenosine A_1 receptor agonist, alleviates neuropathic pain in mice through functional glial and microglial changes without affecting motor or cardiovascular functions. Molecules 17:13712–13726
28. Fedele DE, Li T, La JQ et al (2006) Adenosine A_1 receptors are crucial in keeping an epileptic focus localized. Exp Neurol 200:184–190
29. Sebastião AM, Ribeiro JA (2009) Adenosine receptors and the central nervous system. In: Wilson CN, Mustafa SJ (eds) Adenosine receptors in health and disease, vol 193. Springer, Berlin, pp 471–534
30. Peleli M, Carlstrom M (2017) Adenosine signaling in diabetes mellitus and associated cardiovascular and renal complications. Mol Aspects Med 55:62–74
31. Kiesman WF, Elzein E, Zablocki J (2009) A_1 adenosine receptor antagonists, agonists, and allosteric enhancers. In: Wilson CN, Mustafa SJ (eds) Adenosine receptors in health and disease, vol 193. Springer, Berlin, pp 25–58
32. Hasko G, Pacher P (2008) A_{2A} receptors in inflammation and injury: lessons learned from transgenic animals. J Leukoc Biol 83:447–455

33. Wilson CN, Nadeem A, Spina D et al (2009) Adenosine receptors and asthma. In: Wilson CN, Mustafa SJ (eds) Adenosine receptors in health and disease, vol 193. Springer, Berlin, pp 329–362
34. Golzar Y, Doukky R (2014) Regadenoson use in patients with chronic obstructive pulmonary disease: the state of current knowledge. Int J COPD 9:129–137
35. Boknik P, Drzewiecki K, Eskandar J et al (2019) Evidence for arrhythmogenic effects of A_{2A}-adenosine receptors. Front Pharmacol 10:1051
36. Dunwiddie TV, Masino SA (2001) The role and regulation of adenosine in the central nervous system. Annu Rev Neurosci 24:31–55
37. Boison D, Singer P, Shen HY et al (2012) Adenosine hypothesis of schizophrenia-opportunities for pharmacotherapy. Neuropharmacology 62:1527–1543
38. Loram LC, Harrison JA, Sloane EM et al (2009) Enduring reversal of neuropathic pain by a single intrathecal injection of adenosine A_{2A} receptor agonists: a novel therapy for neuropathic pain. J Neurosci 29:14015–14025
39. Knezevic NN, Cicmil N, Knezevic I et al (2015) Discontinued neuropathic pain therapy between 2009-2015. Expert Opin Investig Drugs 24:1631–1646
40. Knapp CM, Foye MM, Cottam N et al (2001) Adenosine agonists CGS 21680 and NECA inhibit the initiation of cocaine self-administration. Pharmacol Biochem Behav 68:797–803
41. Mohamed RA, Agha AM, Abdel-Rahman AA et al (2016) Role of adenosine A_{2A} receptor in cerebral ischemia reperfusion injury: signaling to phosphorylated extracellular signal-regulated protein kinase (pERK1/2). Neuroscience 314:145–159
42. Yamada K, Kobayashi M, Kanda T (2014) Involvement of adenosine A_{2A} receptors in depression and anxiety. Int Rev Neurobiol 119:373–393
43. Faivre E, Coelho JE, Zornbach K et al (2018) Beneficial effect of a selective adenosine A_{2A} receptor antagonist in the APPswe/PS1dE9 mouse model of Alzheimer's disease. Front Mol Neurosci 11:235
44. Popoli P, Blum D, Martire A et al (2007) Functions, dysfunctions and possible therapeutic relevance of adenosine A_{2A} receptors in Huntington's disease. Prog Neurobiol 81:331–348
45. De Lera RM, Lim YH, Zheng J (2014) Adenosine A_2A receptor as a drug discovery target. J Med Chem 57:3623–3650
46. Nazario LR, da Silva RS, Bonan CD (2017) Targeting adenosine signaling in Parkinson's disease: from pharmacological to non-pharmacological approaches. Front Neurosci 11:658
47. Ribeiro JA, Sebastião AM, De Mendonça A (2002) Adenosine receptors in the nervous system: pathophysiological implications. Prog Neurobiol 68:377–392
48. Li X, Kang H, Liu X et al (2012) Effect of adenosine A_{2A} receptor antagonist ZM241385 on amygdala-kindled seizures and progression of amygdala kindling. J Huazhong Univ Sci Technolog Med Sci 32:257–264
49. Sek K, Mølck C, Stewart GD et al (2018) Targeting adenosine receptor signaling in cancer immunotherapy. Int J Mol Sci 19:3837
50. Congreve M, Brown GA, Borodovsky A et al (2018) Targeting adenosine A_{2A} receptor antagonism for treatment of cancer. Expert Opin Drug Discovery 13:997–1003
51. Antonioli L, Csoka B, Fornai M et al (2014) Adenosine and inflammation: what's new on the horizon? Drug Discov Today 19:1051–1068
52. Linden J (2006) New insights into the regulation of inflammation by adenosine. J Clin Investig 116:1835–1837
53. Gao ZG, Balasubramanian R, Kiselev E et al (2014) Probing biased/partial agonism at the G protein-coupled A_{2B} adenosine receptor. Biochem Pharmacol 90:297–306
54. Van Den Berge M, Hylkema MN, Versluis M et al (2007) Role of adenosine receptors in the treatment of asthma and chronic obstructive pulmonary disease: recent developments. Drugs R D 8:13–23
55. Mustafa SJ, Nadeem A, Fan M et al (2007) Effect of a specific and selective A_{2B} adenosine receptor antagonist on adenosine agonist AMP and allergen-induced airway responsiveness and cellular influx in a mouse model of asthma. J Pharmacol Exp Ther 320:1246–1251

56. Eckle T, Grenz A, Laucher S et al (2008) A_{2B} adenosine receptor signaling attenuates acute lung injury by enhancing alveolar fluid clearance in mice. J Clin Investig 118:3301–3315
57. Hoegl S, Brodsky KS, Blackburn MR et al (2015) Alveolar epithelial A_{2B} adenosine receptors in pulmonary protection during acute lung injury. J Immunol 195:1815–1824
58. Haskó G, Csóka B, Németh ZH et al (2009) A_{2B} adenosine receptors in immunity and inflammation. Trends Immunol 30:263–270
59. Maas JE, Wan TC, Figler RA et al (2010) Evidence that the acute phase of ischemic preconditioning does not require signaling by the A_{2B} adenosine receptor. J Mol Cell Cardiol 49:886–893
60. Dubey RK, Gillespie DG, Osaka K et al (1996) Adenosine inhibits growth of rat aortic smooth muscle cells. Possible role of A_{2B} receptor. Hypertension 27:786–793
61. Eisenstein A, Patterson S, Ravid K (2015) The many faces of the A_{2B} adenosine receptor in cardiovascular and metabolic diseases. J Cell Physiol 230:2891–2897
62. Németh ZH, Bleich D, Csóka B et al (2007) Adenosine receptor activation ameliorates type 1 diabetes. FASEB J 21:2379–2388
63. Sun Y, Huang P (2016) Adenosine A_{2B} receptor: from cell biology to human diseases. Front Chem 4:37
64. Cekic C, Sag D, Li Y et al (2012) Adenosine A_{2B} receptor blockade slows growth of bladder and breast tumors. J Immunol 188:198–205
65. Walker BA, Jacobson MA, Knight DA et al (1997) Adenosine A_3 receptor expression and function in eosinophils. Am J Respir Cell Mol Biol 16:531–537
66. Gao Z, Li BS, Day YJ et al (2001) A_3 adenosine receptor activation triggers phosphorylation of protein kinase B and protects rat basophilic leukemia 2H3 mast cells from apoptosis. Mol Pharmacol 59:76–82
67. Martin L, Pingle SC, Hallam DM et al (2006) Activation of the adenosine A_3 receptor in RAW 264.7 cells inhibits lipopolysaccharide-stimulated tumor necrosis factor-alpha release by reducing calcium-dependent activation of nuclear factor-kappaB and extracellular signal-regulated kinase 1/2. J Pharmacol Exp Ther 316:71–78
68. Lee JY, Jhun BS, Oh YT et al (2006) Activation of adenosine A_3 receptor suppresses lipopolysaccharide-induced TNF-alpha production through inhibition of PI 3-kinase/Akt and NF-kappaB activation in murine BV2 microglial cells. Neurosci Lett 396:1–6
69. Cohen S, Barer F, Bar-Yehuda S et al (2014) A_3 adenosine receptor allosteric modulator induces an anti-inflammatory effect: in vivo studies and molecular mechanism of action. Mediators Inflamm 3:708746
70. David M, Akerman L, Ziv M et al (2012) Treatment of plaque-type psoriasis with oral CF101: data from an exploratory randomized phase 2 clinical trial. J Eur Acad Dermatol Venereol 26: 361–367
71. Yang H, Avila MY, Peterson-Yantorno K et al (2005) The cross-species A_3 adenosine-receptor antagonist MRS 1292 inhibits adenosine-triggered human nonpigmented ciliary epithelial cell fluid release and reduces mouse intraocular pressure. Curr Eye Res 30:747–754
72. Okamura T, Kurogi Y, Hashimoto K et al (2004) Structure-activity relationships of adenosine A_3 receptor ligands: new potential therapy for the treatment of glaucoma. Bioorg Med Chem Lett 14:3775–3779
73. Zhong H, Shlykov SG, Molina JG et al (2003) Activation of murine lung mast cells by the adenosine A_3 receptor. J Immunol 171:338–345
74. Borea PA, Varani K, Vincenzi F et al (2015) The A_3 adenosine receptor: history and perspectives. Pharmacol Rev 67:74–102
75. Headrick JP, Lasley RD (2009) Adenosine receptors and reperfusion injury of the heart. In: Wilson CN, Mustafa SJ (eds) Adenosine receptors in health and disease, vol 193. Springer, Berlin, pp 189–214
76. Auchampach JA, Ge ZD, Wan TC et al (2003) A_3 adenosine receptor agonist IB-MECA reduces myocardial ischemia-reperfusion injury in dogs. Am J Physiol Heart Circ Physiol 285: H607–H613

77. Ge ZD, Peart JN, Kreckler LM et al (2006) Cl-IB-MECA [2-chloro-N6-(3-iodobenzyl)adenosine-5'-N-methylcarboxamide] reduces ischemia/reperfusion injury in mice by activating the A_3 adenosine receptor. J Pharmacol Exp Ther 319:1200–1210
78. Black RGJ, Guo Y, Ge ZD et al (2002) Gene dosage-dependent effects of cardiac-specific overexpression of the A_3 adenosine receptor. Circ Res 91:165–172
79. Park JG, Jeong SJ, Yu J et al (2018) LJ-1888, a selective antagonist for the A_3 adenosine receptor, ameliorates the development of atherosclerosis and hypercholesterolemia in apolipoprotein E knock-out mice. BMB Rep 51:520–525
80. Rothermel BA, Hill JA (2008) Adenosine A_3 receptor and cardioprotection: enticing, enigmatic, elusive. Circulation 118:1691–1693
81. Pugliese AM, Coppi E, Volpini R et al (2007) Role of adenosine A_3 receptors on CA1 hippocampal neurotransmission during oxygen-glucose deprivation episodes of different duration. Biochem Pharmacol 74:768–779
82. Wittendorp MC, Boddeke HWGM, Biber K (2004) Adenosine A_3 receptor-induced CCL2 synthesis in cultured mouse astrocytes. Glia 46:410–418
83. Etherington LAV, Frenguelli BG (2004) Endogenous adenosine modulates epileptiform activity in rat hippocampus in a receptor subtype-dependent manner. Eur J Neurosci 19: 2539–2550
84. Durante M, Squillace S, Lauro F et al (2021) Adenosine A_3 agonists reverse neuropathic pain via T cell-mediated production of IL10. J Clin Invest 131:e139299
85. Roseti C, Palma E, Martinello K et al (2009) Blockage of A_{2A} and A_3 adenosine receptors decreases the desensitization of human GABAA receptors microtransplanted to Xenopus oocytes. Proc Natl Acad Sci U S A 106:15927–15931
86. Mabley J, Soriano F, Pacher P et al (2003) The adenosine A_3 receptor agonist, N6-(3-iodobenzyl)-adenosine-5'-N-methyluronamide, is protective in two murine models of colitis. Eur J Pharmacol 466:323–329
87. Lee HT, Ota-Setlik A, Xu H et al (2003) A_3 adenosine receptor knockout mice are protected against ischemia- and myoglobinuria-induced renal failure. Am J Physiol Renal Physiol 284: F267–F273
88. Madi L, Ochaion A, Rath-Wolfson L et al (2004) The A_3 adenosine receptor is highly expressed in tumor versus normal cells: potential target for tumor growth inhibition. Clin Cancer Res 10:4472–4479
89. Gessi S, Merighi S, Varani K et al (2008) The A_3 adenosine receptor: an enigmatic player in cell biology. Pharmacol Ther 117:123–140
90. Fishman P, Bar-Yehuda S, Vagman L (1998) Adenosine and other low molecular weight factors released by muscle cells inhibit tumor cell growth. Cancer Res 58:3181–3187
91. Fishman P, Bar-Yehuda S, Ohana G et al (2000) Adenosine acts as an inhibitor of lymphoma cell growth: a major role for the A_3 adenosine receptor. Eur J Cancer 36:1452–1458
92. Zhang X, Zhang M, Laties AM et al (2006) Balance of purines may determine life or death of retinal ganglion cells as A_3 adenosine receptors prevent loss following P2X7 receptor stimulation. J Neurochem 98:566–575
93. Gessi S, Merighi S, Varani K et al (2007) Adenosine receptors in colon carcinoma tissues and colon tumoral cell lines: focus on the A_3 adenosine subtype. J Cell Physiol 211:826–836
94. Ohana G, Bar-Yehuda S, Arich A et al (2003) Inhibition of primary colon carcinoma growth and liver metastasis by the A_3 adenosine receptor agonist CF101. Br J Cancer 89:1552–1558
95. Madi L, Bar-Yehuda S, Barer F et al (2003) A_3 adenosine receptor activation in melanoma cells: association between receptor fate and tumor growth inhibition. J Biol Chem 278:42121–42130
96. Fishman P, Bar-Yehuda S, Ardon E et al (2003) Targeting the A_3 adenosine receptor for cancer therapy: inhibition of prostate carcinoma cell growth by A_3AR agonist. Anticancer Res 23:2077–2083
97. Francis JE, Cash WD, Psychoyos S et al (1988) Structure-activity profile of a series of novel triazoloquinazoline adenosine antagonists. J Med Chem 31:1014–1020

98. Sarges R, Howard HR, Browne RG et al (1990) 4-Amino[1,2,4]triazolo[4,3-*a*]quinoxalines. A novel class of potent adenosine receptor antagonists and potential rapid-onset antidepressants. J Med Chem 33:2240–2254
99. Jacobson KA, van Galen PJM, Williams M (1992) Adenosine receptors: pharmacology, structure-activity relationships, and therapeutic potential. J Med Chem 35:407–422
100. Gatta F, Del Giudice MR, Borioni A et al (1993) Synthesis of imidazo[1,2-*c*]pyrazolo[4,3-*e*] 1,2,4-triazolo[1,5-*c*]pyrimidines and 1,2,4-triazolo[5,1-*i*]purines: new potent adenosine A_2 receptor antagonists. Eur J Med Chem 28:569–576
101. Baraldi PG, Manfredini S, Simoni D et al (1994) Synthesis of new pyrazolo[4,3-e]1,2,4-triazolo[1,5-*c*] pyrimidine and 1,2,3-triazolo[4,5-*e*]1,2,4-triazolo[1,5-*c*] pyrimidine displaying potent and selective activity as A_{2A} adenosine receptor antagonists. Bioorg Med Chem Lett 4: 2539–2544
102. Baraldi PG, Cacciari B, Spalluto G et al (1996) Pyrazolo[4,3-e]1,2,4-triazolo[1,5-c]pyrimidine derivatives: potent and selective A_{2A} adenosine antagonists. J Med Chem 39:1164–1171
103. Ongini E (1997) SCH 58261: a selective A_{2A} adenosine receptor antagonists. Drug Dev Res 42:63–70
104. Baraldi PG, Cacciari B, Spalluto G et al (1998) Design, synthesis and biological evaluation of a second generation of pyrazolo[4,3-e]1,2,4-triazolo[1,5-c]pyrimidines as potent and selective A_{2A} adenosine receptor antagonists. J Med Chem 41:2126–2133
105. Baraldi PG, Cacciari B, Romagnoli R et al (1999) Pyrazolo[4,3-e]1,2,4-triazolo[1,5-c]pyrimidines derivatives as highly potent and selective human A_3 adenosine receptor antagonists. J Med Chem 42:4473–4478
106. Baraldi PG, Cacciari B, Moro S et al (2001) Fluorosulfonyl- and bis-(β-chloroethyl)aminophenylamino functionalized pyrazolo[4,3-e]1,2,4-triazolo[1,5-c]pyrimidines derivatives: irreversible antagonists at the human A_3 adenosine receptor and molecular modelling studies. J Med Chem 44:2735–2742
107. Baraldi PG, Cacciari B, Romagnoli R et al (2002) 7-Substituted 5-Amino-2-(2-furyl)pyrazolo [4,3-e]1,2,4-triazolo[1,5-c]pyrimidines as A_{2A} adenosine receptor antagonists: a study on the importance of modifications at the side chain on the activity and solubility. J Med Chem 45: 115–126
108. Baraldi PG, Fruttarolo F, Tabrizi MA et al (2003) Design synthesis and biological evaluation of c^9- and C^2-substituted pyrazolo[4,3-e]1,2,4-triazolo[1,5-c]pyrimidines as new A_{2A} and A_3 adenosine receptor antagonists. J Med Chem 46:1229–1241
109. Neustadt BR, Hao J, Lindo N et al (2007) Potent, selective and orally active adenosine A_{2A} receptor antagonists: arylpiperazine derivatives of pyrazolo[4,3-e]1,2,4-triazolo[1,5-c]pyrimidines. Bioorg Med Chem Lett 17:1376–1380
110. Cheong SL, Dolzhenko A, Kachler S et al (2010) The significance of 2-furyl ring substitution with a 2-(para-substituted) aryl group in a new series of pyrazolo-triazolo-pyrimidines as potent and highly selective hA_3 adenosine receptors antagonists: new insights into structure-affinity relationship and receptor-antagonist recognition. J Med Chem 53:3361–3375
111. Harris JM, Neustadt BR, Zhang H et al (2011) Potent and selective adenosine A_{2A} receptor antagonists: [1,2,4]-triazolo[4,3-c]pyrimidin-3-ones. Bioorg Med Chem Lett 21:2497–2501
112. Baraldi PG, Saponaro G, Romagnoli R et al (2012) Water-soluble pyrazolo[4,3-e][1,2,4] triazolo[1,5-c]pyrimidines as human A_3 adenosine receptor antagonists. J Med Chem 55: 5380–5390
113. Federico S, Redenti S, Sturlese M et al (2015) The influence of the 1-(3-trifluoromethyl-benzyl)-1H-pyrazole-4-yl moiety on the adenosine receptors affinity profile of pyrazolo[4,3-e] [1,2,4]triazolo[1,5-c]pyrimidine derivatives. PloS One 10:e0143504
114. Redenti S, Ciancetta A, Pastorin G et al (2016) Pyrazolo[4,3-e][1,2,4]triazolo[1,5-c]pyrimidines and structurally simplified analogs. Chemistry and SAR profile as adenosine receptor antagonists. Curr Top Med Chem 16:3224–3257
115. Hodgson RA, Bertorelli R, Varty GB et al (2009) Characterization of the potent and highly selective A_{2A} receptor antagonist preladenant and SCH 412348 [7-[2-[4-2,4-difluorophenyl]-

1-piperazinyl]ethyl-2-(2-furanyl)-7H-pyrazolo[4,3-e][1,2,4]triazolo[1,5-c]pyrimidin-5-amine] in rodent models of movement disorders and depression. J Pharmacol Exp Ther 330: 294–303
116. Hauser R, Stocchi F, Rascol O et al (2014) Phase-3 clinical trials of adjunctive therapy with preladenant, an adenosine 2a antagonist, in patient with Parkinson's disease. Neurology 82: P7.087
117. Stocchi F, Rascol O, Hauser R et al (2014) Phase-3 clinical trials of adjunctive therapy with preladenant, an adenosine 2a antagonist, in patient with Parkinson's disease. Neurology 82 (S7):004
118. Zocchi C, Ongini E, Conti A et al (1996) The non-xanthine heterocyclic compound SCH58261 is a new potent and selective A_{2A} adenosine receptor antagonist. J Pharmacol Exp Ther 276: 398–404
119. Fredholm BB, Lindstrom K, Dionisotti S et al (1998) [^{3}H]-SCH58261, a selective adenosine A_{2A} receptor antagonist, is a useful ligand in autoradiographic studies. J Neurochem 70:1210–1216
120. Svenningsson P, Nomikos GG, Ongini E et al (1997) Antagonism of adenosine A_{2A} receptors underlies the behavioural activating effect of caffeine and is associated with reduced expression of mRNA for NGFI-A and NGFI-B in caudate-putamen and nucleus accumbens. Neuroscience 79:753–764
121. Bertorelli R, Ferri N, Adami M et al (1996) Effects of selective agonists and antagonists for A_1 and A_{2A} adenosine receptors on sleep-waking patterns in rats. Drug Dev Res 37:65–72
122. Monopoli A, Casati C, Lozza G et al (1998) Cardiovascular pharmacology of the A_{2A} adenosine receptor antagonist, SCH 58261, in the rat. J Pharmacol Exp Ther 285:9–15
123. Wardas J (2003) Synergistic effect of SCH 58261, an adenosine A_{2A} receptor antagonist, and L-DOPA on the reserpine-induced muscle rigidity in rats. Pol J Pharmacol 55:155–164
124. Wardas J, Konieczny J, Lorenc-Koci E (2001) SCH 58261, an A_{2A} adenosine receptor antagonist, counteracts parkinsonian-like muscle rigidity in rats. Synapse 41:160–171
125. Yong L, Xiaohong L, Shouyue H et al (2015) Effect of SCH442416 on glutamate uptake in retinal Muller cells at increased hydrostatic pressure. Mol Med Rep 12:3993–3997
126. Orrú M, Quiroz C, Guitart X et al (2011) Pharmacological evidence for different populations of postsynaptic adenosine A_{2A} receptors in the rat striatum. Neuropharmacology 61:967–974
127. Baraldi PG, Borea PA (2000) New potent and selective human adenosine A_3 receptor antagonists. Trends Pharmacol Sci 21:456–459
128. Varani K, Merighi S, Gessi S et al (2000) [^{3}H]MRE 3008F20: a novel antagonist radioligand for the pharmacological and biochemical characterization of human A_3 adenosine receptors. Mol Pharmacol 57:968–975
129. Claff T, Klapschinski TA, Tiruttani Subhramanyam UK et al (2022) Single stabilizing point mutation enables high-resolution co-crystal structures of the adenosine A_{2A} receptor with preladenant conjugates. Angew Chem Int Ed 61:e202115545
130. Sajjadi FG, Firestein GS (1993) cDNA cloning and sequence analysis of the human A_3 adenosine receptor. Biochim Biophys Acta 1179:105–107
131. Salvatore CA, Jacobson MA, Taylor HE (1993) Molecular cloning and characterization of the human A_3 adenosine receptor. Proc Natl Acad Sci U S A 90:10365–10369
132. Linden J, Taylor HE, Robeva AS (1993) Molecular cloning and functional expression of a sheep A_3 adenosine receptor with widespread tissue distribution. Mol Pharmacol 44:524–532
133. Zhao ZH, Ravid S, Ravid K (1995) Chromosomal mapping of the mouse A_3 adenosine receptor gene, adora3. Genomics 30:118–119
134. Hill RJ, Oleynek JJ, Hoth CF (1997) Cloning, expression and pharmacological characterization of rabbit A_1 and A_3 receptors. J Pharmacol Exp Ther 280:122–128
135. Muller CE (2003) Medicinal chemistry of adenosine A_3 receptor ligands. Curr Top Med Chem 3:445–462

136. Kim YC, Ji XD, Jacobson KA (1996) Derivatives of the triazoloquinazoline adenosine antagonist (GCS15943) are selective for the human A_3 receptor subtype. J Med Chem 39: 4142–4148
137. Baraldi PG, Cacciari B, Spalluto G et al (1996) Novel N6-(substituted-phenylcarbamoyl) adenosine-5′-uronamides as potent agonists for A_3 adenosine receptor. J Med Chem 39: 802–806
138. Baraldi PG, Cacciari B, Romagnoli R et al (2000) Pyrazolo[3,4-e]1,2,4-triazolo[1,5-c]pyrimidine derivatives as highly potent and selective human A_3 adenosine receptor antagonists: influence of the chain at the N8 pyrazole nitrogen. J Med Chem 43:4768–4780
139. Baraldi PG, Cacciari B, Moro S et al (2002) Synthesis, biological activity and molecular modeling investigation of new pyrazolo[4,3-e]1,2,4-triazolo[1,5-c]pyrimidine derivatives as human A_3 adenosine receptor antagonists. J Med Chem 45:770–780
140. Bolcato C, Cusan C, Pastorin G et al (2008) Pyrazolo-triazolo-pyrimidines as adenosine receptor antagonists: effect of the N-5 bond type on the affinity and selectivity at the four adenosine receptor subtypes. Purinergic Signal 4:39–46
141. Maconi A, Pastorin G, Da Ros T et al (2002) Synthesis, biological properties and molecular modeling investigation of the first potent, selective and water soluble human A_3 adenosine receptor antagonist. J Med Chem 45:3579–3582
142. Pastorin G, Da Ros T, Bolcato C et al (2006) Synthesis and biological studies of a new series of 5-heteroarylcarbamoylaminopyrazolo[4,3-*e*]1,2,4-triazolo[1,5-*c*]pyrimidines as human A_3 adenosine receptor antagonists. Influence of the heteroaryl substituent on binding affinity and molecular modeling investigation. J Med Chem 49:1720–1729
143. Federico S, Ciancetta A, Sabbadin D et al (2012) Exploring the directionally of 5-substitutions in a new series of 5-alkylaminopyrazolo[4,3-*e*]1,2,4-triazolo[1,5-*c*]pyrimidine as a strategy to design novel human A_3 adenosine receptor antagonists. J Med Chem 55:9654–9668
144. Federico S, Margiotta E, Moro S et al (2021) Potent and selective A_3 adenosine receptor antagonists bearing aminoesters as heterobifunctional moieties. RCS Med Chem 12:254–262
145. Cheong SL, Federico S, Venkatesan G et al (2013) The A_3 adenosine receptor as multifaceted therapeutic target: pharmacology, medicinal chemistry and in silico approaches. Med Res Rev 33:235–335
146. Baraldi PG, Cacciari B, Romagnoli R et al (2001) Pyrazolo[3,4-*e*]1,2,4-triazolo[1,5-*c*]pyrimidine derivatives as adenosine receptor ligands: a starting point for searching A_{2B} adenosine receptor antagonists. Drug Res Rev 53:225–235
147. Pastorin G, Da Ros T, Spalluto G et al (2003) Pyrazolo[3,4-*e*]1,2,4-triazolo[1,5-*c*]pyrimidine derivatives as adenosine receptor antagonists. Influence of the N5 substituent on the affinity at the human A_3 and A_{2B} adenosine receptor subtypes: a molecular modeling investigation. J Med Chem 46:4287–4296
148. Michielan L, Federico S, Terfloth L et al (2009) Exploring potency and selectivity receptor antagonist profiles using a multilabel classification approach: the human adenosine receptors as a key study. J Chem Inf Model 49:2820–2286
149. Beauglehole AR, Baker SP, Scammells PJ (2002) New irreversible adenosine A1 antagonists based on FSCPX. Bioorg Med Chem Lett 12:3179–3182
150. Glukhova A, Thal DM, Nguyen AT et al (2017) Structure of the adenosine A_1 receptor reveals the basis for subtype selectivity. Cell 168:867–877.e13
151. Weichert D, Gmeiner P (2015) Covalent molecular probes for class A G protein-coupled receptors: advances and applications. ACS Chem Biol 10:1376–1386
152. Shryock JC, Snowdy S, Baraldi PG et al (1998) A_{2A}-adenosine receptor reserve for coronary vasodilation. Circulation 98:711–718
153. Grunbeck A, Sakmar TP (2013) Probing G protein-coupled receptor—ligand interactions with targeted photoactivatable cross-linkers. Biochemistry 52:8625–8632
154. Muranaka H, Momose T, Handa C, Ozawa T (2017) Photoaffinity labeling of the human A_{2A} adenosine receptor and cross-link position analysis by mass spectrometry. ACS Med Chem Lett 8:660–665

155. Matthews PM, Rabiner EA, Passchier J, Gunn RN (2012) Positron emission tomography molecular imaging for drug development. Br J Clin Pharmacol 73:175–186
156. Honer M, Gobbi L, Martarello L, Comley RA (2014) Radioligand development for molecular imaging of the central nervous system with positron emission tomography. Drug Discov Today 19:1936–1944
157. Baraldi PG, Cacciari B, Dionisotti S et al (1996) Synthesis of the tritium labeled SCH 58261, a new non-xanthine A_{2A} adenosine receptor antagonist. J Label Compd Radiopharm 38:725–732
158. Dionisotti S, Ongini E, Zocchi C et al (1997) Characterization of human A_{2A} adenosine receptors with the antagonist radioligand [^{3}H]-SCH 58261. Br J Pharmacol 121:353–360
159. El Yacoubi M, Ledent C, Parmentier M et al (2001) In vivo labelling of the adenosine A_{2A} receptor in mouse brain using the selective antagonist [^{3}H]SCH 58261. Eur J Neurosci 14: 1567–1570
160. Svenningsson P, Hall H, Gö G et al (1997) Distribution of adenosine receptors in the postmortem human brain: an extended autoradiographic study. Synapse 27:322–335
161. Zocchi C, Ongini E, Ferrara S et al (1996) Binding of the radioligand [^{3}H]-SCH 58261, a new non-xanthine A_{2A} adenosine receptor antagonist, to rat striatal membranes. Br J Pharmacol 117:1381–1386
162. Bryant R, McGuinness D, Turek-Etienne T et al (2004) WGA-coated yttrium oxide beads enable an imaging-based adenosine 2A receptor binding scintillation proximity assay suitable for high throughput screening. Assay Drug Dev Technol 2:290–299
163. Todde S, Moresco RM, Simonelli P et al (2000) Design, radiosynthesis, and biodistribution of a new potent and selective ligand for in vivo imaging of the adenosine A_{2A} receptor system using positron emission tomography. J Med Chem 43:4359–4362
164. Moresco RM, Todde S, Belloli S et al (2005) In vivo imaging of adenosine A_{2A} receptors in rat and primate brain using [^{11}C]SCH442416. Eur J Nucl Med Mol Imaging 32:405–413
165. Varani K, Vincenzi F, Tosi A et al (2010) A_{2A} adenosine receptor overexpression and functionality, as well as TNF-α levels, correlate with motor symptoms in Parkinson's disease. FASEB J 24:587–598
166. Khanapur S, Paul S, Shah A et al (2014) Development of [^{18}F]-labeled pyrazolo[4,3-*e*]-1,2,4-triazolo[1, 5-*c*]pyrimidine (SCH442416) analogs for the imaging of cerebral adenosine A_{2A} receptors with positron emission tomography. J Med Chem 57:6765–6780
167. Bhattacharjee AK, Lang L, Jacobson O et al (2011) Striatal adenosine A_{2A} receptor-mediated positron emission tomographic imaging in 6-hydroxydopamine-lesioned rats using [^{18}F]-MRS5425. Nucl Med Biol 38:897–906
168. Schröder S, Lai TH, Toussaint M et al (2020) PET imaging of the adenosine A_{2A} receptor in the rotenone-based mouse model of Parkinson's disease with [^{18}F]FESCH synthesized by a simplified two-step one-pot radiolabeling strategy. Molecules 25:1633
169. Lai TH, Toussaint M, Teodoro R et al (2021) Improved in vivo PET imaging of the adenosine A_{2A} receptor in the brain using [^{18}F]FLUDA, a deuterated radiotracer with high metabolic stability. Eur J Nucl Med Mol Imaging 48:2727–2736
170. Rosenberg AJ, Cheung Y, Liu F et al (2022) Automated synthesis of (R)-[^{18}F]MH.MZ on the iPhase Flexlab reaction platform. J Label Compd Radiopharm 65:223–229
171. Hesk D, Borges S, Dumpit R et al (2017) Synthesis of ^{3}H, $^{2}H_4$, and ^{14}C-MK 3814 (preladenant). J Label Compd Radiopharm 60:194–199
172. Tavares AADS, Batis J, Barret O et al (2013) In vivo evaluation of [^{123}I]MNI-420: a novel single photon emission computed tomography radiotracer for imaging of adenosine 2A receptors in brain. Nucl Med Biol 40:403–409
173. Tavares AAS, Batis JC, Papin C et al (2013) Kinetic modeling, test–retest, and dosimetry of ^{123}I-MNI-420 in humans. J Nucl Med 54:1760–1767
174. Barret O, Hannestad J, Alagille D et al (2014) Adenosine 2A receptor occupancy by tozadenant and preladenant in rhesus monkeys. J Nucl Med 55:1712–1718

175. Barret O, Hannestad J, Vala C et al (2015) Characterization in humans of ^{18}F-MNI-444, a PET radiotracer for brain adenosine 2A receptors. J Nucl Med 56:586–591
176. Yang X, Heitman LH, Jzerman AP, van der Es D (2021) Molecular probes for the human adenosine receptors. Purinergic Signal 17:85–108
177. Baraldi PG, Cacciari B, Romagnoli R et al (2000) Synthesis and preliminary biological evaluation of [^{3}H]-MRE 3008-F20: the first high affinity radioligand antagonist for the human A_3 adenosine receptors. Bioorg Med Chem Lett 10:209–211
178. Gessi S, Varani K, Merighi S et al (2004) Expression of A_3 adenosine receptors in human lymphocytes: up-regulation in T cell activation. Mol Pharmacol 65:711–719
179. Gessi S, Cattabriga E, Avitabile A et al (2004) Elevated expression of A_3 adenosine receptors in human colorectal cancer is reflected in peripheral blood cells. Clin Cancer Res 10:5895–5901
180. Comeo E, Kindon ND, Soave M et al (2020) Subtype-selective fluorescent ligands as pharmacological research tools for the human adenosine A_{2A} receptor. J Med Chem 63: 2656–2672
181. Kecskés M, Kumar TS, Yoo L et al (2010) Novel Alexa Fluor-488 labeled antagonist of the A_{2A} adenosine receptor: application to a fluorescence polarization-based receptor binding assay. Biochem Pharmacol 80:506–511. https://doi.org/10.1016/j.bcp.2010.04.027
182. Kumar TS, Mishra S, Deflorian F et al (2011) Molecular probes for the A_{2A} adenosine receptor based on a pyrazolo[4,3-*e*][1,2,4]triazolo[1,5-*c*]pyrimidin-5-amine scaffold. Bioorg Med Chem Lett 21:2740–2745. https://doi.org/10.1016/j.bmcl.2010.11.082
183. Duroux R, Ciancetta A, Mannes P et al (2017) Bitopic fluorescent antagonists of the A_{2A} adenosine receptor based on pyrazolo[4,3-*e*] [1,2,4]triazolo[1,5-*c*] pyrimidin-5-amine functionalized congeners. Med Chem Commun 8:1659–1667
184. Gao ZG, Toti KS, Campbell R et al (2020) Allosteric antagonism of the A_{2A} adenosine receptor by a series of bitopic ligands. Cell 9:1200
185. Ciruela F, Fernández-Dueñas V, Jacobson KA (2015) Lighting up G protein-coupled purinergic receptors with engineered fluorescent ligands. Neuropharmacology 98:58–67
186. Federico S, Margiotta E, Paoletta S et al (2019) Pyrazolo[4,3-*e*][1,2,4]triazolo[1,5-*c*]pyrimidines to develop functionalized ligands to target adenosine receptors: fluorescent ligands as an example. Med Chem Commun 10:1094–1108
187. Kozma E, Kumar TS, Federico S et al (2012) Novel fluorescent antagonist as a molecular probe in A_3 adenosine receptor binding assays using flow cytometry. Biochem Pharmacol 83: 1552–1561
188. Federico S, Margiotta E, Moro S et al (2020) Conjugable A_3 adenosine receptor antagonists for the development of functionalized ligands and their use in fluorescent probes. Eur J Med Chem 186:111886
189. Jacobson KA (2013) Structure-based approaches to ligands for G-protein-coupled adenosine and P2Y receptors, from small molecules to nanoconjugates. J Med Chem 56:3749–3767
190. Pineux F, Federico S, Klotz K-N et al (2020) Targeting G protein-coupled receptors with magnetic carbon nanotubes: the case of the A_3 adenosine receptor. ChemMedChem 15:1909–1920
191. Yan W, Ling L, Wu Y et al (2021) Structure-based design of dual-acting compounds targeting adenosine A_{2A} receptor and histone deacetylase as novel tumor immunotherapeutic agents. J Med Chem 64:16573–16597

Top Med Chem (2023) 41: 89–100
https://doi.org/10.1007/7355_2022_155

Published online: 25 November 2022

Adenosine Receptor Ligands as Potential Therapeutic Agents for Impaired Wound Healing and Fibrosis

Flavia Varano, Daniela Catarzi, Erica Vigiani, Sara Calenda, and Vittoria Colotta

Contents

Abstract Wound healing is a complex process that involves a series of events including inflammation, tissue formation, angiogenesis, and tissue remodeling. Such events must be well coordinated with each other, and highly regulated. Otherwise, the wounds do not heal, incurring into chronic wounds, or the healing process, excessively activated, leads to fibrosis. Adenosine, a widely distributed nucleoside, is involved in every phase of the wound healing process through the activation of its receptors (A_1, A_{2A}, A_{2B}, and A_3 adenosine receptors ARs). Thus, targeting ARs may represent a new therapeutic approach to treating chronic wounds or fibrosis, the major pathologies related to impaired wound healing.

Keywords Adenosine, Adenosine receptors, Chronic ulcers, Fibrosis, Wound healing

F. Varano (✉), D. Catarzi, E. Vigiani, S. Calenda, and V. Colotta
Dipartimento di Neuroscienze, Psicologia, Area del Farmaco e Salute del Bambino, Sezione di Farmaceutica e Nutraceutica, Universita'degli Studi di Firenze, Florence, Italy
e-mail: flavia.varano@unifi.it

1 Introduction

Wound healing is one of the most complicated processes that occurs in the human body and is made up of four partially overlapping phases: hemostasis, inflammation, proliferation, and remodeling [1].

For some people, the process does not work, and wounds do not heal in a fair amount of time or at all. Patients with wound healing disorders are often old and suffer from multiple co-morbidities such as diabetes and cardiovascular disease. Compromised wound healing is associated with pain, disability, loss of productiveness, and depression. Patients suffering from abnormalities of wound healing are reported to be about 8.2 million only in the USA, and the estimated annual cost to treat wound complications is around 28 billion US dollars [2, 3]. Moreover, chronic ulcers have been related to increased mortality; in fact, 5-year mortality data indicate that diabetic ischemic ulcers are more lethal than some types of cancer [4]. Despite a clear unmet clinical need, there are currently few FDA-approved therapies for the treatment of chronic ulcers such as venous leg ulcers and diabetic foot ulcers. These include becaplermin, a recombinant human platelet-derived growth factor (rhPDFG-BB), and tissue-engineered living human skin substitutes, Apligraf and Dermagraft [5–7].

Although tissue repair is initially beneficial, the wound healing process may lead to excess matrix production, scarring, and fibrosis with consequent organ dysfunction [8]. Fibrosis is responsible for a wide range of clinically important diseases including idiopathic pulmonary fibrosis, liver cirrhosis, cardiovascular fibrosis, hypertrophic scarring, keloid formation, radiation fibrosis, and scleroderma [9–14]. Scleroderma, also known as systemic sclerosis, is a rare disease that causes hardening of the skin, and it can also affect blood vessels, internal organs, and the digestive tract. There is no cure for scleroderma, available treatments can only ease symptoms, slow disease progression improving patients' quality of life. Ionizing radiation therapy of common tumors, such as head, neck, and breast cancers, must often be stopped due to the appearance of radiation dermatitis in the short term and dermal fibrosis in the long term. Finally, disfiguring scars are a major cause of physical and psychological disability after trauma, resulting in more than 170,000 scar revision surgeries each year, with a significant healthcare cost.

A variety of different factors regulate the wound healing process, ranging from growth factors to small molecules released at the wounded site. One of such molecules is adenosine, an endogenous purine nucleoside that is generated extracellularly from the serial dephosphorylation of ATP to AMP, by nucleoside triphosphate dephosphorylase (CD39), followed by dephosphorylation to adenosine by 5′-ectonucleotidase (CD73) [15]. Extracellular adenosine is then degraded to inosine by the adenosine deaminase enzyme (ADA). Adenosine mediates its pathophysiological actions through the activation of adenosine receptors (ARs) which belong to the G-protein-coupled receptors family and comprise four subtypes, namely A_1, A_{2A}, A_{2B}, and A_3 ARs. Activation of ARs modulates a variety of effector systems, although, classically, the adenosine-mediated signaling has been subdivided based on its ability to raise or lower cAMP levels. In fact, the A_1 and A_3 ARs preferentially couple to $G_{i/o}$ proteins to inhibit adenylyl cyclase thus decreasing cAMP levels, whereas the A_{2A} and A_{2B} ARs stimulate adenylyl cyclase activity through activation of G_s proteins, thus increasing cAMP levels.

2 The Wound Healing Process

Wounds repair process begins with the "hemostasis phase" that is characterized by vascular conscription and fibrin clot formation by platelets. Then, the clot and surrounding wound tissue release pro-inflammatory cytokines and growth factors such as transforming growth factor (TGF-β), platelet-derived growth factor (PDGF), and fibroblast growth factor (FGF). Once bleeding is stopped, inflammatory cells such as neutrophils, macrophages, and lymphocytes migrate into the wound thus promoting the "inflammatory phase" [1, 16]. Neutrophils are able, through phagocytosis activity, to both eliminate any bacteria and degrade necrotic tissue. Moreover, neutrophils release inflammatory mediators such as tumor necrosis factor (TNF-α), interleukins IL-1β and IL-6, and vascular endothelial growth factor (VEGF) which increases the vessels permeability and promotes the growth of new blood vessels [16, 17].

Macrophages play multiple roles in wound healing. In the early stage, they enter the lesion area and contribute to wound cleansing by phagocytosis. Then macrophages undergo a phenotypic transition to a reparative state that stimulates keratinocytes, fibroblasts, and the process of angiogenesis thus promoting tissue regeneration [18].

The "proliferative phase" of healing is characterized by a broad activation of fibroblasts, macrophages, and endothelial cells. Wound closure, matrix deposition, and angiogenesis take place during this phase. Fibroblasts are the main ones responsible for replacing the fibrin-rich platelet-derived matrix with granulation tissue. Fibroblasts, driven by signaling molecules including TFG-β and PDGF, degrade the provisional matrix and replace it with a granulation tissue rich in fibronectin, proteoglycans, and type III collagen, an immature form of collagen protein. This immature granulation tissue acts as a scaffold supporting both the formation of new blood vessels and the deposition of mature extracellular matrix (ECM) [19, 20].

Macrophages during this phase play a significant role in angiogenesis. New blood vessels are created to meet the metabolic demands of the highly proliferative healing tissue allowing nutrients and factors involved in the wound repair process to enter the injury site [21].

"Remodeling" is the last stage of wound healing and can last up to 2 years after injury. During this phase, the formation of granulation tissue stops due to apoptosis. Fibroblasts are the major cell type responsible for wound remodeling. At this stage, the components of ECM undergo some changes. Type III collagen, which was produced in the proliferative phase, is now replaced by type I collagen. Which increases the tensile strength of the forming scar. However, the integrity and architecture of scar ECM never fully return to that of unwounded tissue. Subsequently, myofibroblasts cause contractions of the wound and help to reduce the surface of the developing scar. In addition, the angiogenetic processes decrease, the blood flow of the wound is reduced, and the metabolic activity of the wound slows down and finally stops [19, 20].

It is important to underline that many factors, both endogenous and exogenous, can lead to impaired wound healing process. Among endogenous factors, aging is the most determinant causing a temporal delay in each phase of the process. Nutritional status is a crucial exogenous factor. In particular, a condition of malnutrition can compromise the activity of the immune system, causing healing disorders and infectious complications. Of course, the general health status of people can affect the speed and quality of healing. In particular, diabetic individuals, which total hundreds of millions around the world, are characterized by impaired healing of the skin wounds and are prone to developing chronic non-healing ulcers in the lower extremities known as diabetic foot ulcers [22].

3 Adenosine in Wound Healing

During normal physiological conditions, the extracellular adenosine concentration is maintained low (30–300 nM) and constant. In contrast, under conditions of cellular or tissue necrosis, stress, or injury, its concentration rises dramatically to micromolar range, as a result of ATP catabolism. All the cells involved in the healing process, macrophages, epidermal cells, fibroblasts, and endothelial cells, express the various AR subtypes, although different cells and even the same cellular type, such as endothelial cells, express different AR subtypes [23, 24].

Adenosine is already involved in the early stages of the wound healing process, represented by the hemostatic and inflammatory phase. Numerous studies show that adenosine is a powerful regulator of the inflammatory process [25]. It has been observed that adenosine reduces inflammatory functions mainly through the activation of the A_{2A} AR (Fig. 1). Studies with genetically modified animals showed that the absence of A_{2A} ARs increased inflammatory damage [26, 27]. Adenosine acting at the A_{2A} AR subtype can reduce the activation of neutrophils and macrophages, the differentiation and proliferation of T cells, the activity of NK cells and the production of pro-inflammatory cytokines. It has also been observed that, through activation of the A_{2A} AR, adenosine induces the differentiation of macrophages into M2-type that promote wound healing through the release of factors such as VEGF. At the same time adenosine and A_{2A} AR agonists strongly downregulate TNFα expression, thus promoting an angiogenic switch, shifting macrophages from an inflammatory to an angiogenic phenotype [28–30].

Furthermore, adenosine acting at all ARs has been recognized as a potent stimulator of angiogenesis which is a pivotal event in the wound healing process because injured tissue needs oxygen and nutrients to be efficiently repaired (Fig. 1).

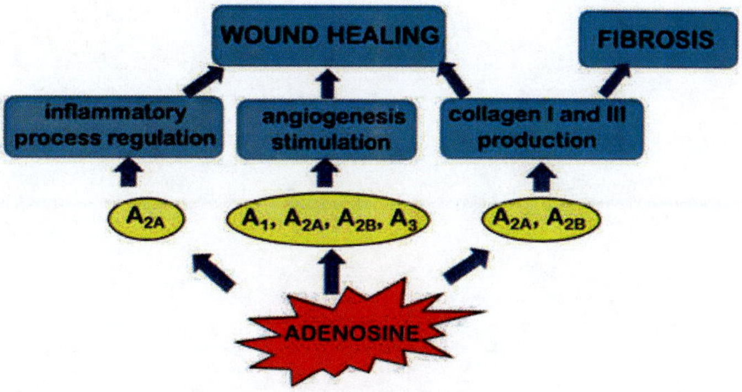

Fig. 1 Adenosine and ARs in wound healing and fibrosis

Studies in animal models showed that treatment with CGS 21680, a selective A_{2A} AR agonist, promoted blood vessel formation and accelerated wound closure in wild-type mice, while having no effect on A_{2A} AR knockout mice [31]. The principal pro-angiogenic actions of adenosine are due to its ability to control the production of pro-angiogenic substances from vascular and immune cells within the microenvironment of injured tissues. In fact, A_1 AR mediated angiogenesis involves the release of VEGF from monocytes [32], while activation of the A_{2B} subtype induces wound angiogenesis by upregulation in human microvascular endothelial cells of VEGF and endothelial NO synthase (eNOS) [33]. Adenosine, via the A_{2A} and A_{2B} ARs, stimulates angiogenesis by modulating thrombospondin-1 expression in human macrophages [34]. Moreover, adenosine through A_{2B} and A_3 ARs also promotes the production of VEFG, IL-8, and angiopoietin-1 from mast cells [35], and A_1, A_{2A}, and A_{2B} AR selective agonists contribute to proliferation and wound healing in human EAhy926 endothelial cells [36].

Finally, activation of the A_{2A} AR induces dermal fibroblast to produce collagen I and III, and to downregulate matrix metalloproteinases (MMP 9, 2, and 14) which are responsible for collagen disruption, thus evidencing a role of adenosine also in the tissue proliferation phase of the wound healing process. In fact, the blockade or deletion of this receptor subtype protects mice from bleomycin-induced dermal fibrosis which is a murine model of scleroderma [37, 38] (Fig. 1).

In synthesis, adenosine receptor agonists, selective or not, may represent new and interesting therapeutic agents for the treatment of impaired wound healing which is one of the main concerns, especially for diabetic patients (up to 28% of diabetic foot ulcers can cause amputation). Topical application of A_{2A} AR agonists accelerates the healing of skin wounds in healthy and diabetic rats, but not in A_{2A} AR knockout mice [31, 39]. CGS21680 and sonedenoson (MRE-0094) (Fig. 2), two well-known A_{2A} AR agonists, induce wound closure by a mechanism that depends on tissue plasminogen activator [40] and promote angiogenesis in wild-type mice. They heal wounds faster than beclapermin gel which is the only growth factor approved for

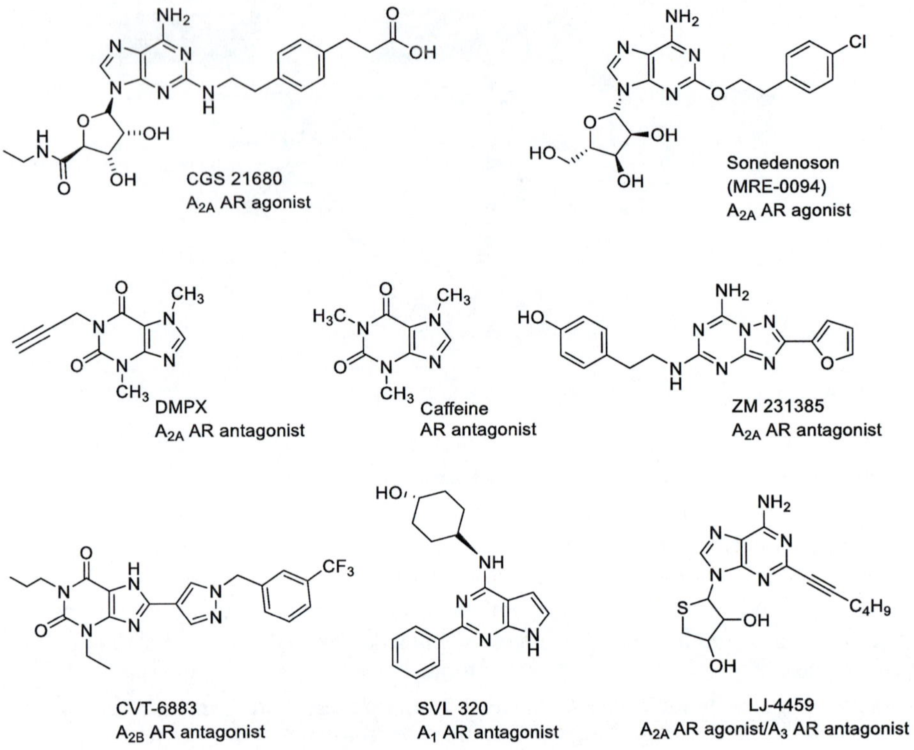

Fig. 2 AR ligands evaluated in impaired wound healing and fibrosis

clinical use [31, 39]. Sonedenoson administered as a topical gel was tested as a potential drug for diabetic foot ulcers, but phase 2 clinical trial had poor enrollment and was terminated [41]. Despite this, encouraging data come from more recent studies which highlighted the tissue repair, anti-ischemic and anti-inflammatory properties of polydeoxyribonucleotide (PDRN) which is a proprietary and registered DNA-derived drug extracted from the sperm of trout bred for human food purposes. In vivo experiments demonstrated that PDRN promoted wound healing of diabetic foot ulcers better than placebo. All data obtained from these studies indicated that the most relevant mechanism of action of PDRN for promoting tissue repair was the A_{2A} AR activation. In fact, DMPX (Fig. 2), a potent and selective A_{2A} AR antagonist, was able to revert the PDRN-induced effects on wound healing.

Indeed, it has been speculated that PDRN may represent a pro-drug able to generate active deoxyribonucleotides, nucleosides, and bases that interact with the A_{2A} AR [42, 43].

4 Adenosine and Fibrosis

The observation that adenosine, acting at its receptors, stimulates the formation of ECM has suggested the possibility that excessive activation of ARs could lead to excess matrix production, scarring, and fibrosis of the skin and other organs, such as the liver, heart, lungs, and kidneys (Fig. 1).

The A_{2A} AR is expressed on human hepatic stellate cells, which are the fibroblasts of the liver, and its occupancy induces collagen I and III production by these cells. A_{2A} AR knockout mice, unlike wild-type littermates, are protected from developing hepatic fibrosis following administration of the known hepatic fibrosis-inducing agents CCl_4 or thioacetamide, as well as wild-type mice treated with caffeine (Fig. 2), a nonselective AR antagonist, or with ZM231385 (Fig. 2) a highly selective A_{2A} AR antagonist [44].

The pivotal role of adenosine and A_{2A} AR in dermal fibrosis has been demonstrated in ADA deficient mice, a murine model in which both circulating and tissue adenosine levels are extremely high. Indeed, ADA is the key enzyme that catabolized adenosine into inosine. These genetically modified mice spontaneously develop dermal fibrosis which is diminished by treatment with an A_{2A} AR antagonist [45]. Also, expression of A_{2B} AR is significantly elevated in bleomycin-induced fibrosis, a well-established mouse model of dermal fibrosis, thus suggesting both a pro-fibrotic role for this receptor subtype too and a potential therapeutic use of A_{2B} AR antagonists to treat dermal fibrosis [46].

Pharmacological blockade of A_{2A} AR with ZM241385 decreased size and enhanced the tensile strength of the scar in a murine model that mimics human scarring. These data show that A_{2A} AR antagonists can be used to reduce scarring, also improving the collagen composition of the healed wound [47]. Moreover, topical application of ZM241385 or A_{2A} AR deletion prevents radiation-induced dermal injury in a murine model of irradiation-induced fibrosis. Therefore, A_{2A} AR antagonism may be a useful strategy for preventing or ameliorating radiation dermatitis which is a common side effect of radiotherapy treatment of cancers [48].

Adenosine deaminase deficient mice also develop pulmonary inflammation and fibrosis that are significantly reduced by treatment with CVT-6883 (Fig. 2), a selective A_{2B} AR antagonist. This result indicates that A_{2B} AR signaling contributes to the pro-inflammatory and pro-fibrotic activities present in chronic lung diseases, such as severe asthma, COPD, and pulmonary fibrosis [49, 50].

On the contrary, adenosine A_{2B} AR stimulation was shown to suppress collagen production in cardiac fibroblasts, suggesting a protective role of this receptor in myocardial fibrosis [51]. Reduction of cardiac fibrosis in a model of uremic cardiomyopathy is also obtained by the use of SVL320 (Fig. 2) a highly selective A_1 AR antagonist [52].

The effects of ARs signaling in the development and progression of renal fibrosis indicate that, while A_{2A} AR stimulation protects against inflammation, A_3 and A_{2B} AR activation promotes renal fibrosis. In fact, LJ-4459 (Fig. 2) [53], a dual-acting ligand that is an A_{2A} AR agonist and an A_3 AR antagonist, prevents the progression of tubule-interstitial fibrosis [54, 55].

From the above, the important role of adenosine in the pathophysiology of tissue fibrosis is evidenced. In fact, adenosine stimulates fibrosis in the skin, lungs, liver, and kidneys, while inhibiting myocardial fibrosis. Thus, adenosine receptor antagonism may represent a novel therapeutic strategy for the treatment of fibrotic diseases. A2A receptor antagonists could be used for the therapy of dermal and hepatic fibrosis, while A_{2B} receptor blockade potentially leads to pulmonary fibrosis control.

5 Conclusions

Adenosine plays important roles in inflammation, angiogenesis, and matrix production, which are critical for wound healing and tissue repair. There are a few FDA-approved therapies for the treatment of impaired healing which remains an unsolved clinical issue. Taking this into account, the development of AR agonists, either selective for the A_{2A} receptor subtype or with a combined action at all the ARs, could lead to new drugs representing interesting alternative agents to beclapermin a growth factor derived drug, approved by the FDA for the treatment of chronic ulcers. In fact, the use of beclapermin has been drastically reduced due to its high production costs, its short storage stability along with an increased risk to malignant infections. On the other hand, AR agonists are in general small synthetic molecules of greater stability and lower cost compared to growth factors derived drugs.

Moreover, adenosine plays a direct role in promoting fibrosis in the skin, lungs, and liver, whereas inhibiting cardiac fibrosis. Targeting of ARs to diminish fibrosis is currently under investigation, with the primary objective to find new therapies for these difficult clinical settings.

Compliance with Ethical Standards **Conflict of Interest**: The Author declares that they have no conflict of interest.

Funding: Original research of our team is funded by an intramural grant from the University of Florence (Fondi Ateneo Ricerca 2022).

Ethical Approval: This chapter does not contain any studies with human participants or animals performed by the authors.

References

1. Rodrigues M, Kosaric N, Bonham CA, Gurtner GC (2019) Wound healing: a cellular perspective. Physiol Rev 99:665–706
2. Sen CK (2019) Human wounds and its burden an updated compendium of estimates. Adv Wound Care 8:39–48

3. Nussbaum SR, Carter MJ, Fife CE, DaVanzo J, Haught R, Nusgart M et al (2018) An economic evaluation of the impact, cost, and medicare policy implication of chronic nonhealing wounds. Value Health 21:27–32
4. Armstrong DG, Wrobel J, Robbins JM (2007) Guest editorial: are diabetes-related wounds and amputations worse than cancer? Int Wound J 4:286–287
5. Smiell JM, Weiman J, Steed DL, Perry BH, Sampson AR, Schwab BH (1999) Efficacy and safety of becaplermin (recombinant human platelet-derived growth factor-BB) in patients with nonhealing, lower extremity diabetic ulcers: a combined analysis of four randomized studies. Wound Repair Regen 7:335–346
6. Zaulyanov L, Kirsner RS (2007) A review of a bi-layered living cell treatment (Apligraf) in the treatment of venous leg ulcers and diabetic foot ulcers. Clin Interv Aging 2:93–98
7. Hart CE, Loewen-Rodriguez A, Lessem J (2012) Dermagraft: use in the treatment of chronic wounds. Adv Wound Care 1:138–141
8. Wynn TA (2008) Cellular and molecular mechanisms of fibrosis. J Pathol 214:199–210
9. King TE, Pardo A, Selman M (2011) Idiopathic pulmonary fibrosis. Lancet 378:1949–1961
10. Gines P, Krag A, Abraldes JG, Sola E, Fabrellas N, Kamath PS (2021) Liver cirrhosis. Lancet 398:1359–1376
11. Hinderer S, Schenke-Layland K (2019) Cardiac fibrosis – a short review of causes and therapeutic strategies. Adv Drug Deliv Rev 146:77–82
12. Wang Z-C, Zhao W-Y, Cao Y, Liu Y-Q, Sun Q, Shi P, Cai J-Q, Shen XZ, Tan W-Q (2020) The roles of inflammation in keloid and hypertrophic scars. Front Immunol 11:603187
13. Straub JM, New J, Hamilton CD, Lomiska C, Shnayder Y, Thomass SM (2015) Radiation-induced fibrosis: mechanism and implications for therapy. J Cancer Res Clin Oncol 141:1985–1994
14. Denton CP, Khanna D (2017) Systemic sclerosis. Lancet 390:1685–1699
15. Borea PA, Gessi S, Merighi S, Vincenzi F, Varani K (2018) Pharmacology of adenosine receptors: the state of the art. Physiol Rev 98:1591–1625
16. Chesko DM, Wilgus TA (2022) Immune cells in cutaneous wound healing: a review of functional data from animal models. Int J Mol Sci 23:2444
17. Werner S, Grose R (2003) Regulation of wound healing by growth factors and cytokines. Physiol Rev 83:835–870
18. Koh TJ, Dipietro LA (2011) Inflammation and wound healing: the role of the macrophage. Expert Rev Mol Med 13:E23
19. Xue M, Jackson CJ (2013) Extracellular matrix reorganization during wound healing and its impact on abnormal scarring. Adv Wound Care 4:119–136
20. Li B, Wang JHC (2011) Fibroblasts and myofibroblasts in wound healing: force generation and measurement. J Tissue Viability 20:108–120
21. Honnegowda TM, Kumar P, Udupa EG, Kumar S, Kumar U, Rao P (2015) Role of angiogenesis and angiogenic factors in acute and chronic wound healing. Plast Aesthet Res 2:243–249
22. Guo S, DiPietro LA (2010) Factors affecting wound healing. J Dent Res 89:219–229
23. Foektistov I, Biaggioni I, Cronstein BN (2009) Adenosine receptors in wound healing, fibrosis and angiogenesis. Handb Exp Pharmacol 193:383–397
24. Feoktisstov I, Goldstein AE, Ryzhov S, Zeng D, Belardinelli L, Voyno-Yasenetskaya T, Biaggioni I (2002) Differential expression of adenosine receptors in human endothelial cells. Role of A2B receptors in angiogenic factor regulation. Circ Res 90:531–538
25. Pasquini S, Contri C, Borea PA, Vincenzi F, Varani K (2021) Adenosine and inflammation: here, there and everywhere. Int J Mol Sci 22:7685
26. Haskò G, Pacher P (2008) A2A receptors in inflammation and injury: lessons learned from transgenic animals. J Leukoc Biol 83:447–455
27. Haskò G, Cronstein B (2013) Regulation of inflammation by adenosine. Front Immunol 4:85
28. Pinhal-Enfield G, Ramanathan M, Haskò G, Vogel SN, Salzman AL, Boons GJ, Leibovich SJ (2003) An angiogenic switch in macrophages involving synergy between toll-like receptors 2, 4, 7, and 9 and adenosine A2A receptors. Am J Pathol 163:711–721

29. Macedo L, Pinhal-Enfield G, Alshits V, Elson G, Cronstein BN, Leibovich SJ (2007) Wound healing is impaired in MyD88-deficient mice. A role for MyD88 in the regulation of wound healing by adenosine A2A receptors. Am J Pathol 171:1774–1788
30. Leibovich SJ, Chen JF, Pinhal-Enfield G, Belem PC, Elson G, Rosania A, Ramanathan M, Montesinos C, Jacobson M, Schwarzschild MA, Fink JS, Cronstein B (2002) Synergistic up-regulation of vascular endothelial growth factor expression in murine macrophages by adenosine A2A receptor agonists and endotoxin. Am J Pathol 160:2231–2244
31. Montesinos MC, Desai A, Chen JF, Yee H, Schwarzschild MA, Fink JS, Cronstein BN (2002) Adenosine promotes wound healing and mediates angiogenesis in response to tissue injury via occupancy of A2A receptors. Am J Pathol 160:2009–2018
32. Auchampach JA (2007) Adenosine receptors and angiogenesis. Circ Res 101:1075–1077
33. Du X, Ou X, Song T, Zhang W, Cong F, Zhang S, Xiong Y (2015) Adenosine A2B receptor stimulates angiogenesis by inducing VEGF and eNOS in human microvascular endothelial cells. Exp Biol Med 240:1472–1479
34. Ernens I, Bousquenaud M, Lenoir B, Devaux Y, Wagner DR (2015) Adenosine stimulates angiogenesis by up-regulating production of trombospondin-1 by macrophages. J Leukoc Biol 97:9–18
35. Feoktistov I, Ryzhov S, Goldstein AE, Biaggioni I (2003) Mast cell-mediated stimulation of angiogenesis. Cooperative interaction between A2B and A3 adenosine receptors. Circ Res 92: 485–492
36. Bonyanian Z, Walker M, Du Toit E, Rose'Meyer RB (2019) Multiple adenosine receptor subtypes stimulate wound healing in human EA.hy926 endothelial cells. Purinergic Signal 15:357–366
37. Chan ESL, Fernandez P, Merchant AA, Montesinos MC, Trzaska S, Desai A, Tung CF, Khoa DN, Pillinger MH, Reiss AB, Tomic-Canic M, Chen JF, Schwarzschild MA, Cronstein BN (2006) Adenosine A2A receptors in diffuse dermal fibrosis. Pathogenic role in human dermal fibroblasts and in a murine model of scleroderma. Arthritis Rheum 54:2632–2642
38. Shaikh G, Zhang J, Perez-Aso M, Mediero A, Cronstein B (2016) Adenosine A2A receptor promotes collagen type III synthesis via β-catenin activation in human dermal fibroblasts. Br J Phamacol 173:3279–3291
39. Victor-Vega C, Desai A, Montesinos MC, Cronstein BN (2002) Adenosine A2A receptor agonists promote more rapid wound healing than recombinant human platelet-derived growth factor (Becaplermin gel). Inflammation 26:19–24
40. Montesinos MC, Desai-Merchant A, Cronstein BN (2015) Promotion of wound healing by agonist of adenosine A2A receptor is dependent on tissue plasminogen activator. Inflammation 38:2036–2041
41. Jacobson KA, Tosh DK, Jain S, Gao ZG (2019) Historical and current adenosine receptor agonists in preclinical and clinical development. Front Cell Neurosci 13:124
42. Squadrito F, Bitto A, Altavilla D, Arcoraci V, De Caridi G, De Feo ME, Corrao S, Pallio G, Sterrantino C, Minutoli L, Saitta A, Vaccaro M, Cucinotta D (2014) The effect of PDRN, an adenosine receptor A2A agonist, on the healing of chronic diabetic foot ulcers: results of a clinical trial. J Clin Endocrinol Metab 99:E746–E753
43. Galeano M, Pallio G, Irrera N, Mannino F, Bitto A, Altavilla D, Vaccaro M, Squadrito G, Arcoraci V, Colonna MR, Lauro R, Squadrito F (2021) Polydeoxyribonucleotide: a promising biological platform to accelerate impaired skin wound healing. Pharmaceuticals 14:1103
44. Chan ESL, Montesinos MC, Fernandez P, Desai A, Delano DL, Yee H, Reiss AB, Pillinger MH, Chen JF, Schwarzschild MA, Friedman SL, Cronstein BN (2006) Adenosine A2A receptors play a role in the pathogenesis of hepatic cirrhosis. Br J Pharmacol 148:1144–1155
45. Fernandez P, Trzaska S, Wilder T, Chiriboga L, Blackburn MR, Cronstein BN, Chan ESL (2008) Pharmacological blockade of A2A receptors prevents dermal fibrosis in a model of elevated tissue adenosine. Am J Pathol 172:1675–1682

46. Karmouty-Quintana H, Molina JG, Philip K, Bellocchi C, Gudenkauf B, Wu M, Chen NY, Collum SD, Ko J, Agarwal SK, Assassi S, Zhong H, Blackburn MR, Weng T (2018) The antifibrotic effect of A2B adenosine receptor antagonism in a mouse model of dermal fibrosis. Arthritis Rheumatol 70:1673–1684
47. Perez-Aso M, Chiriboga L, Cronstein BN (2012) Pharmacological blockade of adenosine A2A receptors diminishes scarring. FASEB J 26:4254–4263
48. Perz-Aso M, Mediero A, Low YC, Levine J, Cronstein BN (2016) Adenosine A2A receptor plays an important role in radiation-induced dermal injury. FASEB J 30:457–465
49. Sun CX, Zhong H, Mohsenin A, Morschl E, Chunn JL, Molina BL, Zeng D, Blackburn MR (2006) Role of A2B adenosine receptor signaling in adenosine-dependent pulmonary inflammation and injury. J Clin Invest 116:2173–2182
50. Philip K, Weng Mills T, Davies J, Chen NY, Karmouty-Quintana H, Luo F, Molina JG, Amione-Guerra J, Sinha N, Guha A, Eltzschig HK, Blackburn MR (2017) HIF1A up-regulates the ADORA2B receptor on alternatively activated macrophages and contributes to pulmonary fibrosis. FASEB J 31:4745–4758
51. Wakeno M, Minamino T, Seguchi O, Okazaki H, Tsukamoto O, Okada K, HirataA FM, Asanuma H, Kim J, Komamura K, Takashima S, Mochizuki N, Kitakaze M (2006) Long-term stimulation of adenosine A2B receptors begun after myocardial infarction prevents cardiac remodeling in rats. Circulation 114:1923–1932
52. Kalk P, Eggert B, Relle K, Godes M, Heiden S, Sharkovska Y, Fischer Y, Ziegler D, Bielenberg GW, Hocher B (2007) The adenosine A1 receptor antagonist SVL320 reduces myocardial fibrosis in rats with 5/6 nephrectomy without affecting blood pressure. Br J Pharmacol 151:1025–1032
53. Hou X, Majik MS, Kim K, Pyee Y, Alexander V, Chung HJ, Lee HW, Chandra G, Lee JH, Park S, Choi WJ, Kim HO, Phan K, Gao ZG, Jacobson KA, Choi S, Lee SK, Jeong LS (2012) Structure-activity relationships of truncated C2- or C8-substituted adenosine derivatives as dual acting A2A and A3 adenosine receptor ligands. J Med Chem 55:342–356
54. Pak ES, Jeong LS, Hou X, Tripathi SK, Lee J, Ha H (2021) Dual actions of A2A and A3 adenosine receptor ligand prevents obstruction-induced kidney fibrosis in mice. Int J Mol Sci 22:5667
55. Roberts VS, Cowan PJ, Alexander SI, Robson SC, Dwyer KM (2014) The role of adenosine receptors A2A and A2B signaling in renal fibrosis. Kidney Int 86:685–692

Top Med Chem (2023) 41: 101–142
https://doi.org/10.1007/7355_2023_162

Published online: 19 August 2023

Adenosine A_{2A} Receptor Antagonists: Chemistry, SARs, and Therapeutic Potential

Andrea Spinaci, Michela Buccioni, Cui Chang, Diego Dal Ben, Beatrice Francucci, Catia Lambertucci, Rosaria Volpini, and Gabriella Marucci

Contents

Abstract Many physiological processes are modulated by the naturally occurred nucleoside adenosine after interaction with four G-protein-coupled receptor subtypes named A_1, A_{2A}, A_{2B}, and A_3. All adenosine receptors are widely distributed in the body. The A_{2A} subtype is present in the central nervous system in the dopamine-rich regions, in glutamatergic and GABAergic pathways intrinsic to the hippocampus, and in the cortex, where it is also co-expressed with dopamine D_2 and cannabinoid CB_1 receptors. The A_{2A} adenosine receptor distribution also regards numerous peripheral tissues such as blood vessels, endothelial, lymphoid, smooth muscle cells, and neurons. Since it is an emerging receptor implicated in many pathologies and its blockade could be beneficial in neuroinflammation, neurodegenerative disorders, and cancer, this review focuses the attention on the chemical structures of selective A_{2A} adenosine receptor antagonists, divided into different

A. Spinaci, M. Buccioni, C. Chang, D. Dal Ben, B. Francucci, C. Lambertucci, R. Volpini (✉), and G. Marucci
Medicinal Chemistry Unit, School of Pharmacy, University of Camerino, Camerino, MC, Italy
e-mail: rosaria.volpini@unicam.it

chemical classes. One section is dedicated to the representation of published crystallographic structures and molecular modeling studies. Furthermore, the use of A_{2A} adenosine receptor antagonists for the treatment of neurodegenerative disorders, as well as their potential as neuroprotective agents, will be discussed. Finally, due to the involvement of the A_{2A} adenosine receptor in activated immune cells, it will be taken into consideration the potential application of A_{2A} adenosine receptor antagonists in the anti-cancer therapy.

Keywords A_{2A} adenosine receptor antagonists, A_{2A} adenosine receptor crystal structure, Cancer, Neurodegeneration, Neuroinflammation, Non-xanthine derivatives, Xanthine derivatives

1 Introduction

Adenosine (Ado) is a natural nucleoside, which is widespread in mammalian organism tissues where it modulates a series of physiological processes through the interaction with four G-protein-coupled receptor subtypes named A_1, A_{2A}, A_{2B}, and A_3. Ado receptors (ARs) are distributed in all tissue and are often co-expressed in the same cell type [1]. In particular, the localization of A_{2A}ARs, by using pharmacological and immunohistochemical tools, was found, in the central nervous system (CNS), in the dopamine-rich regions, in glutamatergic and GABAergic pathways intrinsic to the hippocampus, and in the cortex [2]. In the CNS, A_{2A}ARs are co-expressed with dopamine D_2 and cannabinoid CB1 receptors [3, 4].The distribution of A_{2A}ARs is not limited to the medium-sized spiny neurons in the basal ganglia. In fact, PCR and northern blot analysis demonstrated that the A_{2A}AR gene is also expressed in numerous other tissues, namely blood vessels, endothelial cells, lymphoid cells, smooth muscle cells, and several neurons [5]. From its first cloning as an orphan receptor and deorphanization around the 1990s to today, at least 66 crystal and cryo-EM structures of A_{2A}ARs are known [6, 7], which allowed to depict the structural basis for the binding of agonists and antagonists. The therapeutic potential of this receptor ligands led to an intense medicinal chemistry efforts and many selective Ado receptor agonists and antagonists were discovered [8–10]. Commonly, the agonist structures are related to the natural ligand Ado while the antagonists belong to different chemical classes [11]. As the A_{2A}AR is an emerging receptor implicated in many pathologies and its blockade could be useful in neuroinflammation [12, 13], neurodegenerative disorders [14, 15], and cancer [16, 17], this review draws the attention on the chemical structures of selective A_{2A}AR antagonists. The high therapeutic potential underlying the modulation of the A_{2A}AR has prompted researchers to design and develop increasingly potent and selective ligands to treat various clinical disorders. The first A_{2A}AR antagonists resulted from the purines adenine and xanthine often combined with potentially reactive chemical moieties, such as the electron-rich ring furan and styrene group.

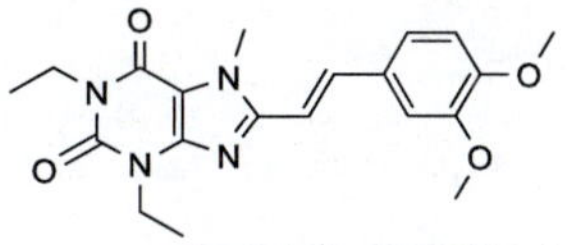
Istradefylline, KW-6002, 1
KiA_{2A}= 13 (r), 36 (h)

Fig. 1 Structure and binding affinity of istradefylline (Ki nM)

The most important example is undoubtedly istradefylline (**1**, also called KW-6002, Fig. 1), which contains a xanthine ring and a 3,4-dimethoxystyryl substituent in the 8-position [18, 19]. It has been approved in Japan (2013), and then in the USA (2019), as a co-therapy for the treatment of Parkinson's disease in association with levodopa/carbidopa to reduce off-time episodes and dyskinesias [20, 21]. Compounds with similar structure turned out to be poorly soluble in water and unstable, in a short time, due to photo-isomerization. In an attempt to get over these limitations, a large series of molecules possessing purine and non-purine heterocyclic scaffolds with considerable structural diversity were synthesized. Hence, the present review takes into consideration research on A_{2A}AR antagonists from the first molecules possessing an alkyl xanthine structure to the more potent and selective heterocyclic compounds unrelated to xanthines. Then, a section will be dedicated at depicting the published crystallographic structures of the A_{2A}ARs and their interaction with antagonists. Furthermore, the use of Ado A_{2A}AR antagonists for the treatment of neurodegenerative disorders, as well as their potential as neuroprotective agents, will be discussed. Finally, due to the involvement of the A_{2A}AR in activated immune cells, it will be taken into consideration the potential application of A_{2A}AR antagonists in the anti-cancer therapy.

2 A_{2A} Adenosine Receptor Antagonists Evaluated in Clinical Trials

Many A_{2A}AR antagonists have been investigated in clinical trials for a variety of neurological diseases and cancers. A number of clinical trials regarding the natural occurring xanthine caffeine, one of the first AR antagonists discovered, have been already completed and some are still ongoing. At present, 1261 studies are present in the USA clinical trial portal (access to clinicaltrials.gov 26.10.2022) involving caffeine for its SNC psychostimulant and exercise and cognition effects, in the symptomatic Alzheimer's disease even if its utility in some cases is debated. A comprehensive study report on caffeine has been recently published by Jacobson K.A. et al. [22]. Preladenant, also called SCH-420814 (**2**, Fig. 2) [23], reached the Phase III clinical trials for Parkinson disease (PD), but in contrast to istradefylline, it was discontinued due to lack of efficacy [24]. The A_{2A}AR antagonist tozadenant (**3**, SYN115, Fig. 2) [10] entered in Phase III of clinical evaluation for Parkinson's disease but was discontinued due to deaths from drug-induced agranulocytosis [21].

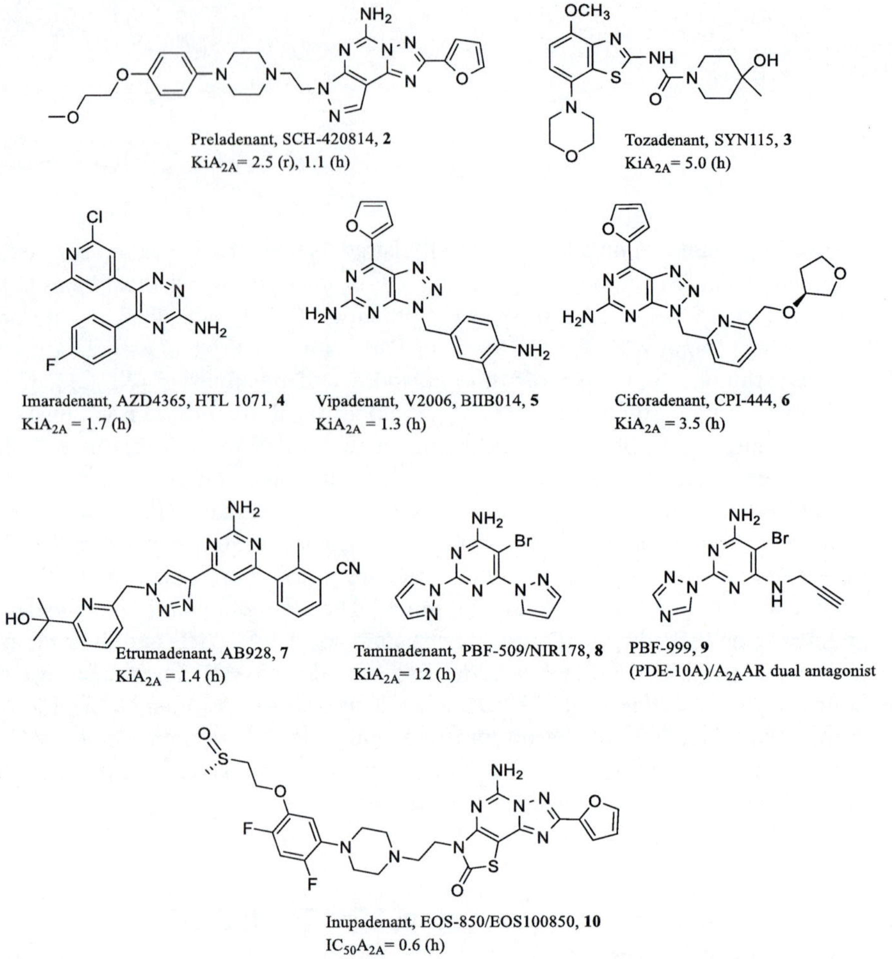

Fig. 2 A_{2A}AR antagonists evaluated in clinical trials. Binding affinity are in Ki nM and functional studies are in IC_{50} nM

The potent triazene A_{2A}AR antagonist, AZD4365, also called HTL 1071 or imaradenant, (**4**, Fig. 2) [25] is in early clinical phase trial for cancer and it is showing encouraging results in enhancing antitumor immunity. Recently it was studied the encapsulation of AZD4635, with other active principles, into nanoparticles for intelligent drug delivery and the achievement of an excellent synergistic effect in photothermal therapy, chemotherapy, and immunotherapy [26].

The triazolopyrimidine vipadenant (**5**, BIIB014/V2006, Fig. 2) [27] was found efficacious in two phase II studies as an antiparkinsonian drug in association with levodopa and as monotherapy [28]; however, it was discontinued in June 2010 for toxicity problems (www.vernalis.com/media-centre/latest-releases/2010-releases/584). Another triazolopyrimidine ciforadenant (**6**, CPI444, Fig. 2) [29] is under

Phase I/II trials in treatment-refractory renal cell cancer [30]. Etrumadenant (**7**, AB928, Fig. 2), an A_{2A}/A_{2B}AR mixed antagonist, is in clinical trials for cancer immunotherapy co-administered with doxorubicin. The strategy is based on the assumption that A_{2B}AR activation leads to immunosuppressive effects in cancer microenvironment and preliminary studies showed promising results [31]. Taminadenant (**8**, PBF509/NIR178, Fig. 2) is in Phase I for the treatment of advanced non-small cell lung cancer (NSCLC). It is well tolerated in patients, causing controllable adverse effects, and produces an appreciable reduction in tumor mass [32]. The dual A_{2A}AR antagonist/phosphodiesterase 10 (PDE-10) inhibitor PBF-999 (**9**, Fig. 2) has recently completed the phase I study for the treatment of solid tumors (ClinicalTrials.gov Identifier: NCT03786484). Other A_{2A}AR antagonists are in phase I clinical trial studies to fight cancer or PD. One of these is inupadenant (**10**, EOS-850/EOS100850, Fig. 2; ClinicalTrials.gov Identifier: NCT03873883) which shows good human tolerability and initial evidence of clinical efficacy over solid tumors in patients not responding to standard treatment options [33].

3 A_{2A} Adenosine Receptor Antagonists

In the last decades, several heterocyclic classes of compounds, which can be divided into xanthine and non-xanthine derivatives, have been synthesized as A_{2A}AR antagonists. As above mentioned, xanthines represent the prototypical group of antagonists, which have been modified to give a comprehensive collection of derivatives endowed with subtype selectivity. On the other hand, many other scaffolds, substituted in different positions and including tricyclic, bicyclic, and monocyclic derivatives, have been identified as selective A_{2A}AR antagonists.

3.1 Xanthine Derivatives

As above mentioned, the xanthine caffeine and theophylline (**11** and **12**, Fig. 3) were the first A_{2A}AR antagonists reported although with an affinity in the micromolar range and a low selectivity versus the other AR subtypes [9]. Theophylline is in therapy as anti-asthmatic drug while caffeine is used as an adjuvant in analgesic combinations. It was recently demonstrated that caffeine enhances the activity of non-steroidal analgesic drugs by pharmacodynamic interaction [34]. The 8-styrylxanthine analogue of caffeine, istradefylline (**1**, Fig. 1), was among the first A_{2A}AR antagonists reported and the most representative compound of the xanthine family. As previously reported, 8-styrylxanthines are photosensitive and in solution the styryl group isomerizes from trans to cis conformation with consequent loss of activity. The simultaneous substitution in the 1,3,7 positions of the 8-styryl xanthine nucleus has given rise to interesting compounds, among which the

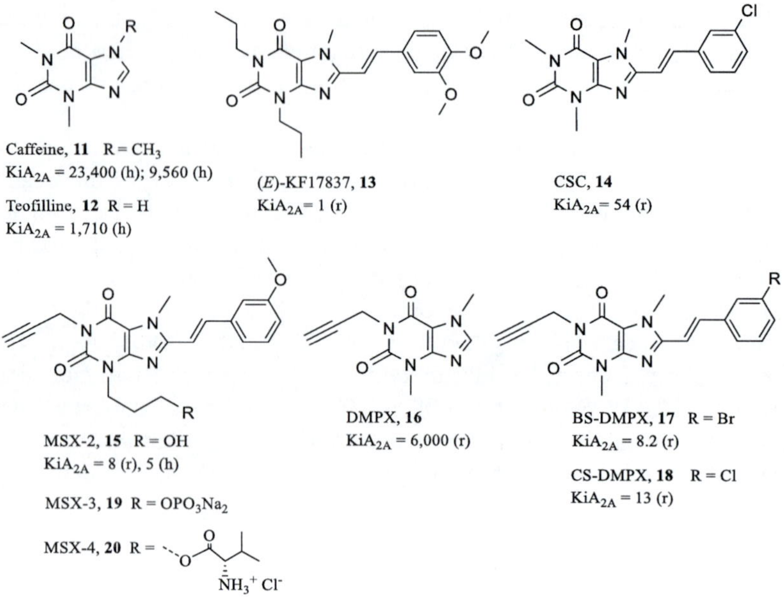

Fig. 3 Structures and binding affinity of xanthine derivatives (Ki nM)

(*E*)-8-(3,4-dimethoxystyryl)-1,3-dipropyl-7-methylxanthine (**13**, (*E*)-KF17837; $KiA_{2A} = 1$ nM at rat receptor; selectivity $A_1/A_{2A} = 62$, Fig. 3) resulted in one of the first A_{2A} antagonists showing high selectivity versus the A_1 subtype [35, 36]. Other A_{2A} receptor-selective 8-styrylxanthines extensively studied comprise CSC (**14**, 8-(3-chlorostyryl)caffeine; Ki = 54 nM on rat receptor) [37] and its congener MSX-2 (**15**, 3-(3-hydroxypropyl)-8-(*m*-methoxystyryl)-1-propargylxanthine) (Ki = 8 nM and 5 nM on rat brain striatal membranes and human recombinant A_{2A}AR, respectively). Interestingly CSC and its derivatives resulted to be inhibitors of monoamine oxidase-B (MAO-B) in addition to A_{2A}AR [38, 39]. The dual mechanism of action makes these drugs particularly suitable for the treatment of PD [2]. In fact, since MAO-B appears to be one of the main catabolic pathways of dopamine in the striatum, the block of this enzyme slows down the elimination of dopamine and increases its stay in the synapse. On the other hand, A_{2A}AR blockade causes an increase in dopamine affinity versus the D2 dopamine receptor and clinical studies revealed that A_{2A}AR antagonists improve motor dysfunctions of PD by reducing side effects such as dyskinesia. MSX2 belongs to the series of propargylxanthines, an extensively studied class of A_{2A}AR antagonists, among which the 3,7-dimethyl-1-propargylxanthine (**16**, DMPX) is considered the first synthetic A_{2A}AR antagonist endowed with xanthine core even if it showed low affinity and selectivity [40].

High selective derivatives of DMPX were developed by Muller et al. Among these, particular attention deserves the compounds 8-(m-bromostyryl)-3,7-dimethyl-1-propargylxanthine (**17**, BS-DMPX) and 8-(m-methoxystyryl)-3,7-dimethyl-1-

propargylxanthine (**18**, CS-DMPX) which, like the lead compound, have a poor water solubility. A useful approach to increase water solubility was the preparation of phosphate or amino acid prodrugs, in which the polar group/moiety is cleaved off in vivo to release the active compound [22, 41–43]. In particular, from MSX-2 derived two interesting compounds, the phosphate prodrug MSX-3, **19**, and the L-valine ester prodrug MSX-4, **20**. Binding studies confirmed that MSX-3 is readily cleaved by phosphatases to MSX-2, showing a Ki value equal to that obtained for its parent compound ($KiA_{2A} = 8$ nM, on rat brain striatal membranes). Both MSX-3 and MSX-4 are commonly used tools in research, particularly in in vivo studies [44–46].

3.2 *Non-Xanthine Derivatives*

3.2.1 Triazolo-Quinazoline and -Quinoxaline Derivatives

First examples of heterocycles derivatives different from xanthines were the triazoloquinazoline CGS15943 (**21**), which was discovered in a search for benzodiazepine receptor modulators [47, 48], and the triazoloquinoxaline CP-66,713 (**23**) [49] and their congeners **22** and **24** (Fig. 4). These tricyclic compounds were found to possess very high A_{2A}AR affinity at rat and human receptors with Ki or IC_{50} in the nM and sub nM range, although they resulted poorly selective.

About 10 years later, a series of triazoloquinoxaline-1-one derivatives were reported; among them, the 4-amino-6-(benzylamino)-2-phenyl-[1,2,4]triazolo[4,3-*a*]quinoxalin-1(2*H*)-one (**25**) was found to possess high A_{2A}AR affinity and selectivity. Removal of the N-benzyl group shifted the selectivity versus the A_1AR receptor [50]. A subsequent study on 1,2,4-triazolo[4,3-*a*]quinoxalin-1-one derivatives demonstrated that replacement of the benzylamino group phenyl ring with other heterocycles like furan, thiophene, or pyridine led to compounds which maintained high A_{2A}AR affinity although resulted less selective especially versus the A_1AR [51].

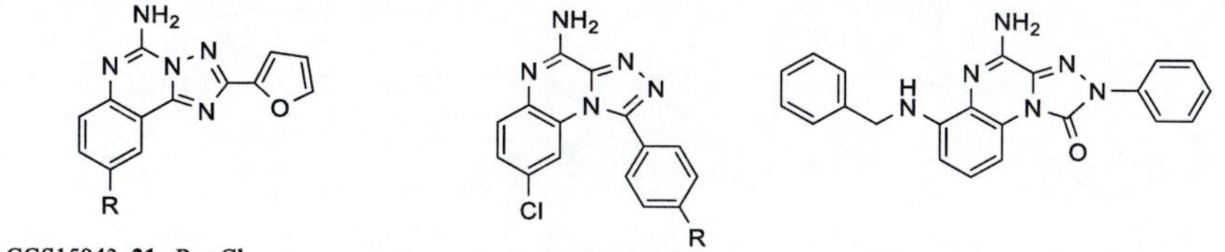

CGS15943, **21** R = Cl
$IC_{50}A_{2A}$ = 3.3 (r), KiA_{2A} = 0.43 (h)
22 R = H
IC_{50} = 27 (r)

CP-66713, **23** R = H
$IC_{50}A_{2A}$ = 21 (r)
24 R = Cl
$IC_{50}A_{2A}$ = 33 (r)

25
KiA_{2A} = 6.5 (h)

Fig. 4 Structures and binding affinity of triazoloquinazoline and triazoloquinoxalines (Ki or IC_{50} nM)

3.2.2 Pyrazolo-, Imidazolo-, and Thiazolo-Triazolopyrimidine Derivatives

Isosteric substitution of CGS15943 phenyl ring with a pyrazole nucleus led to A_{2A}AR antagonists which maintained high affinity but were rather unselective. An example is represented by 8FB-PTP (**26**; Fig. 5), which opens the way to the search for new A_{2A}AR antagonists with pyrazolotriazolopyrimidine scaffold [52]. Then, a number of potent and selective compounds were synthesized allowing to depict structure–activity relationships (SAR), which revealed the importance of the free amino group at the 5-position and the furane ring in the 2-position for the affinity and selectivity. Furthermore, substitution of the pyrazole ring in 7-position was more favorable for the A_{2A} selectivity than that in 8-position as well as the presence of the pyrazole ring with respect to the triazole moiety [10, 53]. In particular, the presence at the 7-position of the pyrazole nucleus of a phenylalkyl chain improved the A_{2A} affinity and selectivity, especially when an amino, hydroxy, and a methoxy group substituted the phenyl para position. These compounds **27-31** were found among the most potent and selective A_{2A}AR antagonists with affinity in the nM and sub-nM range (Fig. 5). In particular, the 2-(furan-2-yl)-7-(3-(4-methoxyphenyl)propyl)-7*H*-pyrazolo[4,3-*e*][1,2,4]triazolo[1,5-*c*]pyrimidin-5-amine (**30**, SCH-442416, $KiA_{2A} = 0.048$ nM) results endowed with the highest affinity and a selectivity of about 23,000-fold versus the A_1 subtype and greater than 200,000-fold versus the other human AR subtypes. Later report did not confirm this extremely high affinity, which however remained in the low nM range [54]. Unfortunately, attempts to increase the water solubility of these molecules by replacement of the phenyl para substituent with a carboxylic or a more polar sulfonic group decreased the A_{2A} binding affinity [55]. Subsequent modifications of these compounds led to the discovery of a series of pyrazolotriazolopyrimidines bearing at the 7-position of the pyrazole ring various arylpiperazine groups. Among them, the orally active

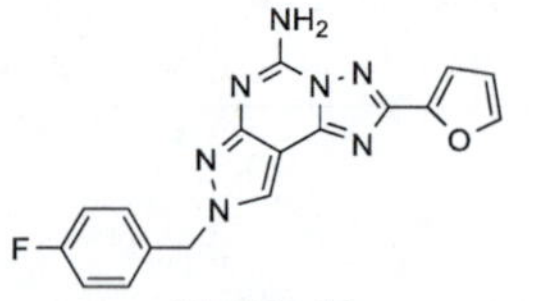

8FB-PTP, **26**
$KiA_{2A} = 1.2$ (r)

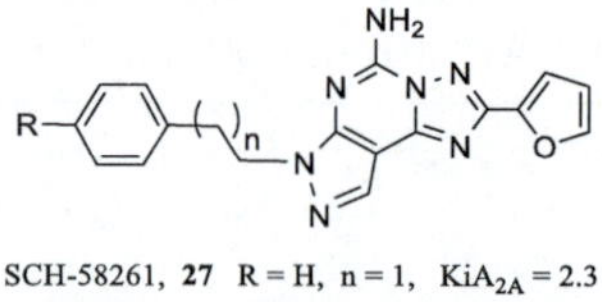

SCH-58261, **27** R = H, n = 1, $KiA_{2A} = 2.3$ (r)
SCH-63390, **28** R = H, n = 2, $KiA_{2A} = 2.4$ (r)
29 R = OH, n = 2, $KiA_{2A} = 0.94$ (r)
SCH-442416, **30** R = OCH_3, n = 2, $KiA_{2A} = 0.048$ (h)
31 R = NH_2, n = 2, $KiA_{2A} = 0.22$ (h)

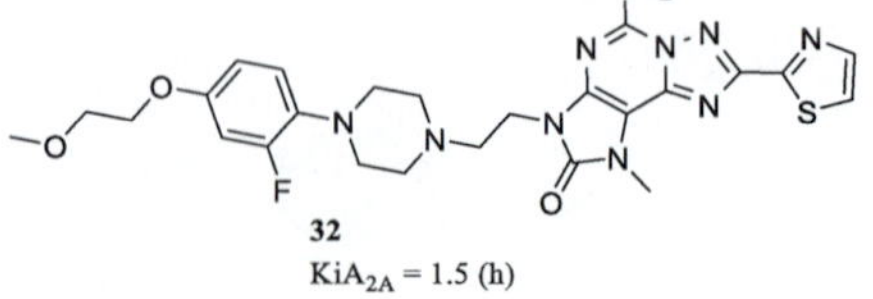

32
$KiA_{2A} = 1.5$ (h)

Fig. 5 Structures and binding affinity of pyrazolo- and imidazolo-triazolopyrimidines (Ki, nM)

Fig. 6 Structure and binding affinity of thiazolotriazolopyrimidines (Ki nM)

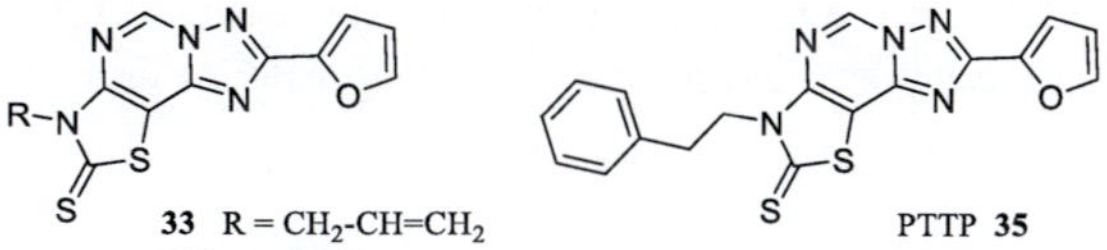

33 R = CH_2-CH=CH_2
KiA_{2A} = 0.016

34 R = 3-I-Ph
KiA_{2A} = 0.0082

PTTP **35**
KiA_{2A} = 6.3

methoxyethoxy ether derivative, preladenant (**2**, SCH-420814, Fig. 2), showed high affinity at both rat and human A$_{2A}$AR with Ki of 2.5 nM and 1.1 nM, respectively, and very good selectivity versus the other AR subtypes and a panel of 59 receptors, enzymes, and ion channels together with favorable pharmacokinetic properties in rats. These characteristics make it a widely studied A$_{2A}$AR antagonists in in vivo animal studies and clinical trials [23, 56].

Recently, Iteos Therapeutics patented a series of thiazolotriazolopyrimidines, bearing a carbonyl group in the thiazole ring, which have been designed as new A$_{2A}$AR antagonists with low CNS penetrance and potentially useful in the treatment of peripheral cancers. Among them, inupadenant (**10**, Fig. 2) showed very high activity at A$_{2A}$AR with an IC$_{50}$ of 0.6 nM in the inhibition of cAMP production [57]. A novel series of [1,2,4]triazolo[5,1-*f*]purin-2-one derivatives by Advinus Therapeutics Ltd. were reported to display functional antagonism at the A$_{2A}$AR with good selectivity over other AR subtypes. Among reported compounds, **32** (Fig. 5) was the most active with a Ki of 1.5 nM at human A$_{2A}$ARs measured in binding experiments. This compound showed good in vitro and in vivo pharmacokinetics properties and demonstrated good effectiveness in two models of Parkinson's disease (haloperidol-induced catalepsy and 6-OHDA lesioned rat models) and depression (tail suspension and forced swim tests mice models) [58]. The design, synthesis, and in silico molecular docking of new thiazolotriazolopyrimidine derivatives as potent and selective A$_{2A}$AR antagonists was undertaken by Mishra et al. [59]. In silico molecular docking studies revealed that all compounds exhibited strong interaction with A$_{2A}$AR, which was confirmed by binding studies at human A$_{2A}$ARs stably expressed in HEK293 cells [59]. In fact, the thiazolotriazolopyrimidine derivatives **33** and **34** (KiA_{2A} = 0.016 nM and 0.0082 nM, respectively, Fig. 6) demonstrated very high binding affinity and selectivity for the A$_{2A}$AR. **33** was able to counteract the haloperidol-induced hypo-locomotor behavior (catalepsy and akinesia) in PD animal model producing motor stimulant effects.

The synthesis and biological evaluation of 8-(furan-2-yl)-3-phenethylthiazolo [5,4-*e*] [1,2,4] triazolo[1,5*c*]pyrimidine-2(3*H*)-thione (**35**, PTTP, Fig. 6) were performed by Kumari et al. [60]. This compound showed a good affinity for A$_{2A}$AR (KiA_{2A} = 6.3 nM) and selectivity versus the A$_1$AR (A$_1$/A$_{2A}$ = 4,603). Docking analysis revealed that PTTP and SCH-58261 shared almost similar binding motif inside the transmembrane regions and extracellular loops (ECL2) at the human A$_{2A}$AR co-crystallized with ZM241385. cAMP functional study, in HEK293T, cells demonstrated that this compound is an antagonist with a potency better than Scheme

58261. In addition, in PD animal model, the pre-treatment with **35** attenuated catalepsy and akinesia without significant neurotoxicity in rotarod test.

3.2.3 Triazolo-Triazine, -Pyrimidine, and -Pyrazine Derivatives

The tricyclic scaffold of the triazoloquinazolines was simplified by researchers of the Zeneca group, which synthesized a series of bicyclic compounds including the triazolotriazine 4-(2-((7-amino-2-(furan-2-yl)-[1,2,4]triazolo[1,5-*a*][1,3,5]triazin-5-yl)amino)ethyl)phenol (**36**, ZM 241385, Fig. 7), which showed high A_{2A}AR affinity and good selectivity ($KiA_{2A} = 1.78$ nM and 16.6 nM at rat and human receptor, respectively) [61, 62]. Indeed, later studies demonstrated ZM241385 to be an inverse agonist at the A_{2A}AR [63].

This compound, which resulted in one of the most potent A_{2A}AR ligands reported, showed a good activity also at the A_{2B}AR subtype and has rapidly become the reference compound for the study of the A_{2A}AR subtype, also in the tritiated form [64]. A number of modifications of the ZM241385 *p*-hydroxyphenylethylamino chain were reported but none of the resulted compounds showed higher affinity [65].

Substitution of the furyl ring with an ethoxy group decreased the A_{2A}AR affinity of about 100 times (**37**, $KiA_{2A} = 178$ nM), while removal of the 2-amino chain was detrimental for the A_{2A}AR affinity (**38**, $KiA_{2A} = 2{,}029$ nM) [66]. In 2009, researchers of Vernalis (R&D) Ltd. reported the synthesis of a series of functionalized triazolo[4,5-*d*]pyrimidine derivatives as A_{2A}AR antagonists. Among them, vipadenant (**5**, BIIB014/V2006, Fig. 2) showed to possess high A_{2A}AR affinity ($Ki = 1.3$ nM) and good selectivity versus the other AR subtypes, especially A_3ARs [27]. Furthermore, this compound demonstrated very good

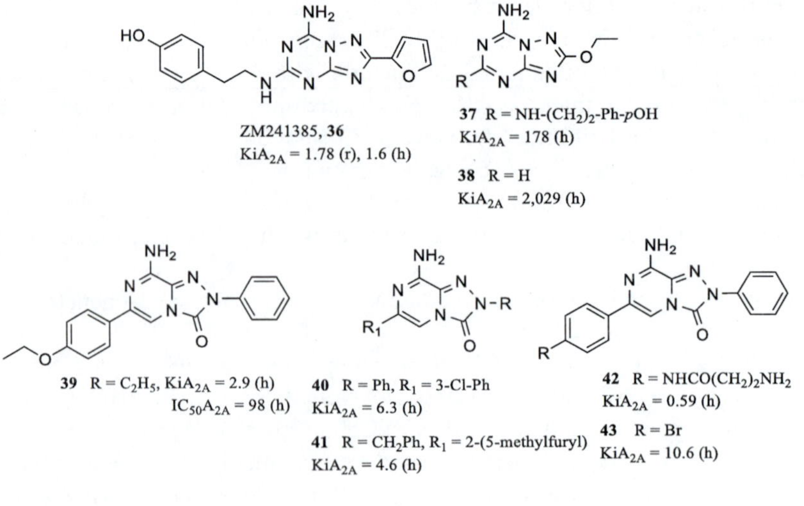

ZM241385, **36** $KiA_{2A} = 1.78$ (r), 1.6 (h)

37 R = NH-$(CH_2)_2$-Ph-*p*OH $KiA_{2A} = 178$ (h)

38 R = H $KiA_{2A} = 2{,}029$ (h)

39 R = C_2H_5, $KiA_{2A} = 2.9$ (h) $IC_{50}A_{2A} = 98$ (h)

40 R = Ph, R_1 = 3-Cl-Ph $KiA_{2A} = 6.3$ (h)

41 R = CH_2Ph, R_1 = 2-(5-methylfuryl) $KiA_{2A} = 4.6$ (h)

42 R = NHCO$(CH_2)_2NH_2$ $KiA_{2A} = 0.59$ (h)

43 R = Br $KiA_{2A} = 10.6$ (h)

Fig. 7 Structures of triazolotriazines and triazolopyrimidines. Binding affinity are in Ki nM preclinical pharmacokinetic properties and was found orally active and very potent in animal models of Parkinson's disease [67]. These characteristics allowed clinical evaluation of the compound in collaboration with Biogen Idec but, as already mentioned, despite positive clinical results, it was discontinued for toxicity problems. Modification of the vipadenant benzylic chain led to the analogue ciforadenant (**6**, CPI-444, Fig. 2), bearing at the 3-position of the triazolopyrimidine nucleus a substituted pyridine [68]. This compound showed high A_{2A}AR affinity (Ki = 3.5 nM) and a better selectivity versus the A_{2B} subtype (more than 400 times) with respect to vipadenant and was able, through the A_{2A}AR blockade, to neutralize the Ado signaling and to enable antitumor immunity in preclinical models of cancers [29]. Recently, a number of 8-amino-6-aryl-1,2,4-triazolo[4,3-*a*]pyrazin-3-ones derivatives were reported to possess very high affinity for the human A_{2A}AR combined, in some cases, with an extremely high selectivity. Compound **39** (Fig. 7) being the most relevant of the first series with a Ki = 2.9 nM. It was also tested to evaluate its ability to inhibit cAMP production with an IC_{50} of 98 nM [69]. Successively, the same research group reported the synthesis of other derivatives of the same scaffold and demonstrated that, on the basis of substituents present in 2- and 6-positions, it was possible to obtain potent and dual A_1/A_{2A}AR ligands with Ki values in the low nM range on both subtypes (**40** and **41**, Fig. 7). In a neuroprotection assay, the compounds were evaluated for their ability in counteracting β-amyloid peptide (Aβ)-induced toxicity on SH-SY5Y cells. Both of them significantly prevented toxicity induced by Aβ peptide or restored viability to control level highlighting their potential as new neuroprotective agents in Alzheimer's disease (AD) [70]. The introduction of a functionalized alkyl-amino chain at the phenyl in 6-position furnished the 3-amino-*N*-(4-(8-amino-3-oxo-2-phenyl-2,3-dihydro-[1,2,4]triazolo[4,3-*a*]pyrazin-6-yl)phenyl)propenamide (**42**), which exhibited a Ki = 0.59 nM and a selectivity more than 800-fold versus all the other ARs [71]. Its bromo derivative **43** (KiA_{2A} = 10.6 nM, Fig. 7) was very efficacious as neuroprotective agent in the in vitro model of AD previously described [70].

3.2.4 Thiazolo- and Thieno-Pyrimidine Derivatives

A series of N^5-substituted-2-(2-furanyl)thiazolo[5,4-*d*]pyrimidine-5,7-diamine derivatives was synthesized and pharmacologically characterized by Varano et al. [72] The introduction of a methoxy substituent at the 2- or 3-position of the phenyl ring of N^5-benzyl-2-(2-furanyl)[1,3]thiazolo[5,4-*d*]pyridine-5,7-diamine (**44,** KiA_{2A} = 22 nM, KiA_1 = 43 nM, and KiA_3 = 37 nM; Fig. 8), which is an unselective compound with nM affinity, led to derivatives 2-(2-furanyl)-N^5-(2-methoxybenzyl)[1,3]thiazolo[5,4-*d*]pyrimidine-5,7-diamine (**45**, Fig. 8) and 2-(2-furanyl)-N^5-(3-methoxybenzyl)[1,3]thiazolo[5,4-*d*]pyrimidine-5,7-diamine (**46**, Fig. 8), respectively.

These two compounds showed a peculiar behavior with two different affinity values for the A_{2A}AR; a high affinity Ki value in the femtomolar range (KiH = 3.55

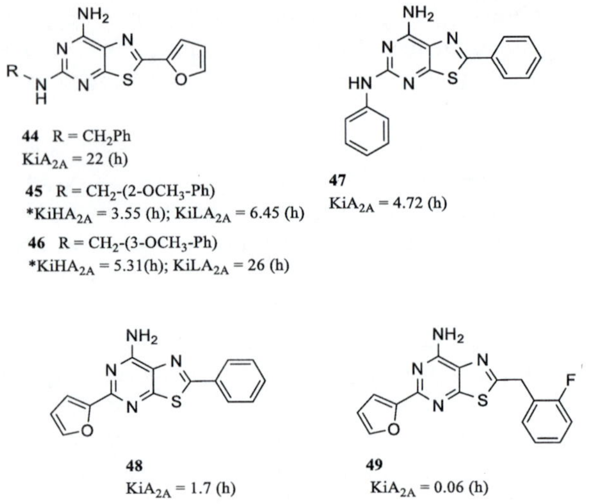

Fig. 8 Structure and binding affinity of thiazolopyrimidines (Ki nM; KiH fM). * $KiHA_{2A}$ and $KiLA_{2A}$ are referred to high affinity and low affinity calculated in a biphasic response curve obtained in binding studies

and 5.31 fM, respectively) and a low affinity Ki value in the nanomolar order (KiL = 6.45 and 26 nM, respectively). Moreover, **45** and **46** were able to inhibit cAMP in the basal condition behaving as inverse agonists (IC_{50} = 1.9 and 8.3 pM, respectively). The same authors synthesized similar analogues bearing different substituents at 2- and 5- position of the thiazolopyrimidine core [73]. Generally, these compounds showed high affinity toward the A_1AR and A_{2A}AR behaving as dual A_1AR/A_{2A}AR antagonists/inverse agonists. The N^5-(2-methoxybenzyl)-2-phenyl-thiazolo[5,4-*d*]pyrimidine-5,7-diamine (**47**, Fig. 8) presented the highest affinity (KiA_1 = 10.2 nM; KiA_{2A} = 4.72 nM) and behaved as a potent A_1/A_{2A} antagonist/inverse agonist ($IC_{50}A_1$ = 13.4 nM; $IC_{50}A_{2A}$ = 5.34 nM). Then, other 7-amino-thiazolo[5,4-*d*]pyrimidines, bearing different substituents at 2-, 5-, and 7-position of the thiazolopyrimidine scaffold, were designed, synthesized, and biologically evaluated [74]. Results showed that affinity and selectivity of the 7-amino-thiazolo [5,4-*d*]pyrimidine derivatives toward the AR subtypes can be modulated by the nature of the groups linked at the different positions of the bicyclic scaffold. The 5-(furan-2-yl)-2-phenylthiazolo[5,4-*d*]pyrimidin-7-amine (**48**, KiA_{2A} = 1.7 nM Fig. 8) showed an nM affinity. Furthermore, also sub-nM binding affinities were found in compounds with a 7-amino-2-arylmethyl-thiazolo[5,4-*d*]pyrimidine structure [75]. The 2-(2-fluorobenzyl)-5-(furan-2yl)-thiazolo[5,4-*d*]pyrimidin-7-amine (**49**, K_iA_1 = 1.9 nM; K_iA_{2A} = 0.06 nM; Fig. 8) was evaluated for its antidepressant-like activity in in vivo studies in mice, showing an effect comparable to that of the reference compound amitriptyline. The synthesis of molecules bearing thieno[2,3-*d*]pyrimidine scaffold led to compounds with weak affinity toward A_1ARs, but good binding affinity at A_{2A} and A_3ARs [76]. Compound showing the best A_{2A}AR affinity was the 5,6-dimethyl-2-(benzamide)thieno[2,3-*d*]pyrimidin-4 (3*H*)-one (**50**, KiA_1 > 30 μM, KiA_{2A} = 650 nM, and KiA_3 = 124 nM; Fig. 9).

Compounds bearing the thieno[2,3-*d*]pyrimidine scaffold, substituted at the 6-position with various aryl and heteroaryl moieties, were also reported by Shook

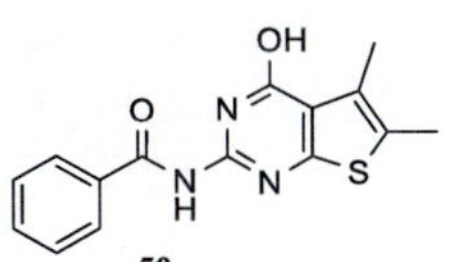

50
$KiA_{2A} = 650$

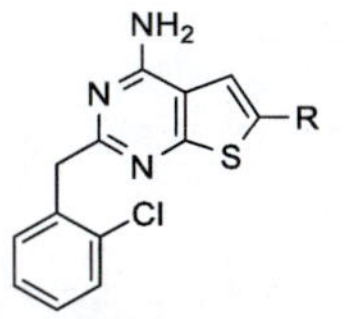

51 R= methylfuran $KiA_{2A} = 2.9$
52 R= 3-cyanophenyl $KiA_{2A} = 1.4$

Fig. 9 Structures and binding affinity of thieno-pyrimidines (Ki nM)

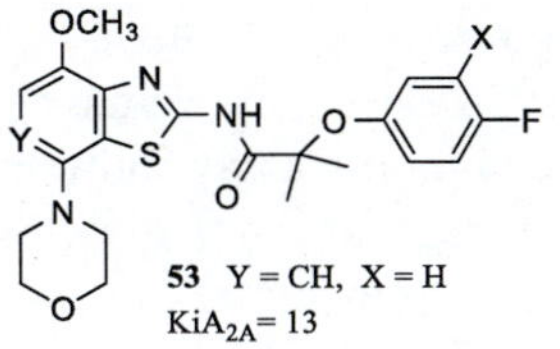

53 Y = CH, X = H
KiA_{2A}= 13
54 Y = CH, X = F
KiA_{2A}= 6.4
55 Y = N, X = H
KiA_{2A}= 3.4

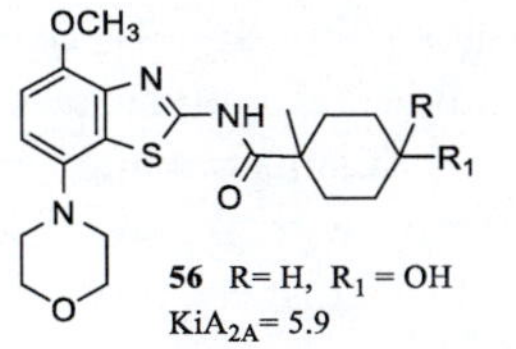

56 R= H, R_1 = OH
KiA_{2A}= 5.9
57 R= CH_3, R_1 = OH
KiA_{2A}= 8.6
58 R= H, $R_1 = N(C_2H_5)_2$
KiA_{2A}= 2.1

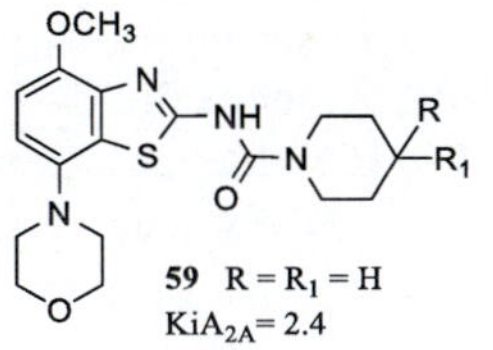

59 R = R_1 = H
KiA_{2A}= 2.4
60 R = H, R_1 = F
KiA_{2A}= 4.9
61 R = H, R_1 = CH_2-F
KiA_{2A}= 3.6
62 R = OH, R_1 = CH_2-F
KiA_{2A}= 2.8

Fig. 10 Structure and binding affinity of benzothiazole and pyrimidothiazole derivatives (Ki nM)

et al. and proved to possess enhanced A_{2A}AR affinity [77]. These compounds were potent A_{2A} antagonists having good in vivo efficacy in the mouse catalepsy model. Unfortunately, they showed a short half-life, limiting their use. Among these compounds are the chlorinebenzylthieno[2,3-*d*]pyrimidine derivatives, which showed good in vitro potency when a methylfuran (**51**, Fig. 9) and a 3-cyanophenyl (**52**, Fig. 9) substituted the scaffold 6-position.

3.2.5 Benzo-, Pyrimido-, and Oxazolo-Thiazolo Derivatives

The best of this class of molecules is tozadenant (**3**, SYN-115, Fig. 2), having a benzothiazole- based scaffold, which was identified by Hoffmann-La Roche [78]. It is a potent and selective A_{2A}AR antagonist with good ADME properties; nevertheless, as reported above, even if entered in Phase III of clinical evaluation for PD, it was discontinued due to deaths from drug-induced agranulocytosis. On this basis, Basu et al. reported the design, synthesis, and biological evaluation of novel series of 2-substituted benzothiazole and thiazolo[5,4-c]pyridine derivatives aimed to discover novel, potent, and selective A_{2A}AR antagonists with desirable physicochemical, pharmacokinetic, and pharmacological properties [79]. The results of the binding assay allowed to select three compounds as potent and selective A_{2A}AR antagonists (**53-55**, Fig. 10).

These three compounds were also assessed for their stability in liver microsomes. They did not show any CYP liability against major CYPs (<5% inhibition at 3 and 10 μM for compound **53** and **54**; <10% inhibition at 10 and 30 μM for compound

Fig. 11 Structures and binding affinity of benzoxazoles (Ki nM)

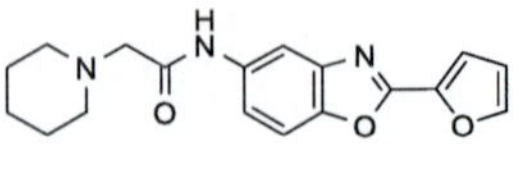

63
KiA_{2A}= 1,000 (h)

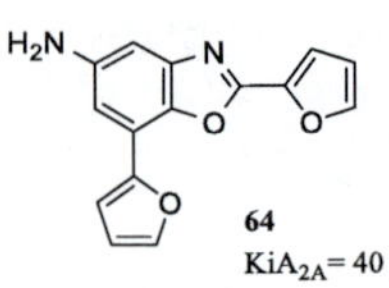

64
KiA_{2A}= 40 (h)

55). Among them, compound **55** showed better pharmacokinetic profile compared to **53** and **54**. Unfortunately, all three compounds showed poor oral bioavailability, which could be due to low water solubility. To overcome these problems, a new series of cyclohexane-carboxamide derivatives was synthesized by replacing the phenyl ring with cyclohexane. The results obtained from binding assay showed that derivatives **56-58** (Fig. 10) interact with A_{2A}ARs with high affinity, showing Ki values of 5.9, 8.6, and 2.1 nM, respectively, with a functional potency in sub-nanomolar range (Ki = 1.1, 0.25, and 0.20 nM, respectively). They were soluble in aqueous medium at all pH and stable to action of liver microsomal enzymes showing good bioavailability. In particular, **58** displayed a much longer half-life than compound **56**. Molecular docking at the active site of the A_{2A}AR was performed to rationalize biological results. In addition, compound **58** was evaluated in in vivo PD model and was found to be able to potentiate the levodopa-induced contralateral rotations. The design and synthesis of a series of tozadenant analogues to obtain potent A_{2A}AR ligands were undertaken also by Renk and coworkers [80]. The piperidine ring functionalization of the A_{2A}AR antagonist tozadenant (**3**, Fig. 2) provided derivatives with high affinity at the A_{2A}AR and good selectivity over the A_1AR. Among them the analogue **59** (Fig. 10) showed the highest potency (Ki = 2.4 nM). Three compounds (**60-62**, Fig. 10), containing a fluorine atom, showed good affinity and selectivity at A_{2A}AR and were chosen for radiolabeling with fluorine-18 and further evaluation in in vitro autoradiography. Since the in vitro studies were performed in rat, the affinity at rat A_{2A}ARs was evaluated using homogenates from rat brain corpora striata, which revealed Ki values roughly 4–5 times higher than the values obtained for the human receptor. Their binding at A_{2A}AR was further confirmed by radiolabeling with ^{18}F and in vitro autoradiography in rat brain slices, which showed almost exclusive striatal binding and complete displacement by the A_{2A}AR antagonist ZM 241385. In conclusion, these compounds represent potential candidates for the visualization of the A_{2A}AR and open the way to novel therapeutic treatments of neurodegenerative disorders or cancer. Duroux et al. reported a series of 2-furylbenzoxazoles as potential A_{2A}AR antagonists. Two "hits" molecules were identified with interesting pharmacokinetic properties, but were found to bind the A_{2A}AR with Ki values in the microM range (**63**, Fig. 11) [81].

In order to improve the A_{2A}AR affinity, the benzoxazole scaffold was further investigated [82]. Modulation on the C-7 position of the benzoxazole ring by adding a furane significantly improved binding affinity toward A_{2A}ARs. Compound **64** displayed the highest binding affinity for A_{2A}AR (Ki = 40 nM) with high antagonist activity (IC_{50} = 70.6 nM;). Moreover, some of these ligands showed interesting drug metabolism and pharmacokinetic properties, especially high-water solubility. In conclusion, the benzoxazole ring can be considered a highly effective scaffold for the design of new A_{2A}AR antagonists.

3.2.6 Purine and Adenine Derivatives

The first purine derivatives identified as A_{2A}AR antagonists were adenine derivatives (6-aminopurines) designed on the basis that modification or removal of the ribose moiety of the natural agonist Ado usually decreases or abolishes activity at ARs [83, 84]. In 1998, Cristalli et al. published the synthesis of a series of 9-ethyladenines and among them the 8-bromo-9-ethyladenine (**65**; Fig. 12) showed a good affinity at hA_{2A}AR with Ki = 52 nM, and a moderate selectivity versus the other AR subtypes [85]. In the search for new ligands endowed with better affinity, a series of adenine derivatives modified at 2-, 8-, and/or 9-positions were later synthesized [86–89]. Among them compounds ANR 94 and ANR 152 (**66** and **67**, Fig. 12) showed the better binding profile at the hA_{2A}AR and then were evaluated in animal models of PD. The 8-ethoxy-9-ethyladenine (**66**) showed the highest activity in ameliorating motor deficits and anti-inflammatory and neuroprotective properties [90, 91]. In the effort to obtain compounds useful as tools for neurodegenerative disease and taking into account the structure–activity relationships data, adenine analogues of ZM241385 (**36**, Fig. 7) have been designed and the 4-(2-((9-ethyl-8-(furan-2-yl)-adenyl)amino)ethyl)phenol (**68**, KiA_{2A} = 1.8 nM, Fig. 12) was found to be a high affinity A_{2A}AR antagonist with low nM Ki. Biological studies on the BV-2 microglial cell line demonstrated that it was able to reduce proinflammatory mediators and increase anti-inflammatory cytokines demonstrating its useful potential in neurodegenerative diseases characterized by strong neuroinflammation such as PD

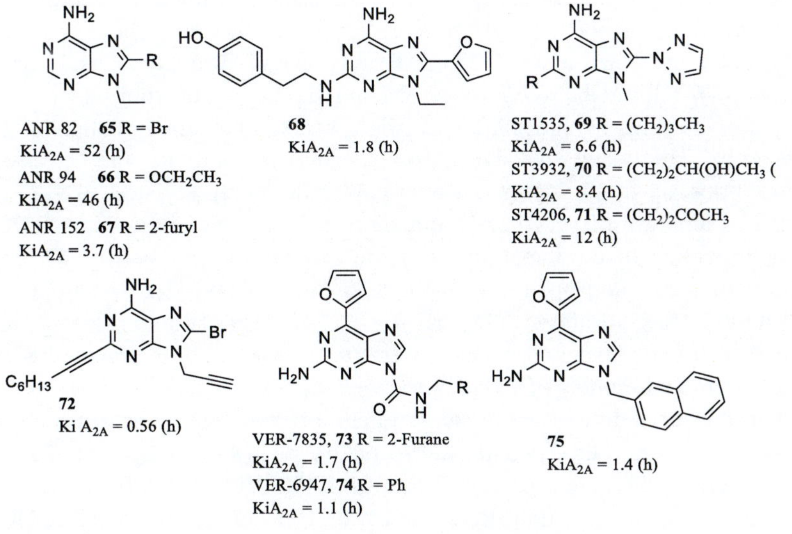

Fig. 12 Structure and binding affinity of purines and adenines (Ki nM)

[66]. Other adenine derivatives reported were characterized by an 8-triazolyl moiety. The prototypical A_{2A} antagonist of this series was ST1535 (**69**, Ki = 6.6 nM, Fig. 12) which showed a low nM affinity at the A_{2A} and low selectivity versus the A_1AR subtype [92]. ST-1535 has been tested in different in vivo models of PD, where it showed to potentiate the effects of a threshold dose of LEVODOPA. The subsequent characterization of the metabolites produced in vivo highlighted the formation, among others, of compounds **70** and **71** (Fig. 12), which showed good affinity on the A_{2A} receptor (**70**, KiA_{2A} = 8.4 nM, **71**, KiA_{2A} = 12 nM) in binding assays. These compounds were also endowed with long-lasting activity at the CNS, contributing to the in vivo long duration of action of ST-1535 [93]. Endo et al. reported the synthesis of 9-propargyladenine derivatives, which displayed good activities. The 8-bromo derivative **72** (Fig. 12), with a sub-nM Ki at A_{2A}AR, showed a comparable activity of istradefylline in an in vivo model of PD [94].

In an attempt to explain the side effects associated with the use of the antimalarial mefloquine, it has been discovered the ability of this drug to inhibit the A_{2A}AR. This finding paved the way for the design of a new series of compounds with non-xanthine structure and led to the development of new A_{2A}AR antagonists endowed with unique structure. Among these compounds, there are the isomers of adenine 2-aminopurines derivatives (VER-7835, **73,** VER-6947, **74**, and **75**, Fig. 12), which showed low nM affinities and moderate selectivity with respect to the other AR subtypes [19, 95]. In an in vivo assay, VER-7835 and VER-6947 potently reverted locomotor deficit induced by haloperidol with doses comparable with istradefylline, demonstrating their potential as anti-PD agents [19].

3.2.7 Thiazole and Triazole Derivatives

An high-throughput screening of extensive combinatorial libraries identified a series of 2-amino-5-benzoyl-4-(2-furyl)thiazole derivatives based on the 2-aminothiazole core, which represented a novel chemical class of A_{2A}AR antagonists. In particular, compound **76** exhibited a KiA_{2A} = 12 nM and more than 500-fold selectivity versus the A_1 subtype (Fig. 13) [96]. After a while, a screening campaign of about 200,000 compounds, conducted by Denmark researchers, led to the discovery of carboxamides of ethyl 2-amino-4-phenylthiazole-5-carboxylates like **77**, which maintained high affinity at A_{2A}AR, but showed lower selectivity than **76**. A hit-to-lead optimization of the metabolic unstable carboxylate function of **77** led to the synthesis of the cyclopropanecarboxylic acid [5-(5-methyl-[1,2,4]oxa-diazol-3-yl)-4-phenyl-thiazol-2-yl]-amide (**78**, KiA_{2A} = 13 nM), demonstrating that the heteroaromatic replacement of the ester group was well tolerated. **78** showed efficacy in animal models of PD [97].

Other thiazolyl derivatives were synthesized from the same researchers and, among them, were the potent and selective A_{2A}AR antagonist **79** and its phosphonooxymethylene prodrug **80** (Lu AA47070, Fig. 13), which was endowed with a very high increased aqueous solubility and was fully converted into **79** in vivo [98]. In a rodent model of PD, Lu AA47070 was able to counteract the effects of

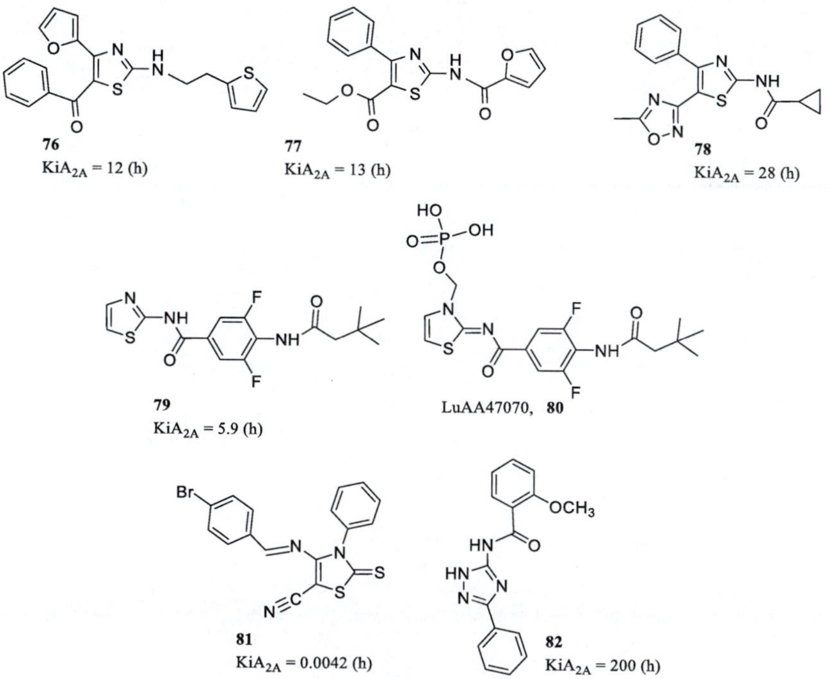

Fig. 13 Structures and binding affinity of thiazoles and triazoles (Ki nM)

pimozide and haloperidol, two D2 dopamine receptor antagonists, suggesting that this compound could have beneficial effects for PD and some depression motivational symptoms [99]. In 2013, a series of 2-thioxothiazole derivatives was reported as high affinity A_{2A}AR antagonists, some of them showing sub-nM affinity and high selectivity versus the A_1AR subtype. Among them, the (*E*)-4-((4-bromobenzylidene)amino)-3-phenyl-2-thioxo-2,3-dihydrothiazole-5-carbonitrile (**81**, Fig. 13) showed a Ki = 0.0042 nM and a selectivity A_1/A_{2A} more than 40,000-fold and was found efficacious in an in vivo PD model [100]. Other five-membered heterocycle derivatives, belonging to the 1,2,4-triazole chemical class, were described by Carlsson et al. as A_{2A}AR antagonists. Among them, **82** was found to possess good A_{2A}AR affinity but was rather unselective [101].

3.2.8 Pyridine and Pyrimidine Derivatives

On the basis of a pharmacophore model, a new series of 2-amino-3-cyanopyridine derivatives was synthesized and reported as A_{2A}AR antagonists. Different substitution at the pyridine core revealed that the presence of a five-membered heterocycle bearing an H-bond acceptor favored the interaction with the A_{2A}AR. Examples of such derivatives are compounds **83** and **84** (Fig. 14), which exhibited low nM A_{2A}AR affinity, although resulted slightly selective especially versus the A_1 subtype

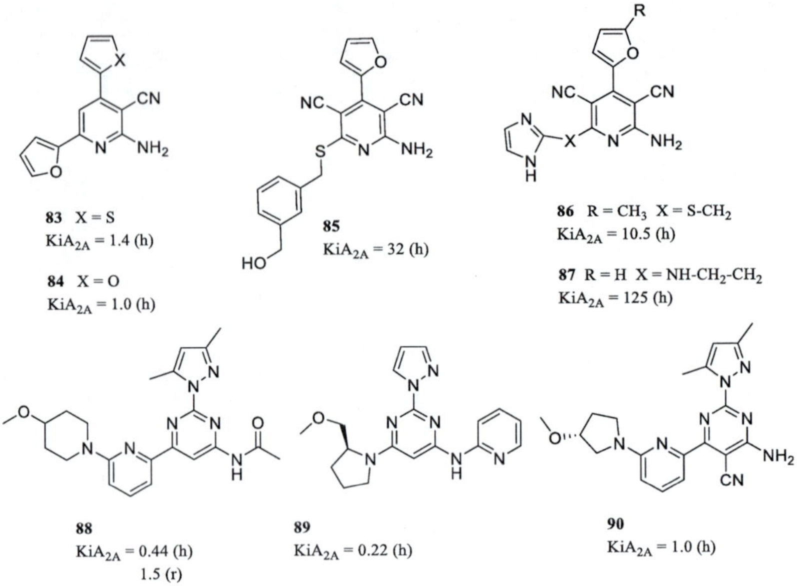

83 X = S
KiA_{2A} = 1.4 (h)

84 X = O
KiA_{2A} = 1.0 (h)

85
KiA_{2A} = 32 (h)

86 R = CH_3 X = S-CH_2
KiA_{2A} = 10.5 (h)

87 R = H X = NH-CH_2-CH_2
KiA_{2A} = 125 (h)

88
KiA_{2A} = 0.44 (h)
1.5 (r)

89
KiA_{2A} = 0.22 (h)

90
KiA_{2A} = 1.0 (h)

Fig. 14 Structures and binding affinity of pyridine and pyrimidine derivatives (Ki nM)

[102]. Similar compounds bearing a further cyan group at the pyridine core were recently described. The 2-amino-4-(furan-2-yl)-6-((3-(hydroxymethyl)benzyl)thio) pyridine-3,5-dicarbonitrile (**85**, Fig. 14) was reported as an A_{2A}AR antagonist, which behaved as inverse agonist at A_1AR with $K_i = 31$ and 0.98 nM at A_{2A} and A_1AR, respectively [103]. Very recently, the same authors described dicyanopyridine derivatives as A_{2A} and A_1 AR dual ligands. Among them, **86** and **87** were endowed with $KiA_{2A} = 10.5$ nM and $KiA_1 = 10.1$ nM and $KiA_{2A} = 125$ nM and $KiA_1 = 174$ nM, respectively (Fig. 14) [104].

A number of water soluble pyrimidine-acetamide derivatives were reported by Almiral and Neurocrine Biosciences with good affinity for the A_{2A}AR [105]. Modification of the pyrimidine substituents improved the compound pharmacokinetic properties and binding affinity leading to derivatives **88** and **89**, which showed sub-nM A_{2A}AR affinity (**88**, KiA_{2A} of 0.44 nM at human receptor and 1.5 nM at rat receptor; **89**, $KiA_{2A} = 0.22$ nM, at human receptor; Fig. 14). Furthermore, when i.p. administered, **89** readily crossed the blood-brain barrier (BBB) and was efficacious in reversing the haloperidol-induced catalepsy, an in vivo model of PD [106, 107]. Then, the same authors reported new A_{2A}AR antagonists based on a 4-amino-5-carbonitrile pyrimidine scaffold. In particular, the (*R*)-4-amino-2-(3,5-dimethyl-1*H*-pyrazol-1-yl)-6-(6-(3-methoxypyrrolidin-1-yl)pyridin-2-yl) pyrimidine-5-carbonitrile (**90**, Fig. 14) exhibited high potency at A_{2A}AR with a Ki = 1.0 nM and an IC_{50} of 1.1 nM in a functional assay at the same receptor. Furthermore, the compound resulted less sensible than **88** and **89** to the cytochrome P450 metabolism and demonstrated very good efficacy in the haloperidol-induced catalepsy when orally administered [108]. In 2011 researchers of Palobiopharma

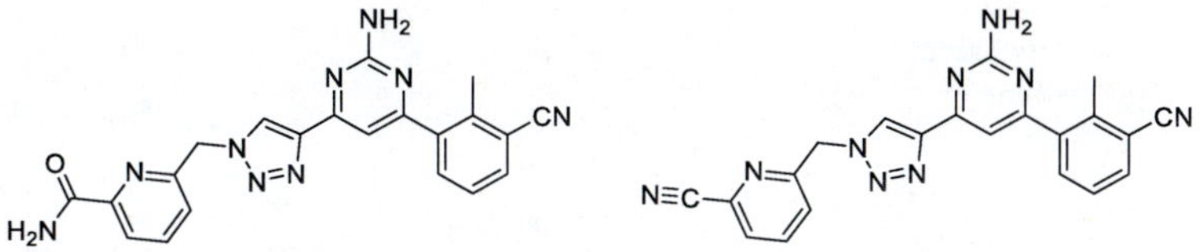

91
$IC_{50}A_{2A} = 6.0$ (h)

92
$IC_{50}A_{2A} = 24$ (h)

Fig. 15 Structures and functional activity on cAMP assay of pyrimidines (IC_{50} nM)

S. L. patented a series of new 4-aminopyrimidine derivatives like taminadenant (**8**, Fig. 2) endowed with high affinity and selectivity for the A_{2A}AR [109]. This derivative was out-licensed to Novartis in 2015 and is actually in clinical trials in some different tumor types. Simplification of 8 imidazole ring gave PBF-999 (compound **9**, Fig. 2), which resulted to be a dual A_{2A}AR antagonist/phosphodiesterase 10 (PDE-10) inhibitor useful especially for the treatment of "cold tumors," as colon cancers (https://www.palobiofarma.com/pipeline-2/).

Etrumadenant represents the lead compound of a novel series of potent and selective A_{2A}AR and A_{2B}AR dual antagonist endowed with high affinity (**7**, $KiA_{2A} = 1.4$ nM and $KiA_{2B} = 2.0$ nM, respectively, Fig. 2). This derivative does not cross the BBB and inhibited NECA-mediated A_{2A}AR activation with a potency of <10 nM. It showed to be orally bioavailable and inhibits Ado in suppressing activation of human CD4 or CD8 T cells, as well as monocyte-derived dendritic cells, in in vitro experiments [31, 110, 111]. Very recently, analogue of this compound endowed with a triazole-pyrimidine-methylbenzonitrile core was reported as potent A_{2A}AR and A_{2B}AR dual antagonists endowed with comparable and even higher activity, at A_{2B}ARs, than etrumadenant itself in functional studies. Examples of such derivatives are compounds **91** and **92** (Fig. 15) [112].

3.2.9 Pyrazine and Triazine Derivatives

In 2005, a number of pyrazine derivatives were reported as dual A_{2A}/A_1AR antagonists endowed with Ki in the sub-nM range. These compounds resulted slightly selective for the A_{2A}AR [113].

Among them, the 5-(5-amino-3-(4-fluorophenyl)pyrazin-2-yl)-1-isopropylpyridin-2(1*H*)-one (**93**, ASP5854, $KiA_{2A}AR = 1.8$ nM and $KiA_1AR = 9.0$) was found orally active in animal models of PD and cognition [114]. This last effect was related to the ability to block A_1AR subtype (https://clinicaltrials.gov/ct2/show/NCT00783276). Replacement of **93** fluorophenyl group with a thiophene ring resulted in an improvement of the A_{2A}AR affinity leading **94**, an A_{2A}AR antagonist with sub-nM affinity (Ki = 0.84 nM) [113]. A patent from Heptares Therapeutics Limited reported a series of 1,2,4-triazine derivatives as A_{2A}AR antagonists endowed with high affinity [115]. Among them, the 6-(2-methyl-6-(trifluoromethyl)-pyridin-4-yl)-5-phenyl-1,2,4-triazin-3-amine (**95**, $KiA_{2A} = 3.5$ nM, Fig. 16) was

Fig. 16 Structures and binding affinity of pyrazine and triazine derivatives (Ki nM)

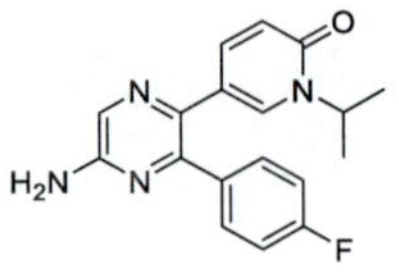

ASP5854, **93**
KiA_{2A} = 1.8 (h)

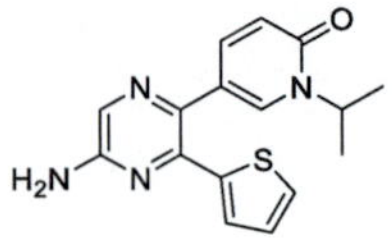

94
KiA_{2A} = 0.84 (h)

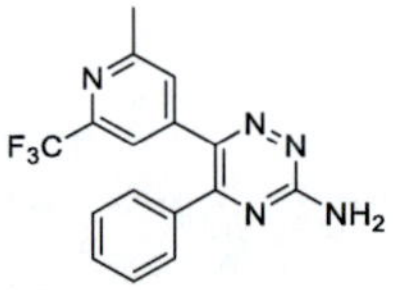

95
KiA_{2A} = 3.5 (h)

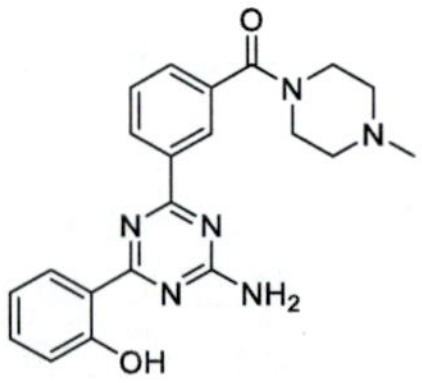

96
KiA_{2A} = 1.0 (h)

found with low nM affinity and moderate selectivity versus the A_1 subtype (about 10-fold). SAR revealed that the presence of a phenyl ring at the 1,2,4-triazine 5-position favored the interaction with the A_{2A}AR. The further introduction of a fluorine atom in this aromatic ring increased the A_{2A}AR selectivity. Replacement of the 6-pyridil substituent with a morpholine or phenoxy group led to a further increase of selectivity but decreased the A_{2A}AR affinity [116]. Among 1,2,4-triazine derivatives, imaradenant (**4**, Ki = 1.7 nM, Fig. 2) is in clinical trial for cancer [22]. Another study reported also 1,3,5-triazines as potent A_{2A}AR antagonists identified through a virtual screening. In this series, the presence of the phenol group is very important for A_{2A}AR affinity. The most representative compound of this series is **96** (KiA_{2A} = 1.0 nM, Fig. 16). **96** has good physicochemical and drug-like properties, including high oral bioavailability and was very potent in in vivo model of PD [117].

4 Structural Features of A_{2A} Structure and Its Interaction with Antagonists

The structure of the A_{2A}AR presents the typical features of a rhodopsin-like G-Protein-Coupled Receptor, with an N-terminal domain exposed into the extracellular environment, seven transmembrane (TM) domains called TM1-7, three extracellular and three intracellular loops (EL1-3 and IL1-3, respectively), and an intracellular C-terminal domain. While in the '90s and in the early 2000s the structural and functional roles of the A_{2A}AR residues were only predicted through the development of theoretical models, the first crystal structure of this protein (published in 2008 and reported into the protein data bank with the code 3EML) definitely clarified the architecture of the receptor and its ligand binding site [6]. This structure was obtained in complex with ZM241385 (**36**) and gave a detailed picture of the receptor–ligand interaction from a molecular point of view, highlighting the central role of the amino acids Glu168, Phe169, Asn253, and Leu249 in the stabilization of the ligand within the binding cavity and in polar interactions with the ligand itself. This X-ray structure was obtained by engineering the receptor structure with the insertion of a T4 lysozyme moiety within the receptor IL3 segment, to make the receptor more stable in the experimental conditions during the crystallization process. Since then, a number of experimental 3D structures of the A_{2A}AR were obtained, again by X-ray crystallography or, more recently, also by electron cryomicroscopy. To date, 66 experimental structures of the A_{2A}AR are available in the protein data bank. While some of these structures were obtained in complex with agonists (like the endogenous ligand Ado [118], but also its derivatives NECA [118], UK-432097 [119], CGS21680 [120], and the non-nucleoside agonist LUF5833 [121]), even bound to an engineered G protein (pdb code 6GDG [122]), the great majority of the published A_{2A}AR experimental structures are complexed with antagonists. Several structures of this receptor were obtained in complex with ZM241385 by both crystallography and electron cryomicroscopy and by operating in various experimental conditions, with various approaches to stabilize the receptor structure; beside the above-cited insertion of a T4 lysozyme moiety within the receptor IL3 segment, other receptor enginerizations were made (i.e., various mutations to obtain a thermostable structure, the so-called thermostabilized receptor [123], or by replacing the IL3 loop with apocytochrome b(562)RIL, to obtain the so-called A_{2A}AR-BRIL or A_{2A}AR-StaR2-bRIL structures [124, 125]).

Regarding the ligand binding site of the A_{2A}AR, Fig. 17 shows the complex of this protein with ZM241385 (pdb code 5NM4 [126]) and the residues involved in ligand interaction. The deep region of the binding cavity (see surface representation in Fig. 18) is narrow and presents mainly hydrophobic residues, even if some polar interaction is still possible. The central region (see surface representation in Fig. 18) presents three key residues like Phe168, Glu169, and Asn253. The phenyl ring of Phe168 makes a π-stacking interaction with the aromatic central core of the co-crystallized ligands. This feature may be seen not only with ZM241385 (Fig. 17) but also with other antagonists and also with agonists. Another residue, Leu249, helps in the stabilization of the ligand scaffold within the cavity. Asn253 presents an amide function within its side chain, hence it is able to make both H-bond donor and acceptor interactions. This feature is clearly visible in the A_{2A}AR-ZM241385 complex, in which this amino acid provides such double interaction with the ligand scaffold and exocyclic amine group. An additional polar interaction is made by Glu169, whose carboxylic group provides electron-rich oxygen atoms for a polar interaction again with the exocyclic amine function of the ligand. All these non-polar and polar contacts constitute a sort of pattern for all the ligand–receptor interactions observed at the A_{2A}AR experimental structures. The central region of the binding cavity is large, and a typical feature of antagonist-bound experimental structures of the A_{2A}AR is the presence of water molecules working as bridges between the ligand and the surrounding protein residues. Such feature is not so evident for the nucleoside agonist-bound structures of the A_{2A}AR, where the above

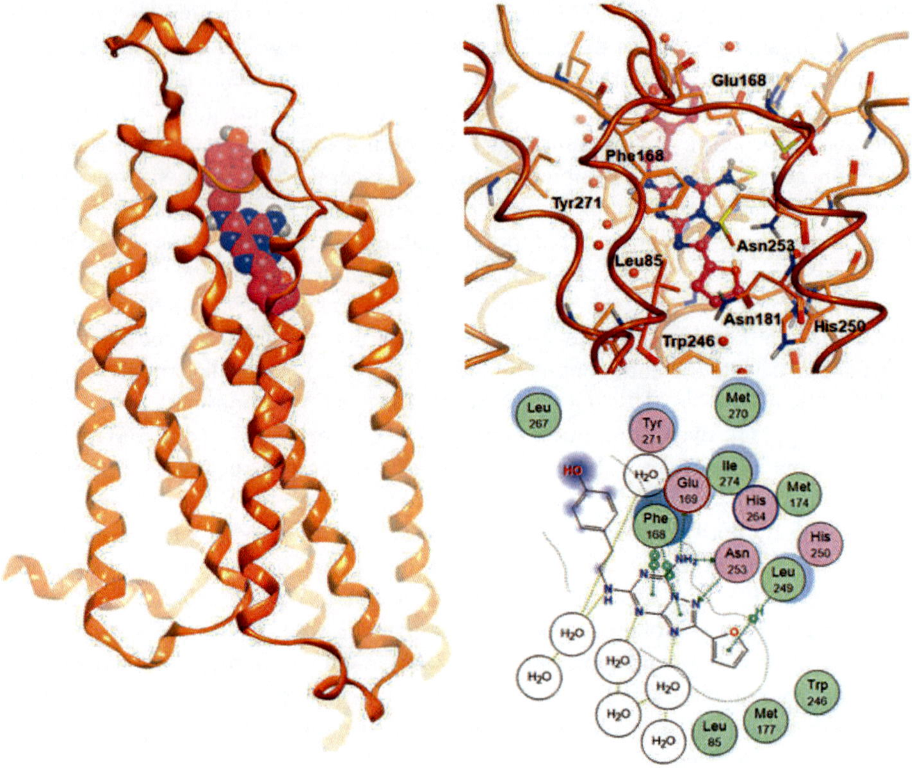

Fig. 17 Crystal structure of the A_{2A}AR in complex with ZM241385 (pdb code 5NM4). Left: global view of the receptor and space filling-style view of the ligand. Top right: detailed view of the ligand–target interaction with the key residues indicated (water molecules are showed as red spheres). Bottom right: schematic plot of the ligand–target interaction for the same complex

cited water molecules are partially replaced by the sugar moiety of the agonists. The *p*-hydroxy-phenylethylamino group of ZM241385 is oriented toward the extracellular environment and is located in a wider region of the receptor cavity (external region, see surface representation in Fig. 18). This group makes some polar and non-polar interactions with the protein, with the contribution of co-crystallized water molecules.

Figure 18 shows antagonists of the A_{2A}AR present in complex with the receptor in its experimental 3D structures, with the respective pdb codes. The chemical structures are colored according to the location of their molecule fragments in the deep, central, and external regions (please refer to the surface representation in Fig. 18). It can be observed that all these molecules present an aromatic core structure, always present in the central region of the cavity. An additional substituent to be inserted in the deep region of the cavity is almost always an aromatic ring. While a typical substituent pointing toward the external region of the binding site is made by a small flexible chain followed by one or more rings (i.e., ZM241385,

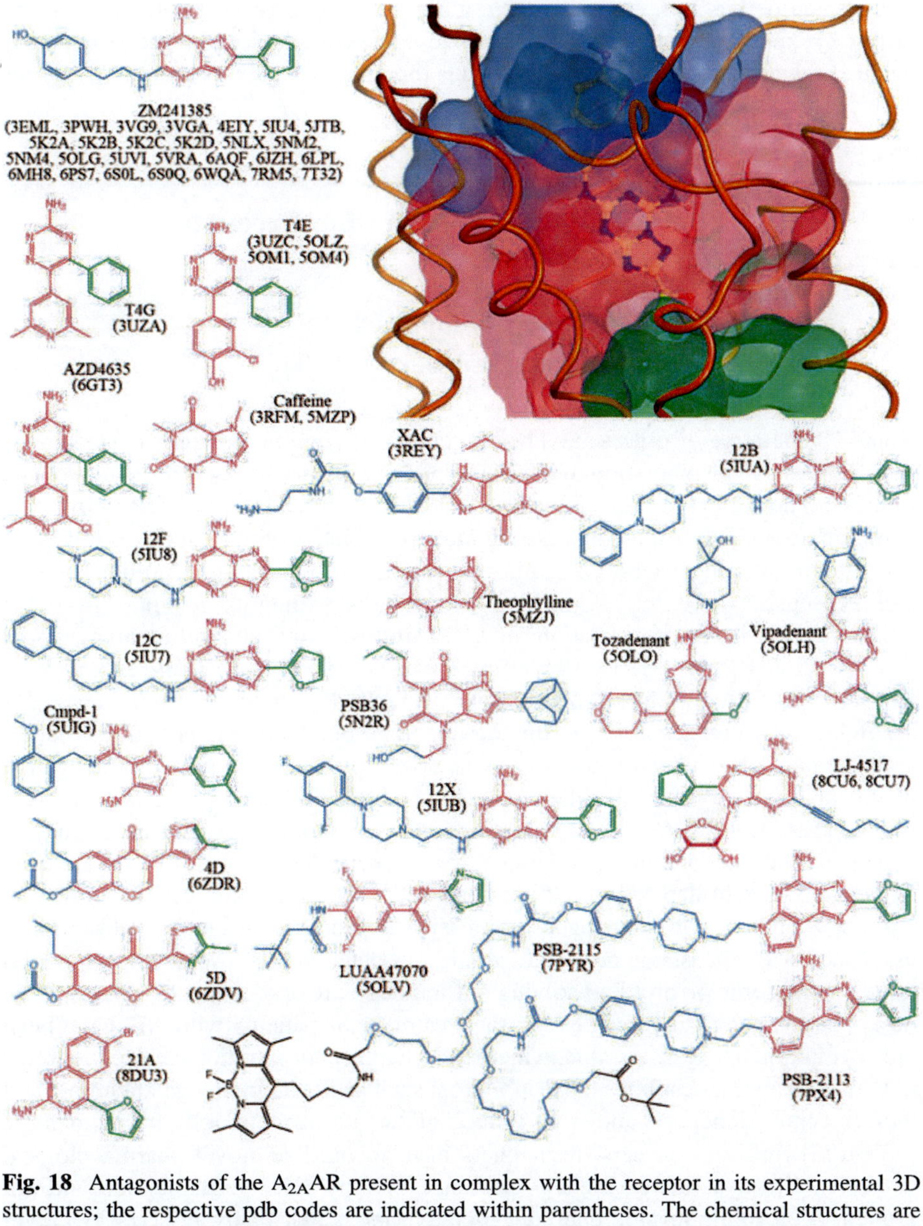

Fig. 18 Antagonists of the $A_{2A}AR$ present in complex with the receptor in its experimental 3D structures; the respective pdb codes are indicated within parentheses. The chemical structures are colored according to the location of their respective molecular fragments in the deep, central, and external regions of the binding site, which are represented as surfaces in the top-right section of the figure: deep region, green; central region, purple; external region, blue ($A_{2A}AR$ in complex with ZM241385, pdb code 5NM4, taken as reference). The black colored fragments are not present in the pdb files

Preladenant derivatives, etc.), this moiety may consist of a variety of combinations of flexible and rigid moieties of various sizes. The complexity of these groups may be due to the searching of selectivity versus the other AR subtypes and not only the affinity for the A_{2A}AR.

5 Potential Application of A_{2A} Adenosine Receptor Antagonists

5.1 A_{2A} Adenosine Receptor Antagonists and Neurodegenerative Disorders

Neurodegenerative disorders are a heterogeneous group of conditions characterized by a progressive degeneration of central or peripheral nervous systems structure and function. They affect millions of people worldwide, with AD and PD being the most common neurodegenerative diseases in modern societies with high life expectancy. The number of people with dementia was estimated to stand at 55 million in 2019 and expected to rise to 139 million in 2050, according to a report from the Alzheimer's Disease Association in 2022 [https://www.alz.org/research?gclid=CjwKCAjwqJSaBhBUEiwAg5W9p2KWH7gSedyUvN5VLmyDAB5p50t3Eu95opzYZL5PJgejfg0M4zL0ZBoCDekQAvD_BwE]; nearly 10 million peoples are living with PD, according to the Parkinson's Foundation [127]. AD is a multi-factorial neurodegenerative disease characterized by abnormal protein aggregation of Aβ plaques and neurofibrillary tangles consisting of hyperphosphorylated tau protein [128, 129]. In AD there is a prolonged activation of microglia and astrocytes and increased production of inflammatory markers, like IL-1β [130, 131]. Both aggregated protein deposition and immune-cell over-stimulation can affect neuron structure and function, causing a progressive appearance of amnesic disabilities, associated with aphasia, apraxia, and agnosia. Related to this disease are behavioral issues, consisting of problem-solving difficulties and orientation disorders. Until now, pharmacological strategies for the treatment of patients with AD have been only symptomatic. They are able to increase synaptic concentrations of acetylcholine (ACh), by using the cholinesterase inhibitors such as rivastigmine, galantamine, and more recently donepezil and/or to reduce glutamate excitotoxicity by memantine [132, 133]. In the last years, researchers have focused on novel pharmacological strategies aimed at increasing the clearance of amyloid β (Aβ) peptide and tau protein, or at inhibiting the neurodegenerative molecular pathways [134, 135]. In 2021, the FDA controversially approved the use of the anti-Aβ monoclonal antibody aducanumab in mild cognitive impairment and early-stage Alzheimer's disease [136, 137]. Experimental results, in both cells and animal models, have highlighted the possibility of targeting A_{2A}AR to contrasting dementia-related neurodegeneration. Results of electrophysiological studies on pyramidal neurons have showed that activation of these ARs increases evoked *N*-methyl-D-aspartate receptor (NMDAR) currents [138]. In addition, the increase of Ado in early β-amyloid-based AD models showed a detrimental effect on cognition. This effect was not observed in A_{2A}AR knock-out mouse models demonstrating the potential efficacy of A_{2A}AR antagonists [139]. The A_{2A}AR plays a role also in tau phosphorylation; it was observed that caffeine, a well-known A_{2A}AR antagonist, decreases hippocampus tau hyperphosphorylation. Administration of selective A_{2A}AR antagonists in animal models has been shown to improve memory deficits and mitigate disease-associated Aβ toxicity or tau hyperphosphorylation [15]. Moreover, the reduction of memory performance in rats, induced by administration of the human $A\beta_{1-42}$ fragment, was reversed by SCH58261 and KW6002 [140]. Animal studies demonstrated that Aβ-induced neurotoxicity is mainly mediated by A_{2A}AR activity, since ZM241385 attenuated Aβ-induced neuronal death and further reduced synaptic loss [141]. Consequently, drugs targeting A_{2A}AR are of particular interest, given the potential role of these receptors in inflammatory processes, as well as in tauopathy and Aβ pathologies. PD is another pathology belonging to the family of neurodegenerative disorders, in which A_{2A}AR plays a crucial role. In PD, the neuronal degeneration is observed especially in the ventralis region of the pars compacta, substantia nigra, and locus coeruleus. Neurons have eosinophilic inclusion bodies called Lewy bodies, composed of the presynaptic protein α-synuclein. The etiology of PD is still unknown, although participation of environmental toxins, oxidative stress, and free radicals is postulated. The principal pathophysiological event in PD is the progressive damage of dopaminergic neurons in the substantia nigra, which leads to a reduction in the dopamine concentration in striatum. Typical symptoms of this disease such as bradykinesia, muscular rigidity, and tremor are consequences of this reduction. Current therapies for PD are based principally on dopamine substitutes such as levodopa, dopamine agonists, enzymatic inhibitors. This therapy shows some limitations such as a loss of efficacy and significant side effects like dyskinesia, disorders, hallucinations, impulse control disorders, etc. [142]. The non-dopaminergic approaches are based on NMDA antagonist amantadine and anticholinergic drugs like benzhexol but both with undesirable adverse effects [143]. More recently, pimavanserin, a 5-HT_{2A} receptor antagonist/inverse agonist, was approved for the psychosis in PD treatment [144]. The efficacy of this drug is lower than clozapine but it showed minor side effects for managing psychosis with respect to clozapine itself. Moreover, pimavanserin decreases the risk of falls in individuals with PD [145]. The undesirable side effects of dopamine replacement therapy have promoted many research studies of non-dopaminergic approaches and particular attention was dedicated to A_{2A}AR receptor antagonists since A_{2A}AR is co-expressed with dopamine D_2 receptors in the striatum. More in detail, the interconnection of these two receptors forms an A_{2A}AR–D_2R heteromer. The effect on motor function by these receptors is opposite, since the stimulation of A_{2A}AR inhibits the signaling of dopamine D_2 receptors, and contrariwise, the inactivation of A_{2A}AR increases the responsiveness of dopamine D_2 receptors facilitating dopamine D_2-mediated responses [146]. In addition, in the long-term levodopa therapy, the co-administration of A_{2A}AR antagonists decreases dyskinesia increasing the therapeutic effect of levodopa [147]. The compound 8-ethoxy-9-ethyladenine (ANR 94) proved to be useful in the reversal of haloperidol catalepsy and the 6-OHDA model of contralateral turning behavior, two rat models of Parkinson's disease. Interestingly, this compound produces contralateral turning in rats primed with levodopa when it is also administered alone, without concomitant levodopa administration contrary to other known A_{2A}AR antagonists such as Scheme 58261 or istradefylline. This finding suggests that the compound could be effective as a monotherapy, when dopamine receptors are repeatedly stimulated by levodopa [88, 90]. In addition, ANR 94 protects nigrostriatal dopaminergic neurons from degeneration and counteracted neuroinflammatory processes by contrasting astroglial (glial fibrillary acidic protein, GFAP) and microglial (CD11b) activation in a sub-chronic 1-methyl-4-phenyl-1,2,3,6-tetrahydropyridine (MPTP) mouse model of PD [91]. These results strongly support the A_{2A}AR antagonists as a real opportunity for improving PD treatment. On this basis, A_{2A}AR antagonists caused a balanced improvement of motor and non-motor symptoms with an acceptable adverse event profile. Initially, vipadenant was tested only on restricted human trials and, unfortunately, it was discontinued due to side effects [148–150]. Then, istradefylline [151–155], preladenant [56], and tozadenant [156] were enrolled for clinical trials but only istradefylline has successfully reached commercialization [24]. To date, istradefylline has received approval in Japan and, more recently, by FDA in the USA for the treatment of adult PD patients with OFF time episodes due to recurrent assumption of levodopa (Nourianz https://www.accessdata.fda.gov/drugsatfda_docs/label/2019/022075s000lbl.pdf; accessed on 10 November 2022).

5.2 A_{2A} *Adenosine Receptor Antagonists and Neuroprotection*

Cellular responses of CNS to injury, infection, and neurodegenerative processes, such as ischemia, epilepsy, PD, and AD, comprise a wide range of biological effects that cause diseases, which, despite having different etiologies, have neuroinflammation as a common basis. This biological process is due to the activation of neuroimmune cells from the brain astrocytes and microglia, leading to abnormal proinflammatory cytokines (CK) [157–160]. Remarkably, the A_{2A}AR has a relevant function in activated microglia, which was found surrounding the pathological features of AD [161]. Activation of microglial A_{2A}AR enhances the production of proinflammatory mediators [162]. These results are in agreement with the finding that the microglial receptor blockade reduces neuroinflammation and, more importantly, may lead to an improvement of cognitive impairment [163]. During the harmful brain conditions, A_{2A}AR is up-regulated especially in microglia cells and its blockade shows brain neuroprotection in studied animals [164]. The blockade of A_{2A}AR alleviates the long-term burden of brain disorders in different neurodegenerative conditions through its control on neuronal cell death. Experimental study in the forebrain A_{2A}AR knock-out mice combined with flow cytometric analyses and intracerebroventricular injection of istradefylline demonstrated that the selective deletion of A_{2A}ARs in forebrain neurons abolished the motor stimulant effects of the A_{2A}AR antagonist KW-6002 but did not affect acute MPTP neurotoxicity. Intracerebroventricular administration of KW-6002 into forebrain A_{2A}AR knock-out mice reinstated protection against acute MPTP-induced dopaminergic neurotoxicity and attenuated MPTP-induced striatal microglial and astroglial activation. Microglial activation induced by MPTP was also attenuated by the A_{2A}AR antagonist KW-6002. Furthermore, flow cytometry analysis showed that the reduction in microglial activation by istradefylline pre-treatment is largely seen in the subpopulation of microglia with large cell body size, which indicates a full activation. This suggests that KW-6002 mainly attenuates the progression of microglial activation to fully activated status. In conclusion, A_{2A}AR in forebrain neurons is responsible for motor stimulation, and A_{2A}AR in glial cells is among the factors involved in the neuroprotective effects observed in the acute MPTP model of PD. This finding underlines the role of A_{2A}AR in different cellular elements to control neurodegeneration and/or locomotion opening the real possibility of targeting the A_{2A}ARs using different dose treatments. In fact, A_{2A}AR antagonists exert neuroprotection against ischemic and excitotoxic brain injuries at a dose approximately 100-fold lower than those needed to modulate motor effects [165]. In mixed glial cells and in an animal model of neuroinflammation, the compound 8-chloro-9-ethyl-2-phenethoxyadenine (KiA_{2A} = 0.75 nM; KiA_1 = 17 nM and KiA_3 = 227 nM, $IC_{50}A_{2A}$ = 251 nM) possesses anti-inflammatory properties, counteracting neuroinflammation and by its antioxidant activity [13, 166]. In addition, A_{2A}ARs are mainly located in both presynaptic and postsynaptic brain synapses [167]. Presynaptic A_{2A}AR controls the release of glutamate and plays a major role in arresting the inhibition of A_1AR-mediated synaptic transmission [168, 169]. The A_1AR-mediated inhibition prevents the enhancing synaptic efficacy upon increasing neuronal activity. Postsynaptic A_{2A}AR controls NMDA receptors [170, 171]. For this reason, the A_{2A}AR antagonists reduce long-term potentiation at different excitatory synapses [171]. As mentioned before, the peculiarity of A_{2A}AR is its remarkable ability to up-regulate its expression and density upon harmful conditions, in both neurons [172] and glial cells [164]. Very recently, in a study of purine and triazolotriazine-based A_{2A}AR antagonists the compound 9-ethyl-8-furyl-2-(4-hydroxyphenyl)-2-ethylaminoadenine (**68**) showed a good activity on reducing proinflammatory mediators and on increasing anti-inflammatory cytokines in microglia cells. This A_{2A}AR antagonist could represent a promising compound useful in neurodegenerative diseases [66].

5.3 *A_{2A} Adenosine Receptor Antagonists and Cancer*

During the last decade, despite the significant improvement in cancer treatment with radiotherapy/chemotherapy and targeted drug therapy, which lead to a longer survival period, the mortality rate of patients with solid metastatic tumors has not significantly changed. In addition, available treatments have numerous side effects for patients. An innovative strategy to combat cancer diseases is now cancer immunotherapy [173]. The immune system is absolutely critical in fighting cancer; in fact, the approach to treat many tumors with the immunotherapy represents an innovation, since it targets unfitting responses of the immune system in the human body rather than cancerous cells [174]. In addition, it is now well established that an over-expression of some receptors, like ARs, occurs in various tumors and their modulation could represent a therapeutic strategy [62, 175, 176]. A_{2A}AR modulation is one of the few novel approaches with a great potential in the treatment of patients, especially those presenting tumors refractory to other therapies. The human immune system is activated by damage pathogens and it could be stimulated by a range of mechanisms to protect the body against an excess inflammatory response or an autoimmune process [177]. Unfortunately, these circumvents are sometimes bypassed by tumors. More than 100 monoclonal antibodies were approved by FDA to modulate the human immune system, but the cost of therapy limits their use (The Antibody Society. In: Approved antibodies. https://www.antibodysociety.org; Accessed on 10 May 2023). The A_{2A}AR is widely distributed on the surface of most immune cells. The gene CD73 promotes the Ado increase activating A_{2A}AR signal pathway in tumor cells, which in turn leads to the activation of a signaling cascade including Rap1, P110, and PIP3. After that, the PIK3/AKT signaling pathway is turned on. This activation promotes tumor cell epithelial-mesenchymal transition and anti-apoptosis, which promotes tumor cell growth and metastasis [178]. Under normal conditions, the extracellular level of Ado is very low (<1 μM). Conversely, transient or chronic hypoxia is established in the tumor microenvironment (TME), which increases the level of extracellular Ado at concentrations of about 10 μM. The high level of Ado triggers an increase in cAMP-protein kinase A signaling concomitant with immune regulation [178]. During hypoxia there is a release of adenine nucleotides as ATP, ADP, and AMP resulting also in HIF-1α overexpression. This supports the transcription of genes like CD39 and CD73, which represent the major source of extracellular Ado in the TME [16]. Activation of A_{2A}AR mediates the primary immunosuppression in the TME. Several studies demonstrated that elevated intracellular cAMP in T cells, after A_{2A}AR activation, interferes with level of interleukin-2 (IL-2), tumor necrosis factor (TNFα), and interferon-γ (IFN-γ) and compromises CD8+ T cell infiltration [179]. There are some methods to block or reduce the immunosuppression caused through the hypoxia induced by A_{2A}AR pathway. It is possible to block hypoxia by using agents able to interact with CD39, CD73, and PKA, or A_{2A}AR antagonists [180]. Preliminary clinical studies reported that some agents, including the anti-CD73 antibodies BMS-986179 and CPI-006 [181], are efficacious to modulate the hypoxia in the TME. On the other hand, the therapeutic potential of A_{2A}AR blockade, using small molecules, was validated by commercially available istradefylline. Some A_{2A}AR antagonists are in different phase of clinical trials for solid tumor such as taminadenant (**8**, Fig. 2) [32], inupadenant (**10**, Fig. 2) [33], ciforadenant (**6**, Fig. 2) [29], and the A_{2A}AR/A_{2B}AR dual antagonist etrumadenant (**7**, Fig. 2) [31] showing good tolerability with low side effects.

In particular, as reported above, taminadenant was well tolerated in patients with advanced NSCLC in a phase 1 clinical trial when administered alone or in combination with spartalizumab, an anti-PD-1 Ab. The maximum tolerated dose of taminadenant alone was 480 mg twice a day, and 240 mg twice a day plus spartalizumab. Efficacy was neither a primary nor secondary endpoint; however, some clinical benefits were noted regardless of prior immunotherapy. Very recently, preliminary antitumor studies in a phase Ib study with co-administration of taminadenant and spartalizumab in non-small cell lung carcinoma showed clinical benefit with tumor shrinkage (Clinical trial identification NCT03207867: Protocol number: CNIR178X2201; release date: October 22, 2020). Ongoing clinical studies showed a favorable safety profile of inupadenant with preliminary evidence of clinical activity in multiple tumor types, including patients who have exhausted standard treatment options [33]. The standard treatment for non-squamous mNSCLC patients is a platinum-based doublet chemotherapy regimen: carboplatin plus pemetrexed (C + P). Study A2A-005, actually in process, is evaluating the efficacy of inupadenant in combination with C + P as a second-line therapy in adult patients with non-squamous mNSCLC [182]. Ciforadenant is under evaluation for treatment-refractory renal cell cancer. Now it is in phase 1b/2 trial in combination with Ipilimumab and Nivolumab in First-line Advanced Renal Cell Carcinoma [https://clinicaltrials.gov/ct2/show/NCT05501054]. In addition, a study of SRF617 (human antibody designed to inhibit the enzymatic activity of CD39) with etrumadenant and Zimberelimab (human anti-PD-1 monoclonal antibody) in patients with metastatic castration-resistant prostate cancer is undergoing [https://clinicaltrials.gov/ct2/show/NCT05177770]. In conclusion, the monotherapy with A_{2A}AR antagonists could represent an innovative strategy for cancer therapy. On the other hand, especially in refractory tumors, the combination of A_{2A}AR blockers with other chemotherapeutic drugs can effectively improve the antitumor effect and significantly increase the A_{2A}AR antagonist application.

6 Concluding Remarks

Over the past decades, the A_{2A}AR has become an attractive target for various clinical disorders. This high therapeutic potential has prompted researchers to design and synthesize potent and selective ligands and the development of compounds able to modulate the activity of A_{2A}ARs is a rapidly growing area of research, with an important impact on the drug discovery process. In this review, the structure, distribution, and signal transduction pathways of A_{2A}ARs, together with the SARs of some of the most interesting A_{2A}AR antagonists, have been discussed, combined with the corresponding pharmacological roles and therapeutic potential. A_{2A}AR antagonists have typically been divided into xanthine-based and non-xanthine derivatives. One of the earliest antagonists known is caffeine, a weak psychostimulant that plays also a neuroprotective role and whose effects are mediated by the blockade of the A_{2A}AR. It is present in many food products, especially in the coffee and energetic beverages, which led to large assumption, even if it is considered a psychoactive drug. There are about 1,000 studies listed in ClinicalTrials.gov in which caffeine is studied. In addition, numerous reports, often contradictory, claim other health effects of caffeine, for example, its benefits in liver disease, diabetes, cancer, and CNS disorders.

From a medicinal chemistry point of view, this review describes some common chemical scaffolds that can be highlighted among the plethora of molecules synthesized. Here, we present the state of the art about the current advancement of A_{2A}AR antagonists in drug development. For each of the identified templates, we have indicated the general mono- or multitarget profile, the most representative molecules, and corresponding advantages and disadvantages. During the reviewing process of literature, examples of dual-acting ligands (A_{2A}/A_1 or A_{2B}/ARs) have been found and reported. A caffeine derivative like the 8-styrylxanthine istradefylline was among the first A_{2A}AR antagonists reported and now it is commercially available in Japan and in the USA as a co-therapy for the treatment of PD in association with levodopa/carbidopa to reduce off-time episodes and dyskinesias. Although only this compound has reached the market, many other A_{2A}AR antagonists are in various phases of clinical trials for different pathological status. For example, preladenant, tozadenant, and vipadenant have undergone testing in PD. The A_{2A}AR is an emerging target also for cancer treatment. In fact, the ability of A_{2A}AR antagonists to counteract the Ado immunosuppression in TME makes these receptors an innovative target in cancer immunotherapy, with a great potential in the treatment of patients especially those presenting tumors refractory to other therapies. Ciforadenant, etrumadenant, taminadenant, and PBF-999 are in clinical trials for cancer immunotherapy. The discovery of potent and selective A_{2A}AR antagonists opens new boundaries for the elucidation of specific and selective therapeutic potentials. The characterization of intracellular pathways involving A_{2A}ARs indicates that the modulation of these signaling pathways could lead to considerable advances in the management of a number of disease states. It is clear that A_{2A}AR antagonists play an important role in neurodegenerative diseases, such as PD. The combined therapy showed major effects than different single-drug effects, with minor side consequences. Finally, dual-acting ligands could also open remarkable therapeutic perspectives for adenosine antagonists. The optimization of a single molecule to make it able to interact with multiple targets is a major and stimulating challenge to be addressed in order to improve the clinical utility for the treatment of complex multifactorial disorders.

References

1. IJzerman AP, Jacobson KA, Muller CE, Cronstein BN, Cunha RA (2022) International Union of Basic and Clinical Pharmacology. CXII: adenosine receptors: a further update. Pharmacol Rev 74(2):340–372
2. Mori A, Chen JF, Uchida S, Durlach C, King SM, Jenner P (2022) The pharmacological potential of adenosine A_{2A} receptor antagonists for treating Parkinson's disease. Molecules 27(7):2366

3. Carriba P, Ortiz O, Patkar K, Justinova Z, Stroik J, Themann A, Muller C, Woods AS, Hope BT, Ciruela F, Casado V, Canela EI, Lluis C, Goldberg SR, Moratalla R, Franco R, Ferre S (2007) Striatal adenosine A(2A) and cannabinoid CB(1) receptors form functional heteromeric complexes that mediate the motor effects of cannabinoids. Neuropsychopharmacology 32(11): 2249–2259
4. Ferre S, Ciruela F, Borycz J, Solinas M, Quarta D, Antoniou K, Quiroz C, Justinova Z, Lluis C, Franco R, Goldberg SR (2008) Adenosine A_1-A_{2A} receptor heteromers: new targets for caffeine in the brain. Front Biosci 13:2391–2399
5. Chen JF (2014) Adenosine receptor control of cognition in normal and disease. Int Rev Neurobiol 119:257–307
6. Jaakola VP, Griffith MT, Hanson MA, Cherezov V, Chien EY, Lane JR, Ijzerman AP, Stevens RC (2008) The 2.6 angstrom crystal structure of a human A_{2A} adenosine receptor bound to an antagonist. Science 322(5905):1211–1217
7. Carpenter B, Lebon G (2017) Human adenosine A_{2A} receptor: molecular mechanism of ligand binding and activation. Front Pharmacol 8:898
8. Muller CE, Jacobson KA (2011) Recent developments in adenosine receptor ligands and their potential as novel drugs. Biochim Biophys Acta 1808(5):1290–1308
9. Lambertucci C, Marucci G, Catarzi D, Colotta V, Francucci B, Spinaci A, Varano F, Volpini R (2022) A_{2A} adenosine receptor antagonists and their potential in neurological disorders. Curr Med Chem. Accepted
10. Cacciari B, Spalluto G, Federico S (2018) A_{2A} adenosine receptor antagonists as therapeutic candidates: are they still an interesting challenge? Mini Rev Med Chem 18(14):1168–1174
11. Jacobson KA, IJzerman AP, Muller CE (2021) Medicinal chemistry of P2 and adenosine receptors: common scaffolds adapted for multiple targets. Biochem Pharmacol:114311
12. Ferrante A, De Simone R, Ajmone-Cat MA, Minghetti L, Popoli P (2018) Adenosine receptors and neuroinflammation. In: The adenosine receptors. The receptors. Humana Press, Cham
13. Marti Navia A, Dal Ben D, Lambertucci C, Spinaci A, Volpini R, Marques-Morgado I, Coelho JE, Lopes LV, Marucci G, Buccioni M (2020) Adenosine receptors as neuroinflammation modulators: role of A_1 agonists and A_{2A} antagonists. Cell 9(7):1739
14. Pinna A, Wardas J, Domenici MR, Popoli P, Cossu G, Morelli M (2017) Control of motor function by adenosine A_{2A} receptors in Parkinson's and Huntington's disease. In: Blum D, Lopes L (eds) Adenosine receptors in neurodegenerative diseases. Elsevier Inc., pp 187–213
15. Merighi S, Borea PA, Varani K, Vincenzi F, Jacobson KA, Gessi S (2022) A_{2A} adenosine receptor antagonists in neurodegenerative diseases. Curr Med Chem 29(24):4138–4151
16. Zhang J, Yan W, Duan W, Wuthrich K, Cheng J (2020) Tumor immunotherapy using A_{2A} adenosine receptor antagonists. Pharmaceuticals (Basel) 13(9):237
17. Franco R, Rivas-Santisteban R, Navarro G, Reyes-Resina I (2021) Adenosine receptor antagonists to combat cancer and to boost anti-cancer chemotherapy and immunotherapy. Cell 10(11):2831
18. Knutsen LJ, Weiss SM (2001) KW-6002 (Kyowa Hakko Kogyo). Curr Opin Investig Drugs 2(5):668–673
19. Weiss SM, Benwell K, Cliffe IA, Gillespie RJ, Knight AR, Lerpiniere J, Misra A, Pratt RM, Revell D, Upton R, Dourish CT (2003) Discovery of nonxanthine adenosine A_{2A} receptor antagonists for the treatment of Parkinson's disease. Neurology 61(11 Suppl 6):S101–S106
20. Berger AA, Winnick A, Welschmeyer A, Kaneb A, Berardino K, Cornett EM, Kaye AD, Viswanath O, Urits I (2020) Istradefylline to treat patients with Parkinson's disease experiencing “Off” episodes: a comprehensive review. Neurol Int 12(3):109–129
21. Chen JF, Cunha RA (2020) The belated US FDA approval of the adenosine A_{2A} receptor antagonist istradefylline for treatment of Parkinson's disease. Purinergic Signal 16(2):167–174
22. Jacobson KA, Gao ZG, Matricon P, Eddy MT, Carlsson J (2022) Adenosine A_{2A} receptor antagonists: from caffeine to selective non-xanthines. Br J Pharmacol 179(14):3496–3511
23. Neustadt BR, Hao J, Lindo N, Greenlee WJ, Stamford AW, Tulshian D, Ongini E, Hunter J, Monopoli A, Bertorelli R, Foster C, Arik L, Lachowicz J, Ng K, Feng KI (2007) Potent,

selective, and orally active adenosine A_{2A} receptor antagonists: arylpiperazine derivatives of pyrazolo[4,3-*e*]-1,2,4-triazolo[1,5-*c*]pyrimidines. Bioorg Med Chem Lett 17(5):1376–1380
24. LeWitt PA, Aradi SD, Hauser RA, Rascol O (2020) The challenge of developing adenosine A (2A) antagonists for Parkinson disease: Istradefylline, preladenant, and tozadenant. Parkinsonism Relat Disord 80:S54–S63
25. Borodovsky A, Barbon CM, Wang Y, Ye M, Prickett L, Chandra D, Shaw J, Deng N, Sachsenmeier K, Clarke JD, Linghu B, Brown GA, Brown J, Congreve M, Cheng RK, Dore AS, Hurrell E, Shao W, Woessner R, Reimer C, Drew L, Fawell S, Schuller AG, Mele DA (2020) Small molecule AZD4635 inhibitor of $A_{2A}R$ signaling rescues immune cell function including CD103(+) dendritic cells enhancing anti-tumor immunity. J Immunother Cancer 8(2):e000417
26. Wang Z, Yu L, Wang Y, Wang C, Mu Q, Liu X, Yu M, Wang KN, Yao G, Yu Z (2022) Dynamic adjust of non-radiative and radiative attenuation of AIE molecules reinforces NIR-II imaging mediated photothermal therapy and immunotherapy. Adv Sci (Weinh) 9(8):e2104793
27. Gillespie RJ, Bamford SJ, Botting R, Comer M, Denny S, Gaur S, Griffin M, Jordan AM, Knight AR, Lerpiniere J, Leonardi S, Lightowler S, McAteer S, Merrett A, Misra A, Padfield A, Reece M, Saadi M, Selwood DL, Stratton GC, Surry D, Todd R, Tong X, Ruston V, Upton R, Weiss SM (2009) Antagonists of the human A(2A) adenosine receptor. 4. Design, synthesis, and preclinical evaluation of 7-aryltriazolo[4,5-d]pyrimidines. J Med Chem 52(1):33–47
28. Pinna A (2009) Novel investigational adenosine A_{2A} receptor antagonists for Parkinson's disease. Expert Opin Investig Drugs 18(11):1619–1631
29. Willingham SB, Ho PY, Hotson A, Hill C, Piccione EC, Hsieh J, Liu L, Buggy JJ, McCaffery I, Miller RA (2018) $A_{2A}R$ antagonism with CPI-444 induces antitumor responses and augments efficacy to anti-PD-(L)1 and anti-CTLA-4 in preclinical models. Cancer Immunol Res 6(10):1136–1149
30. Iacovelli R, Ciccarese C, Procopio G, Astore S, Cannella MA, Maratta MG, Rizzo M, Verzoni E, Porta C, Tortora G (2022) Current evidence for second-line treatment in metastatic renal cell carcinoma after progression to immune-based combinations. Cancer Treat Rev 105: 102379
31. Seitz L, Jin L, Leleti M, Ashok D, Jeffrey J, Rieger A, Tiessen RG, Arold G, Tan JBL, Powers JP, Walters MJ, Karakunnel J (2019) Safety, tolerability, and pharmacology of AB928, a novel dual adenosine receptor antagonist, in a randomized, phase 1 study in healthy volunteers. Invest New Drugs 37(4):711–721
32. Chiappori AA, Creelan B, Tanvetyanon T, Gray JE, Haura EB, Thapa R, Barlow ML, Chen Z, Chen DT, Beg AA, Boyle TA, Castro J, Morgan L, Morris E, Aregay M, Hurtado FK, Manenti L, Antonia S (2022) Phase I study of Taminadenant (PBF509/NIR178), an adenosine 2A receptor antagonist, with or without Spartalizumab (PDR001), in patients with advanced non-small cell lung cancer. Clin Cancer Res 28(11):2313–2320
33. Buisseret L, Rottey S, De Bono JS, Migeotte A, Delafontaine B, Manickavasagar T, Martinoli C, Wald N, Rossetti M, Gangolli EA, Wiegert E, McIntyre N, Lager JJ, Machiels J-PH (2021) Phase 1 trial of the adenosine A_{2A} receptor antagonist inupadenant (EOS-850): update on tolerability, and antitumor activity potentially associated with the expression of the A2A receptor within the tumor. J Clin Oncol 39(15):2562
34. Weiser T, Weigmann H (2019) Effect of caffeine on the bioavailability and pharmacokinetics of an acetylsalicylic acid-paracetamol combination: results of a phase I study. Adv Ther 36(3): 597–607
35. Nonaka H, Ichimura M, Takeda M, Nonaka Y, Shimada J, Suzuki F, Yamaguchi K, Kase H (1994) KF17837 ((E)-8-(3,4-dimethoxystyryl)-1,3-dipropyl-7-methylxanthine), a potent and selective adenosine A2 receptor antagonist. Eur J Pharmacol 267(3):335–341
36. Shimada J, Suzuki F, Nonaka H, Ishii A, Ichikawa S (1992) (E)-1,3-dialkyl-7-methyl-8-(3,4,5-trimethoxystyryl)xanthines: potent and selective adenosine A2 antagonists. J Med Chem 35(12):2342–2345

37. Jacobson KA, Gallo-Rodriguez C, Melman N, Fischer B, Maillard M, van Bergen A, van Galen PJ, Karton Y (1993) Structure-activity relationships of 8-styrylxanthines as A2-selective adenosine antagonists. J Med Chem 36(10):1333–1342
38. Petzer JP, Castagnoli Jr N, Schwarzschild MA, Chen JF, Van der Schyf CJ (2009) Dual-target-directed drugs that block monoamine oxidase B and adenosine A(2A) receptors for Parkinson's disease. Neurotherapeutics 6(1):141–151
39. Stossel A, Schlenk M, Hinz S, Kuppers P, Heer J, Gutschow M, Muller CE (2013) Dual targeting of adenosine A(2A) receptors and monoamine oxidase B by 4H-3,1-benzothiazin-4-ones. J Med Chem 56(11):4580–4596
40. Daly JW, Padgett WL, Shamim MT (1986) Analogues of caffeine and theophylline: effect of structural alterations on affinity at adenosine receptors. J Med Chem 29(7):1305–1308
41. Sauer R, Maurinsh J, Reith U, Fulle F, Klotz KN, Muller CE (2000) Water-soluble phosphate prodrugs of 1-propargyl-8-styrylxanthine derivatives, A(2A)-selective adenosine receptor antagonists. J Med Chem 43(3):440–448
42. Vollmann K, Qurishi R, Hockemeyer J, Muller CE (2008) Synthesis and properties of a new water-soluble prodrug of the adenosine A_{2A} receptor antagonist MSX-2. Molecules 13(2): 348–359
43. Jacobson KA, Muller CE (2016) Medicinal chemistry of adenosine, P2Y and P2X receptors. Neuropharmacology 104:31–49
44. Johnson SM, Vasdev RMS, Miller MM, Baker TL, Watters JJ (2021) Adenosine A2a receptors modulate TrkB receptor-dependent respiratory plasticity in neonatal rats. Respir Physiol Neurobiol 294:103743
45. Guo F, Wang X, Guo Y, Wan W, Cui Y, Wang J, Liu W (2022) Shenfu administration improves cardiac fibrosis in rats with myocardial ischemia-reperfusion through adenosine A_{2A} receptor activation. Hum Exp Toxicol 41:9603271221077684
46. Santerre JL, Nunes EJ, Kovner R, Leser CE, Randall PA, Collins-Praino LE, Lopez Cruz L, Correa M, Baqi Y, Muller CE, Salamone JD (2012) The novel adenosine A(2A) antagonist prodrug MSX-4 is effective in animal models related to motivational and motor functions. Pharmacol Biochem Behav 102(4):477–487
47. Francis JE, Cash WD, Psychoyos S, Ghai G, Wenk P, Friedmann RC, Atkins C, Warren V, Furness P, Hyun JL et al (1988) Structure-activity profile of a series of novel triazoloquinazoline adenosine antagonists. J Med Chem 31(5):1014–1020
48. Baraldi PG, Cacciari B, Romagnoli R, Spalluto G, Moro S, Klotz KN, Leung E, Varani K, Gessi S, Merighi S, Borea PA (2000) Pyrazolo[4,3-e]1,2,4-triazolo[1,5-c]pyrimidine derivatives as highly potent and selective human A_3 adenosine receptor antagonists: influence of the chain at the N(8) pyrazole nitrogen. J Med Chem 43(25):4768–4780
49. Sarges R, Howard HR, Browne RG, Lebel LA, Seymour PA, Koe BK (1990) 4-amino[1,2,4] triazolo[4,3-a]quinoxalines. A novel class of potent adenosine receptor antagonists and potential rapid-onset antidepressants. J Med Chem 33(8):2240–2254
50. Colotta V, Catarzi D, Varano F, Cecchi L, Filacchioni G, Martini C, Trincavelli L, Lucacchini A (1999) 4-Amino-6-benzylamino-1,2-dihydro-2-phenyl-1,2,4-triazolo [4,3-alpha]-quinoxalin-1-one: a new A_{2A} adenosine receptor antagonist with high selectivity versus A_1 receptors. Arch Pharm (Weinheim) 332(2):39–41
51. Colotta V, Catarzi D, Varano F, Filacchioni G, Martini C, Trincavelli L, Lucacchini A (2003) Synthesis of 4-amino-6-(hetero)arylalkylamino-1,2,4-triazolo[4,3-a]quinoxalin-1-one derivatives as potent A(2A) adenosine receptor antagonists. Bioorg Med Chem 11(24):5509–5518
52. Gatta F, Del Giudice MR, Borioni A, Borea PA, Dionisotti S, Ongini E (1993) Synthesis of imidazo[1,2-c]pyrazolo[4,3-e]pyrimidines, pyrazolo[4,3-e]1,2,4-triazolo[1,5-c]pyrimidines and 1,2,4-triazolo[5,1-i]purines: new potent adenosine A2 receptor antagonists. Eur J Med Chem 28:569–577
53. Baraldi PG, Cacciari B, Spalluto G, Pineda de las Infantas y Villatoro MJ, Zocchi C, Dionisotti S, Ongini E (1996) Pyrazolo[4,3-e]-1,2,4-triazolo[1,5-c]pyrimidine derivatives: potent and selective A(2A) adenosine antagonists. J Med Chem 39(5):1164–1171

54. Kecskes M, Kumar TS, Yoo L, Gao ZG, Jacobson KA (2010) Novel Alexa Fluor-488 labeled antagonist of the A(2A) adenosine receptor: application to a fluorescence polarization-based receptor binding assay. Biochem Pharmacol 80(4):506–511
55. Cacciari B, Pastorin G, Spalluto G (2003) Medicinal chemistry of A_{2A} adenosine receptor antagonists. Curr Top Med Chem 3(4):403–411
56. Hauser RA, Stocchi F, Rascol O, Huyck SB, Capece R, Ho TW, Sklar P, Lines C, Michelson D, Hewitt D (2015) Preladenant as an adjunctive therapy with levodopa in Parkinson disease two randomized clinical trials and lessons learned. JAMA Neurol 72(12): 1491–1500
57. Crosignani S, Gomes B, Houthuys E (2018) 2-Oxo-thiazole derivatives as A_{2A} inhibitors and compounds for use in the treatment of cancers WO2018178338A1
58. Basu S, Barawkar DA, Ramdas V, Naykodi M, Shejul YD, Patel M, Thorat S, Panmand A, Kashinath K, Bonagiri R, Prasad V, Bhat G, Quraishi A, Chaudhary S, Magdum A, Meru AV, Ghosh I, Bhamidipati RK, Raje AA, Madgula VLM, De S, Rouduri SR, Palle VP, Chugh A, Hariharan N, Mookhtiar KA (2017) Discovery of potent and selective A_{2A} antagonists with efficacy in animal models of Parkinson's disease and depression. ACS Med Chem Lett 8(8): 835–840
59. Mishra CB, Barodia SK, Prakash A, Senthil Kumar JB, Luthra PM (2010) Novel 8-(furan-2-yl)-3-substituted thiazolo [5,4-e][1,2,4] triazolo[1,5-c] pyrimidine-2(3H)-thione derivatives as potential adenosine A(2A) receptor antagonists. Bioorg Med Chem 18(7):2491–2500
60. Kumari N, Mishra CB, Prakash A, Kumar N, Mongre R, Luthra PM (2014) 8-(Furan-2-yl)-3-phenethylthiazolo[5,4-e][1,2,4]triazolo[1,5-c]pyrimidine-2(3H)-t hione as novel, selective and potent adenosine A(2A) receptor antagonist. Neurosci Lett 558:203–207
61. Poucher SM, Keddie JR, Singh P, Stoggall SM, Caulkett PW, Jones G, Coll MG (1995) The in vitro pharmacology of ZM 241385, a potent, non-xanthine A2a selective adenosine receptor antagonist. Br J Pharmacol 115(6):1096–1102
62. Jacobson KA, Gao ZG (2006) Adenosine receptors as therapeutic targets. Nat Rev Drug Discov 5(3):247–264
63. Dore AS, Robertson N, Errey JC, Ng I, Hollenstein K, Tehan B, Hurrell E, Bennett K, Congreve M, Magnani F, Tate CG, Weir M, Marshall FH (2011) Structure of the adenosine A(2A) receptor in complex with ZM241385 and the xanthines XAC and caffeine. Structure 19(9):1283–1293
64. Cristalli G, Muller CE, Volpini R (2009) Recent developments in adenosine A_{2A} receptor ligands. Handb Exp Pharmacol 193:59–98
65. de Zwart M, Vollinga RC, Beukers MW, Sleegers DF, Von Frijtag Drabbe Künzel J, de Groote M, IJzerman AP (1999) Potent antagonists for the human adenosine A(2B) receptor. Derivatives of the triazolotriazine adenosine receptor antagonist ZM241385 with high affinity. Drug Dev Res 48:95–103
66. Spinaci A, Lambertucci C, Buccioni M, Dal Ben D, Graiff C, Barbalace MC, Hrelia S, Angeloni C, Tayebati SK, Ubaldi M, Masi A, Klotz KN, Volpini R, Marucci G (2022) A_{2A} adenosine receptor antagonists: are triazolotriazine and purine scaffolds interchangeable? Molecules 27(8):2386
67. Shah U, Hodgson R (2010) Recent progress in the discovery of adenosine A(2A) receptor antagonists for the treatment of Parkinson's disease. Curr Opin Drug Discov Devel 13(4): 466–480
68. Bamford SJ, Gillespie RJ, Todd RS (2021) Triazolo[4,5-*d*] pyramidine derivatives and their use as purine receptor antagonists. US20210115054A1
69. Falsini M, Squarcialupi L, Catarzi D, Varano F, Betti M, Dal Ben D, Marucci G, Buccioni M, Volpini R, De Vita T, Cavalli A, Colotta V (2017) The 1,2,4-triazolo[4,3-*a*]pyrazin-3-one as a versatile scaffold for the design of potent adenosine human receptor antagonists. Structural investigations to target the A_{2A} receptor subtype. J Med Chem 60(13):5772–5790
70. Falsini M, Catarzi D, Varano F, Dal Ben D, Marucci G, Buccioni M, Volpini R, Di Cesare Mannelli L, Ghelardini C, Colotta V (2019) Novel 8-amino-1,2,4-triazolo[4,3-a]pyrazin-3-one

derivatives as potent human adenosine A_1 and A_{2A} receptor antagonists. Evaluation of their protective effect against beta-amyloid-induced neurotoxicity in SH-SY5Y cells. Bioorg Chem 87:380–394
71. Falsini M, Catarzi D, Varano F, Ceni C, Dal Ben D, Marucci G, Buccioni M, Volpini R, Di Cesare ML, Lucarini E, Ghelardini C, Bartolucci G, Menicatti M, Colotta V (2019) Antioxidant-conjugated 1,2,4-triazolo[4,3-a]pyrazin-3-one derivatives: highly potent and selective human A_{2A} adenosine receptor antagonists possessing protective efficacy in neuropathic pain. J Med Chem 62(18):8511–8531
72. Varano F, Catarzi D, Vincenzi F, Betti M, Falsini M, Ravani A, Borea PA, Colotta V, Varani K (2016) Design, synthesis, and pharmacological characterization of 2-(2-Furanyl)thiazolo [5,4-*d*]pyrimidine-5,7-diamine derivatives: new highly potent A_{2A} adenosine receptor inverse agonists with antinociceptive activity. J Med Chem 59(23):10564–10576
73. Varano F, Catarzi D, Falsini M, Vincenzi F, Pasquini S, Varani K, Colotta V (2018) Identification of novel thiazolo[5,4-*d*]pyrimidine derivatives as human A_1 and A_{2A} adenosine receptor antagonists/inverse agonists. Bioorg Med Chem 26(12):3688–3695
74. Varano F, Catarzi D, Falsini M, Dal Ben D, Buccioni M, Marucci G, Volpini R, Colotta V (2019) Novel human adenosine receptor antagonists based on the 7-amino-thiazolo[5,4-d] pyrimidine scaffold. Structural investigations at the 2-, 5- and 7-positions to enhance affinity and tune selectivity. Bioorg Med Chem Lett 29(4):563–569
75. Varano F, Catarzi D, Vigiani E, Dal Ben D, Buccioni M, Marucci G, Di Cesare ML, Lucarini E, Ghelardini C, Volpini R, Colotta V (2021) Design and synthesis of novel thiazolo[5,4-d]pyrimidine derivatives with high affinity for both the adenosine A_1 and A_{2A} receptors, and efficacy in animal models of depression. Pharmaceuticals (Basel) 14(7):657
76. Deb PK, Mailavaram R, Chandrasekaran B, Kaki VR, Kaur R, Kachler S, Klotz KN, Akkinepally RR (2018) Synthesis, adenosine receptor binding and molecular modelling studies of novel thieno[2,3-d]pyrimidine derivatives. Chem Biol Drug Des 91(4):962–969
77. Shook BC, Chakravarty D, Barbay JK, Wang A, Leonard K, Alford V, Powell MT, Rassnick S, Scannevin RH, Carroll K, Wallace N, Crooke J, Ault M, Lampron L, Westover L, Rhodes K, Jackson PF (2013) Substituted thieno[2,3-d]pyrimidines as adenosine A_{2A} receptor antagonists. Bioorg Med Chem Lett 23(9):2688–2691
78. Flohr A, Moreau JL, Poli SM, Riemer C, Steward L (2005) 4-Hydroxy-4-methyl-piperidine-1-carboxylic acid (4-methoxy-7-morpholin-4-yl-benzothiazol-2-yl)-amide. US-20050261289
79. Basu S, Barawkar DA, Thorat S, Shejul YD, Patel M, Naykodi M, Jain V, Salve Y, Prasad V, Chaudhary S, Ghosh I, Bhat G, Quraishi A, Patil H, Ansari S, Menon S, Unadkat V, Thakare R, Seervi MS, Meru AV, De S, Bhamidipati RK, Rouduri SR, Palle VP, Chug A, Mookhtiar KA (2017) Design, synthesis of novel, potent, selective, orally bioavailable adenosine A_{2A} receptor antagonists and their biological evaluation. J Med Chem 60(2): 681–694
80. Renk DR, Skraban M, Bier D, Schulze A, Wabbals E, Wedekind F, Neumaier F, Neumaier B, Holschbach M (2021) Design, synthesis and biological evaluation of Tozadenant analogues as adenosine A_{2A} receptor ligands. Eur J Med Chem 214:113214
81. Duroux R, Renault N, Cuelho JE, Agouridas L, Blum D, Lopes LV, Melnyk P, Yous S (2017) Design, synthesis and evaluation of 2-aryl benzoxazoles as promising hit for the A_{2A} receptor. J Enzyme Inhib Med Chem 32(1):850–864
82. Duroux R, Agouridas L, Renault N, El Bakali J, Furman C, Melnyk P, Yous S (2018) Antagonists of the adenosine A_{2A} receptor based on a 2-arylbenzoxazole scaffold: investigation of the C5- and C7-positions to enhance affinity. Eur J Med Chem 144:151–163
83. Olsson RA, Khouri EM, Bedynek Jr JL, McLean J (1979) Coronary vasoactivity of adenosine in the conscious dog. Circ Res 45(4):468–478
84. Bruns RF (1980) Adenosine receptor activation in human fibroblasts: nucleoside agonists and antagonists. Can J Physiol Pharmacol 58(6):673–691
85. Camaioni E, Costanzi S, Vittori S, Volpini R, Klotz KN, Cristalli G (1998) New substituted 9-alkylpurines as adenosine receptor ligands. Bioorg Med Chem 6(5):523–533

86. Lambertucci C, Vittori S, Mishra RC, Dal Ben D, Klotz KN, Volpini R, Cristalli G (2007) Synthesis and biological activity of trisubstituted adenines as A_{2A} adenosine receptor antagonists. Nucleosides Nucleotides Nucleic Acids 26(10–12):1443–1446
87. Lambertucci C, Antonini I, Buccioni M, Dal Ben D, Kachare DD, Volpini R, Klotz KN, Cristalli G (2009) 8-Bromo-9-alkyl adenine derivatives as tools for developing new adenosine A_{2A} and A_{2B} receptors ligands. Bioorg Med Chem 17(7):2812–2822
88. Volpini R, Dal Ben D, Lambertucci C, Marucci G, Mishra RC, Ramadori AT, Klotz KN, Trincavelli ML, Martini C, Cristalli G (2009) Adenosine A_{2A} receptor antagonists: new 8-substituted 9-ethyladenines as tools for in vivo rat models of Parkinson's disease. ChemMedChem 4(6):1010–1019
89. Thomas A, Buccioni M, Dal Ben D, Lambertucci C, Marucci G, Santinelli C, Spinaci A, Kachler S, Klotz KN, Volpini R (2016) The length and flexibility of the 2-substituent of 9-ethyladenine derivatives modulate affinity and selectivity for the human A_{2A} adenosine receptor. ChemMedChem 11(16):1829–1839
90. Pinna A, Volpini R, Cristalli G, Morelli M (2005) New adenosine A_{2A} receptor antagonists: actions on Parkinson's disease models. Eur J Pharmacol 512(2–3):157–164
91. Pinna A, Tronci E, Schintu N, Simola N, Volpini R, Pontis S, Cristalli G, Morelli M (2010) A new ethyladenine antagonist of adenosine A(2A) receptors: behavioral and biochemical characterization as an antiparkinsonian drug. Neuropharmacology 58(3):613–623
92. Minetti P, Tinti MO, Carminati P, Castorina M, Di Cesare MA, Di Serio S, Gallo G, Ghirardi O, Giorgi F, Giorgi L, Piersanti G, Bartoccini F, Tarzia G (2005) 2-n-Butyl-9-methyl-8-[1,2,3]triazol-2-yl-9H-purin-6-ylamine and analogues as A_{2A} adenosine receptor antagonists. Design, synthesis, and pharmacological characterization. J Med Chem 48(22): 6887–6896
93. Piersanti G, Bartoccini F, Lucarini S, Cabri W, Stasi MA, Riccioni T, Borsini F, Tarzia G, Minetti P (2013) Synthesis and biological evaluation of metabolites of 2-n-butyl-9-methyl-8-[1,2,3]triazol-2-yl-9H-purin-6-ylamine (ST1535), a potent antagonist of the A_{2A} adenosine receptor for the treatment of Parkinson's disease. J Med Chem 56(13):5456–5463
94. Endo K, Deguchi K, Matsunaga H, Tomaya K, Yamada K (2014) 8-Substituted 2-alkynyl-N (9)-propargyladenines as A_{2A} adenosine receptor antagonists. Bioorg Med Chem 22(12): 3072–3082
95. Kiselgof E, Tulshian DB, Arik L, Zhang H, Fawzi A (2005) 6-(2-Furanyl)-9H-purin-2-amine derivatives as A_{2A} adenosine antagonists. Bioorg Med Chem Lett 15(8):2119–2122
96. Cole AG, Stauffer TM, Rokosz LL, Metzger A, Dillard LW, Zeng W, Henderson I (2009) Synthesis of 2-amino-5-benzoyl-4-(2-furyl)thiazoles as adenosine A(2A) receptor antagonists. Bioorg Med Chem Lett 19(2):378–381
97. Sams AG, Mikkelsen GK, Larsen M, Torup L, Brennum LT, Schroder TJ, Bang-Andersen B (2010) Hit-to-lead optimization of a series of carboxamides of ethyl 2-amino-4-phenylthiazole-5-carboxylates as novel adenosine A_{2A} receptor antagonists. Bioorg Med Chem Lett 20(17):5241–5244
98. Sams AG, Mikkelsen GK, Larsen M, Langgard M, Howells ME, Schroder TJ, Brennum LT, Torup L, Jorgensen EB, Bundgaard C, Kreilgard M, Bang-Andersen B (2011) Discovery of phosphoric acid mono-{2-[(E/Z)-4-(3,3-dimethyl-butyrylamino)-3,5-difluoro-benzoylimino]-thiazol-3-ylmethyl} ester (Lu AA47070): a phosphonooxymethylene prodrug of a potent and selective hA(2A) receptor antagonist. J Med Chem 54(3):751–764
99. Collins LE, Sager TN, Sams AG, Pennarola A, Port RG, Shahriari M, Salamone JD (2012) The novel adenosine A_{2A} antagonist Lu AA47070 reverses the motor and motivational effects produced by dopamine D2 receptor blockade. Pharmacol Biochem Behav 100(3):498–505
100. Mishra CB, Sharma D, Prakash A, Kumari N, Kumar N, Luthra PM (2013) Design and synthesis of (4E)-4-(4-substitutedbenzylideneamino)-3-substituted-2,3-dihydro-2-thioxothiazole-5-carbonitrile as novel A_{2A} receptor antagonists. Bioorg Med Chem 21(19): 6077–6083

101. Carlsson J, Tosh DK, Phan K, Gao ZG, Jacobson KA (2012) Structure-activity relationships and molecular modeling of 1,2,4-triazoles as adenosine receptor antagonists. ACS Med Chem Lett 3(9):715–720
102. Mantri M, de Graaf O, van Veldhoven J, Goblyos A, von Frijtag Drabbe Kunzel JK, Mulder-Krieger T, Link R, de Vries H, Beukers MW, Brussee J, Ijzerman AP (2008) 2-Amino-6-furan-2-yl-4-substituted nicotinonitriles as A_{2A} adenosine receptor antagonists. J Med Chem 51(15):4449–4455
103. Betti M, Catarzi D, Varano F, Falsini M, Varani K, Vincenzi F, Pasquini S, di Cesare ML, Ghelardini C, Lucarini E, Dal Ben D, Spinaci A, Bartolucci G, Menicatti M, Colotta V (2019) Modifications on the amino-3,5-dicyanopyridine core to obtain multifaceted adenosine receptor ligands with antineuropathic activity. J Med Chem 62(15):6894–6912
104. Catarzi D, Varano F, Vigiani E, Calenda S, Melani F, Varani K, Vincenzi F, Pasquini S, Mennini N, Nerli G, Dal Ben D, Volpini R, Colotta V (2022) 4-Heteroaryl substituted amino-3,5-dicyanopyridines as new adenosine receptor ligands: novel insights on structure-activity relationships and perspectives. Pharmaceuticals (Basel) 15(4):478
105. Slee DH, Zhang X, Moorjani M, Lin E, Lanier MC, Chen Y, Rueter JK, Lechner SM, Markison S, Malany S, Joswig T, Santos M, Gross RS, Williams JP, Castro-Palomino JC, Crespo MI, Prat M, Gual S, Diaz JL, Wen J, O'Brien Z, Saunders J (2008) Identification of novel, water-soluble, 2-amino-N-pyrimidin-4-yl acetamides as A_{2A} receptor antagonists with in vivo efficacy. J Med Chem 51(3):400–406
106. Zhang X, Tellew JE, Luo Z, Moorjani M, Lin E, Lanier MC, Chen Y, Williams JP, Saunders J, Lechner SM, Markison S, Joswig T, Petroski R, Piercey J, Kargo W, Malany S, Santos M, Gross RS, Wen J, Jalali K, O'Brien Z, Stotz CE, Crespo MI, Diaz JL, Slee DH (2008) Lead optimization of 4-acetylamino-2-(3,5-dimethylpyrazol-1-yl)-6-pyridylpyrimidines as A_{2A} adenosine receptor antagonists for the treatment of Parkinson's disease. J Med Chem 51(22): 7099–7110
107. Zheng J, Yang Z, Li X, Li L, Ma H, Wang M, Zhang H, Zhen X, Zhang X (2014) Optimization of 6-heterocyclic-2-(1H-pyrazol-1-yl)-N-(pyridin-2-yl)pyrimidin-4-amine as potent adenosine A_{2A} receptor antagonists for the treatment of Parkinson's disease. ACS Chem Nerosci 5(8): 674–682
108. Yang Z, Li L, Zheng J, Ma H, Tian S, Li J, Zhang H, Zhen X, Zhang X (2016) Identification of a new series of potent adenosine A_{2A} receptor antagonists based on 4-amino-5-carbonitrile pyrimidine template for the treatment of Parkinson's disease. ACS Chem Nerosci 7(11): 1575–1584
109. Camacho Gomez JA, Castro-Palomino Laria JC (2011) 4-Aminopyrimidine derivatives and their as as adenosine A_{2A} receptor antagonists WO2011/121418 A9
110. Walters MJ, Piovesan D, Tan J, DiRenzo D, Yin F, Miles D, Leleti MR, Park T, Soriano F, Sharif E, Schindler U, Powers JP (2018) Abstract 5556: combining adenosine receptor inhibition with AB928 and chemotherapy results in greater immune activation and tumor control. Cancer Res 78(13):5556
111. Walters MJ, Tan JB, Becker A, Yi F, Park T, Leleti MR, Rosen B, Sharif E, Debien L, Young S, Lim WH, Garrido-Shaqfeh S, Jaen JC, Powers JP (2017) Characterization of the potent and selective A_{2A}R antagonist AB928 for the treatment of cancer. Cancer Res 77(13): 4572
112. Li Z, Kou L, Fu X, Xie Z, Xu M, Guo L, Lin T, Gong S, Zhang S, Liu M (2022) Design, synthesis, and biological evaluation of triazole-pyrimidine-methylbenzonitrile derivatives as dual A_{2A}/A_{2B} adenosine receptor antagonists. J Enzyme Inhib Med Chem 37(1):1514–1526
113. Yonishi S, Aoki S, Matsushima Y, Akahane A (2005) Preparation of pyrazines as adenosine A_1 and A_{2A} receptor antagonists and their pharmaceutical compositions. WO 2005040151 A1
114. Mihara T, Mihara K, Yarimizu J, Mitani Y, Matsuda R, Yamamoto H, Aoki S, Akahane A, Iwashita A, Matsuoka N (2007) Pharmacological characterization of a novel, potent adenosine A_1 and A_{2A} receptor dual antagonist, 5-[5-amino-3-(4-fluorophenyl)pyrazin-2-yl]-1-isopropylpyridine-2(1H)-one (ASP5854), in models of Parkinson's disease and cognition. J Pharmacol Exp Ther 323(2):708–719

115. Congreve M, Andrews SP, Mason JS, Richardson CM, Brown GA (2011) Preparation of 1,2,4-triazin-3-amine derivatives as A_1 and A_{2A} receptor inhibitors useful in the treatment of diseases. WO 2011095625
116. Congreve M, Andrews SP, Dore AS, Hollenstein K, Hurrell E, Langmead CJ, Mason JS, Ng IW, Tehan B, Zhukov A, Weir M, Marshall FH (2012) Discovery of 1,2,4-triazine derivatives as adenosine A(2A) antagonists using structure based drug design. J Med Chem 55(5): 1898–1903
117. Langmead CJ, Andrews SP, Congreve M, Errey JC, Hurrell E, Marshall FH, Mason JS, Richardson CM, Robertson N, Zhukov A, Weir M (2012) Identification of novel adenosine A(2A) receptor antagonists by virtual screening. J Med Chem 55(5):1904–1909
118. Lebon G, Warne T, Edwards PC, Bennett K, Langmead CJ, Leslie AG, Tate CG (2011) Agonist-bound adenosine A_{2A} receptor structures reveal common features of GPCR activation. Nature 474(7352):521–525
119. Xu F, Wu H, Katritch V, Han GW, Jacobson KA, Gao ZG, Cherezov V, Stevens RC (2011) Structure of an agonist-bound human A_{2A} adenosine receptor. Science 332(6027):322–327
120. Lebon G, Edwards PC, Leslie AG, Tate CG (2015) Molecular determinants of CGS21680 binding to the human adenosine A_{2A} receptor. Mol Pharmacol 87(6):907–915
121. Amelia T, van Veldhoven JPD, Falsini M, Liu R, Heitman LH, van Westen GJP, Segala E, Verdon G, Cheng RKY, Cooke RM, van der Es D, IJzerman AP (2021) Crystal structure and subsequent ligand design of a nonriboside partial agonist bound to the adenosine A_{2A} receptor. J Med Chem 64(7):3827–3842
122. Garcia-Nafria J, Lee Y, Bai X, Carpenter B, Tate CG (2018) Cryo-EM structure of the adenosine A(2A) receptor coupled to an engineered heterotrimeric G protein. Elife 7:e35946
123. Doré AS, Robertson N, Errey JC, Ng I, Hollenstein K, Tehan B, Hurrell E, Bennett K, Congreve M, Magnani F, Tate CG, Weir M, Marshall FH (2011) Structure of the adenosine A_{2A} receptor in complex with ZM241385 and the xanthines XAC and caffeine. Structure 19(9):1283–1293
124. Liu W, Chun E, Thompson AA, Chubukov P, Xu F, Katritch V, Han GW, Roth CB, Heitman LH, IJzerman AP, Cherezov V, Stevens RC (2012) Structural basis for allosteric regulation of GPCRs by sodium ions. Science 337(6091):232–236
125. Segala E, Guo D, Cheng RK, Bortolato A, Deflorian F, Dore AS, Errey JC, Heitman LH, IJzerman AP, Marshall FH, Cooke RM (2016) Controlling the dissociation of ligands from the adenosine A_{2A} receptor through modulation of salt bridge strength. J Med Chem 59(13): 6470–6479
126. Weinert T, Olieric N, Cheng R, Brunle S, James D, Ozerov D, Gashi D, Vera L, Marsh M, Jaeger K, Dworkowski F, Panepucci E, Basu S, Skopintsev P, Dore AS, Geng T, Cooke RM, Liang M, Prota AE, Panneels V, Nogly P, Ermler U, Schertler G, Hennig M, Steinmetz MO, Wang M, Standfuss J (2017) Serial millisecond crystallography for routine room-temperature structure determination at synchrotrons. Nat Commun 8(1):542
127. https://www.parkinson.org. Accessed 3 Nob 2022
128. Selkoe DJ, Hardy J (2016) The amyloid hypothesis of Alzheimer's disease at 25 years. EMBO Mol Med 8(6):595–608
129. Nortley R, Korte N, Izquierdo P, Hirunpattarasilp C, Mishra A, Jaunmuktane Z, Kyrargyri V, Pfeiffer T, Khennouf L, Madry C, Gong H, Richard-Loendt A, Huang W, Saito T, Saido TC, Brandner S, Sethi H, Attwell D (2019) Amyloid beta oligomers constrict human capillaries in Alzheimer's disease via signaling to pericytes. Science 365(6450):eaav9518
130. Griffin WS, Stanley LC, Ling C, White L, MacLeod V, Perrot LJ, White 3rd CL, Araoz C (1989) Brain interleukin 1 and S-100 immunoreactivity are elevated in down syndrome and Alzheimer disease. Proc Natl Acad Sci U S A 86(19):7611–7615
131. Simard AR, Soulet D, Gowing G, Julien JP, Rivest S (2006) Bone marrow-derived microglia play a critical role in restricting senile plaque formation in Alzheimer's disease. Neuron 49(4): 489–502

132. Park KW, Kim EJ, Han HJ, Shim YS, Kwon JC, Ku BD, Park KH, Yi HA, Kim KK, Yang DW, Lee HW, Kang H, Kwon OD, Kim S, Lee JH, Chung EJ, Park SW, Park MY, Yoon B, Kim BC, Seo SW, Choi SH (2017) Efficacy and tolerability of rivastigmine patch therapy in patients with mild-to-moderate Alzheimer's dementia associated with minimal and moderate ischemic white matter hyperintensities: a multicenter prospective open-label clinical trial. PloS One 12(8):e0182123
133. Stoiljkovic M, Horvath TL, Hajos M (2021) Therapy for Alzheimer's disease: missing targets and functional markers? Ageing Res Rev 68:101318
134. Johansson M, Stomrud E, Lindberg O, Westman E, Johansson PM, van Westen D, Mattsson N, Hansson O (2020) Apathy and anxiety are early markers of Alzheimer's disease. Neurobiol Aging 85:74–82
135. Cummings J (2021) Correction to: new approaches to symptomatic treatments for Alzheimer's disease. Mol Neurodegener 16(1):21
136. Long JM, Holtzman DM (2019) Alzheimer disease: an update on pathobiology and treatment strategies. Cell 179(2):312–339
137. Mullard A (2021) FDA approval for Biogen's aducanumab sparks Alzheimer disease firestorm. Nat Rev Drug Discov 20(7):496
138. Mouro FM, Rombo DM, Dias RB, Ribeiro JA, Sebastiao AM (2018) Adenosine A_{2A} receptors facilitate synaptic NMDA currents in CA1 pyramidal neurons. Br J Pharmacol 175(23): 4386–4397
139. Goncalves FQ, Lopes JP, Silva HB, Lemos C, Silva AC, Goncalves N, Tome AR, Ferreira SG, Canas PM, Rial D, Agostinho P, Cunha RA (2019) Synaptic and memory dysfunction in a beta-amyloid model of early Alzheimer's disease depends on increased formation of ATP-derived extracellular adenosine. Neurobiol Dis 132:104570
140. Cunha GM, Canas PM, Melo CS, Hockemeyer J, Muller CE, Oliveira CR, Cunha RA (2008) Adenosine A_{2A} receptor blockade prevents memory dysfunction caused by beta-amyloid peptides but not by scopolamine or MK-801. Exp Neurol 210(2):776–781
141. Cunha RA (2005) Neuroprotection by adenosine in the brain: from A(1) receptor activation to A(2A) receptor blockade. Purinergic Signal 1(2):111–134
142. Armstrong MJ, Okun MS (2020) Diagnosis and treatment of Parkinson disease: a review. JAMA 323(6):548–560
143. Barrett MJ, Sargent L, Nawaz H, Weintraub D, Price ET, Willis AW (2021) Antimuscarinic anticholinergic medications in Parkinson disease: to prescribe or deprescribe? Mov Disord Clin Pract 8(8):1181–1188
144. Isaacson SH, Ballard CG, Kreitzman DL, Coate B, Norton JC, Fernandez HH, Ilic TV, Azulay JP, Ferreira JJ, Abler V, Stankovic S, Study Group (2021) Efficacy results of pimavanserin from a multi-center, open-label extension study in Parkinson's disease psychosis patients. Parkinsonism Relat Disord 87:25–31
145. Layton JB, Forns J, Turner ME, Dempsey C, Bartsch JL, Anthony MS, Danysh HE, Ritchey ME, Demos G (2022) Falls and fractures in patients with Parkinson's disease-related psychosis treated with Pimavanserin vs atypical antipsychotics: a cohort study. Drugs Real World Outcomes 9(1):9–22
146. Pollack AE, Fink JS (1995) Adenosine antagonists potentiate D2 dopamine-dependent activation of Fos in the striatopallidal pathway. Neuroscience 68(3):721–728
147. Kanda T, Jackson MJ, Smith LA, Pearce RK, Nakamura J, Kase H, Kuwana Y, Jenner P (2000) Combined use of the adenosine A(2A) antagonist KW-6002 with L-DOPA or with selective D1 or D2 dopamine agonists increases antiparkinsonian activity but not dyskinesia in MPTP-treated monkeys. Exp Neurol 162(2):321–327
148. Brooks DJ, Papapetropoulos S, Vandenhende F, Tomic D, He P, Coppell A, O'Neill G (2010) An open-label, positron emission tomography study to assess adenosine A_{2A} brain receptor occupancy of vipadenant (BIIB014) at steady-state levels in healthy male volunteers. Clin Neuropharmacol 33(2):55–60

149. Pawsey S, Donaldson K, Warrington S, Tranter E (2013) A phase I single and repeated dose pharmacokinetic study of oral V81444, a selective non-xanthine adenosine A_{2A} receptor antagonist. J Neurol Sci 333(Suppl 1):e135
150. Pinna A (2014) Adenosine A_{2A} receptor antagonists in Parkinson's disease: progress in clinical trials from the newly approved istradefylline to drugs in early development and those already discontinued. CNS Drugs 28(5):455–474
151. Mizuno Y, Hasegawa K, Kondo T, Kuno S, Yamamoto M, Japanese Istradefylline Study Group (2010) Clinical efficacy of istradefylline (KW-6002) in Parkinson's disease: a randomized, controlled study. Mov Disord 25(10):1437–1443
152. Fernandez HH, Greeley DR, Zweig RM, Wojcieszek J, Mori A, Sussman NM, Group USS (2010) Istradefylline as monotherapy for Parkinson disease: results of the 6002-US-051 trial. Parkinsonism Relat Disord 16(1):16–20
153. Stacy M, Silver D, Mendis T, Sutton J, Mori A, Chaikin P, Sussman NM (2008) A 12-week, placebo-controlled study (6002-US-006) of istradefylline in Parkinson disease. Neurology 70(23):2233–2240
154. LeWitt PA, Guttman M, Tetrud JW, Tuite PJ, Mori A, Chaikin P, Sussman NM, Group USS (2008) Adenosine A_{2A} receptor antagonist istradefylline (KW-6002) reduces "off" time in Parkinson's disease: a double-blind, randomized, multicenter clinical trial (6002-US-005). Ann Neurol 63(3):295–302
155. Pourcher E, Fernandez HH, Stacy M, Mori A, Ballerini R, Chaikin P (2012) Istradefylline for Parkinson's disease patients experiencing motor fluctuations: results of the KW-6002-US-018 study. Parkinsonism Relat Disord 18(2):178–184
156. Hauser RA, Olanow CW, Kieburtz KD, Pourcher E, Docu-Axelerad A, Lew M, Kozyolkin O, Neale A, Resburg C, Meya U, Kenney C, Bandak S (2014) Tozadenant (SYN115) in patients with Parkinson's disease who have motor fluctuations on levodopa: a phase 2b, double-blind, randomised trial. Lancet Neurol 13(8):767–776
157. Patraca I, Martinez N, Busquets O, Marti A, Pedros I, Beas-Zarate C, Marin M, Ettcheto M, Sureda F, Auladell C, Camins A, Folch J (2017) Anti-inflammatory role of leptin in glial cells through p38 MAPK pathway inhibition. Pharmacol Rep 69(3):409–418
158. Mendiola AS, Cardona AE (2018) The IL-1beta phenomena in neuroinflammatory diseases. J Neural Transm (Vienna) 125(5):781–795
159. Taylor JM, Moore Z, Minter MR, Crack PJ (2018) Type-I interferon pathway in neuroinflammation and neurodegeneration: focus on Alzheimer's disease. J Neural Transm (Vienna) 125(5):797–807
160. Mubarak M (2019) Tumor necrosis factor alpha: a major cytokine of brain neuroinflammation. In: Behzadi P (ed) Cytokines. IntechOpen
161. Carvalho K, Faivre E, Pietrowski MJ, Marques X, Gomez-Murcia V, Deleau A, Huin V, Hansen JN, Kozlov S, Danis C, Temido-Ferreira M, Coelho JE, Meriaux C, Eddarkaoui S, Gras SL, Dumoulin M, Cellai L, Neuro CEBBB, Landrieu I, Chern Y, Hamdane M, Buee L, Boutillier AL, Levi S, Halle A, Lopes LV, Blum D (2019) Exacerbation of C1q dysregulation, synaptic loss and memory deficits in tau pathology linked to neuronal adenosine A_{2A} receptor. Brain 142(11):3636–3654
162. Catafau AM, Bullich S (2017) Non-amyloid PET imaging biomarkers for neurodegeneration: focus on tau, alpha-synuclein and neuroinflammation. Curr Alzheimer Res 14(2):169–177
163. Rebola N, Simoes AP, Canas PM, Tome AR, Andrade GM, Barry CE, Agostinho PM, Lynch MA, Cunha RA (2011) Adenosine A_{2A} receptors control neuroinflammation and consequent hippocampal neuronal dysfunction. J Neurochem 117(1):100–111
164. Wei CJ, Li W, Chen JF (2011) Normal and abnormal functions of adenosine receptors in the central nervous system revealed by genetic knockout studies. Biochim Biophys Acta 1808(5): 1358–1379
165. Yu L, Shen HY, Coelho JE, Araujo IM, Huang QY, Day YJ, Rebola N, Canas PM, Rapp EK, Ferrara J, Taylor D, Muller CE, Linden J, Cunha RA, Chen JF (2008) Adenosine A_{2A} receptor

antagonists exert motor and neuroprotective effects by distinct cellular mechanisms. Ann Neurol 63(3):338–346
166. Marucci G, Dal Ben D, Lambertucci C, Navia AM, Spinaci A, Volpini R, Buccioni M (2021) Combined therapy of A_1AR agonists and A_{2A}AR antagonists in neuroinflammation. Molecules 26(4):1188
167. Rebola N, Canas PM, Oliveira CR, Cunha RA (2005) Different synaptic and subsynaptic localization of adenosine A_{2A} receptors in the hippocampus and striatum of the rat. Neuroscience 132(4):893–903
168. Lopes LV, Cunha RA, Ribeiro JA (1999) Cross talk between A(1) and A(2A) adenosine receptors in the hippocampus and cortex of young adult and old rats. J Neurophysiol 82(6): 3196–3203
169. Ciruela F, Casado V, Rodrigues RJ, Lujan R, Burgueno J, Canals M, Borycz J, Rebola N, Goldberg SR, Mallol J, Cortes A, Canela EI, Lopez-Gimenez JF, Milligan G, Lluis C, Cunha RA, Ferre S, Franco R (2006) Presynaptic control of striatal glutamatergic neurotransmission by adenosine A_1-A_{2A} receptor heteromers. J Neurosci 26(7):2080–2087
170. Wirkner K, Gerevich Z, Krause T, Gunther A, Koles L, Schneider D, Norenberg W, Illes P (2004) Adenosine A_{2A} receptor-induced inhibition of NMDA and GABAA receptor-mediated synaptic currents in a subpopulation of rat striatal neurons. Neuropharmacology 46(7): 994–1007
171. Rebola N, Lujan R, Cunha RA, Mulle C (2008) Adenosine A_{2A} receptors are essential for long-term potentiation of NMDA-EPSCs at hippocampal mossy fiber synapses. Neuron 57(1): 121–134
172. Rebola N, Porciuncula LO, Lopes LV, Oliveira CR, Soares-da-Silva P, Cunha RA (2005) Long-term effect of convulsive behavior on the density of adenosine A_1 and A_{2A} receptors in the rat cerebral cortex. Epilepsia 46(Suppl 5):159–165
173. Kirkwood JM, Tarhini AA, Panelli MC, Moschos SJ, Zarour HM, Butterfield LH, Gogas HJ (2008) Next generation of immunotherapy for melanoma. J Clin Oncol 26(20):3445–3455
174. Tang J, Shalabi A, Hubbard-Lucey VM (2018) Comprehensive analysis of the clinical immuno-oncology landscape. Ann Oncol 29(1):84–91
175. Merighi S, Mirandola P, Varani K, Gessi S, Leung E, Baraldi PG, Tabrizi MA, Borea PA (2003) A glance at adenosine receptors: novel target for antitumor therapy. Pharmacol Ther 100(1):31–48
176. Gessi S, Merighi S, Sacchetto V, Simioni C, Borea PA (2011) Adenosine receptors and cancer. Biochim Biophys Acta 1808(5):1400–1412
177. Pardoll DM (2012) The blockade of immune checkpoints in cancer immunotherapy. Nat Rev Cancer 12(4):252–264
178. Sun C, Wang B, Hao S (2022) Adenosine-A_{2A} receptor pathway in cancer immunotherapy. Front Immunol 13:837230
179. Clayton A, Al-Taei S, Webber J, Mason MD, Tabi Z (2011) Cancer exosomes express CD39 and CD73, which suppress T cells through adenosine production. J Immunol 187(2):676–683
180. Sitkovsky MV (2020) Lessons from the A2A adenosine receptor antagonist-enabled tumor regression and survival in patients with treatment-refractory renal cell cancer. Cancer Discov 10(1):16–19
181. Luke JJ, Powderly JD, Merchan JR, Barve MA, Hotson AN, Mobasher M, Kwei L, Luciano G, Buggy JJ, Piccione EC, Miller RA (2019) Immunobiology, preliminary safety, and efficacy of CPI-006, an anti-CD73 antibody with immune modulating activity, in a phase 1 trial in advanced cancers. J Clin Oncol 37(15):2505
182. Srivastava S, O'Brien M, Cheema PK, Grohe C, Carcereny E, Girard N, Chiappori AA, Ross S, Rossetti M, Dubois F, Lager J, Velcheti V (2022) EP08.01-021 phase 2 study evaluating inupadenant in combination with chemotherapy in adults with NSCLC who progressed on immunotherapy. J Thorac Oncol 17(9):S346–S347

Top Med Chem (2023) 41: 143–156
https://doi.org/10.1007/7355_2022_153

Published online: 25 November 2022

A_{2B} Adenosine Receptor as a New and Attractive Target to Treat Brain Ischemia or Demyelination

Federica Cherchi, Martina Venturini, Ilaria Dettori, Felicita Pedata, Elisabetta Coppi, and Anna Maria Pugliese

Contents

Abstract Ischemic stroke is one of the most prevalent vascular diseases worldwide that cause a decline in quality of life and a high mortality rate. Cerebral ischemia usually results in damage to white matter regions, especially the corpus callosum and the optic tract, with consequent demyelination. Animal models of stroke demonstrated that white matter injured after ischemia causes both axonal and myelin damage, thereby impacting behaviour and functional outcome after stroke. Oligodendrocyte precursor cells (OPCs) differentiate into mature oligodendrocyte, that is responsible for myelin production also in the adult brain. OPC migration, proliferation, and differentiation are altered after stroke, interfering, and limiting a functional recovery in this condition. Accumulated evidence indicates that after stroke, there is a strikingly release of adenosine, that may activate all its receptors, including the low-affinity A_{2B} subtype (A_{2B}Rs). After ischemia this receptor plays different and often contradictory roles, strictly depending on its cellular localisation (i.e. neuronal vs glial cells) indicating A_{2B}Rs as a novel and encouraging target to improve white matter integrity in stroke and demyelinating pathologies.

Keywords Adenosine signalling, Cerebral ischemia, Myelination, Neuroinflammation, Oligodendrocyte differentiation, Oxygen and glucose deprivation

F. Cherchi (✉), M. Venturini, I. Dettori, F. Pedata, E. Coppi, and A. M. Pugliese (✉)
Division of Pharmacology and Toxicology, Department of Neuroscience, Psychology, Drug Research and Child Health (NEUROFARBA), University of Florence, Florence, Italy
e-mail: federica.cherchi@unifi.it; annamaria.pugliese@unifi.it

1 Introduction

In this chapter, we aimed to compile the recent updates on the role of adenosine A_{2B} receptors ($A_{2B}Rs$) and cellular and molecular mechanisms produced by different $A_{2B}Rs$ ligands in experimental models of cerebral ischemia and demyelination. These receptors, which are known as Gs/q-protein-coupled receptors, are rarely distributed in the central nervous system (CNS) [1]. In the brain, $A_{2B}Rs$ are expressed in the presynaptic site of neurons but are also present in other cells, such as astrocytes, oligodendrocytes and microglia. In pathological situations, like hypoxia and inflammation, the extracellular adenosine concentration significantly increases up to activate all its receptor subtypes, including $A_{2B}Rs$ [2]. Moreover, in the same conditions, significant variations in the level of $A_{2B}Rs$ have been reported in recent years [3, 4], suggesting that $A_{2B}R$ may be a potential therapeutic target in these diseases. Furthermore, $A_{2B}Rs$ are implicated in neuroinflammation and microglial activation, by increasing IL-6 secretion and cell proliferation in murine primary microglial cells [5]. The absence of suitable ligands, that can distinguish between the other adenosine receptor subtypes, has long prevented a pharmacological and physiological characterisation of $A_{2B}Rs$; the most common $A_{2B}Rs$ ligands are shown in Fig. 1. $A_{2B}Rs$ are expressed in glutamatergic CA1 hippocampal terminals [3] where their selective stimulation with BAY60-6583 attenuates the predominant A_1 receptor (A_1R)-mediated inhibitory control of synaptic transmission [6]. Particularly, the facilitatory synaptic control of $A_{2B}Rs$ on glutamatergic neurotransmission was assessed by two groups of researchers in murine acute hippocampal slices using the protocol of paired-pulse facilitation (PPF), an electrophysiological paradigm whose reduction reflects an increase in glutamate

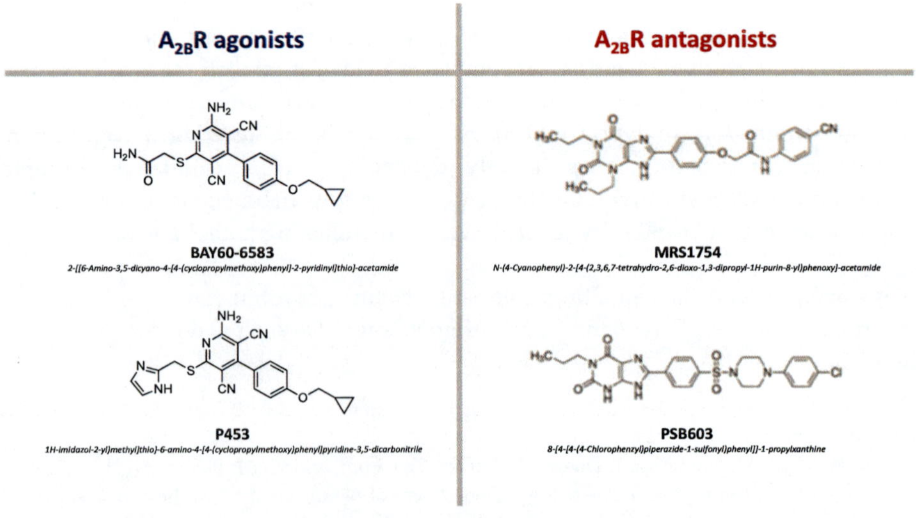

Fig. 1 Chemical structures of commercially available (BAY60-6583, MRS1754 and PSB603) and of the recently synthetised (P453) $A_{2B}R$ ligands

release. The $A_{2B}R$ stimulation decreases PPF, thus enhancing glutamate release, in an A_1R-dependent manner, an effect prevented by the $A_{2B}R$ antagonists MRS1754 and PSB-603 and the A_1R blocker DPCPX [6, 7]. Furthermore, Fusco and colleagues extended results to a newly synthetised BAY60-6583 analogue, the $A_{2B}R$-selective agonist P453 recently described [8], which proved to have a higher affinity than BAY60-6583 [7].

The fact that neither $A_{2B}R$ agonists [6, 7] nor antagonists [9] affect basal hippocampal glutamatergic transmission suggests that the role of $A_{2B}Rs$ might be confined to conditions of synaptic plasticity changes. Nevertheless, the effect of $A_{2B}R$ activation on PPF should not be underestimated as it could be crucial to cognitive performances and mechanisms related to memory and learning [10]. Indeed, it has been shown that compounds that can facilitate a PPF ratio, such as $A_{2B}R$ agonists, may improve cognitive functions and memory in mouse behavioural tests [11]. In confirm, $A_{2B}R^{-/-}$ mice have cognitive deficits, evaluated by using T maze alternation test, a sensitive tool to identify hippocampal dysfunction [12]. BAY60-6583, throughout hippocampal-expressed $A_{2B}Rs$, reversed the cognitive dysfunction observed in these mice, following midazolam exposure [12]. However, beyond neuroprotection exerted by $A_{2B}R$ antagonists acting at the neuro-glial level, it is worth noting that the bulk of evidence in the literature points to a beneficial role exerted by $A_{2B}R$ agonists acting on the same receptor subtype expressed also on peripheral blood cells [13, 14]. Indeed, studies in mice ablated with $A_{2B}Rs$ on bone marrow cells indicated an important contribution of vascular peripheral $A_{2B}Rs$ in attenuating vascular leakage during hypoxia [14]. It was also found that activation of $A_{2B}Rs$ in a model of femoral artery injury is vasoprotective, as it reduces myocardial infarct size in rabbit and mouse hearts when administered before or at the onset of reperfusion [13].

2 Role of $A_{2B}Rs$ in Cerebral Ischemia

Due to the poor selectivity of endogenous adenosine for $A_{2B}R$ and the paucity of identified selective ligands, the involvement of this adenosine receptor subtype in brain ischemia has received little attention so far.

Ischemic stroke is the major cause of long-lasting disabilities worldwide and the second greatest cause of death in industrialised countries [15]. In most cases, it is caused by a transient or permanent decrease in cerebral blood flow, mostly due to the occlusion of a brain artery. Current therapies are restricted to improving blood fluidity, by intravenous administration of tissue plasminogen activator (tPA) within 4–4.5 h of the insult because of its thrombolytic activity and its ability to restore circulation to the brain. The pathology is significant socioeconomic burden and the ineffectiveness of present treatments highlights the urgent need for new therapeutic targets able to prevent or reverse brain damage. The consequent reduction of blood and/or oxygen supply to the brain and consequent depletion of ATP leads to neuronal death caused, at least in the first phase, by excessive glutamate release

[16], thus confirming glutamate as an important driver in the generation of brain damage. At the same time, a significant extracellular adenosine increase, able to activate all subtypes of P1 adenosine receptors, was described. Adenosine release during hypoxic/ischemic conditions suppresses synaptic glutamate release principally via activation of presynaptic A_1Rs and modulating the activity of other subtypes. The excitotoxic, glutamatergic, phase is followed by a subsequent chronic phase of neuroinflammation that emerges hours and days after the insult. The blood-brain barrier (BBB) is damaged as a result of brain ischemia, which also induces early granulocyte recruitment and delayed T cell and macrophage infiltration into the ischemic areas and the boundary zones. This inflammatory scenario is mainly triggered by necrotic cells, reactive oxygen species (ROS) generation, and numerous other factors caused by blood flow interruption in the brain. Once activated, these initiators of inflammation lead to several responses among which is the activation of microglia, the brain's resident immune cells. Microglia then generate more proinflammatory cytokines [17] which in turn lead to adhesion molecule induction in the cerebral vasculature. The most common mechanism of subsequent ischemia injury is now recognised to be chronic neuroinflammation [18]. Therefore, controlling brain inflammation after ischemia is a key strategy in addition to the approved tPA treatment in the early hours following ischemia. Again adenosine, by acting on its central and peripheral receptors, is involved as a front-line endogenous mediator for anti-inflammatory responses [19].

As mentioned, during ischemia adenosine is released in large amounts [2, 20] and has long been known to act predominantly as an endogenous neuroprotectant agent [21, 22]. Adenosine infusion into the ischemic striatum has been shown to significantly ameliorate neurological outcomes and to reduce infarct volume after transient focal cerebral ischemia [23]. This well-recognised neuroprotection by adenosine is principally ascribed to A_1R stimulation [24], as these adenosine receptor subtypes emerged as potent inhibitors of excitatory synaptic transmission both in vitro [25–27] or in vivo [28]. In particular, A_1Rs counteract overstimulation of N-methyl D-aspartate (NMDA) receptors due to excessive glutamate release and consequent intracellular Ca^{2+} overload [29]. Among favourable consequences of A_1R stimulation during ischemic insults are also reduction in cell metabolism and energy consumption [25] and a moderate hypothermia [28, 30]. Unfortunately, the use of adenosine A_1R agonists is hampered by profound central and peripheral side effects, such as bradycardia and sedation [31–33].

Of note, elevated levels of the enzyme adenosine kinase (AK), which recycles and removes adenosine by phosphorylation to form AMP, result in a decrease in the extracellular adenosine and thus in a reduced adenosine receptor activation. As studied extensively in the past, in the adult brain AK is predominantly expressed in astrocytes, which serve as a metabolic sink for the clearance of adenosine; any treatments which reduce AK expression or activity increase adenosine and have neuroprotective action and antiepileptogenic properties [4].

Since it is known that A_{2B}Rs enhance glutamate release in the CA1 hippocampus [6, 7], one of the brain areas more susceptible to ischemic insults, the block of A_{2B}Rs may be neuroprotective, as it counteracts glutamate overload by preserving the inhibitory role of A_1Rs on neurotransmission [6, 7, 9]. This is indeed the case in an in vitro model of brain ischemia reproduced in rat hippocampal slices by oxygen and glucose deprivation (OGD). In this experimental condition, inhibition of A_{2B}R by the prototypical antagonist PSB603 or by MRS1754, prevents irreversible synaptic failure produced by a severe, 7 min, OGD event in CA1 hippocampal slices [9]. In the same study, the Authors also demonstrated that anoxic depolarisation (AD), a clear sign of tissue damage, is completely abolished in slices treated with an A_{2B}R antagonist and exposed to a 7 min OGD insult, while it is significantly delayed in slices exposed to a prolonged OGD insult. These results were accompanied by immunohistochemical analysis revealing that the selective A_{2B}R antagonism also counteracts the reduction of neuronal density found in CA1 stratum pyramidale at 3 h after OGD insults, decreases apoptosis, and maintains activated the mechanistic target of rapamycin (mTOR) levels similar to those of controls, thus sparing neurons from the degenerative effects caused by the injury. Astrocytes significantly proliferate in CA1 stratum radiatum at 3 h after the end of OGD, an effect possibly due to increased glutamate release. A_{2B}R antagonism significantly prevents astrocyte modifications [9]. Of note, neither A_{2B}R antagonist tested protects CA1 neurons from the neurodegeneration induced by exogenous glutamate exposure, indicating that the antagonistic effect is upstream of glutamate release, in line with data indicating a presynaptic effect of A_{2B}Rs on glutamatergic terminals [6, 7]. The increase in glutamate release induced by selective activation of A_{2B}Rs was further strengthened by the experiments performed in CA1 hippocampal region during moderate, 2 min long, OGD. In particular, A_{2B}Rs attenuate the A_1-mediated depression of synaptic transmission induced by in vitro ischemia [7], indicating an opposite effect on glutamate release.

In contrast to the data discussed above and describing the beneficial effect of the A_{2B}R blockade, A_{2B}R stimulation proved to be protective in vivo model of cerebral ischemia. Post-treatment with intravenous BAY60-6583 reduces lesion volume in the absence or presence of tPA and attenuates brain swelling, BBB disruption, and tPA-exacerbated haemorrhagic transformation at 24 h after ischemia induced by transient (2 h) middle cerebral artery occlusion (tMCAo) [34]. Additionally, in the same work, BAY60-6583 mitigates sensorimotor deficits in the presence of tPA and inhibits tPA-enhanced matrix metalloprotease-9 activation, thus decreasing BBB permeability 24 h after ischemia [34]. Protection from BBB permeability after ischemia might protect from blood cell infiltration, that on their turn promote the expansion of the inflammatory response in the ischemic tissue [35].

Additionally, chronic intraperitoneal treatment with BAY60-6583, starting 4 h after focal ischemia induced by tMCAo, protects from neurological deficit as evaluated 24 h from the end of the insult. Seven days after ischemia, BAY60-6583 also protected from ischemic brain damage, neuronal death, microglia activation, and astrocyte alteration [36]. Interestingly, 7 days after ischemia, the agonist decreases tumour necrosis factor alpha (TNF-α) and increases interleukin 10 (IL-10) levels in the blood [36] both valuable markers of brain damage [37]. Accordingly, it was demonstrated that A_{2B}R activation reduces the expression of TNF-α in primary microglia cultures [38] and increases IL-10 production from murine microglial cells [39], with the consequent rescuing of the resting state of microglia. Besides protection due to a direct agonism of $A_{2B}R$ located on rat microglial and/or astrocytic cells, observation that 2 days after tMCAo, BAY60-6583 significantly reduces granulocyte infiltration in the cortex [36] supports the idea that $A_{2B}R$ activation on peripheral endothelial and blood cells is involved in counteracting inflammation of brain parenchyma. This possibility is also supported by the evidence that $A_{2B}R$ knock-out (KO) mice show increased basal levels of TNF-α and expression of adhesion molecules in lymphoid cells, resulting in increased leukocyte rolling and adhesion [14]. Increasing evidence indicates a role for $A_{2B}R$ in the modulation of inflammation and immune responses in distinct pathologies like cancer, diabetes, as well renal, lung, and vascular diseases [40]. Indeed, stroke and inflammation are strictly correlated. Brain ischemia induces profound inflammatory changes in peripheral organs (especially lungs and gut) as early as 2 h after tMCAo in mice as detected by whole-body single-photon emission computed tomography (SPECT)-based imaging protocols [41]. Such peripheral inflammatory changes, in turn, may contribute to poorer recovery after stroke [41]. Data suggest that the $A_{2B}R$ agonists can be proposed as adjuvant therapy to the accepted pharmacological strategy with tPA in brain ischemia.

Taken together, data point towards the possibility that stimulation of $A_{2B}R$ plays a dual time-related role after ischemia. In the early hours after this pathological condition, a strong and sustained increase in cerebral extracellular levels of adenosine able to activate low-affinity $A_{2B}Rs$ in the brain may contribute to expanding excitotoxicity. However, in the hours and days following ischemia, when profound neuroinflammation develops, $A_{2B}Rs$ located on glial, vascular endothelial, and blood cells exert a prevalent immunomodulatory role in attenuating neuroinflammation. Thus, overall, it appears that $A_{2B}Rs$ located on different cell types in the brain and on vascular and blood cells contribute to either recovery or demise of the tissue after the insult and represent an important target for drugs with different therapeutic time windows after ischemia.

Since it is known that an $A_{2B}R$ agonist administered within the first 4 h after brain ischemia is able to reduce neurological deficit measured 24 h after the insult [36], we could hypothesise that a beneficial effect of $A_{2B}R$ activation could be achieved by administration of a selective agonist starting at 1 day after stroke.

Recently indications support the hypothesis that $A_{2B}R$ may be a "sensor" of blood oxygen levels by an additive mechanism involving the release of oxygen from erythrocytes. $A_{2B}Rs$ expressed on erythrocytes elevate their O_2 releasing capacity by increasing 2,3-biphosphoglycerate (2,3-BPG) levels in an AMP-activated protein kinase-dependent manner [42]. Of note, the same authors also reported a significant increase in plasma adenosine concentrations and soluble CD73 activity in 21 healthy humans within 2 h of arrival at 5,260 m altitude [42]. These findings unveiled a novel mechanism of human adaptation to hypoxia and pointed to $A_{2B}Rs$ as possible targets to facilitate O_2 release capacity to peripheral tissues and a potential therapeutic approach for counteracting hypoxia-induced tissue damage. Effects of the selective stimulation or block of $A_{2B}Rs$ in pre-clinical in vivo and in vitro studies of stroke are reported in Table 1.

3 Role of A_{2B}Rs in Demyelinating Conditions

White matter damage is frequently observed in surviving stroke patients and is an important clinical aspect of this pathology [43, 44]. It should be noted that ischemic injury consists not only in grey matter damage but also in white matter alterations. The last is mainly composed of axon bundles wrapped by myelin sheaths, and it plays an important role in neurotransmission by allowing fast signal propagation between different regions of grey matter and providing metabolic support to axons [45, 46]. In addition, white matter is more susceptible to ischemic damage than grey matter because of the lower collateral circulation and blood supply [47]. It has been demonstrated that the prognosis of neurological function after stroke is strictly related to the degree of white matter damage.

Oligodendrocytes (OLs) are glial cells in the CNS, responsible for myelin sheath formation. During development, OLs are generated in the germinal zone from migratory bipolar oligodendrocyte precursor cells (OPCs), which are renowned for the expression of the proteoglycan nerve-glial antigen 2 (NG2) [48]. Most OPCs differentiate into myelinating OLs; however, a pool of immature OPCs, comprising 5–8% of total glial cells, persists within the adult CNS where they represent the major population of proliferating cells [49, 50] and retain the ability to generate new OLs. This process guarantees the generation of new OLs and consequently the myelin turnover, under physiological conditions, [51] and, the generation of new myelin, during normal ageing or pathological conditions. In addition, OLs and OPCs are very vulnerable to conditions common to CNS injury, such as inflammation, oxidative stress, and elevated glutamate levels leading to excitotoxicity. Therefore, OPCs are becoming a promising target in demyelinating disorders, since fostering OPC maturation into myelinating OLs could prevent disease progression and the consequent disability [52]. Postnatal development of oligodendrocytes has been well studied and many processes have been described such as migration, proliferation, and differentiation of precursor cells. However, the extent to which OLs and damaged myelin are replaced following injury to the adult CNS is still unclear. All P1 purinergic receptors are expressed at each stage of oligodendroglial cell maturation [53, 54] and exert a key role in cell development. Furthermore, the expression by OLs of the equilibrative nucleoside transporters ENT1 and ENT2, as well as adenosine degrading enzymes, such as adenosine deaminase (ADA) and AK, suggests that these cells are able to sense and finely tune extracellular adenosine levels [55], thus supporting the notion that purinergic signalling exerts a prominent role in oligodendroglia [56].

It is interesting to note that elevated adenosine levels have been detected in cerebrospinal fluid of multiple sclerosis (MS) patients [57, 58]. Furthermore, an upregulation of A_{2A}R [59] and A_{2B}R [60] has been reported in peripheral blood leukocytes of MS patients and in the CNS [61] and peripheral lymphoid tissues [60, 62] in a mouse model of MS, the experimental autoimmune encephalomyelitis (EAE).

Hence, a crucial role of adenosine, and in particular of $A_{2A}R$ and/or $A_{2B}R$ subtypes, in demyelinating pathologies have been postulated. Under these conditions, excessive signalling by excitatory neurotransmitters like glutamate may be deleterious to neurons and oligodendroglia by exacerbating excitotoxicity and contributing to brain injury. For this reason, the $A_{2B}R$-mediated inhibitory effect on glutamate release described above could be protective in demyelinating lesions. Indeed, as demonstrated by Chen and colleagues [63] and by Wei and co-workers [60] the $A_{2A}R$ and $A_{2B}R$ antagonists, respectively, alleviated the clinical symptoms of EAE and prevented demyelination and CNS damage. Recent data by Liu and co-workers [61] confirmed that $A_{2B}R$ activation seems to participate in EAE-induced damage as BAY60-6583 reverted the protective effects, that is, reduced inflammatory cell infiltration and demyelination, exerted by mesenchymal stem cell therapy in EAE mice. Of note, the above results demonstrating a deleterious role of $A_{2B}Rs$ in demyelinating diseases are in agreement with in vitro data demonstrating that $A_{2B}R$ blockade [64], as well as $A_{2A}R$ antagonism [65], facilitates OPC differentiation in vitro. Furthermore, it was demonstrated that the selective $A_{2B}R$ antagonist PSB0788 stimulated differentiation and development of OPCs in vivo by the maternal inflammation-induced periventricular leukomalacia (PVL) model [66]. In particular, lipopolysaccharide- (LPS-) induced maternal inflammation reduced the expression of myelin basic protein (MBP), which related to the decrease in the numbers of OPCs and mature OLs in neonatal rats and treatment with PSB0788 increased the levels of MBP proteins in the rats' offspring and improved the remyelination [66]. Accordingly, treatment with the selective $A_{2B}R$ antagonist PSB603 greatly ameliorated cerebral ischemic injury by increasing body weight, reducing infarct volume, brain injury, inflammation and contributing to long-term learning memory functional recovery of the PVL mice [67]. Meanwhile, PSB603 treatment suppressed neuron apoptosis as characterised by reducing of caspase-3 level and inhibited microglia activation. Furthermore, PSB603 reduced a direct target of hypoxia, Hypoxia Inducible Factor-1α (HIF-1α), which is a key regulator of adenosine signalling by binding to the $A_{2B}R$ promoter to induce its expression [67].

Therefore, it appears that $A_{2B}Rs$ are crucially involved in OPC maturation. The selective $A_{2B}R$ agonists BAY60-6583 and P453 inhibited the differentiation of purified primary OPC cultures, as demonstrated by the reduced expression of myelin-related proteins such as MBP or myelin-associated glycoprotein (MAG) [64].

Furthermore, $A_{2B}R$ activation reversibly inhibits also tetraethylammonium- (TEA-) sensitive delayed rectifier (IK) and 4-aminopyridine- (4-AP-) sensitive transient (IA) potassium currents [64]. As IK are known to be necessary for OPC maturation [68], this may represent one of the mechanisms by which $A_{2B}Rs$ inhibit myelin production. These results are similar to what was observed in cultured OPCs exposed to the $A_{2A}R$ agonist CGS21680 [55, 65, 69]. Indeed, the adenylate cyclase activator forskolin, applied to patch-clamped cultured OPCs, evoked the same effect as A_2R agonists, i.e., a decrease in IK. Furthermore, a subsequent BAY60-65683 application in the continuous presence of forskolin was devoid of effects [64], thus confirming that cAMP increase is responsible at least for $A_{2B}R$-mediated inhibition of IK and cell maturation. Interestingly, it is present an interplay in cultured OPCs between A_{2B}Rs and sphingosine kinase 1 (SphK1), one of the enzymes involved in the synthesis of sphingosine 1 phosphate (S1P). This signalling sphingolipid, also known as lysosphingolipid, is reported to play a role as a mitogen in OPC cells [70]. The A_{2B}R agonist BAY60-6583 activates SphK1, whereas its silencing, by small interference RNA (siRNA), increases the expression of S1P lyase, the enzyme involved in the irreversible degradation of S1P inside the cells [64]. This observation led to hypothesise that the anti-differentiating effect exerted by A_{2B}R activation in OPCs is mediated by an increase in S1P intracellular levels, as confirmed by the application of the SphK inhibitors VPC96047 or VPC96091, that markedly increased MAG and MBP expression and also significantly reduced IK currents in cultured OPCs [64]. Data about an inhibitory role of A_{2B}R in myelin formation are consistent with recent findings from Manalo et al. [70] who demonstrated that elevated cochlear adenosine levels in $ADA^{-/-}$ mice are associated with sensorineural hearing loss (SNHL) due to cochlear nerve fibre demyelination and mild hair cell loss. Intriguingly, A_{2B}R-specific antagonists administered in $ADA^{-/-}$ mice significantly restored auditory capacity, nerve fibre density, and myelin compaction [70]. The same Authors also provided genetic evidence for A_{2B}R upregulation not only in $ADA^{-/-}$ hearing-impaired mice but also in age-related SNHL [70].

On the other hand, it was demonstrated that BAY60-6583 could facilitate myelin repair, leading to an improvement in cognitive function in a dizocilpine maleate (MK801)-induced mouse model of schizophrenia [71]. This was a result of an increase in cells that were positive for the mature OL markers anti-adenomatous polyposis coli clone ($CC1^+$) and OL transcription factor 2 ($Olig2^+$) [71]. Summary of the selective stimulation or block of A_{2B}Rs in pre-clinical in vivo and in vitro studies of demyelination is showed in Table 1.

4 Conclusions

In conclusion, results underlie that after hypoxia/ischemia, brain injury results from a complex sequence of pathophysiological events that evolve over time: a primary acute mechanism of excitotoxicity and periinfarct depolarisations followed by a secondary brain injury activation triggered by protracted neuroinflammation. Information so far acquired indicates that adenosine A_{2B}Rs located on any cell type of the brain and on vascular and blood cells partake in either salvage or demise of the tissue after a stroke, including protracted demyelination.

Thus, they all represent important targets for drugs having different therapeutic time windows after stroke. The final protective outcome for an agonist versus antagonist compound depends on the time of administration and district of activation of the receptor targeted by the drug.

Table 1 Effects of the selective stimulation or block of A_{2B}Rs in pre-clinical in vivo and in vitro studies of stroke and/or demyelination

	A_{2B}R activation	A_{2B}R block
Animal models	Reduces *lesion volume* and attenuates *brain swelling*, *BBB disruption*, and *tPA-exacerbated haemorrhage* at 24 h after ischemia induced by tMCAo (2 h) [34]	Increases basal levels of *TNF-α* and expression of *adhesion molecules* in lymphoid cells, resulting in increased *leukocyte rolling* and *adhesion* in $A_{2B}R^{-/-}$ mice [14]
	Mitigates *sensorimotor deficits* in the presence of tPA and inhibits tPA-enhanced matrix *metalloprotease-9 activation*, thus decreasing BBB permeability 24 h after ischemia [34]	Alleviates the *clinical symptoms of EAE* and prevents *demyelination* and *CNS damage* [60, 63]
	Protects from *neurological deficit* 1 day after ischemia induced by tMCAo [36]	$A_{2B}R^{-/-}$ mice have *cognitive deficits* [12]
	Protects from *ischemic brain damage*, *neuronal death*, *microglia activation*, and *astrocyte alteration* [36]	Restores *auditory capacity*, *nerve fibre density*, and *myelin compaction* in $ADA^{-/-}$ mice [70]
	Decreases *TNF-α* and increases *IL-10* levels in the blood of tMCAo rats [36]	Promotes *development and differentiation of OPCs* in the maternal inflammation-induced periventricular leukomalacia model [66]
	Facilitates *myelin repair*, leading to the improvement of cognitive function in a dizocilpine maleate (MK-801)-induced mouse model of schizophrenia, since it increased the *number of CC-1$^+$/Olig2$^+$* cells [71]	
Hippocampus	Enhances *glutamate release* in the CA1 hippocampus [6] and *decreases PPF*, thus enhancing glutamate release, in an A_1R-dependent manner [7] in hippocampal slices	Prevents *irreversible synaptic failure* and *AD appearance* produced by a severe, OGD insult in CA1 hippocampal slices [9]
		Delays *AD appearance* in slices undergoing to severe OGD insult [9]
	Attenuates the predominant *A_1R-mediated inhibitory control* of synaptic transmission in hippocampal synaptosome [6]	Counteracts the reduction of *neuronal density*, found in CA1 stratum pyramidale at 3 h after OGD insults, decreases *apoptosis* and maintains activated *mTOR levels* similar to those of controls, prevents *astrocyte modifications* [9]
Glial cells	Reduces the expression of *TNF-α* in primary microglia cultures [72] and increases *IL-10* production from murine microglial cells [39]	Increases the expression of *S1P lyase*, the enzyme catalysing *irreversible S1P degradation,* inside the cells in A_{2B}R-silenced OPC cultures [64]
	Inhibits the *differentiation* of purified primary *OPC* cultures by reducing I_K *and* I_A *conductances* [64]	
	Activates *SphK1*, thus enhancing *S1P production* [64]	

tPA tissue Plasminogen Activator, *BBB* Blood-Brain Barrier, *tMCAo* transient Middle Cerebral Artery occlusion, *$A_{2B}R$* Adenosine A_{2B} Receptor, *TNF-α* Tumour Necrosis Factor – alpha, *EAE* Experimental Autoimmune Encephalomyelitis, *IL-10* InterLeukin-10, *ADA* Adenosine Deaminase, *OPCs* Oligodendrocyte Precursor Cells, *CC-1* anti-adenomatous polyposis Coli Clone, *Olig2* Oligodendrocyte transcription factor 2, *PPF* Paired Pulse Facilitation, *A_1R* Adenosine A_1 receptor, *AD* Anoxic Depolarisation, *OGD* Oxygen and Glucose Deprivation, *mTOR* mechanistic Target of Rapamycin, *SphK1* Sphingosine Kinase 1, *S1P* Sphingosine-1-Phospate

Compliance with Ethical Standards **Conflict of Interest**: The authors declare that they have no conflict of interest.

Funding: The Authors received no financial support.

Ethical Approval: This chapter does not contain any studies with human participants or animals performed by the authors.

References

1. Schulte G, Fredholm BB (2003) Signalling from adenosine receptors to mitogen-activated protein kinases. Cell Signal 15:813–827. https://doi.org/10.1016/s0898-6568(03)00058-5
2. Pedata F, Dettori I, Coppi E et al (2016) Purinergic signalling in brain ischemia. Neuropharmacology 104:105–130. https://doi.org/10.1016/j.neuropharm.2015.11.007
3. Eltzschig HK, Ibla JC, Furuta GT et al (2003) Coordinated adenine nucleotide phosphohydrolysis and nucleoside signaling in posthypoxic endothelium: role of ectonucleotidases and adenosine A2B receptors. J Exp Med 198:783–796. https://doi.org/10.1084/JEM.20030891
4. Boison D, Jarvis MF (2021) Adenosine kinase: a key regulator of purinergic physiology. Biochem Pharmacol 187:114321. https://doi.org/10.1016/J.BCP.2020.114321
5. Merighi S, Bencivenni S, Vincenzi F et al (2017) A 2B adenosine receptors stimulate IL-6 production in primary murine microglia through p38 MAPK kinase pathway. Pharmacol Res 117:9–19. https://doi.org/10.1016/J.PHRS.2016.11.024
6. Gonçalves FQ, Pires J, Pliassova A et al (2015) Adenosine A(2b) receptors control A(1) receptor-mediated inhibition of synaptic transmission in the mouse hippocampus. Eur J Neurosci 41: 878–888. https://doi.org/10.1111/ejn.12851
7. Fusco I, Cherchi F, Catarzi D et al (2019) Functional characterization of a novel adenosine A2B receptor agonist on short-term plasticity and synaptic inhibition during oxygen and glucose deprivation in the rat CA1 hippocampus. Brain Res Bull 151:174–180. https://doi.org/10.1016/j.brainresbull.2019.05.018
8. Betti M, Catarzi D, Varano F et al (2018) The aminopyridine-3,5-dicarbonitrile core for the design of new non-nucleoside-like agonists of the human adenosine A2B receptor. Eur J Med Chem 150:127–139. https://doi.org/10.1016/j.ejmech.2018.02.081
9. Fusco I, Ugolini F, Lana D et al (2018) The selective antagonism of adenosine A2B receptors reduces the synaptic failure and neuronal death induced by oxygen and glucose deprivation in rat CA1 hippocampus in vitro. Front Pharmacol 9:399. https://doi.org/10.3389/fphar.2018.00399
10. Li X, Sun W, An L (2018) Nano-CuO impairs spatial cognition associated with inhibiting hippocampal long-term potentiation via affecting glutamatergic neurotransmission in rats. Toxicol Ind Health 34:409–421. https://doi.org/10.1177/0748233718758233
11. Pereda D, Al-Osta I, Okorocha AE et al (2019) Changes in presynaptic calcium signalling accompany age-related deficits in hippocampal LTP and cognitive impairment. Aging Cell 18: 5. https://doi.org/10.1111/acel.13008
12. Gile J, Oyama Y, Shuff S, Eckle T (2020) A role for the adenosine ADORA2B receptor in midazolam induced cognitive dysfunction. Curr Pharm Des 26:4330. https://doi.org/10.2174/1381612826666200415171622
13. Yang D, Zhang Y, Nguyen HG et al (2006) The A2B adenosine receptor protects against inflammation and excessive vascular adhesion. J Clin Investig 116:1913–1923. https://doi.org/10.1172/JCI27933
14. Eckle T, Faigle M, Grenz A et al (2008) A2B adenosine receptor dampens hypoxia-induced vascular leak. Blood 111:2024–2035. https://doi.org/10.1182/blood-2007-10-117044
15. Boehme AK, Esenwa C, Elkind MSV (2017) Stroke risk factors, genetics, and prevention. Circ Res 120:472–495. https://doi.org/10.1161/CIRCRESAHA.116.308398

16. Rossi DJ, Oshima T, Attwell D (2000) Glutamate release in severe brain ischaemia is mainly by reversed uptake. Nature 403:316–321. https://doi.org/10.1038/35002090
17. Stoll G, Jander S, Schroeter M (1998) Inflammation and glial responses in ischemic brain lesions. Prog Neurobiol 56:149–171. https://doi.org/10.1016/S0301-0082(98)00034-3
18. Wang X, Xuan W, Zhu ZY et al (2018) The evolving role of neuro-immune interaction in brain repair after cerebral ischemic stroke. CNS Neurosci Ther 24:1100–1114. https://doi.org/10.1111/CNS.13077
19. Coppi E, Dettori I, Cherchi F et al (2020) A2b adenosine receptors: when outsiders may become an attractive target to treat brain ischemia or demyelination. Int J Mol Sci 21:9697. https://doi.org/10.3390/ijms21249697
20. Dale N, Pearson T, Frenguelli BG (2000) Direct measurement of adenosine release during hypoxia in the CA1 region of the rat hippocampal slice. J Physiol 526:143. https://doi.org/10.1111/J.1469-7793.2000.00143.X
21. Latini S, Pedata F (2001) Adenosine in the central nervous system: release mechanisms and extracellular concentrations. J Neurochem 79:463–484. https://doi.org/10.1046/j.1471-4159.2001.00607.x
22. Pedata F, Melani A, Pugliese AM et al (2007) The role of ATP and adenosine in the brain under normoxic and ischemic conditions. Purinergic Signal 3:299–310. https://doi.org/10.1007/s11302-007-9085-8
23. Kitagawa H, Mori A, Shimada J et al (2002) Intracerebral adenosine infusion improves neurological outcome after transient focal ischemia in rats. Neurol Res 24:317–323. https://doi.org/10.1179/016164102101199819
24. Kashfi S, Ghaedi K, Baharvand H et al (2017) A 1 adenosine receptor activation modulates central nervous system development and repair. Mol Neurobiol 54:8128–8139. https://doi.org/10.1007/S12035-016-0292-6
25. Greene RW, Haas HL (1991) The electrophysiology of adenosine in the mammalian central nervous system. Prog Neurobiol 36:329–341. https://doi.org/10.1016/0301-0082(91)90005-1
26. Fowler JC, Gervitz LM, Hamilton ME, Walker JA (2003) Systemic hypoxia and the depression of synaptic transmission in rat hippocampus after carotid artery occlusion. J Physiol 550:961. https://doi.org/10.1113/JPHYSIOL.2003.039594
27. Colotta V, Lenzi O, Catarzi D et al (2012) 3-Hydroxy-1H-quinazoline-2,4-dione derivatives as new antagonists at ionotropic glutamate receptors: molecular modeling and pharmacological studies. Eur J Med Chem 54:470–482. https://doi.org/10.1016/j.ejmech.2012.05.036
28. Gourine AV, Dale N, Gourine VN, Spyer KM (2004) Fever in systemic inflammation: roles of purines. Front Biosci 9:1011–1022. https://doi.org/10.2741/1301
29. Andiné P (1993) Involvement of adenosine in ischemic and postischemic calcium regulation. Mol Chem Neuropathol 18:35–49. https://doi.org/10.1007/BF03160021
30. von Lubitz DKJE, Lin RCS, Popik P et al (1994) Adenosine A3 receptor stimulation and cerebral ischemia. Eur J Pharmacol 263:59–67. https://doi.org/10.1016/0014-2999(94)90523-1
31. Sweeney JE, Höhmann CF, Oster-Granite ML, Coyle JT (1989) Neurogenesis of the basal forebrain in euploid and trisomy 16 mice: an animal model for developmental disorders in down syndrome. Neuroscience 31:413–425. https://doi.org/10.1016/0306-4522(89)90384-9
32. Sollevi A (1986) Cardiovascular effects of adenosine in man; possible clinical implications. Prog Neurobiol 27:319–349. https://doi.org/10.1016/0301-0082(86)90005-5
33. von Lubitz DKJE, Lin RCS, Melman N et al (1994) Chronic administration of selective adenosine A1 receptor agonist or antagonist in cerebral ischemia. Eur J Pharmacol 256:161–167. https://doi.org/10.1016/0014-2999(94)90241-0
34. Li Q, Han X, Lan X et al (2017) Inhibition of tPA-induced hemorrhagic transformation involves adenosine A2b receptor activation after cerebral ischemia. Neurobiol Dis 108:173–182. https://doi.org/10.1016/J.NBD.2017.08.011
35. Haskó G, Linden J, Cronstein B, Pacher P (2008) Adenosine receptors: therapeutic aspects for inflammatory and immune diseases. Nat Rev Drug Discov 7:759–770. https://doi.org/10.1038/nrd2638

36. Dettori I, Gaviano L, Ugolini F et al (2020) Protective effect of adenosine A2B receptor agonist, BAY60-6583, against transient focal brain ischemia in rat. Front Pharmacol 11:1639. https://doi.org/10.3389/FPHAR.2020.588757
37. Jickling GC, Sharp FR (2011) Blood biomarkers of ischemic stroke. Neurotherapeutics 8:349–360. https://doi.org/10.1007/S13311-011-0050-4
38. Merighi S, Borea PA, Stefanelli A et al (2015) A2a and a2b adenosine receptors affect HIF-1α signaling in activated primary microglial cells. Glia 63:1933–1952. https://doi.org/10.1002/glia.22861
39. Pacher P, Virág L, Haskó Balázs Koscsó G et al (2012) Adenosine augments IL-10 production by microglial cells through an A2B adenosine receptor-mediated process. Am Assoc Immnol 188:445–453. https://doi.org/10.4049/jimmunol.1101224
40. Borea PA, Gessi S, Merighi S et al (2018) Pharmacology of adenosine receptors: the state of the art. Physiol Rev 98:1591–1625. https://doi.org/10.1152/physrev.00049.2017
41. Szigeti K, Horváth I, Veres DS et al (2015) A novel SPECT-based approach reveals early mechanisms of central and peripheral inflammation after cerebral ischemia. J Cereb Blood Flow Metab 35:1921–1929. https://doi.org/10.1038/JCBFM.2015.174
42. Liu H, Zhang Y, Wu H et al (2016) Beneficial role of erythrocyte adenosine A2B receptor-mediated AMP-activated protein kinase activation in high-altitude hypoxia. Circulation 134: 405–421. https://doi.org/10.1161/CIRCULATIONAHA.116.021311
43. Yoshioka H, Niizuma K, Katsu M et al (2011) Consistent injury to medium spiny neurons and white matter in the mouse striatum after prolonged transient global cerebral ischemia. J Neurotrauma 28:649. https://doi.org/10.1089/NEU.2010.1662
44. Wang LE, Tittgemeyer M, Imperati D et al (2012) Degeneration of corpus callosum and recovery of motor function after stroke: a multimodal magnetic resonance imaging study. Hum Brain Mapp 33:2941–2956. https://doi.org/10.1002/HBM.21417
45. Saab AS, Tzvetavona ID, Trevisiol A et al (2016) Oligodendroglial NMDA receptors regulate glucose import and axonal energy metabolism. Neuron 91:119–132. https://doi.org/10.1016/j.neuron.2016.05.016
46. Simons M, Nave KA (2016) Oligodendrocytes: myelination and axonal support. Cold Spring Harb Perspect Biol 8:a020479. https://doi.org/10.1101/cshperspect.a020479
47. Wang Y, Liu G, Hong D et al (2016) White matter injury in ischemic stroke. Prog Neurobiol 141:45. https://doi.org/10.1016/J.PNEUROBIO.2016.04.005
48. Sakry D, Trotter J (2016) The role of the NG2 proteoglycan in OPC and CNS network function. Brain Res 1638:161–166. https://doi.org/10.1016/j.brainres.2015.06.003
49. Dawson MRL, Polito A, Levine JM, Reynolds R (2003) NG2-expressing glial progenitor cells: an abundant and widespread population of cycling cells in the adult rat CNS. Mol Cell Neurosci 24:476–488. https://doi.org/10.1016/S1044-7431(03)00210-0
50. Domingues HS, Cruz A, Chan JR et al (2018) Mechanical plasticity during oligodendrocyte differentiation and myelination. Glia 66:5–14. https://doi.org/10.1002/glia.23206
51. Bergles DE, Richardson WD (2016) Oligodendrocyte development and plasticity. Cold Spring Harb Perspect Biol 8. https://doi.org/10.1101/cshperspect.a020453
52. Skaper SD (2019) Oligodendrocyte precursor cells as a therapeutic target for demyelinating diseases. Prog Brain Res:119–144. https://doi.org/10.1016/bs.pbr.2019.03.013
53. Stevens B, Porta S, Haak LL et al (2002) Adenosine: a neuron-glial transmitter promoting myelination in the CNS in response to action potentials. Neuron 36:855–868. https://doi.org/10.1016/S0896-6273(02)01067-X
54. Fields RD (2004) Volume transmission in activity-dependent regulation of myelinating glia. Neurochem Int 45:503–509. https://doi.org/10.1016/j.neuint.2003.11.015
55. Cherchi F, Pugliese AAM, Coppi E (2021) Oligodendrocyte precursor cell maturation: role of adenosine receptors. Neural Regen Res 16:1686. https://doi.org/10.4103/1673-5374.306058
56. Burnstock G, Fredholm B, Verkhratsky A (2011) Adenosine and ATP receptors in the brain. Curr Top Med Chem 11:973–1011. https://doi.org/10.2174/156802611795347627

57. Lazzarino G, Amorini A, Eikelenboom M et al (2010) Cerebrospinal fluid ATP metabolites in multiple sclerosis. Mult Scler 16:549–554. https://doi.org/10.1177/1352458510364196
58. Amorini AM, Petzold A, Tavazzi B et al (2009) Increase of uric acid and purine compounds in biological fluids of multiple sclerosis patients. Clin Biochem 42:1001–1006. https://doi.org/10.1016/J.CLINBIOCHEM.2009.03.020
59. Vincenzi F, Corciulo C, Targa M et al (2013) Multiple sclerosis lymphocytes upregulate A2A adenosine receptors that are antiinflammatory when stimulated. Eur J Immunol 43:2206–2216. https://doi.org/10.1002/eji.201343314
60. Wei W, Du C, Lv J et al (2013) Blocking a 2B adenosine receptor alleviates pathogenesis of experimental autoimmune encephalomyelitis via inhibition of IL-6 production and Th17 differentiation. J Immunol 190:138–146. https://doi.org/10.4049/jimmunol.1103721
61. Liu Y, Ma Y, Du B et al (2020) Mesenchymal stem cells attenuated blood-brain barrier disruption via downregulation of Aquaporin-4 expression in EAE mice. Mol Neurobiol 57: 3891–3901. https://doi.org/10.1007/s12035-020-01998-z
62. Ingwersen J, Wingerath B, Graf J et al (2016) Dual roles of the adenosine A2a receptor in autoimmune neuroinflammation. J Neuroinflammation 13:48. https://doi.org/10.1186/s12974-016-0512-z
63. Chen Y, Zhang ZX, Zheng LP et al (2019) The adenosine A2A receptor antagonist SCH58261 reduces macrophage/microglia activation and protects against experimental autoimmune encephalomyelitis in mice. Neurochem Int 129:104490. https://doi.org/10.1016/j.neuint.2019.104490
64. Coppi E, Cherchi F, Fusco I et al (2020) Adenosine A2B receptors inhibit K+ currents and cell differentiation in cultured oligodendrocyte precursor cells and modulate sphingosine-1-phosphate signaling pathway. Biochem Pharmacol 177:113956. https://doi.org/10.1016/j.bcp.2020.113956
65. Coppi E, Cellai L, Maraula G et al (2013) Adenosine A2A receptors inhibit delayed rectifier potassium currents and cell differentiation in primary purified oligodendrocyte cultures. Neuropharmacology 73:301–310. https://doi.org/10.1016/j.neuropharm.2013.05.035
66. Li Y, Wang D, Li Z, Ouyang Z (2022) PSB0788 ameliorates maternal inflammation-induced periventricular leukomalacia-like injury. Bioengineered 13:10224–10234. https://doi.org/10.1080/21655979.2022.2061296
67. Wang J, Wang D, Zheng X et al (2022) A 2B adenosine receptor inhibition ameliorates hypoxic-ischemic injury in neonatal mice via PKC/Erk/Creb/HIF-1α signaling pathway. Brain Res 1782:147837. https://doi.org/10.1016/J.BRAINRES.2022.147837
68. Gallo V, Zhou JM, McBain CJ et al (1996) Oligodendrocyte progenitor cell proliferation and lineage progression are regulated by glutamate receptor-mediated K+ channel block. J Neurosci 16:2659–2670. https://doi.org/10.1523/jneurosci.16-08-02659.1996
69. Coppi E, Cellai L, Maraula G et al (2015) Role of adenosine in oligodendrocyte precursor maturation. Front Cell Neurosci 9:155. https://doi.org/10.3389/fncel.2015.00155
70. Manalo JM, Liu H, Ding D et al (2020) Adenosine A2B receptor: a pathogenic factor and a therapeutic target for sensorineural hearing loss. FASEB J 34:15771–15787. https://doi.org/10.1096/fj.202000939R
71. Ma Q, Wang D, Li Y et al (2022) Activation of A_{2B} adenosine receptor protects against demyelination in a mouse model of schizophrenia. Exp Ther Med 23:1–9. https://doi.org/10.3892/ETM.2022.11323
72. Merighi S, Borea PA, Gessi S (2015) Adenosine receptors and diabetes: focus on the A2B adenosine receptor subtype. Pharmacol Res 99:229–236. https://doi.org/10.1016/j.phrs.2015.06.015

Top Med Chem (2023) 41: 157–178
https://doi.org/10.1007/7355_2023_161

Published online: 23 June 2023

A_3 Adenosine Receptor Ligands: From Discovery to Clinical Trials

Kenneth A. Jacobson, Paola Oliva, and R. Rama Suresh

Contents

Abstract Numerous classes of selective agonists, antagonists, and allosteric modulators of the A_3 adenosine receptor (AR) have been reported. The structure–affinity relationships and selectivity of A_3AR ligands compared to the other three ARs have been described. Furthermore, prodrugs of some of these ligands have been reported, and their efficacy was demonstrated in animal models. In recent years, a structural approach has been adopted for the discovery of novel A_3AR ligands, either to improve the potency of known ligands or to discover novel chemotypes for this receptor. Some A_3AR ligands are being developed as therapeutic candidate molecules for a variety of chronic (e.g., psoriasis, NASH, pain, cancer, glaucoma) and acute (stroke, cardiac ischemia) conditions. The number of current clinical trials for A_3AR ligands has now grown, with A_3AR agonists accounting for the majority.

K. A. Jacobson (✉), P. Oliva, and R. R. Suresh
Molecular Recognition Section, Laboratory of Bioorganic Chemistry, National Institute of Diabetes and Digestive and Kidney Diseases, National Institutes of Health, Bethesda, MD, USA
e-mail: kennethj@niddk.nih.gov

Both the historical development of this field and the current state of the clinical trial-directed medicinal chemistry are included in this chapter.

Keywords Adenosine, Agonist, Antagonist, GPCR, Nucleosides, Receptor

1 Introduction

The A_3 adenosine receptor (AR) was the last of the four ARs to be discovered (1992) [1], and the introduction of selective agonists and antagonists began during the time period following the discovery of a considerable number of A_1AR and A_{2A}AR ligands [2]. Thus, the early structure–activity relationship (SAR) studies of this receptor were able to make use of many available compounds known to bind to other ARs. A_3AR ligands continue to be generated and refined for affinity and selectivity. Some A_3AR ligands are being developed as therapeutic candidate molecules for a variety of chronic (e.g., psoriasis, NASH, pain, cancer, and glaucoma) [3–7] and acute (stroke, cardiac ischemia) conditions [8]. A_3AR agonists have found a greater degree of translation onto a clinical path than antagonists, although it initially appeared to be the opposite case [9, 10]. Currently, there are a number of ongoing clinical trials for A_3AR ligands, including multiple A_3AR agonists, three A_3AR antagonists, and another A_3AR modulator, as listed in clinicaltrials.gov and other sources (Table 1) [9, 22]. How this GPCR became the focus of extensive medicinal chemistry and translational studies is described below.

1.1 Historical Sketch: A_3AR Discovery

Cloning and early discovered biological effects. The A_3AR was the last of the four ARs to be cloned. In fact, the A_3AR sequence designated "tgpcr1" was reported first by Meyerhoff et al. in 1991 [23]. It was described as a "novel G-protein coupled receptor" without any indication of its native ligand. It was cloned from a rat testis cDNA library and was noted to increase with sexual maturation. The sequence of this orphan receptor resembled known sequences of biogenic amine receptors and the canine RDC7, another orphan GPCR later shown to be the A_1AR. Subsequently, the rat A_3AR was identified as a bona fide adenosine receptor by Gary Stiles and coworkers at Duke University [24], and the sheep A_3AR was cloned by the Steve Reppert lab at Harvard Medical School together with Joel Linden of the University of Virginia and the human A_3AR by Marlene Jacobson at Merck with Joel Linden [25, 26]. The earlier reported effects of adenosine and R-N^6-(phenylisopropyl)-adenosine (R-PIA) to cause mediator release in rat basophilic (mast cell-like) cells [27] were later identified as being mediated by the same A_3AR as was later cloned in several species. The availability of selective agonists and antagonists for the AR

Table 1 Current and former clinical trials of A_3AR ligands (source: clinicaltrials.gov, a website to consult for current status, as well as Ref. [11] and other sources)

Drug	Action	Condition or disease	Phase, NCT number (and name) [reference]
Agonists			
Adenosine	AR agonist	Perioperative pain	2, 00298636 [9]
Piclodenoson (IB-MECA, CF-101)[a]	A_3AR agonist	Rheumatoid arthritis	2, 00556894; 2, 01034306; 3, 02647762 [9]
		Psoriasis	2, 00428974; 2/3, 01265667 [4] 3, 03168256 (COMFORT) [11]
		Osteoarthritis	2, 00837291 [12]
		Glaucoma	2, 01033422 [13]
		Uveitis	2, 01905124 [14]
		Dry eye disease	2, 00349466; 3, 01235234 [9]
		Covid-19	2, 04333472 [15]
Namodenoson (Cl-IB-MECA, CF-102)[a]	A_3AR agonist	HCC (hepatocellular carcinoma)	1/2, 00790218; 2, 02128958; 3, 05201404 (LIVERATION) [11, 16]
		Chronic HCV	1/2, 00790673 [9]
		NASH (non-alcoholic steatohepatitis)	2, 02927314 [17]; 2, 04697810 [11]
AST-004[b]	A_3AR/A_1AR agonist	Stroke	1 [18]
HU010[c,d]	A_3AR agonist	Stroke	2
Antagonists			
PBF-677[d,e]	A_3AR antagonist	Glaucoma	1, 02639975
		Ulcerative colitis	2, 03773952, ADENOIBD
PBF-1650[d,e]	A_3AR antagonist	Psoriasis	1, 03798236, ADENOIMMUNE
FM101[f]	A_3AR antag/ part. Agonist	Safety study	1, 03879928 [4]
		NASH	2, 04710524 [19]
		Glaucoma	1/2, 04585100 [20]
Allosteric modulator			
NMT-006[d,g] (JNJ-10,450,232)	A_3AR modulator	Postoperative pain	2 [21]

[a]Can Fite Biopharma
[b]Astrocyte Pharma
[c]Huons
[d]Structure not disclosed
[e]Palobiofarma SL
[f]Future Medicines
[g]Neumentum

subtypes, including A_3AR, has facilitated extensive pharmacological studies, leading to hypotheses that A_3AR modulators could potentially be applied therapeutically.

2 A_3AR Agonists

2.1 *Early Medicinal Chemistry of A_3AR Agonists*

The Jacobson lab at NIH in collaboration with the Stiles lab began the search for the first selective agonists soon after the rat A_3AR cloning and identification. The agonist radioligand used initially in the Stiles for A_3AR binding lab was $[^{125}I]$I-APNEA, while Linden et al. used $[^{125}I]$I-ABA (Fig. 1) [24–26]. However, both radioligands were originally introduced to study the A_1AR, and their A_1AR affinity is actually greater than their A_3AR affinity. Nevertheless, these were important tools in the initial stages of nucleoside screening at the A_3AR using membranes from A_3AR-overexpressing mammalian cell lines. Many off-the-shelf known adenosine agonists, such as potent and non-selective AR agonist *N*-ethylcarboxamidoadenosine (NECA, Fig. 1) and many A_1AR-selective N^6-substituted adenosines, were tested in rat A_3AR binding [28]. A key conclusion was that a NECA-like 5′-amide group was more conducive to A_3AR affinity than the 5′-CH_2OH group of adenosine. This was also the first report that inosine binds to ARs, albeit at high concentrations (rat A_3AR $K_i = 45$ μM), and the hybrid agonist NECI (Fig. 1, rat A_3AR $K_i = 5.0$ μM) was synthesized with a 6-hydroxyl group rather than the 6-amino group of NECA.

Furthermore, some N^6-substitutions of adenosine were favorable for A_3AR affinity despite similar groups providing known A_1AR versus $A_{2A}AR$ selectivity. It was also discovered that some of the previously classified A_1AR- and $A_{2A}AR$-selective agonists had considerable previously undetected A_3AR affinity that made them invalid as selective pharmacological agents. In the series of N^6-phenyl, N^6-benzyl, and N^6-2-phenylethyl, it was N^6-benzyl that provided the highest rat A_3AR affinity within the homologous series ($K_i = 120$ nM) relative to the A_1AR and $A_{2A}AR$ [28]. When the N^6-benzyl group was combined with the 5′-N-ethyluronamide group of NECA, the product N^6-benzyl-NECA was slightly selective (rat A_3AR $K_i = 6.8$ nM, sevenfold versus rat A_1AR) [28]. This A_3AR agonist was the first to become commercially available, but its only marginal selectivity led to errors of interpretation when applied as a pharmacological probe. Functional group substitution further enhanced the selectivity, leading to the identification of the 3-iodo-benzyl group as providing moderate A_3AR selectivity when combined with a 5′-N-methylamide, i.e., IB-MECA.

The first administration of an A_3AR agonist in vivo was reported in 1993 [29]. IB-MECA was found to cause locomotor depression in mice. The site of action of IB-MECA to induce the behavioral depression was proposed to be in the brain, but it was later shown that much of the locomotor depression by a range of adenosine

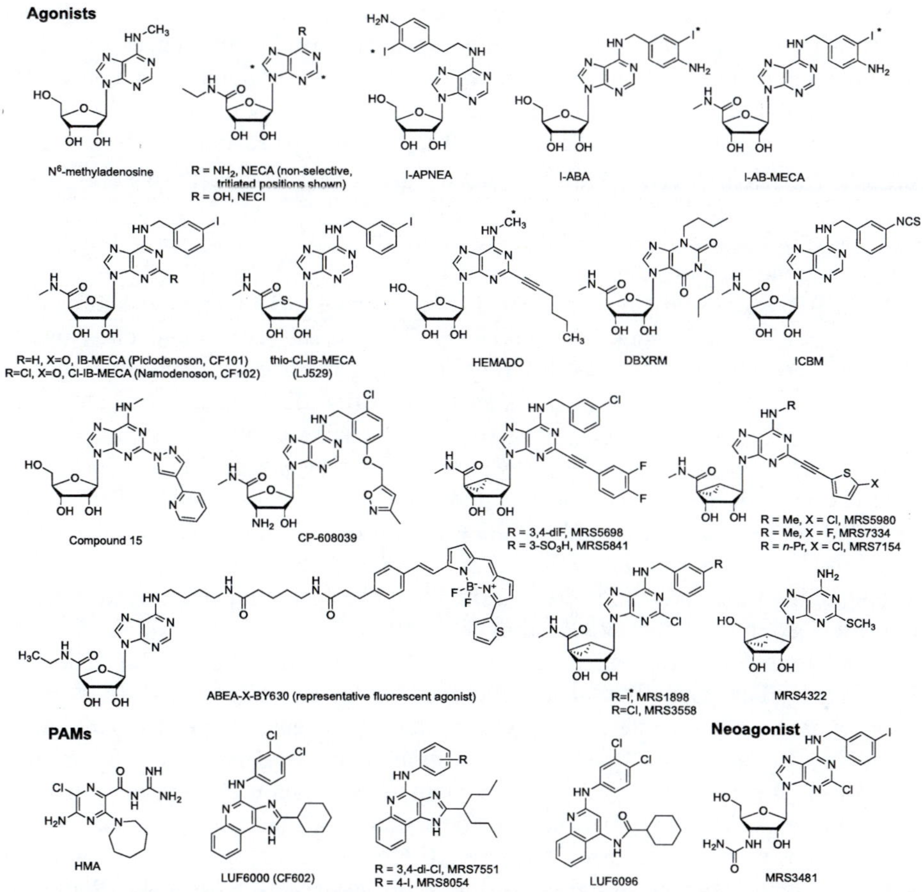

Fig. 1 Representative AR agonists and allosteric modulators that have been useful in studying the A_3AR. The * symbol indicates a radioisotopic label. MRS3481 is a tailored agonist that is selective for a A_3AR neoceptor

agonists, even presumed A_1AR- and $A_{2A}AR$-selective agonists, originates from peripheral action at the mast cell A_3AR to release histamine [30]. However, even the 1993 paper [29], IB-MECA was also observed to cause scratching in the mice typical of histamine release, which was suppressed by coadministration of a H_1 histamine receptor antagonist. The cerebroprotective properties of chronically administered IB-MECA in the forebrain ischemia model in gerbils were described early in the process of A_3AR agonist discovery [31], and the envisioned clinical application to stroke of the anti-ischemic properties of A_3AR agonists motivated the discovery of additional agonists with even higher affinity and selectivity than IB-MECA. The efficacy of A_3AR agonists in cardiac ischemia models in the 1990s [32–34] followed the discovery of their utility in stroke.

A refinement of the IB-MECA structure as a rat A_3AR-selective agonist was the introduction of a 2-chloro group in 1994, i.e., producing Cl-IB-MECA (Fig. 1)

[35, 36]. The ratio of rat A_3AR selectivity versus A_1AR was reported to be ~50-fold for IB-MECA and ~200-fold for Cl-IB-MECA. The species differences in affinity were not reported in other species until later when the cloned human and mouse ARs became accessible. In fact, at the mouse A_3AR, the affinity of IB-MECA significantly exceeds that of Cl-IB-MECA [37]. The AR subtype selectivity of IB-MECA and Cl-IB-MECA was later confirmed at the human A_3AR [37]. Species differences in A_3AR affinity of both orthosteric and allosteric ligands are prominent [37–40], especially for non-nucleoside antagonists, which must be carefully chosen for use as probes in pharmacological experiments.

An SAR early study by IJzerman and colleagues [41] identified xanthine-7-ribosides as A_1AR ligands. Extending the SAR of this compound class to the newly discovered A_3AR, DBXRM (Fig. 1) was identified as a selective agonist [42], with maximal efficacy (E_{max}) similar to Cl-IB-MECA in a functional assay [43]. Thus, this was the first report of non-adenosine derivatives activating the A_3AR. The full agonism of DBXRM was assured by the presence of the 5′-*N*-methyluronamide group, as xanthine derivatives themselves would otherwise be expected to be antagonists and xanthine-7-ribosides partial agonists. Given that 1,3-dialkylxanthines had relatively low rat A_3AR affinity as antagonists, the higher affinity of the xanthine-7-ribosides was driven by the anchoring effect of an attached riboside deep in the orthosteric binding site [29].

The introduction of the first somewhat A_3AR-selective radioligand, [^{125}I]I-AB-MECA (Fig. 1), was in 1994 [44]. It incorporated the structure of IB-MECA along with a *p*-amino group on the N^6-benzyl substituent, conveniently placed to introduce an iodine isotope at the preferred 3 position. However, the selectivity of this radioligand for A_3AR versus A_1AR was minimal, so when used in primary tissues, an unlabeled A_1AR antagonist had to be present to block its binding to that subtype [38]. An early attempt to use [^{125}I]I-AB-MECA to identify A_3AR sites by brain autoradiography resulted in the visualization of the densely present A_1AR instead [45]. The more selective agonist Cl-IB-MECA was prepared in the tritiated form (labeled on the *N*-Me group) by the Jacobson lab at NIH, but its high non-specific binding to cell membranes precluded its use as an A_3AR radioligand (unpublished). Another agonist radioligand for the human A_3AR was [^{3}H]HEMADO (Fig. 1), reported by Klotz et al. in 1999 [46]. However, this tracer was not suitable for use in rodent species due to its greatly reduced affinity compared to the human A_3AR.

Affinity probes of the A_3AR, such as irreversibly binding A_3AR agonist ICBM, an isothiocyanate (Fig. 1), and a 1*H*,3*H*-pyrido[2,1-*f*]purine-2,4-dione antagonist compound 17b (LUF7602, Fig. 2), have been reported [48, 49]. The residue that was covalently modified by compound 17b was identified as Y7.36 [49]. Fluorescent agonists and antagonists have been valuable tool compounds for drug screening and characterization of the A_3AR. We include in Figs. 1 and 2 representative structures, ABEA-X-BY630 and XAC-X-BY630 (BODIPY conjugates), respectively, but many other fluorescent A_3AR ligands having a range of affinities and selectivities have been reported [50, 51].

Efforts in the pharmaceutical sector to identify selective A_3AR agonists resulted in compound 15, a 2-biaryl-substituted adenosine, and CP-608039 (Fig. 1), among

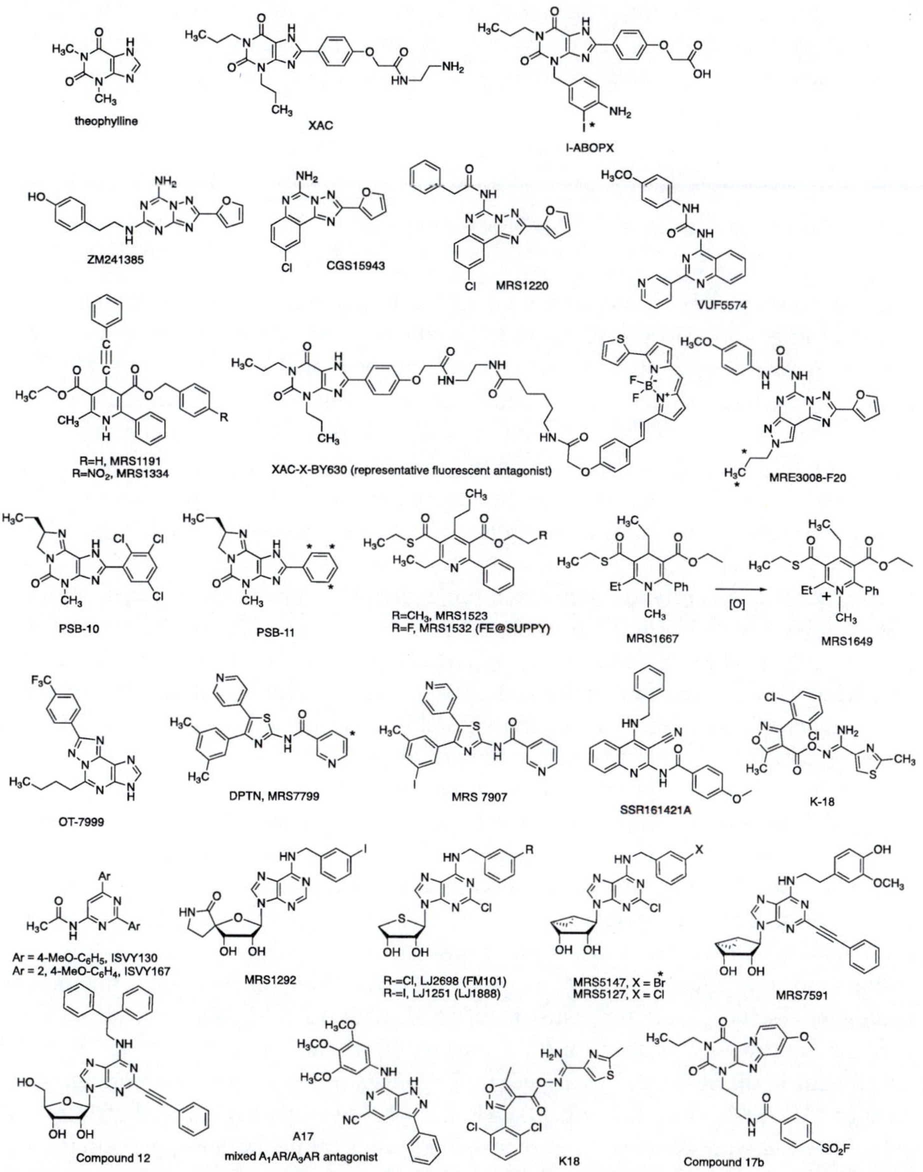

Fig. 2 Representative AR antagonists, consisting of both non-nucleoside heterocycles and strategically modified nucleoside derivatives that have been useful in studying the A_3AR. The * symbol indicates a radioisotopic label. Uncharged prodrug MRS1667 of low aqueous solubility is oxidized to water-soluble cationic active form MRS1649 [47]

others [52, 53]. Curiously, the 3′-amino-3′-deoxy ribose moiety of CP-608039 was not a general modification for diverse nucleosides that preserved A_3AR binding [39]. A neoceptor scheme to selectively activate strategically mutated A_3ARs, to form an electrostatic complementarity to a mutant A_3AR, was based on the low affinity of certain 3′-amino-3′-deoxy derivatives at the wild-type A_3AR [39]. The complementary and H272E mutant A_3AR-specific agonist was MRS3841, which was termed a neoagonist (Fig. 1).

2.2 Recently Reported A_3AR Agonists

Consideration of the ribose ring conformation resulted in agonists of enhanced A_3AR affinity and selectivity. North (N)-methanocarba adenosine derivatives have been extensively explored as A_3AR-selective agonists (2,000-present) [54–59]. This rigid ring system replaces the flexible furanose ring of native nucleosides and nucleotides with a bicyclo[3.1.0]hexane moiety to pre-establish a fixed conformation. This bicyclic ring system, depending on the point of ring fusion, maintains a North (N) conformation, which is highly preferred over the opposite ring twist (South, (S)) at the A_3AR. The enhancing effect of the (N)-methanocarba modification on both A_3AR affinity and selectivity, compared to the corresponding ribose analogue, was recently demonstrated in a direct comparison of hypermodified adenosine derivatives in both series [59]. Fluoro-thienyl derivative MRS7334 (Fig. 1) displayed exceptionally high A_3AR affinity ($K_i = 0.28$ nM) and selectivity. MRS1898 (Fig. 1) was radioiodinated, resulting in a useful A_3AR-selective agonist radioligand [60]. Also, the corresponding 3-chloro analogue MRS3558 was found to be protective in a model of ischemic lung damage [61].

Highly selective agonists have been obtained using a combination of 2-arylalkynyl groups with other A_3AR-enhancing groups [57]. A 2-arylethynyl group on adenine was found by Cristalli and colleagues to enhance A_3AR in the ribose series [62]. Thus, this is a general modification that can be applied to various classes of nucleoside agonists of the A_3AR. This modification has led to the introduction of MRS5698, MRS5980, and MRS7154 (Fig. 1), all of which also contain an extended C2-arylethynyl group [57, 59]. Many of the (N)-methanocarba analogues, including the three just mentioned, have found application in reducing chronic neuropathic pain in animal models [63]. Sulfonation of the C2 or N^6 substituent in the series of C2-arylethynyl agonists of the A_3AR provided agonists, such as MRS5841 (Fig. 1, human A_3AR $K_i = 1.9$ nM), which retained affinity and selectivity but were prevented from crossing biological membranes, particularly the blood–brain barrier (BBB) [64]. This selectivity for the peripheral compartment was utilized to analyze separate contributions to the A_3AR agonist activity in a chronic pain model.

Although the receptor region surrounding the 5′ position is sterically limited, 5′-C-ethyl-tetrazolyl derivatives (not shown) were reported to have considerable A_3AR affinity, but with low efficacy [65].

3 A_3AR Antagonists

3.1 *Early Medicinal Chemistry of A_3AR Antagonists*

Since the rat A_3AR had an extremely low affinity for the usual simple alkylxanthines (mixed A_1/A_2 antagonists), such as caffeine and theophylline, the prototypical adenosine antagonists, an alternative strategy was required to identify the first A_3AR antagonists. A small chemical library (~2000 compounds) was assembled by hand at the Jacobson NIH lab, prior to the availability of commercial chemical libraries [66–70]. These mostly heterocyclic compounds were screened at a standard concentration (10 μM) for inhibition of [^{125}I]I-AB-MECA binding at the rat (and later human) A_3AR. Hit compounds in this exercise were 1,4-dihydropyridines, flavonoids, and later pyridines [70]. Several hit compounds from library screening reported by Merck in abstract form were not published in peer-reviewed publications [71].

Potent tricyclic antagonists selective for the human A_3AR began with the discovery that acylation of CGS15943 increases human A_3AR selectivity, leading to MRS1220 (Fig. 2, $K_i = 0.65$ nM). This derivative contains a phenylacetyl group at the exocyclic NH of potent A_{2A}AR antagonist CGS15943 [72]. The hint that acylation of this amine was tolerated at the A_3AR was the result of testing in rat A_3AR binding ($K_i = 82.3$ nM) of the NH-Boc-protected derivative of CGS15943 [72]. However, the use of MRS1220 is limited to primate A_3AR, because it is nearly inactive at the mouse and rat A_3ARs [37]. Canine and sheep A_3ARs also tend to bind heterocyclic antagonists with similarly high affinity as in primates [1]. OT-7999 is an A_3AR antagonist that contains a modified xanthine scaffold, e.g., 3*H*-[1,2,4]triazolo [5,1-*i*]purine.

Among the first selective non-nucleoside human A_3AR antagonists reported (from SAR modification of experimental screening hits) were 1,4-dihydropyridine (DHP) and pyridine derivatives, including potent DHPs MRS1191 ($K_i = 31$ nM) and MRS1334 ($K_i = 2.69$ nM) and pyridine MRS1523 ($K_i = 43$ nM) at the human A_3AR, respectively (Fig. 2) [70]. MRS1523 is further moderately selective at the mouse and rat A_3ARs and for that reason has been useful in vivo [73]. For DHP and pyridine derivatives, the addition of 4-arylethynyl and 6-aryl groups enhanced the selectivity for ARs and prevented L-type Ca^{2+} channel binding. A relatively new chemical series of A_3AR antagonists led to the introduction of DPTN (MRS7799) and its analogue MRS7907 (Fig. 2), which potently inhibit both primate and rodent A_3ARs [40, 74]. Other diverse chemotypes to serve as human A_3AR antagonists were identified by library screening, including virtual screening.

An alternative approach to finding antagonists was to identify nucleoside derivatives that were bound to but failed to fully activate the A_3AR. The first example was an N^6-benzyl derivative that was bound to but did not activate the rat A_3AR [75]. This theme was expanded with the discovery that certain modifications of the ribose moiety reduced efficacy but not A_3AR affinity. Initially, this effect was observed upon curtailing the H-bonding ability of the *N*-alkyl uronamide group of

NECA-like agonists, by either *N*,*N*-dialkylation or truncation of the 5′-carbon. Thus, IB-MECA analogues (Fig. 2, MRS1292, MRS5147, MRS5127, MRS7591), including 4′-thio analogues (LJ 2698 and LJ1251), could be converted to antagonists or low-efficacy A_3AR agonists by this approach [76–78]. The advantage of this strategy to obtain antagonists, compared to library screening of diverse heterocycles, was that nucleoside ligands tended to preserve affinity at both human and rat A_3ARs. Eventually, this efficacy-lowering phenomenon was expanded to other substituents on the nucleoside scaffold, including at the N^6 and C2 positions of adenine. For example, Spinaci et al. recently reported various bulky N^6 and C2 substituents (see Fig. 2, compound 12) that greatly reduce the efficacy of simple adenosine derivatives, resulting in compounds that demonstrated anticancer activity [79]. Nucleoside-derivative A_3AR antagonists often display a residual low level of agonists – therefore, they are better described as partial agonists of low efficacy. This was recently demonstrated for 5′-truncated (N)-methanocarba derivatives that contained N^6-dopamine derived substituents, which effectively increased mouse A_3AR affinity, e.g., MRS7591 (mouse A_3AR $K_i = 17.8$ nM) [76].

Radiolabeled A_3AR antagonists (labeled with ^{125}I, ^{3}H, ^{76}Br, or ^{18}F) as receptor probes have been prepared from I-ABOPX, PSB-11, MRE3008-F20, MRS5147, and MRS7799 (see Fig. 2 for the sites of radioisotope incorporation) [22, 80–82].

3.2 *Recently Reported A_3AR Antagonists*

A_3AR antagonists have been examined preclinically as potential anti-inflammatory, anti-asthmatic, anti-glaucoma, anticancer, and anti-ischemic therapies [1, 2, 20]. Mikus et al. described a novel human A_3AR antagonist SSR161421A (Fig. 2) that has in vivo anti-asthmatic activity in sensitized guinea pigs [83]. Sotelo and coworkers have studied 2,6-diaryl-2-acetamidopyridines as A_3AR antagonists [84] and introduced ISVY130 and ISVY167 (Fig. 2) as novel, selective A_3AR antagonists [85, 86]. 3-(Dichlorophenyl)-isoxazole derivative K18 (Fig. 2) was disclosed recently as a novel A_3AR antagonist intended for treating inflammatory pathologies [87]. Kolocouris and coworkers have reported novel mixed A_1AR/A_3AR antagonists, such as A17 (Fig. 2) [88].

4 A_3AR Prodrugs

A prodrug is a biologically weak or inactive molecule that gets metabolized to an active pharmaceutical ingredient (drug) by enzymatic or chemical reactions in vivo. Prodrug strategies optimize the physicochemical and pharmaceutical properties of the drugs incompatible with the development of the desired delivery routes or limit the formulation options [89]. The FDA approved approximately 30 prodrugs from

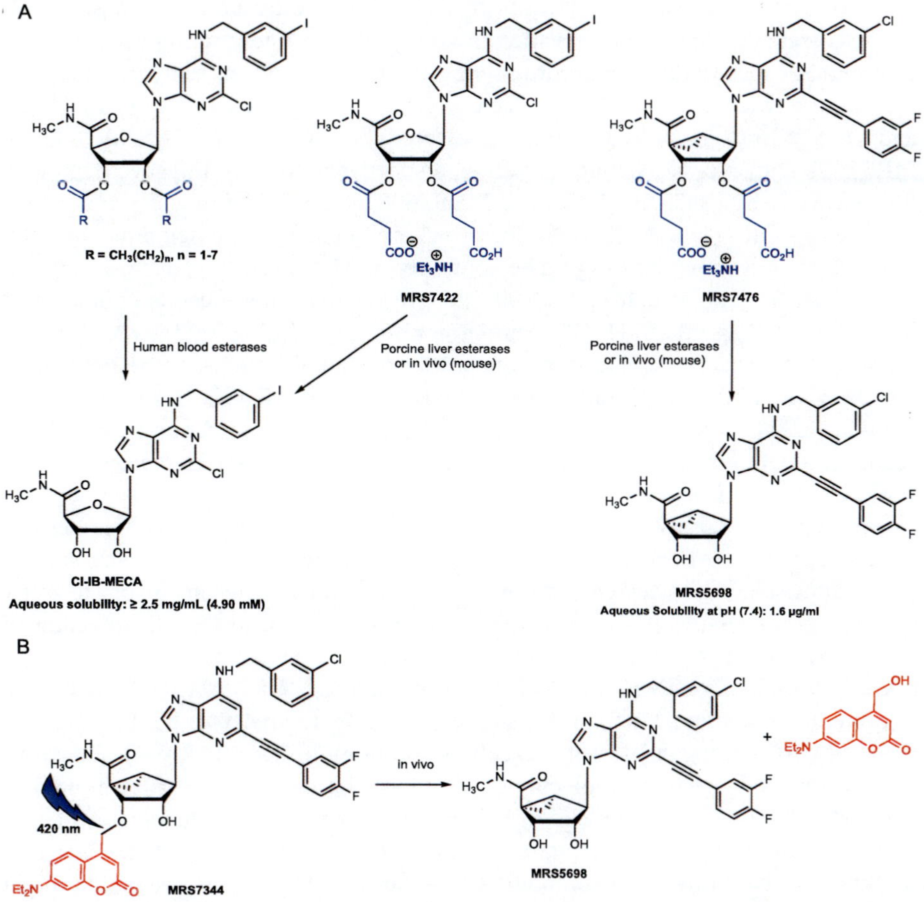

Fig. 3 AR agonist prodrug schemes, designed for cleavage by esterases (**a**) or by light (**b**, photopharmacology) [90, 91]. The prodrug masking groups are in blue (enzymatic removal) or red (photochemical removal)

2008 to 2018, and most recently in 2020, the antiviral drug Remdesivir for treating SARS-CoV-2 infection.

Many A_3AR agonists and antagonists are adenine nucleoside derivatives, which often have poor bioavailability and low aqueous solubility. The free hydroxy groups at the nucleoside ligands' $2'$ and $3'$ positions were derivatized using appropriate fatty acid anhydride or acid chloride to get the aliphatic mono- or diester prodrugs (Fig. 3) [92]. This dual acylation precludes binding of the ribose moiety to ARs in general, because H-bonding of both $2'$ and $3'$-hydroxyl groups to the receptor is required and the ribose subpocket is highly limited sterically [40, 59, 93]. The lipophilic prodrugs were intended to have increased half-life of the active parent compound and potentially more favorable bioavailability. Upon in vitro incubation with human whole blood, the ester prodrugs were cleaved by esterases to generate the active drug in 2 h

(Fig. 3). The in vitro studies showed that the *O*-acyl ester prodrugs mask A_3AR agonists and antagonists for activation in vivo [90]. The ester prodrugs are reminiscent of other successful ester prodrug schemes [94]. Thus, we recently designed and synthesized two water-soluble *O*-acyl carboxylic acid reversible prodrugs MRS7422 and MRS7476 (Fig. 3), by treating the selective A_3AR agonists Cl-IB-MECA and MRS5698 with excess succinic anhydride, respectively. The prodrugs were found to liberate active drugs Cl-IB-MECA and MRS5698 upon incubation with porcine liver esterases in vitro (Fig. 3). The prodrugs had significantly increased aqueous solubility compared to parent drugs. The solubility of MRS7476 in water was approximately 2.5 mg/ml, which is >1,000-fold greater than the aqueous solubility of MRS5698 (1.6 μg/ml at pH 7.4). Furthermore, orally administered MRS7476 was fully efficacious in reversing mouse neuropathic pain (chronic constriction injury model, 3 μmol/kg, p.o.), with a longer duration than MRS7422. In addition, compound MRS7476 (5 mg/kg, p.o., twice daily) induced liver protection in the STAM mouse model against non-alcoholic steatohepatitis (NASH), indicated by the NAFLD activity score [90].

Prodrug schemes were also devised for A_3AR antagonists. Pyridine-derived A_3AR antagonists [69] have poor aqueous solubility. However, water-soluble pyridinium salts were designed from the 1,4-dihydropyridine precursors to overcome solubility issues [47]. Furthermore, SAR studies led to the identification of 2,4-diethyl-1-methyl-3-(ethylsulfanylcarbonyl)-5-ethyloxycarbonyl-6-phenylpyridinium iodide (MRS1649, Fig. 2, human A_3AR $K_i = 219$ nM). Although pyridinium compound MRS1649 is slightly less potent in receptor binding than the reported parent antagonists, it has excellent solubility in water (43 mM). However, as a permanently charged molecule it would not be expected to cross biological membranes. It could be generated rapidly in vivo through the biochemical oxidation of the 1,4-dihydropyridne derivative MRS1667 [47]. Thus, the 3,5-diacyl-1,2,4-trialkyl-1,4-dihydropyridine compounds may be considered as prodrugs that might have better PK properties than their corresponding pyridinium antagonists. The oxidation-activated prodrugs are reminiscent of other redox prodrug schemes involving pyridines [95].

Another type of prodrug fits the approach known as photopharmacology, in which an inactive masked drug molecule is converted to the active drug by visible light irradiation. An example of this scheme applied to A_3AR agonists was MRS7344 (Fig. 3b) [91], which is an *O*-linked coumarin derivative of highly A_3AR-selective agonist MRS5698 [37]. Two other photo-masked structures were proposed, but MRS7344 was the only one prepared. When administered systemically, MRS7344 was cleaved by blue light in a site-specific manner on the mouse ear skin, to protect against psoriasis [91].

In conclusion, both biochemically reversible and photocleavable prodrugs for the A_3AR have been introduced. In vivo activity of masked A_3AR agonists was demonstrated in mouse models of three disease conditions (neuropathic pain, NASH, and psoriasis). Thus, these are novel tools for studying the pharmacological activity of A_3AR agonists in vivo that are suitable for further study.

5 A_3AR Allosteric Modulators

Several classes of heterocyclic positive allosteric modulators (PAMs) of the hA_3AR have been reported (Fig. 1). The initial identification of lead compounds by Zhan-Guo Gao and Jacobson at NIH, in collaboration with A. P. IJzerman of Leiden University, consisted of examining known AR ligands to see if any were able to increase the level of agonist radioligand binding to human A_3AR receptor in membranes [96]. A series of 1*H*-imidazo[4,5-c]quinolin-4-amine derivatives was found to enhance A_3AR agonist maximal efficacy and reduce the dissociation rate of agonist radioligand dramatically at the human but not rat or mouse A_3AR, both measures being characteristic of PAMs. Earlier members of this series were reported as A_1AR antagonists prior to the discovery of the PAM activity of certain derivatives found in extensive A_3AR antagonist SAR studies. Systematic functional group substitution resulted in the prototypical A_3AR PAMs LUF6000 and a bioisosteric amide derivative LUF6096 (Fig. 1). Recently the SAR of 1*H*-imidazo[4,5-c] quinolin-4-amine PAMs of the A_3AR was expanded, resulting in MRS7551 and MRS8054, which contain a heptan-4-yl group instead of the cyclohexyl ring of LUF6000 [80].

6 A_3AR Molecular Modeling

As of early 2023, an experimentally determined A_3AR structure has not been reported, and homology modeling has proven useful to predict the interaction of diverse ligands with the receptor orthosteric binding site. The homology modeling was initially based on the rhodopsin structure [97]; then the subsequent determination of many $A_{2A}AR$ and A_1AR structures, including active structures revealed by cryogenic electron microscopy (cryo-EM) [98, 99], provided templates that were more homologous, at least based on sequence and conserved functionality [57–59, 62]. The predicted A_3AR structures were correlated with mutagenesis results, which increased confidence in the modeling approach. Recent examples of using homology modeling of the human and mouse A_3ARs to understand ligand SAR were reported by Dal Ben et al. [62], Tosh et al. [59, 76], and Suresh et al. [40].

Conserved features across the ARs have been helpful in predicting the agonist binding mode at the A_3AR. The riboside moiety in nucleosidic AR ligands was shown by X-ray crystallography [e.g., PBD IDs 3QAK, 2YDO, 2YDV] to be in a deep hydrophilic region of the $A_{2A}AR$ surrounded by hydrophobicity in more shallow regions of the binding site that are complementary to the potency enhancement seen with bulky alkyl-aryl substitution of the N^6 and C2 positions. The sequence and presumed structure of the A_3AR features a mostly conserved placement of the ribose moiety in its hydrophilic pocket, although one (TM6) of the two key His residues present in this region of the $A_{2A}AR$ (H6.52, using Ballesteros-Weinstein nomenclature [100]), is lacking in the A_3AR, where that position is Asn.

Based on modeling, conserved interactions of the ribose moiety of agonists when A_3AR-bound include T3.36, S7.42, and H7.43. N6.55 is a key conserved residue interacting with the heterocyclic moiety of A_3AR agonists and antagonists [40, 59, 93]. The use of molecular dynamics simulation of the ligand-bound A_3AR has proven useful to correlate models with experimentally determined SAR [59]. Often, the structural plasticity of the receptor is proposed in order to accommodate some agonists, especially those that have rigid or bulky substituents while preserving other important ligand–receptor interactions. Curiously, the extended C2-arylethynyl group present in numerous adenosine-derived A_3AR agonists [57, 59] can dock inside the orthosteric binding site only after a reorganization of the relative position of transmembrane domain helix 2 (TM2) in the TM helical bundle [57, 76]. An outward displacement of the top portion of this helix was predicted using a hybrid model combining the orientation of TM2 in the A_1AR, with the position of the remaining helices corresponding to their position in the $A_{2A}AR$. In the A_3AR, the TM2 position is not confined by disulfide bonds, as it is in the $A_{2A}AR$, possibly explaining the A_3AR selectivity of C2-arylethynyl agonists.

7 Clinical Trials of A_3AR Ligands

A_3AR ligands are being developed as therapeutic candidate molecules for a variety of chronic (e.g., psoriasis, NASH, pain, cancer, glaucoma) and acute (stroke, cardiac ischemia) conditions (Table 1). The number of current and concluded clinical trials for A_3AR ligands has now surpassed those of agonists of the other AR subtypes. The two A_3AR agonists IB-MECA and Cl-IB-MECA have been in clinical trials for more than a decade, for autoimmune inflammatory diseases and liver diseases, respectively [11]. A previous clinical trial of IB-MECA for osteoarthritis was discontinued, but recent evidence points to its efficacy in an animal model [101]. IB-MECA has advanced to a pivotal Phase 3 trial for plaque psoriasis and Cl-IB-MECA is recruiting in a Phase 3 trial for hepatocellular carcinoma, and a Phase 2 trial for non-alcoholic steatohepatitis (NASH). Agonist MRS4322 (AST-004) that contains a (N)-methanocarba ring is in a Phase 1 trial for stroke [18]. NMT-006 (structure not disclosed) is described as an A_3AR modulator that is effective in models of postsurgical pain [21].

Other promising therapeutic concepts have not yet reached clinical trials. For example, intraocular implants that deliver Cl-IB-MECA prevent glutamate-induced damage to the ischemic retina [102]. PAM LUF6000 is being evaluated preclinically as a potential topical treatment of erectile dysfunction [103]. Clinical trials of A_3AR agonists for chronic pain, a promising target on its own or in combination with opioids, have not yet begun.

Although many preclinical studies have indicated the value of A_3AR agonists for therapeutics, A_3AR antagonists, including PBF-677 and PBF-1650 (structure not disclosed), or partial agonists, i.e., FM101 [20], have also progressed to clinical trials (Table 1).

8 Conclusions

Early SAR studies of orthosteric ligands, both agonists and antagonists, of the A_3AR have led to numerous translational studies. Thousands of A_3AR agonists, antagonists, and allosteric modulators have been reported since the cloning of the receptor. Species differences are prominent and determine the choice of pharmacological probes for a given experiment. In vivo disease models have shown the efficacy of many of these ligands, which has led to numerous clinical trials, including Phase 2 and 3 trials of A_3AR agonists.

Acknowledgments We thank the NIDDK Intramural Res. Program (ZIADK031117) for support.

Compliance with Ethical Standards Conflict of Interest The author declares that they have no conflict of interest.

Ethical Approval Not applicable.

Competing Interests The authors declare no competing interests of a financial or personal nature.

Conflict of Interest Kenneth A. Jacobson declares that he/she has no conflict of interest. Paola Oliva declares that he/she has no conflict of interest. R. Rama Suresh declares that he/she has no conflict of interest.

Authors' Contributions K.A.J., P.O. and R. R. S. wrote, edited, and reviewed the main manuscript text.

Funding This work was supported by the Intramural Research Program of the National Institutes of Health National Institute of Diabetes and Digestive and Kidney Diseases for support [ZIADK031117 to K.A.J].

Availability of Data and Materials Not applicable.

References

1. Borea PA, Varani K, Vincenzi F, Baraldi PG, Tabrizi MA, Merighi S et al (2015) The A_3 adenosine receptor: history and perspectives. Pharmacol Rev 67(1):74–102
2. Jacobson KA, Tosh DK, Gao ZG, Yu J, Suresh RR, Rao H, Romagnoli R, Baraldi PG, Tabrizi MA (2018) Chapter 7. Medicinal chemistry of the A_3 adenosine receptor. In: Varani K (ed) The adenosine receptors, the receptors, vol 34. Springer, pp 169–198. https://doi.org/10.1007/978-3-319-90808-3_7
3. Merighi S, Battistello E, Giacomelli L, Varani K, Vincenzi F, Borea PA et al (2019) Targeting A_3 and A_{2A} adenosine receptors in the fight against cancer. Expert Opin Ther Targets:1–10. https://doi.org/10.1080/14728222.2019.1630380
4. David M, Gospodinov DK, Gheorghe N et al (2016) Treatment of plaque-type psoriasis with oral CF101: data from a phase II/III multicenter, randomized, controlled trial. J Drugs Dermatol 15(8):931–938

5. Safadi R, Braun M, Francis A et al (2021) Randomised clinical trial: a phase 2 double-blind study of namodenoson in non-alcoholic fatty liver disease and steatohepatitis. Aliment Pharmacol Ther 54:1405–1415. https://doi.org/10.1111/apt.16664
6. Jung S-M, Peyton L, Essa H, Choi D-S (2022) Adenosine receptors: emerging non-opioids targets for pain medications. Neurobiol Pain 11:100087. https://www.sciencedirect.com/science/article/pii/S2452073X22000046
7. Spinozzi E, Baldassarri C, Acquaticci L et al (2021) Adenosine receptors as promising targets for the management of ocular diseases. Med Chem Res 30:353–370. https://doi.org/10.1007/s00044-021-02704-x
8. Chen G-J, Harvey BK, Shen H, Chou J, Victor A, Wang Y (2006) Activation of adenosine A_3 receptors reduces ischemic brain injury in rodents. J Neurosci Res 84:1848–1855. https://doi.org/10.1002/jnr.21071
9. Jacobson KA, Tosh DK, Jain S, Gao ZG (2019) Historical and current adenosine receptor agonists in preclinical and clinical development. Front Cell Neurosci 13:124. https://doi.org/10.3389/fncel.2019.00124
10. Baraldi PG, Tabrizi MA, Gessi S, Borea PA (2008) Adenosine receptor antagonists: translating medicinal chemistry and pharmacology into clinical utility. Chem Rev 108(1):238–263. https://doi.org/10.1021/cr0682195
11. Fishman P (2022) Drugs targeting the A_3 adenosine receptor: human clinical study data. Molecules 27:3680. https://doi.org/10.3390/molecules27123680
12. Bar-Yehuda S, Rath-Wolfson L, Del Valle L, Ochaion A, Cohen S, Patoka R, Zozulya G, Barer F, Atar E, Pina-Oviedo S, Perez-Liz G, Castel D, Fishman P (2009) Induction of an anti-inflammatory effect and prevention of cartilage damage in rat knee osteoarthritis by CF101 treatment. Arthritis Rheum 60(10):3061–3071. https://doi.org/10.1002/art.24817
13. Fishman P, Cohen S, Bar-Yehuda S (2013) Targeting the A_3 adenosine receptor for glaucoma treatment (review). Mol Med Rep 7(6):1723–1725. https://doi.org/10.3892/mmr.2013.1413
14. Bar-Yehuda S, Luger D, Ochaion A, Cohen S, Patokaa R, Zozulya G, Silver PB, de Morales JM, Caspi RR, Fishman P (2011) Inhibition of experimental auto-immune uveitis by the A_3 adenosine receptor agonist CF101. Int J Mol Med 28(5):727–731. https://doi.org/10.3892/ijmm.2011.753
15. Cohen S, Fishman P (2019) Targeting the A_3 adenosine receptor to treat cytokine release syndrome in cancer immunotherapy. Drug Des Devel Ther 13:491–497
16. Stemmer SM, Manojlovic NS, Marinca MV, Petrov P, Cherciu N, Ganea D, Ciuleanu TE, Pusca IA, Beg MS, Purcell WT et al (2021) Namodenoson in advanced hepatocellular carcinoma and Child–Pugh B cirrhosis: randomized placebo-controlled clinical trial. Cancers 13:187. https://doi.org/10.3390/cancers13020187
17. Safadi R, Braun M, Francis A, Milgrom Y, Massarwa M, Hakimian D, Hazou W, Issachar A, Harpaz Z, Farbstein M, Itzhak I, Lev-Cohain N, Bareket-Samish A, Silverman MH, Fishman P (2021) Randomised clinical trial: A phase 2 double-blind study of namodenoson in non-alcoholic fatty liver disease and steatohepatitis. Aliment Pharmacol Ther 54(11–12): 1405–1415. https://doi.org/10.1111/apt.16664
18. Bozdemir E, Vigil FA, Chun SH, Espinoza L, Bugay V, Khoury SM, Holstein DM, Stoja A, Lozano D, Tunca C, Sprague SM, Cavazos JE, Brenner R, Liston TE, Shapiro MS, Lechleiter JD (2021) Neuroprotective roles of the adenosine A_3 receptor agonist AST-004 in mouse model of traumatic brain injury. Neurotherapeutics 18:2707–2721. https://doi.org/10.1007/s13311-021-01113-7
19. Shon C, Choi SY, Park JS, Park CW, Lee J, Roh YS (2021) Preclinical characterization of FM101, a first-in-class A_3 adenosine receptor modulator for the treatment of non-alcoholic steatohepatitis. Eur Assoc Stud Liver Abstr. https://www.postersessiononline.eu/173580348_eu/congresos/ILC2019/aula/-FRI_344_ILC2019.pdf
20. Park CW, Han CT, Sakaguchi Y, Lee J, Youn HY (2020) Safety evaluation of FM101, an A_3 adenosine receptor modulator, in rat, for developing as therapeutics of glaucoma and hepatitis. EXCLI J 19:187–200. https://doi.org/10.17179/excli2019-2058

21. Steigerwald I, Pergolizzia J, Argoff C (2020) Efficacy and safety of NTM-006 in a randomized, double-blind, placebo- and active-controlled trial in moderate to severe pain after third molar extraction. Meaningful and sustained analgesia over 24 h demonstrated for a single dose (1000 mg) of NTM- 006, an NCE with a novel non-opioid/non-NSAID mechanism of action. (2020) PAINWeek Abstract Book 2020. Postgrad Med 132(Suppl 1):1–76. https://doi.org/10.1080/00325481.2020.1824967. abstract 22, pp 19–21
22. IJzerman AP, Jacobson KA, Müller CE, Cronstein BN, Cunha RA (2022) International Union of Basic and Clinical Pharmacology. CXII: adenosine receptors – a further update. Pharmacol Rev 74:340–372
23. Meyerhof W, Müller-Brechlin R, Richter D (1991) Molecular cloning of a novel putative G-protein coupled receptor expressed during rat spermiogenesis. FEBS Lett 284(2):155–160. https://doi.org/10.1016/0014-5793(91)80674-r
24. Zhou QY, Li C, Olah ME, Johnson RA, Stiles GL, Civelli O (1992) Molecular cloning and characterization of an adenosine receptor: the A_3 adenosine receptor. Proc Natl Acad Sci U S A 89(16):7432–7326. https://doi.org/10.1073/pnas.89.16.7432
25. Linden J, Taylor HE, Robeva AS, Tucker AL, Stehle H, Rivkees SA, Fink JS, Reppert SM (1993) Molecular cloning and functional expression of a sheep A_3 adenosine receptor with widespread tissue distribution. Mol Pharmacol 44(3):524–532
26. Salvatore CA, Jacobson MA, Taylor HE, Linden J, Johnson RG (1993) Molecular cloning and characterization of the human A_3 adenosine receptor. Proc Natl Acad Sci U S A 90(21): 10365–10369. https://doi.org/10.1073/pnas.90.21.10365
27. Ali H, Cunha-Melo JR, Saul WF, Beaven MA (1990) Activation of phospholipase C via adenosine receptors provides synergistic signals for secretion in antigen-stimulated RBL-2H3 cells. Evidence for a novel adenosine receptor. J Biol Chem 265(2):745–753
28. van Galen PJM, van Bergen AH, Gallo-Rodriguez C, Melman N, Olah ME, IJzerman AP, Stiles GL, Jacobson KA (1994) A binding site model and structure-activity relationships for the rat A_3 adenosine receptor. Mol Pharmacol 45:1101–1111
29. Jacobson KA, Nikodijevic O, Shi D, Gallo-Rodriguez C, Olah ME, Stiles GL, Daly JW (1993) A role for central A_3-adenosine receptors: mediation of behavioral depressant effects. FEBS Lett 336:57–60
30. Carlin JL, Jain S, Gizewski E, Wan TC, Tosh DK, Xiao C, Auchampach JA, Jacobson KA, Gavrilova O, Reitman ML (2017) Hypothermia in mouse is caused by adenosine A_1 and A_3 receptor agonists and AMP via three distinct mechanisms. Neuropharmacology 114:101–113
31. von Lubitz DKJE, Lin RC-S, Popik P, Carter MF, Jacobson KA (1994) Adenosine A_3 receptor stimulation and cerebral ischemia. Eur J Pharmacol 263:59–67
32. Liu GS, Richards SC, Olsson RA, Mullane K, Walsh RS, Downey JM (1994) Evidence that the adenosine A_3 receptor may mediate the protection afforded by preconditioning in the isolated rabbit heart. Cardiovasc Res 28:1057–1061
33. Liang BT, Jacobson KA (1998) A physiological role of the adenosine A_3 receptor: sustained cardioprotection. Proc Natl Acad Sci U S A 95:6995–6999
34. Ross Tracey W, Magee W, Masamune H, Kennedy SP, Knight DR, Allan Buchholz R, Hill RJ (1997) Selective adenosine A_3 receptor stimulation reduces ischemic myocardial injury in the rabbit heart. Cardiovasc Res 33(2):410–415. https://doi.org/10.1016/S0008-6363(96)00240-4
35. Gallo-Rodriguez C, Ji X-D, Melman N, Siegman BD, Sanders LH, Orlina J, Fischer B, Pu Q-L, Olah ME, van Galen PJM, Stiles GL, Jacobson KA (1994) Structure-activity relationships of N^6-benzyladenosine-5′-uronamides as A_3-selective adenosine agonists. J Med Chem 37:636–646
36. Kim HO, Ji X-D, Siddiqi SM, Olah ME, Stiles GL, Jacobson KA (1994) 2-Substitution of N^6-benzyladenosine-5′-uronamides enhances selectivity for A_3-adenosine receptors. J Med Chem 37:3614–3621
37. Gao ZG, Auchampach JA, Jacobson KA (2023) Species dependence of A_3 adenosine receptor pharmacology and function. Purinergic Signal:1. https://doi.org/10.1007/s11302-022-09910-1

38. Ji X-D, von Lubitz D, Olah ME, Stiles GL, Jacobson KA (1994) Species differences in ligand affinity at central A_3-adenosine receptors. Drug Dev Res 33:51–59
39. Jacobson KA, Gao ZG, Liang BT (2007) Neoceptors: reengineering GPCRs to recognize tailored ligands. Trends Pharmacol Sci 28:111–116
40. Suresh RR, Gao ZG, Salmaso V, Chen E, Campbell RG, Poe RB, Liston TE, Jacobson KA (2022) Selective A_3 adenosine receptor antagonist radioligand for human and rodent species. ACS Med Chem Lett 13(4):623–631
41. van Galen PJM, Adriaan P, IJzerman AP, Soudijn W (1990) Xanthine-7-ribosides as adenosine A_1 receptor antagonists: further evidence for Adenosine's anti mode of binding. Nucleosides Nucleotides 9(2):275–291. https://doi.org/10.1080/07328319008045138
42. Kim HO, Ji X-D, Melman N, Olah ME, Stiles GL, Jacobson KA (1994) Structure activity relationships of 1,3-dialkylxanthine derivatives at rat A_3-adenosine receptors. J Med Chem 37: 3373–3382
43. Park KS, Hoffmann C, Kim HO, Padgett WL, Daly JW, Brambilla R, Motta C, Abbracchio MP, Jacobson KA (1998) Activation and desensitization of rat A_3-adenosine receptors by selective adenosine derivatives and xanthine-7-ribosides. Drug Dev Res 44:97–105
44. Olah ME, Gallo-Rodriguez C, Jacobson KA, Stiles GL (1994) ^{125}I-4-Aminobenzyl-5′-N-methylcarboxamidoadenosine, a high affinity radioligand for the rat A_3 adenosine receptor. Mol Pharmacol 45:978–982
45. Rivkees SA, Thevananther S, Hao H (2000) Are A_3 adenosine receptors expressed in the brain? Neuroreport 11(5):1025–1030
46. Klotz KN, Falgner N, Kachler S, Lambertucci C, Vittori S, Volpini R, Cristalli G (2007) [^{3}H] HEMADO – a novel tritiated agonist selective for the human adenosine A_3 receptor. Eur J Pharmacol 556(1–3):14–18. https://doi.org/10.1016/j.ejphar.2006.10.048
47. Xie R, Li AH, Ji X-D, Melman N, Olah ME, Stiles GL, Jacobson KA (1999) Selective adenosine A_3 receptor antagonists: water soluble 3,5-diacylpyridinium salts and their oxidative generation from dihydropyridine precursors. J Med Chem 42:4232–4238
48. Ji X-D, Gallo-Rodriguez C, Jacobson KA (1994) A selective agonist affinity label for A_3 adenosine receptors. Biochem Biophys Res Commun 203:570–576
49. Yang X, van Veldhoven JPD, Offringa J et al (2019) Development of covalent ligands for G protein-coupled receptors: a case for the human adenosine A_3 receptor. J Med Chem 62(7): 3539–3552. https://doi.org/10.1021/acs.jmedchem.8b02026
50. Stoddart LA, Vernall AJ, Briddon SJ, Kellam B, Hill SJ (2015) Direct visualisation of internalization of the adenosine A_3 receptor and localization with arrestin3 using a fluorescent agonist. Neuropharmacology 98:68–77. https://doi.org/10.1016/j.neuropharm.2015.04.013
51. Toti KS, Campbell RG, Lee H, Salmaso V, Suresh RR, Gao ZG, Jacobson KA (2022) Fluorescent A_{2A} and A_3 adenosine receptor antagonists as flow cytometry probes. Purinergic Signal. https://doi.org/10.1007/s11302-022-09873-3
52. DeNinno MP, Masamune H, Chenard LK, DiRico KJ, Eller C, Etienne JB, Tickner JE, Kennedy SP, Knight DR, Kong J, Oleynek JJ, Tracey WR, Hill RJ (2003) 3′- -Aminoadenosine-5′-uronamides: discovery of the first highly selective agonist at the human adenosine A_3 receptor. J Med Chem 46:353–355
53. Elzein et al (2004) 2-Pyrazolyl-N^6-substituted adenosine derivatives as high affinity and selective adenosine A_3 receptor agonists. J Med Chem 47:4766–4773
54. Jacobson KA, Ji X, Li A-H, Melman N, Siddiqui MA, Shin K-J, Marquez VE, Ravi RG (2000) Methanocarba analogues of purine nucleosides as potent and selective adenosine receptor agonists. J Med Chem 43:2196–2203. http://pubs.acs.org. https://doi.org/10.1021/jm9905965
55. Tchilibon S, Joshi BV, Kim S-K, Duong HT, Gao Z-G, Jacobson KA (2005) (N)-methanocarba 2,N^6-disubstituted adenine nucleosides as highly potent and selective A_3 adenosine receptor agonists. J Med Chem 48(6):1745–1758. https://doi.org/10.1021/jm049580r
56. Melman A, Gao Z-G, Kumar D, Wan TC, Gizewski E, Auchampach JA, Jacobson KA (2008) Design of (N)-methanocarba adenosine 5′-uronamides as species-independent A_3 receptor-

selective agonists. Bioorg Med Chem Lett 18:2813–2819. https://doi.org/10.1016/j.bmcl.2008.04.001
57. Tosh DK, Finley A, Paoletta S, Moss SM, Gao Z-G, Gizewski ET, Auchampach JA, Salvemini D, Jacobson KA (2014) In vivo phenotypic screening for treating chronic neuropathic pain: modification of C2-arylethynyl group of conformationally constrained A_3 adenosine receptor agonists. J Med Chem 57(23):9901–9914. https://doi.org/10.1021/jm501021n
58. Tosh DK, Ciancetta A, Warnick E, O'Connor R, Chen Z, Gizewski E, Crane S, Gao Z-G, Auchampach JA, Salvemini D, Jacobson KA (2016) Purine (N)-methanocarba nucleoside derivatives lacking an exocyclic amine as selective A_3 adenosine receptor agonists. J Med Chem 59(7):3249–3263. https://doi.org/10.1021/acs.jmedchem.5b01998
59. Tosh DK, Salmaso V, Rao H, Campbell R, Bitant A, Gao Z-G, Auchampach JA, Jacobson KA (2020) Direct comparison of (N)-methanocarba and ribose-containing 2-arylalkynyladenosine derivatives as A_3 receptor agonists. ACS Med Chem Lett 11:1935–1941. https://doi.org/10.1021/acsmedchemlett.9b00637
60. Gao ZG, Teng B, Wu H, Joshi BV, Griffiths GL, Jacobson KA (2009) Synthesis and pharmacological characterization of [^{125}I]MRS1898, a high affinity, selective radioligand for the rat A_3 adenosine receptor. Purinergic Signal 5:31–37
61. Matot I, Einav S, Weininger CF, Pearl RG, Abramovich R, Joshi BV, Jacobson KA (2008) Lung injury following in-vivo reperfusion: outcome at 27 hours post-reperfusion. Anesthesiology 109:269–278
62. Dal Ben D, Buccioni M, Lambertucci C, Kachler S, Falgner N, Marucci G, Thomas A, Cristalli G, Volpini R, Klotz K-N (2014) Different efficacy of adenosine and NECA derivatives at the human A_3 adenosine receptor: insight into the receptor activation switch. Biochem Pharmacol 87(2):321–331. https://doi.org/10.1016/j.bcp.2013.10.011
63. Coppi E, Cherchi F, Lucarini E, Ghelardini C, Pedata F, Jacobson KA, Di Cesare Mannelli L, Pugliese AM, Salvemini D (2021) Uncovering the mechanisms of A_3 adenosine receptor-mediated pain control. Int J Mol Sci 22:7952
64. Paoletta S, Tosh DK, Finley A, Gizewski E, Moss SM, Gao ZG, Auchampach JA, Salvemini D, Jacobson KA (2013) Rational design of sulfonated A_3 adenosine receptor-selective nucleosides as pharmacological tools to study chronic neuropathic pain. J Med Chem 56:5949–5963
65. Petrelli R, Torquati I, Kachler S, Luongo L, Maione S, Franchetti P, Grifantini M, Novellino E, Lavecchia A, Klotz K-N, Cappellacci L (2015) 5′-C-ethyl-tetrazolyl-N^6-substituted adenosine and 2-chloro-adenosine derivatives as highly potent dual acting A_1 adenosine receptor agonists and A_3 adenosine receptor antagonists. J Med Chem 58(5):2560–2566. https://doi.org/10.1021/acs.jmedchem.5b00074
66. Ji XD, Melman N, Jacobson KA (1996) Interactions of flavonoids and other phytochemicals with adenosine receptors. J Med Chem 39:781–788
67. Jacobson KA, Moro S, Manthey JA, West PL, Ji X-D (2002) Interaction of flavones and other phytochemicals with adenosine receptors. Adv Exp Med Biol 505:163–171
68. Jiang J-L, van Rhee AM, Chang L, Patchornik A, Evans P, Melman N, Jacobson KA (1997) Structure activity relationships of 4-phenylethynyl-6-phenyl-1,4-dihydropyridines as highly selective A_3 adenosine receptor antagonists. J Med Chem 40:2596–2608
69. Li AH, Moro S, Melman N, Ji X-D, Jacobson KA (1998) Structure activity relationships and molecular modeling of 3,5-diacyl-2,4-dialkylpyridine derivatives as selective A_3 adenosine receptor antagonists. J Med Chem 41:3186–3201
70. Jacobson KA, Park KS, Jiang J-L, Kim YC, Olah ME, Stiles GL, Ji XD (1997) Pharmacological characterization of novel A_3 adenosine receptor-selective antagonists. Neuropharmacology 36:1157–1165
71. Jacobson M, Chakravarty PK, Johnson RG, Norton R (1996) Novel selective non-xanthine A_3 adenosine receptor antagonists. Drug Dev Res 37:131

72. Kim YC, de Zwart M, Chang L, Moro S, von Frijtag Drabbe Künzel JK, Melman N, IJzerman AP, Jacobson KA (1998) Derivatives of the triazoloquinazoline adenosine antagonist (CGS15943) having high potency at the human A_{2B} and A_3 receptor subtypes. J Med Chem 41:2835–2841
73. Gao ZG, Suresh RR, Jacobson KA (2021) Pharmacological characterization of DPTN and other selective A_3 adenosine receptor antagonists. Purinergic Signal 17:737–746. https://doi.org/10.1007/s11302-021-09823-5
74. Miwatashi S, Arikawa Y, Matsumoto T, Uga K, Kanzaki N, Imai YN, Ohkawa S (2008) Synthesis and biological activities of 4-phenyl-5-pyridyl-1,3-thiazole derivatives as selective adenosine A_3 antagonists. Chem Pharm Bull(Tokyo) 56(8):1126–1137
75. Jacobson KA, Siddiqi SM, Olah ME, Ji XD, Melman N, Bellamkonda K, Meshulam Y, Stiles GL, Kim HO (1995) Structure-activity relationships of 9-alkyladenine and ribose-modified adenosine derivatives at rat A_3 adenosine receptors. J Med Chem 38:1720–1735
76. Tosh DK, Salmaso V, Rao H, Bitant A, Fisher CL, Lieberman DI, Vorbrüggen H, Reitman ML, Gavrilova O, Gao ZG, Auchampach JA, Jacobson KA (2020) Truncated (N)-methanocarba nucleosides as partial agonists at mouse and human A_3 adenosine receptors: affinity enhancement by N^6-(2-phenylethyl) substitution. J Med Chem 63:4334–4348
77. Wang Z, Do CW, Avila MY, Peterson-Yatorno K, Stone RA, Gao ZG, Joshi BV, Besada P, Jeong LS, Jacobson KA, Civan MM (2010) Nucleoside-derived antagonists to A_3 adenosine receptors lower mouse intraocular pressure and act across species. Exp Eye Res 90:146–154
78. Jacobson KA, Gao ZG, Tosh DK, Sanjayan GJ, de Castro S (2010) A_3 adenosine receptor agonists: history and future perspectives, chapter 6. In: Borea PA (ed) A_3 adenosine receptors from cell biology to pharmacology and therapeutics. Springer, pp 93–120
79. Spinaci A, Buccioni M, Dal Ben D, Maggi F, Marucci G, Francucci B, Santoni G, Lambertucci C, Volpini R (2022) A_3 adenosine receptor antagonists with nucleoside structures and their anticancer activity. Pharmaceuticals 15:164. https://doi.org/10.3390/ph15020164
80. Fallot LB, Suresh RR, Fisher CL, Kaufman N, Gao ZG, Auchampach JA, Jacobson KA (2022) Structure activity studies of 1H-imidazo[4,5-c]quinolin-4-amine derivatives as A_3 adenosine receptor positive allosteric modulators. J Med Chem 65(22):15238–15262
81. Ozola V, Thorand M, Diekmann M, Qurishi R, Schumacher B, Jacobson KA, Müller CE (2003) 2-Phenylimidazo[2,1-i]purin-5-ones: structure–activity relationships and characterization of potent and selective inverse agonists at human A3 adenosine receptors. Bioorg Med Chem 11:347–356
82. Varani K, Merighi S, Gessi S, Klotz KN, Leung E, Baraldi PG, Cacciari B, Romagnoli R, Spalluto G, Borea PA (2000) [^{3}H]MRE 3008F20: a novel antagonist radioligand for the pharmacological and biochemical characterization of human A_3 adenosine receptors. Mol Pharmacol 57:968–975
83. Mikus EG, Boér K, Timári G, Urbán-Szabó K, Kapui Z, Szeredi J, Gerber K, Szabó T, Bátori S, Finet M, Arányi P, Galzin AM (2013) Interaction of SSR161421, a novel specific adenosine A_3 receptor antagonist with adenosine A_3 receptor agonists both in vitro and in vivo. Eur J Pharmacol 699(1–3):62–66. https://doi.org/10.1016/j.ejphar.2012.11.046
84. Azuaje J, Jespers W, Yaziji V, Mallo A, Majellaro M, Caamaño O, Loza MI, Cadavid MI, Brea J, Åqvist J, Sotelo E, Gutiérrez-de-Terán H (2017) Effect of nitrogen atom substitution in a_3 adenosine receptor binding: N-(4,6-diarylpyridin-2-yl)acetamides as potent and selective antagonists. J Med Chem 60(17):7502–7511. https://doi.org/10.1021/acs.jmedchem.7b00860
85. Yaziji V, Rodríguez D, Gutiérrez-de-Terán H, Coelho A, Caamaño O, García-Mera X, Brea J, Loza MI, Cadavid MI, Sotelo E (2011) Pyrimidine derivatives as potent and selective A_3 adenosine receptor antagonists. J Med Chem 54(2):457–471. https://doi.org/10.1021/jm100843z
86. Miranda-Pastoriza D, Bernárdez R, Azuaje J, Prieto-Díaz R, Majellaro M, Tamhankar AV, Koenekoop L, González A, Gioé-Gallo C, Mallo-Abreu A, Brea J, Loza MI, García-Rey A, García-Mera X, Gutiérrez-de-Terán H, Sotelo E (2022) Exploring non-orthosteric interactions

with a series of potent and selective A_3 antagonists. ACS Med Chem Lett 13(2):243–249. https://doi.org/10.1021/acsmedchemlett.1c00598

87. Barkan K, Lagarias P, Stampelou M et al (2020) Pharmacological characterisation of novel adenosine A_3 receptor antagonists. Sci Rep 10:20781. https://doi.org/10.1038/s41598-020-74521-y
88. Stampelou M, Suchankova A, Tzortzini E, Dhingra L, Barkan K, Lougiakis N, Marakos P, Pouli N, Ladds G, Kolocouris A (2022) Dual A_1/A_3 adenosine receptor antagonists: binding kinetics and structure–activity relationship studies using mutagenesis and alchemical binding free energy calculations. J Med Chem 65(19):13305–13327
89. Rautio J, Meanwell N, Di L et al (2018) The expanding role of prodrugs in contemporary drug design and development. Nat Rev Drug Discov 17:559–587. https://doi.org/10.1038/nrd.2018.46
90. Suresh RR, Jain S, Chen Z, Tosh DK, Ma Y, Podszun MC, Rotman Y, Salvemini D, Jacobson KA (2020) Design and in vivo activity of A_3 adenosine receptor agonist prodrugs. Purinergic Signal 16:367–377
91. López-Cano M, Filgaria I, Nolen EG, Cabré G, Hernando J, Tosh DK, Jacobson KA, Soler C, Ciruela F (2021) Optical control of adenosine A_3 receptor function in psoriasis. Pharmacol Res 170:105731
92. Besada P, Mamedova LK, Palaniappan KK, Gao ZG, Joshi BV, Jeong LS, Civan MM, Jacobson KA (2006) Nucleoside prodrugs of A_3 adenosine receptor agonists and antagonists. Collect Czechoslov Chem Commun 71:912–928
93. Salmaso V, Jain S, Jacobson KA (2021) Purinergic GPCR transmembrane residues involved in ligand recognition and dimerization. In: Shukla A (ed) Methods cell biology biomolecular interactions, vol 166, pp 133–159
94. Haddad F, Sawalha M, Khawaja Y, Najjar A, Karaman R (2017) Dopamine and levodopa prodrugs for the treatment of Parkinson's disease. Molecules 23(1):40
95. Prokai L, Prokai-Tatri L, Bodor N (2000) Targeting drug to the brain by redox chemical delivery systems. Med Res Rev 20:367–416
96. Du L, Gao ZG, Paoletta S, Wan TC, Barbour S, van Veldhoven JP, IJzerman AP, Jacobson KA, Auchampach JA (2018) Species differences and mechanism of action of A_3 adenosine receptor allosteric modulators. Purinergic Signal 14:59–71
97. Costanzi S, Siegel J, Tikhonova IG, Jacobson KA (2009) Rhodopsin and the others: a historical perspective on structural studies of G protein-coupled receptors. Curr Pharm Des 15:3994–4002
98. Carpenter B, Tate CG (2017) Active state structures of G protein-coupled receptors highlight the similarities and differences in the G protein and arrestin coupling interfaces. Curr Opin Struct Biol 45:124–132. https://doi.org/10.1016/j.sbi.2017.04.01
99. Draper-Joyce CJ, Bhola R, Wang J, Bhattarai A, Nguyen ATN, Cowie-Kent I, O'Sullivan K, Chia LY, Venugopal H, Valant C, Thal DM, Wootten D, Panel N, Carlsson J, Christie MJ, White PJ, Scammells P, May LT, Sexton PM, Danev R, Miao Y, Glukhova A, Imlach WL, Christopoulos A (2021) Positive allosteric mechanisms of adenosine A_1 receptor-mediated analgesia. Nature 597:571–576
100. Ballesteros JA, Weinstein H (1995) Integrated methods for the construction of three dimensional models and computational probing of structure function relations in G protein-coupled receptors. Methods Neurosci 25:366–428
101. Bai H, Zhang Z, Liu L, Wang X, Song X, Gao L (2022) Activation of adenosine A_3 receptor attenuates progression of osteoarthritis through inhibiting the NLRP3/caspase-1/GSDMD induced signalling. J Cell Mol Med 26(15):4230–4243. https://doi.org/10.1111/jcmm.17438
102. Boia R, Dias PAN, Galindo-Romero C, Ferreira H, Aires ID, Vidal-Sanz M et al (2022) Intraocular implants loaded with A_3R agonist rescue retinal ganglion cells from ischemic damage. J Control Release 343:469–481. https://www.sciencedirect.com/science/article/pii/S016836592200075X
103. Itzhak I, Cohen S, Fishman S, Fishman P (2022) A_3 adenosine receptor allosteric modulator CF602 reverses erectile dysfunction in a diabetic rat model. Andrologia:e14498. https://doi.org/10.1111/and.14498

Top Med Chem (2023) 41: 179–222
https://doi.org/10.1007/7355_2022_154

Published online: 25 November 2022

Bifunctional Tools to Study Adenosine Receptors

China Payne, Jon K. Awalt, Lauren T. May, Joel D. A. Tyndall, Manuela Jörg, and Andrea J. Vernall

Contents

C. Payne and A. J. Vernall (✉)
Department of Chemistry, University of Otago, Dunedin, New Zealand
e-mail: andrea.vernall@otago.ac.nz

J. K. Awalt
Monash Institute of Pharmaceutical Sciences, Monash University, Parkville, VIC, Australia

Chemical Biology Division, Walter and Eliza Hall Institute of Medical Research, Parkville, VIC, Australia

L. T. May
Drug Discovery Biology, Monash Institute of Pharmaceutical Sciences, Monash University, Parkville, VIC, Australia

Department of Pharmacology, Monash University, Parkville, VIC, Australia

J. D. A. Tyndall
School of Pharmacy, University of Otago, Dunedin, New Zealand

M. Jörg
Medicinal Chemistry Theme, Monash Institute of Pharmaceutical Sciences, Monash University, Parkville, VIC, Australia

Newcastle University Centre for Cancer, Chemistry – School of Natural & Environmental Sciences, Newcastle University, Newcastle upon Tyne, UK

Abstract Target-based G protein-coupled receptors (GPCRs) drug discovery has historically focused on the development of ligands that either activate or inhibit the signal response of one specific receptor. However, the potential benefits of ligands with multi-target activity have gained more attention, particularly for the development of drugs for complex diseases such as CNS disorders. Furthermore, GPCRs have traditionally been studied in isolation, whereas there is increased evidence of GPCRs existing in dimeric and higher-order oligomeric complexes. These aspects have contributed to the development of a range of novel ligand types, which are able to interact with multiple GPCR monomers, including dual-acting ligands and prodrugs as well as homo- and heterobivalent ligands. This chapter provides an overview on the development of ligands with multi-target activity against at least one receptor subtype of the purinergic receptor family. In addition to summarising reported bifunctional ligands in this field, the important design aspects and the role of computational methods and structural biology in guiding the development of bifunctional ligands are discussed.

Keywords Adenosine receptors, Bifunctional, Bitopic ligands, Bivalent ligands, Dual-acting ligands, G protein-coupled receptors, Heterodimers, Homodimers

1 Introduction

Despite much debate over the last few decades, it is now widely accepted that G protein-coupled receptors (GPCRs) can exist as homodimers, heterodimers, or even higher-order oligomers. Purinergic GPCRs (e.g. adenosine (P1) and P2Y receptors) have been shown to exist as homodimers and also heterodimers with many other GPCRs such as dopamine and β-adrenergic receptors (Table 1).

Adenosine receptors (ARs) are attractive drug targets, due to their roles in the cardiovascular and immune systems, as well as in the central nervous system where ARs have been implicated in diverse conditions from neurodegenerative diseases (such as Alzheimer's Disease) to sleep disorders and schizophrenia [20–23]. The ability to target different (orthosteric and allosteric) ligand binding sites of the receptor, as well as homo- and heterodimers of ARs offers diverse and potentially unique therapeutic avenues different to solely targeting the orthosteric ligand binding site of an AR monomer. ARs are well-known for the prevalence of homo- and hetero-dimerization with other GPCRs [24–26]. There is a pressing need to develop tools to further investigate AR dimers, to explore their role in signalling pathways

Table 1 Detected GPCR dimers containing an adenosine receptor(s), Table adapted from Hiller et al. [1] and George et al. [2]

Receptor dimer	Detection method	Reference
A_1R–A_1R	co-IP, BRET	[3, 4]
A_1R–$A_{2A}R$	co-IP, BiFC, BRET, FRET	[5]
A_1R–A_3R	nd	[6]
A_1R–β_1R	co-IP	[7]
A_1R–β_2R	co-IP	[7]
A_1R–D_1R	co-IP	[8, 9]
A_1R–$mGlu_1R$	co-IP, FRET	[10, 11]
A_1R–$P2Y_1$	co-IP, BRET	[12]
$A_{2A}R$–$A_{2B}R$	BRET, FRET, BiFC	[13]
$A_{2A}R$–CB_1R	BRET, IP	[14]
$A_{2A}R$–D_2R	BiFC, BRET, co-IP	[15–17]
$A_{2A}R$–$mGlu_5R$	co-IP	[18]
A_3R–A_3R	Confocal microscopy	[19]

Receptors: A_1R, adenosine A_1 receptor; $A_{2A}R$, adenosine A_{2A} receptor; $A_{2B}R$, adenosine A_{2B} receptor; A_3R, adenosine A_3 receptor; β_1R, β_1-adrenergic receptor; β_2R, β_2-adrenergic receptor; CB_1R, cannabinoid type 1 receptor; D_1R, dopamine type 1 receptor; D_2R, dopamine type 2 receptor; $mGlu_1R$, glutamate type 1 receptor; $mGlu_5R$, glutamate type 5 receptor; P2Y, purinergic receptor
Detection methods: co-IP, coimmunoprecipitation; BRET, bioluminescence resonance energy transfer; BiFC, bimolecular fluorescence complementation; FRET, fluorescence resonance energy transfer; nd, not determined

and to explore the biological functions in order to provide a better understanding of receptor dimerisation in many diseases and conditions.

Bivalent ligands targeting GPCRs have been the subject of research dating back to at least the 1980s, before dimerisation of GPCRs was proven. In 1982, Erez et al. investigated the distance between opioid receptors by designing ligands that linked two identical antagonist ligands together with spacers of various lengths [27]. They showed that when the appropriate linker length is employed, affinity and subtype selectivity were increased, initiating the broader development of bivalent ligands for GPCRs. A key strategy for targeting GPCR dimers and/or dual targeting of another pharmacological role/action continues to be the design of bivalent, bitopic and/or dual-acting ligands.

1.1 General Design Aspects of Bifunctional Adenosine Receptor Tools

The individual binding moieties of a bivalent ligand are covalently connected by a linker of variable size (also referred to as a spacer). In case of 'traditional' homo- or heterobivalent ligands, which aim to interact with two receptor monomers simultaneously, the spacer is normally much longer than those present in a 'dual acting' or bitopic ligand. Heterobivalent and bitopic ligands usually consist of two individual chemical moieties/individual ligands; however, the difference between the two types of ligands arises in which receptor regions these moieties interact with. Bitopic ligands consist of two moieties linked together, often with a short linker, acting concomitantly to an orthosteric and allosteric site on the same receptor, whereas bivalent probes generally target two orthosteric sites of individual receptors. The design of bifunctional tools targeted to ARs involves consideration of many aspects. Firstly, ligand selection is of the utmost importance.

- Ligands are ideally highly potent and selective for the intended targets (in case of homobivalent ligands, the same ligand can be used twice).
- The chosen ligand(s) must also have a position tolerant of change, from which a linker or spacer may be attached.
- It is crucial that this attached linker causes minimal disruption of the ligand binding to the target receptor in order to maintain ligand selectivity and potency.

A feature common to many bifunctional tools for the AR is the use of chemical moieties based on derivatives of either the endogenous agonist adenosine (**1**) or the non-selective antagonist xanthine (**2**). The structure–activity relationships (SAR) of these AR scaffolds are well understood and have been previously reviewed in depth [28, 29]. The development of adenosine and xanthine-based congeners was pioneered by Kenneth Jacobson in the late 1980s, an approach that has aided in the development of bifunctional and bivalent AR ligands [30, 31]. A congener is defined as a ligand functionalized with a chemically 'useful' functional group for linker or spacer attachment such as an amine or carboxylic acid. Well-known AR congeners include the adenosine amine congener (ADAC) (**3**) xanthine amine congener (XAC) (**4**) and the xanthine carboxylic acid congener (XCC) (**5**) (Fig. 1). Both **3** and **4** were first utilised in 1986 [32], for attachment to carriers and reporter groups by coupling to the terminal reactive amine or carboxylic acid without losing ligand-AR binding of the overall conjugate.

Secondly, the composition of the linker connecting the two ligands is very important.

- The linker must be of an appropriate/optimal length that binding of each ligand to the respective receptor binding sites can occur.
- The chemical composition, in particular the physicochemical properties, of the linker must not disrupt the receptor–ligand interactions and/or result into detrimental interactions with the extracellular loops of the receptors.

The chemical composition of the linker has been investigated since the early work on opioid receptor bivalent ligands by Portoghese and co-workers in the 1980s [33]. Polyethylene glycol (PEG)-based linkers are most commonly employed in bivalent ligands (not limited to AR bivalent ligands); however, there are also many examples of alkyl [34–36], peptide [37, 38] and cyclic chain linkers. PEG-based linkers are often chosen because of intrinsically desirable physicochemical properties, such as low toxicity (including low antigenicity) and high aqueous solubility.

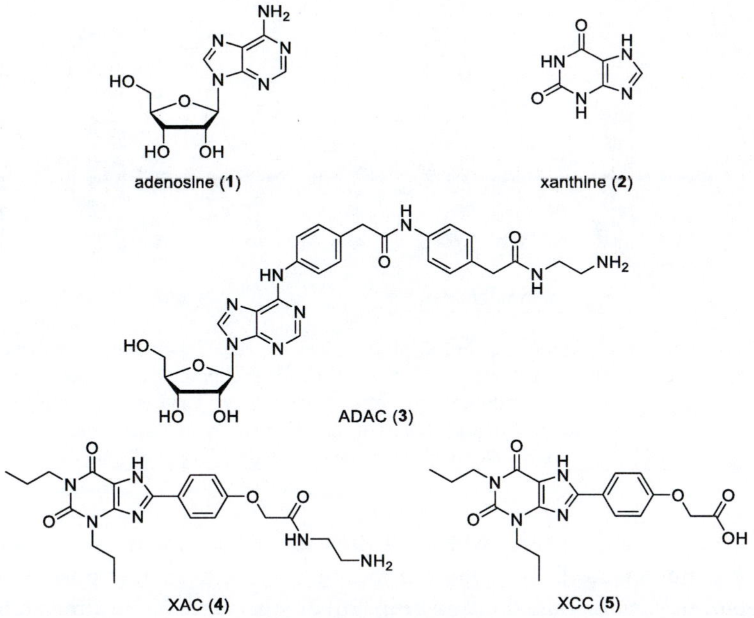

Fig. 1 The AR endogenous agonist adenosine (**1**), non-subtype selective AR antagonist xanthine (**2**), adenosine amine congener ADAC (**3**), xanthine amine congener XAC (**4**) and xanthine carboxylic acid congener XCC (**5**) developed by Jacobson et al. [32]

There is no hard and fast rule about the optimal linker length for all GPCR bifunctional tools, thus linkers like PEG with inherent repeating 'units' also allow for easy systematic variation in linker length to probe optimal distance between targeted ligand binding sites. It has been speculated that a hydrophilic vs hydrophobic linker (for example a PEG- vs. alkyl-based linker) may change the positioning of the linkers and overall mode of bivalent ligand binding. It is thought that hydrophobic linkers, particularly of bivalent ligands, are more likely to bridge the receptors directly through the receptor's hydrophobic transmembrane (TM) domains, whereas hydrophilic linkers are more likely to bridge two receptors via the extracellular space [39].

2 Heterobivalent Ligands

In a 'prototypical' bivalent ligand, each moiety/ligand will be capable of interacting simultaneously with the orthosteric site of two receptor monomers (Fig. 2). The linker length of the bivalent ligand is designed such that the simultaneous interaction of a single bivalent ligand with two receptors is assumed to bridge across a receptor dimer. Bivalent ligands can be further described as either homobivalent, wherein the bivalent ligand simultaneously interacts with two of the same monomers in a

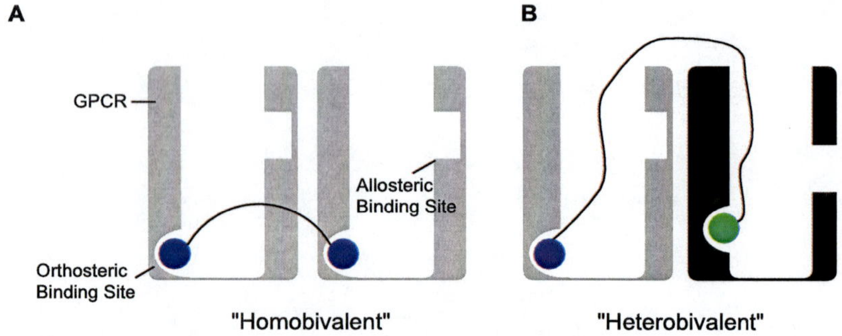

Fig. 2 Depiction of the binding mode of a 'prototypical' bivalent ligand, simultaneously binding across a receptor dimer, further defined as (**a**) homo- or (**b**) heterobivalent, corresponding to binding across a receptor homo- or heterodimeric receptor, respectively. Theoretically, the linker might cross the transmembrane domains (depicted in **a**) or may reach both orthosteric binding sites via the extracellular space (depicted in **b**). Adapted from Lane et al. and Jörg et al. [40, 41]

homodimer or heterobivalent where the bivalent ligand interacts simultaneously across two monomers. In bridging the receptors, the linker may position through the receptor transmembrane domains (Fig. 2a) or span across the extracellular space (Fig. 2b).

When considering the potential binding modes of a bivalent ligand, one must consider not only the intended concomitant binding (Fig. 2) but also several other possibilities (Fig. 3a, b). The so-called 'flip-flop' mode of binding (Fig. 3a) describes a situation where the ligand is only able to bind one of the sites at a time, potentially due to a suboptimal spacer length, and thus 'flip-flops' between them. Alternatively, if only one binding moiety of the bivalent ligand is engaged, two ligands may simultaneously bind, one at each site (Fig. 3b), imparting something analogous to traditional cooperativity of co-administered ligands [40, 42]. This further highlights the importance of selecting a suitable linker length when designing such ligands, otherwise the intended truly simultaneous, bivalent binding mode might not be achieved. To further elucidate the mode of binding, 'monovalent control ligands' (Fig. 3c, ii) are often synthesised alongside bivalent ligands. These can be thought of as a truncated or clipped bivalent ligand (i.e. Fig. 3c, i), containing one individual chemical moiety/ligand covalently connected to a linker (or portion of linker). The binding affinity, potency and/or efficacy of monovalent control ligands (Fig. 3c, ii) can be compared to the fully assembled bivalent ligand and can provide more information about linker influence and binding mode.

The first bivalent ligands targeted to ARs were reported in 2008, months before the first reported crystal structure of an AR – the human adenosine A_{2A} receptor ($A_{2A}R$) [43]. To date, homobivalent ligands for AR homodimers have not been reported, although there are many examples of homobivalent ligands for other GPCRs [44–46], (in particular, the D_2 dopamine receptor (D_2R)) [47–51]. As such, the following section reviews reported heterobivalent ligands, where one active chemical moiety of the bivalent ligand is targeted to an AR.

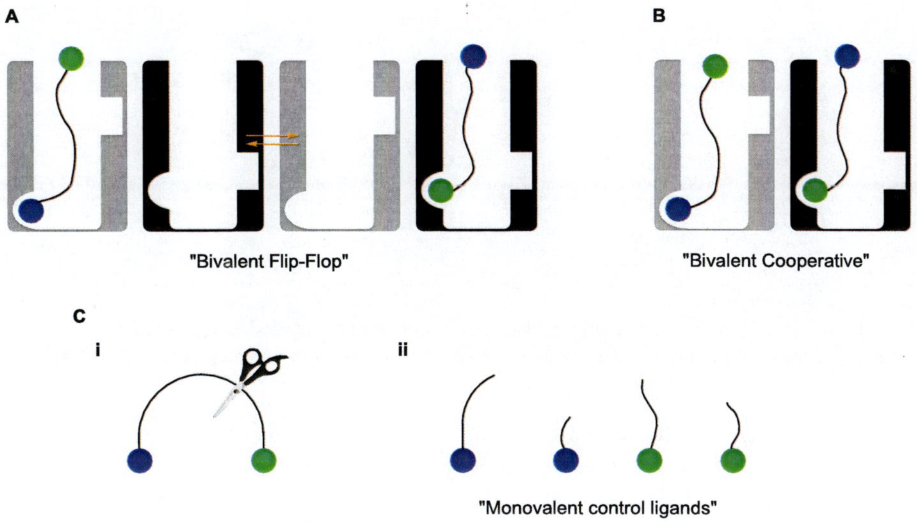

Fig. 3 Depiction of other possible binding modes for bivalent ligands. (**a**) The flip-flop mode depicts a scenario where the bivalent ligand is only able to bind one site at a time and switches between them. (**b**) Cooperative binding involves two molecules of the bivalent ligand engaged with the receptor dimer at the same time. (**c**, ii) 'Monovalent control ligands' are often synthesised alongside bivalent ligands (**c**, i) to probe binding and linker influence and tolerability and can be thought of as a truncated/clipped bivalent ligand (depicted in **c**, i). Adapted from Lane et al. and Jörg et al. [40, 41]

2.1 Heterobivalent A_1R Ligands

In 2008, Karellas and co-workers reported the synthesis of a series of heterobivalent ligands targeting $A_1R - \beta_2$ adrenoceptor (β_2R) heterodimers [34]. Both A_1R and β_2R are involved in the regulation of adenylate cyclase (AC), the activation of β_2R stimulates AC, which is attenuated by the simultaneous activation of A_1R. It was theorised that bivalent ligands targeted to this heterodimer could allow for fine-tuned control of the regulation of AC and provide further insight into the crosstalk of these two receptors. The bivalent ligands **6–11** contained linked A_1R and β_2R ligands, based on adenosine (**1**) and β_2R agonist formoterol, respectively (Fig. 4) [34]. The attachment point of the bridging linker was chosen to be between the N^6-position of adenosine (which is known to tolerate large substituents without loss of binding to A_1R) [52, 53], and the amine sidechain of formoterol. A range of alkyl chain lengths (2 to 8 atoms long) and a PEG linker to confer water solubility were investigated.

The affinities of ligands **5–10** for β_2R and A_1R were determined by displacement of the β_2R radioligand [^{125}I]-(−)iodopindolol and A_1R radioligand [^{3}H]DPCPX in DDT cell membranes, respectively. These assays included 10 μM 5′-guanylyl-imidodiphosphate to maintain the receptor in the agonist low-affinity state. Karellas et al. showed that ligands **6–11** retained activity at both receptors, and that the affinity to A_1R across the series of ligands was relatively insensitive to the linker

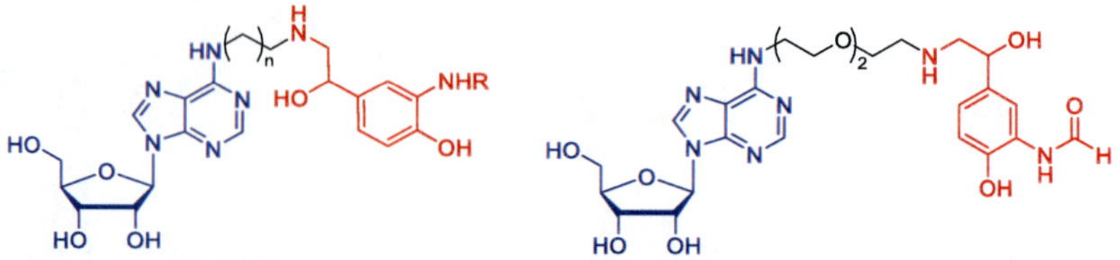

6; R = formyl, n = 1
7; R = H, n = 1
8; R = formyl, n = 3
9; R = H, n = 3
10; R = formyl, n = 5

11

Fig. 4 Bivalent ligands **6–11** developed by Karellas et al. [34] targeted to A_1R–β_2R heterodimers. The A_1R component is based on adenosine (**1**), the β_2R agonist component is based on formoterol

length, however, was more disruptive to β_2R. Ligands **6–11** were also functionally assessed for the ability to stimulate (β_2R) and/or inhibit (A_1R) cAMP accumulation in DDT cells. The butyl and hexyl alkyl-linked ligands (**8** and **10** respectively) produced potent β_2R-mediated stimulation of cAMP with the same maximum response as the full agonist (−)isoproterenol, whereas others (**9** and **11**) had an A_1R-mediated suppression of the maximal stimulatory response suggesting partial β_2R-agonism. Although the pharmacological testing showed ligands **6–11** retained affinity and activity comparable to those of similar individual ligands for each receptor type, the lack of any large increase in affinity and/or linker length correlation suggested that ligands **6–11** were not acting as true simultaneous bivalent ligands. Ligands **6–11** were likely acting in a 'bivalent flip-flop' mode (Fig. 3a) capable of interacting with both A_1R and β_2R, albeit independently. It was rationalised that the linker length may be too short for true bivalent binding bridging the heterodimer. It was also noted that the alkyl-linked ligands (**6–10**) had poor aqueous solubility. Although ligands such as **11** had improved water solubility, the direct attachment of PEG linkers to the adenosine N^6-position had the undesired effect of decreasing both potency and affinity for A_1R.

Subsequently, Barlow et al. [54] investigated longer linker lengths (19–38 atoms) and different linker compositions in a second generation (from first generation Karellas et al. [34]) of $A_1R-\beta_2R$ bivalent ligands. The same adenosine-based ligand was used, again linked via the adenosine N^6-position, but this time to the well-known β_2R agonist salbutamol (**12–17**, Fig. 5). Linkers in this series of bivalent ligands were considerably longer (19–38 atoms in length) than those previously reported by Karellas et al., to improve the possibility of simultaneous binding across the heterodimer. Based on linker results from the Karellas et al. study, bivalent ligands **16** and **17** were designed with alkyl linkers directly attached at N^6 with the PEG component distal to the adenosine moiety, to promote better solubility but not to perturb A_1R affinity as per Karellas et al. [34].

Affinities of the bivalent ligands (**12–17**) were determined by displacement binding assays against [^{125}I]-(−)iodopindolol and [^{3}H]-8-cyclopentyl-1,3-dipropylxanthine (DPCPX) for the β_2R and the A_1R respectively, using β_2R agonist (−)-isoproterenol and A_1R agonist N^{6-}cyclopentyladenosine (CPA) for comparison.

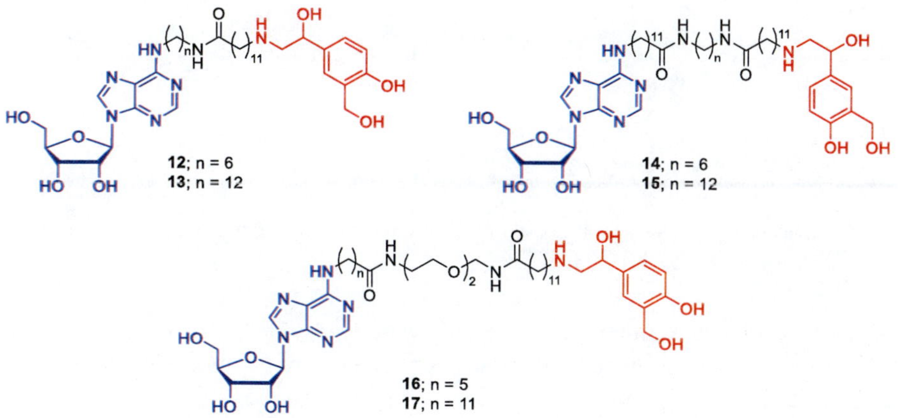

Fig. 5 Bivalent ligands **12–17** developed by Barlow et al. [54]. The A_1R component is based on adenosine (**1**), the β_2R agonist component is based on salbutamol, linked by either alkyl or PEG linkers

Bivalent compounds with solely alkyl-based linkers (**12–15**, Fig. 5) had overall decreased affinities and functional potencies for each receptor compared to the parent agonists and a β_2R monovalent control ligand (not shown). This decrease was more pronounced at β_2R. A pattern emerged with the A_1R, where both affinity and potency of compounds **12–15** decreased with increasing linker length. Previous studies have provided evidence for simultaneous bridging across two receptors by a single bivalent ligand, partly based upon increases in ligand affinity for their targets. However, and in line with the bivalent ligands developed by Karellas et al., a large increase in affinity was not observed for either β_2R or A_1R, indicating that for bivalent ligands **12–15**, which contained only alkyl linkers, simultaneous binding across the heterodimer was likely not occurring.

Bivalent ligands containing a PEG component to the linker (**16**, **17,** Fig. 5) again displayed decreased affinity and potency for each receptor, compared to the corresponding agonists and a β_2R monovalent control ligand. However, increasing the PEG-based linker length from 28 atoms (**16**) to 34 atoms (**17**) resulted in a ninefold increase in affinity at β_2R and a 16-fold increase in affinity at A_1R. In contrast to affinity, the potency (EC_{50}) for inhibiting cAMP accumulation was decreased with increasing linker length, from $EC_{50} = 614 \pm 89$ nM (**16**) to $EC_{50} = 2{,}226 \pm 371$ nM (**17**) at A_1R, far lower than the A_1R adenosine-like agonist CPA ($EC_{50} = 1.6 \pm 0.1$ nM). While it was unlikely any of ligands **12–17** were capable of acting as truly bivalent ligands, they could perhaps be acting via a 'bivalent flip-flop' mode of binding (Fig. 3a). It was concluded from these two studies from the Scammells research group [34, 54] that bivalent ligands such as **16** may be useful tool compounds for investigative studies on the mechanism of receptor crosstalk for opposing AC responses.

In the same year as Barlow et al., Shen et al. [55] also reported the development of A_1R heterodimer bivalent ligands, in this case targeted to the A_1R – dopamine D_1

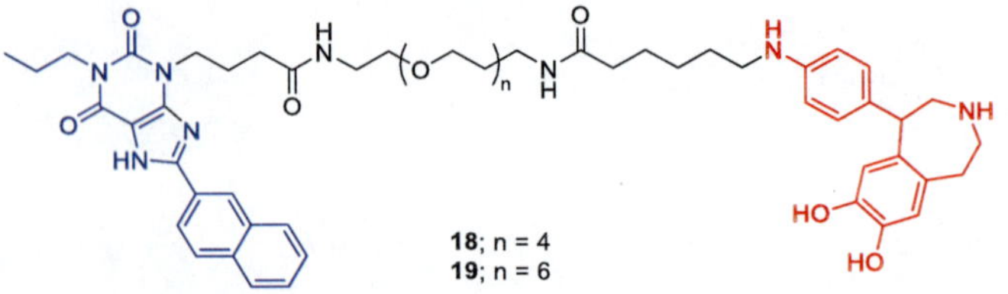

18; n = 4
19; n = 6

Fig. 6 Bivalent ligands **18** and **19**, based on A_1R antagonist naphthylxanthine and D_1R agonist SKF38393 developed by Shen et al. [55]

receptor (D_1R) heterodimer. These ligands were envisioned to be chemical tools to detect heterodimerisation and potentially further explore the biological functions of an A_1R–D_1R heterodimer. Two bivalent ligands (**18**, **19**, Fig. 6) containing a naphthylxanthine-based A_1R antagonist and a 1-phenyl-2,3,4,5-tetrahydro-(1*H*)-3-benzazepine-7,8-diol (SKF38393)-based D_1R agonist linked by differing lengths of combined alkyl-PEG linkers were synthesised, along with the corresponding A_1R monovalent control ligands (not shown).

Established SAR indicated that the N-3 position of the naphthylxanthine moiety [56, 57], and the 4′ position of SKF38393 [58] were tolerant of variation/further substitution; therefore, the linker was attached at these sites. An earlier publication by Kühhorn et al. [47] had reported dopamine D_3 receptor (D_3R) targeted homobivalent ligands linked by 5–8 PEG units, which had showed up to a 70-fold increase of D_3R binding affinity compared to monovalent control ligands, and again demonstrated the utility of the PEG linker for favourable physicochemical properties. This provided the rationale for Shen et al. [55] to design similar bivalent ligands linked via 4- and 6-PEG unit spacers. Bivalent ligands **18** and **19** showed moderate binding to the A_1R and D_1R, similar to that of the original ligands. The binding affinities of **18** and **19** for A_1R were found to be 10–100 times higher compared to the monovalent control ligands.

Evidence for the simultaneous binding of ligands **18** and **19** was shown through fluorescence resonance energy transfer (FRET) experiments [55]. FRET occurs when a donor chromophore, in its electronic excited state, transfers energy to a nearby acceptor chromophore (10–100 Å) through non-radiative dipole-dipole coupling. FRET efficiency values differed significantly between cells co-transfected with A_1YFP and D_1CFP ($P < 0.01$) (YFP = yellow fluorescent protein, CFP = cyan fluorescent protein) when treated with 100 nmol/L **18** ($P < 0.01$) or **19** ($P < 0.05$), or 1 μmol/L **19** ($P < 0.01$), compared to those from cells transfected with either A_1-YFP or D_1-CFP. This supports the idea that the bivalent ligands could bridge and simultaneously bind to each monomer of the A_1R–D_1R heterodimer.

Ligand docking experiments using an A_1R–D_1R heterodimer model further supported **18** and **19** interactions with orthosteric binding sites of both receptors of the heterodimer, indicating that the linkers were of an appropriate length to allow simultaneous binding. The reported minimum distance between the two orthosteric ligand binding sites in this A_1R-D_1R heterodimer model was 48 Å, similar to other reported AR heterodimer models [59, 60]. To date, the two bivalent ligands (**18** and

19) reported by Shen et al. are the most experimentally robust A_1R heterodimer bivalent ligands.

2.2 Heterobivalent $A_{2A}R$ Ligands

In 1991, Ferré et al. first reported crosstalk between the $A_{2A}R$ and D_2R [61]. The heterodimer formed by $A_{2A}R$ and D_2R has become an attractive drug target, owing to potential synergistic effects between $A_{2A}R$ antagonists and D_2R agonists in the treatment of Parkinson's disease (PD) [62–64] or schizophrenia [65]. Interactions between $A_{2A}R$ and D_2R have been demonstrated through co-immunoprecipitation and co-localisation assays, as well as FRET and bioluminescence resonance energy transfer (BRET) techniques in co-transfected cells [16].

The first bivalent ligands for the $A_{2A}R$–D_2R heterodimer were reported in 2009 by Soriano et al. (**20–27**, Fig. 7) [59]. These heterobivalent ligands utilised the $A_{2A}R$ antagonist XCC (**4**, Fig. 1) and the D_2R agonist 6-((4-aminophenethyl)(propyl)-amino)-5,6,7,8-tetrahydronaphthalen-1-ol ((±)-PPHT-NH_2), connected by linkers consisting of trifunctional amino acids and PEG-polyamide units. Before ligand synthesis was undertaken, the optimal linker length was investigated using molecular dynamics (MD) simulations and ligand docking. The estimated maximum distance between the two orthosteric binding sites of the heterodimer was found to be in the range of 40–45 Å. Based on this distance, the shortest predicted linker which would allow a bivalent ligand to simultaneously interact with each receptor was found to be 26 atoms in length. Therefore this study synthesised bivalent ligands

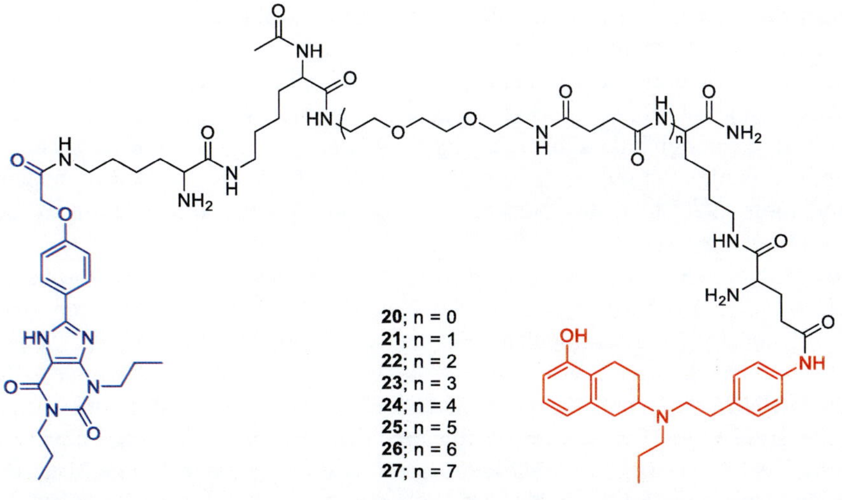

Fig. 7 Heterobivalent ligands (**20–27**) developed by Soriano et al. [59] which contain XCC as the $A_{2A}R$ antagonist and (±)-PPHT-NH_2 as the D_2R agonist and are connected by trifunctional amino acids and PEG-polyamide linkers

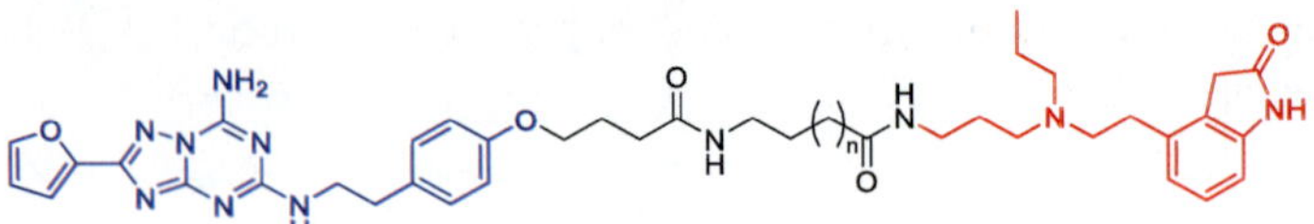

28; n = 1
29; n = 5
30; n = 9

Fig. 8 Heterobivalent ligands 28–30 developed by Jörg et al. [66] comprising of $A_{2A}R$ antagonist ZM241385 and D_2R agonist ropinirole connected by alkyl linkers

(**20**–**27**, Fig. 7) with linkers ranging from 26 to 118 atoms, as well as the corresponding $A_{2A}R$ and D_2R monovalent control ligands (not shown). While the trifunctionalised amino acid-like linkers allowed longer linkers to be synthesised with ease, the drawback is the inability to systematically increase the linker length by smaller atom units in such a manner that the exact distance between the orthosteric ligand binding sites across the dimer can be more precisely investigated.

Competitive radioligand binding experiments, utilising the $A_{2A}R$ radioligand [^{3}H]-ZM241385 were carried out to evaluate the binding of **20**–**27** to $A_{2A}R$, along with the binding of the corresponding $A_{2A}R$ monovalent control ligands. Bivalent ligands **20**–**23** with shorter linkers exhibited higher $A_{2A}R$ affinity than ligands **24**–**27** with longer linkers. Three to seven times higher affinity for $A_{2A}R$ was observed for **20**–**27** (between 50 ± 8 and 80 ± 10 nM) compared to the respective monovalent control ligands in brain striatum and cells co-expressing both $A_{2A}R$ and D_2R. In cells expressing just one receptor type, a change affinity was not observed, indicating that the bivalent ligands **20**–**27** were likely binding to both $A_{2A}R$ and D_2R simultaneously. It was postulated that these results indicated that a heterobivalent ligand bound to $A_{2A}R$ may subsequently bind to D_2R located in the vicinity with a higher affinity than the corresponding D_2R monovalent control ligand. In effect, the 'local concentration' of ligand is higher in a bivalent ligand compared to the monovalent counterpart. The opposite should also be true for optimally linked bivalent ligands – initial binding to D_2R should result in higher affinity for $A_{2A}R$. It was concluded that **20**–**23** (the shorter-linked bivalent ligands) were acting as true heterobivalent ligands, capable of simultaneous interaction with both $A_{2A}R$ and D_2R, representing useful tools to evaluate the presence of $A_{2A}R$–D_2R heterodimers even in native tissue.

In 2015, a series of heterobivalent (Fig. 8), fused and dual-acting (discussed in Sect. 3, Fig. 14) ligands targeted to the $A_{2A}R$–D_2R heterodimer were reported by Jörg et al. [66]. Heterobivalent ligands consisting of the $A_{2A}R$ antagonist 4-(2-([7-amino-2-(furan-2-yl)[1,2,4]triazolo[1,5-*a*][1,3,5]triazin-5-yl]amino)ethyl)-phenol (ZM241385) and the D_2R agonist ropinirole connected by alkyl linkers ranging from 11 to 21 atoms were synthesised (**28**–**30**, Fig. 8). An $A_{2A}R$ monovalent control ligand consisting of ZM241385 connected to the longest linker used in the bivalent ligands was also synthesised (not shown).

Functional assays using cAMP and [^{35}S]GTPγS (for $A_{2A}R$ and D_2R, respectively) were used to compare the activity (IC_{50} and EC_{50}) of heterobivalent ligands

28–30 to each other, to monovalent control ligands, the parent ligand ZM241385 and the literature values of ropinirole, respectively. Heterobivalent ligands **28–30** exhibited moderate potencies at both $A_{2A}R$ and D_2R individually, although somewhat lower than those of the respective parent ligands ZM241385 and ropinirole. An obvious trend between increasing linker length and potency was not observed for either $A_{2A}R$ or D_2R. However, ligand **29** exhibited an IC_{50} value ($A_{2A}R$, IC_{50} ~ 40 nM) similar to that of ZM241385 ($A_{2A}R$, IC_{50} ~ 30 nM), and an EC_{50} value (D_2R, EC_{50} ~ 500 nM) similar to ropinirole (D_2R, EC_{50} ~ 300 nM). Bivalent ligand **30** (with the longest linker) showed a substantial drop in the inhibitory potency at $A_{2A}R$ (IC_{50} ~ 1,000 nM) and drop in efficacy at D_2R (EC_{50} ~ 1,400 nM). The $A_{2A}R$ monovalent control ligand (not shown) showed a 13-fold decrease in potency for $A_{2A}R$; however, this was not surprising based on previous studies that showed the addition of longer linkers to $A_{2A}R$ congeners is often detrimental [41]. It was concluded that while bivalent ligands **28–30** were useful pharmacological tools to probe the $A_{2A}R$–D_2R heterodimer, such ligands have unfavourable physicochemical properties (such as high molecular weight) for a potential CNS drug/probe. As such, the 'fused' and integrated dual-acting compounds (Sect. 3, Fig. 14), which were the main focus of Jörg et al. [66], showed more promise as 'drug-like' compounds.

Recently, Pulido et al. [67] reported a series of heterobivalent ligands for the $A_{2A}R - D_2R$ heterodimer (**31–33**, Fig. 9).

The selected $A_{2A}R$ ligand was based on the selective $A_{2A}R$ antagonist 2-(4-(3-(5-amino-2(furan-2-yl)-7*H*-pyrazolo[4,3-*e*][1,2,4]triazolo[1,5-*c*]pyrimidin-7-yl)propyl)phenoxy)acetic acid (SCH-442,416), which was linked to a D_2R ligand based on N-(p-aminophenethyl)spiperone (NAPS). The derivatisation points in ligands **31–33** from the requisite orthosteric binding moieties were selected due to favourable orientation towards the extracellular side (predicted from docking the derivatised binding moieties into the $A_{2A}R$–D_2R model) and chemical accessibility/ease of derivatization. The linker chemistry was developed in an analogous way to the same research teams' work developing D_2R homobivalent ligands [50] and was once again PEG based, as is with many other bivalent ligands. This linker system differs from others as it also contains a central nitrilotriacetic acid-based scaffold, containing three symmetric carboxylic acids. This allowed for addition of two

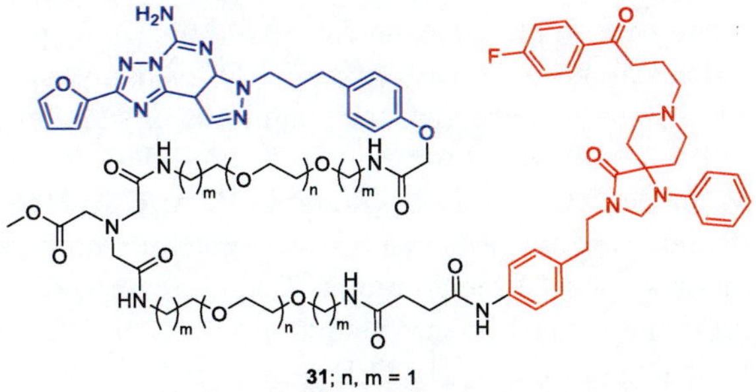

31; n, m = 1
32; n, m = 2
33; n = 4, m = 1

Fig. 9 Heterobivalent ligands developed by Pulido et al. [67], comprising of the $A_{2A}R$ antagonist SCH-442,416 and D_2R antagonist NAPS connected by combined alkyl and PEG linkers ligands, leaving a carboxylic acid (or protected carboxylic acid) handle available for further modification. An earlier report by the same group utilised a similar scaffold for dendritic-type structures, which, depending on the final applications, could be modified to contain peptide or aliphatic chains, or fluorophores [68]. Although Pulido et al. do not use this third possible attachment point for further modification in this report [67], one could imagine using it for attachment of a third orthosteric ligand to probe higher order oligomers or attachment of a reporter/fluorescent probe.

The linker lengths of **31–33** (25, 35 or 43 atoms, respectively) were informed by MD simulations, and informed by the homobivalent ligands previously synthesised for the D_2R homodimer [50], and also the results of Soriano et al. [59] in 2009 that used linkers of a much greater length. MD simulations by Pulido et al. [67] suggested that ligands **32** and **33**, with a linker length of at least 35 atoms, should be capable of simultaneous binding to the orthosteric binding sites of the $A_{2A}R$–D_2R heterodimer. Shorter-linked bivalent ligand **31** was synthesised as a negative control, that, based on the MD simulations, should not be capable of interacting simultaneously across the heterodimer.

Binding affinities of heterobivalent ligands **31–33** and corresponding monovalent control compounds (not shown) for $A_{2A}R$ and D_2R were obtained from [^{3}H]ZM 241385 or [^{3}H]YM-09151-2 radioligand competition-binding assays in both the absence and presence of specific peptides suggested to disrupt the formation of the $A_{2A}R$–D_2R heterodimer. In the more traditional radioligand binding assays (i.e. those that did not contain peptides that disrupt heterodimer formation), ligands **31–33** showed a statistically significant increase in binding affinity to $A_{2A}R$ compared to both the monovalent $A_{2A}R$ control compounds (8-fold, 84-fold and 24-fold, respectively) and to the $A_{2A}R$ parent ligand SCH-442,416 (1.3-fold for **31**, fourfold for **32** and **33**). At D_2R, **31–33** were found to have binding affinities in the low nanomolar to sub-nanomolar range, with only **33** showing an enhanced binding affinity for D_2R when compared to a monovalent control ligand. The presence of peptides that disrupt the $A_{2A}R - D_2R$ heterodimer interface decreased the binding affinity of **33** at $A_{2A}R$. The same assays showed no effect on the binding of **31** and **32**.

To further test the 'true, simultaneous bivalent' character of the bivalent ligands, dynamic mass redistribution (DMR) assays in living cells and BRET experiments were also carried out. Interestingly, **31–33** all showed an increase in BRET relative to untreated cells. The longest bivalent ligand (**33**) increased BRET more than 'negative control' **31**, whereas **32** did not. As expected, monovalent control ligands did not show an increase in BRET. DMR assays showed that bivalent ligands **32** and **33** were able to decrease the magnitude of signalling induced by both CGS21680, a full $A_{2A}R$ agonist, and sumanirole, a full D_2R agonist. This effect was much greater for **33** than **32** at both $A_{2A}R$ and D_2R. As CGS21680 and sumanirole are orthosteric ligands, this suggests that **33** is capable of simultaneous binding to orthosteric sites of an $A_{2A}R$–D_2R heterodimer. Taken together, data suggest only bivalent ligand **33** (with the longest linker of 43 atoms) was capable of acting as a true heterobivalent ligand across the $A_{2A}R$–D_2R heterodimer.

To date, examples of bivalent ligands specifically targeted to $A_{2B}R$ have not been reported. However, due to the high degree of structural similarity between $A_{2A}R$ and $A_{2B}R$, it is possible that bivalent ligands targeted to $A_{2A}R$ may also show some affinity for $A_{2B}R$.

2.3 *Heterobivalent Ligands Targeting A_3R*

In 2000, Jacobson et al. [69] reported bivalent ligands they referred to as 'binary conjugates' targeted to both A_3R and A_1R using a 'functionalized congener' approach (MRS1740 (**34**), MRS1741 (**35**), Fig. 10). Release of adenosine following activation of A_3R and A_1R during ischemic events exerts a potent cardioprotective effect. It was thought that simultaneous activation of A_3R and A_1R could increase this protective effect, allowing for the development of a new class/type of highly effective cardioprotective ligands. To date, these ligands are the only reported bivalent ligands targeted to an A_3R-containing heterodimer.

The agonist IB-MECA (**54**) (Fig. 16) was chosen as the A_3R targeted ligand, covalently linked to the adenosine amine congener ADAC (**3**) (Fig. 1). The linkers for bivalent ligand regioisomers MRS1740 (**34**) and MRS1741 (**35**) (Fig. 10) were thiourea based, differing only in the substitution pattern of a phenyl moiety in the linker. A direct amide-linked analogue with a much shorter linker (which could be considered a 'dual-acting' compound) and a A_3R monovalent control ligand were also synthesised (not shown).

The selectivity of 'binary conjugates' MRS1740 (**34**) and MRS1741 (**35**) at both A_3R and A_1R were measured in radioligand binding assays using rat brain containing A_1R and $A_{2A}R$ and recombinant human A_3R. An A_3R monovalent control ligand was found to be A_3R selective (as was the parent ligand IB-MECA). Both MRS1740 and MRS1741 had high affinity for A_1R ($K_i = 1.15 \pm 0.24$ (**34**) and $K_i = 3.52 \pm 0.48$ nM (**35**)) and A_3R ($K_i = 0.727 \pm 0.261$ (**34**) and $K_i = 1.15 \pm 0.25$ nM (**35**)) vs. $A_{2A}R$ receptors ($K_i = 336 \pm 119$ (**34**) and $K_i = 197 \pm 21$ nM (**35**)) in radioligand binding assays. The ability of **34** and **35** to co-activate both A_3R and A_1R was determined using a recombinant cardiac myocyte

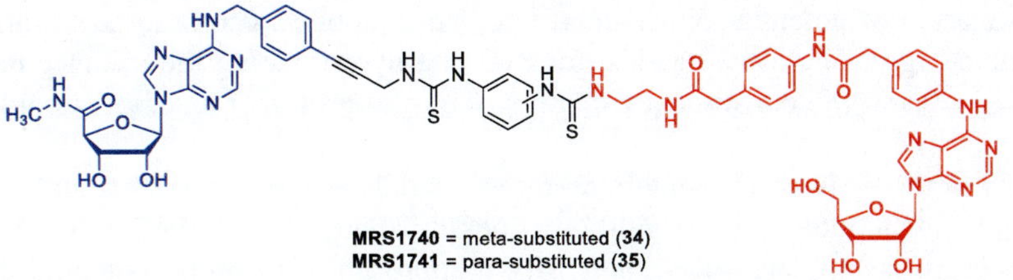

Fig. 10 Binary conjugates MRS1740 and MRS1741 developed by Jacobson et al. [69], based on adenosine amine congener ADAC (3) and an IB-MECA (54) derivative, linked by thiourea-containing linkers

model of cardioprotection, in which a highly potent anti-ischemic effect was observed. This effect was observed to be significantly stronger in myocytes where both A_1R and A_3R were expressed, compared to those only expressing the A_3R. Ligands **33** and **34** each caused a significant reduction in both the percentage of cells killed and the amount of creatine kinase released in myocytes expressing both human A_1/A_3 receptor subtypes following a period of simulated ischemia.

It was theorised by Jacobson et al. [69] that if co-activation of both A_3R and A_1R occurred simultaneously, the response produced by one equivalent of the bivalent ligand should be equal to or higher than that elicited by the combined presence of one equivalent of each constituent agonist. Both **33** and **34** did indeed produce a response in line with this prediction; the cardioprotective effect of either was virtually identical to that caused by a combination of A_1R agonist ADAC and A_3R agonist IB-MECA (**51**) (Fig. 15). It was not discussed by Jacobson et al. whether ligands **34** and **35** were interacting with the heterodimer via a true bivalent binding mode, or an alternative such as the 'flip-flop' or cooperative binding modes (Fig. 3a, b). Techniques such as FRET, BRET and DMR assays, which are now commonplace, could be used to further elucidate the specific binding mode of ligands **34** and **35**.

2.4 *Structural Biology and Molecular Modelling of Bivalent Ligands*

Molecular modelling has contributed to AR drug discovery. Experimental crystal and/or cryo-electron microscopy structures for two (A_1R, $A_{2A}R$) [43, 70, 71] of the four ARs have been reported, both with and without ligands. Homology models for $A_{2B}R$ [72–76] and A_3R [77–80] have been reported by many research groups. These structures have drastically improved the understanding of the structural and functional properties of this receptor family, including particulary with regard to ligand binding and subtype selectivity. Recently, Salmaso et al. [81] thoroughly reviewed the use of in silico techniques such as molecular modelling, docking and MD simulations for monomeric purinergic GPCRs (ARs and P2Y receptors). In silico models and MD simulations of GPCR dimers have become an increasingly important tool for the development of bivalent ligands targeting GPCR dimers and investigation of potential dimer interfaces. These models can be used to rationally aid the design of bivalent ligands, for example by estimating the distance between binding sites of the constituent receptors (the inter-protomer distance) to aid linker length design.

One of the earliest AR-containing dimer models was reported by Canals et al. in 2003 [16], who integrated a rigid-body docking approach with the results of protein engineering, as well as FRET and BRET approaches, to probe potential dimeric interface(s) of $A_{2A}R$-D_2R heterodimers. These models [16] pre-dated any crystal structures of the AR subtypes; therefore, the $A_{2A}R$–D_2R heterodimers were achieved by comparative modelling, using a modified rhodopsin as a template. In 2006, Kim et al. [82] reported an A_3R homodimer model, based on a rhodopsin crystallographic dimer. This model was constructed by building a homology model of the A_3R monomer, followed by superimposition on a dimer model based on a rhodopsin oligomer. This superimposition method has since been utilised for the construction of many other GPCR homo- and heterodimers [83]. In Kim et al. [82] study, eight symmetrical homodimer models were generated for A_3R, subjected to MD simulations (in the absence of a lipid bilayer) and evaluated for contact surface area and shape complementarity. The results of these simulations clearly showed that a symmetrical TM4–5 interface led to an energetically favourable A_3R homodimer with a high degree of shape complementarity at the TM surface contact area. In 2016, Baltoumas et al. [84] evaluated a representative set of GPCR homodimers following Coarse-Grained Molecular Dynamics (CGMD) simulations, with one example being an $A_{2A}R$ homodimer. Like the A_3R homodimer model reported in 2006 [82], a TM4–TM5 interfaced-dimer was observed in the $A_{2A}R$ homodimer model structure [84].

The design of many of the previously discussed bivalent ligands was aided by MD simulations of heterodimer models, such as the A_1R–D_1R targeted bivalent ligands reported by Shen et al. [55]. A heterodimer model was proposed by Shen et al. [55] with the interface of TM-3, −4 and −5 of the A_1R and TM-1, −6 and −7 from D_1R. This model allowed for elucidation of the expected distance between the two orthosteric ligand binding sites of the heterodimer, which was found to be 48 Å. This is similar to the inter-protomer distance reported for other AR heterodimer models and importantly this was shorter than the length of the synthesised bivalent ligands [55] (**18** and **19**, Fig. 6).

The Royo group used molecular modelling and MD simulations [16] to great effect alongside their reported synthesises of $A_{2A}R$–D_2R targeted bivalent ligands [59]. Initial docking simulations by collaborators Canals et al. [16] gave minimised receptor dimer structures that fit into two populations or 'clusters.' In cluster 1, the most populated receptor dimer interface was found to be between TM-5 and/or TM-6 and the *N*-terminal portion of IL-3 of D_2R with TM-4 and the *C*-terminal portion of the C-tail from the $A_{2A}R$, respectively. In a second, less populated cluster (which, however, had high docking scores and resembled the intradimer contact model proposed for rhodopsin), the heterodimer interface is mainly formed by IL-2, TM-4, TM-3, and TM-5, from D_2R, and IL-2, TM-5, TM-3, and TM-4, from $A_{2A}R$. There was only a small contribution to the $A_{2A}R$–D_2R dimer interface from the intracellular and extracellular domains in both clusters. The estimated maximum inter-Cα distance between the two orthosteric ligand binding pockets of the monomeric receptors that participate in the $A_{2A}R$–D_2R heterodimer was found to be about 40–45 Å.

Models of the $A_{2A}R-D_2R$ heterodimer have been refined over the years and in 2018 a paper published by long-time collaborators of the Royo/Franco research team, Navarro et al. [85] reported an $A_{2A}R - D_2R$ heterodimer generated via superimposition of monomers onto the TM4/5 dimeric interface observed in the β1-adrenergic receptor (PDB: 4GPO) [86]. This updated $A_{2A}R - D_2R$ model was

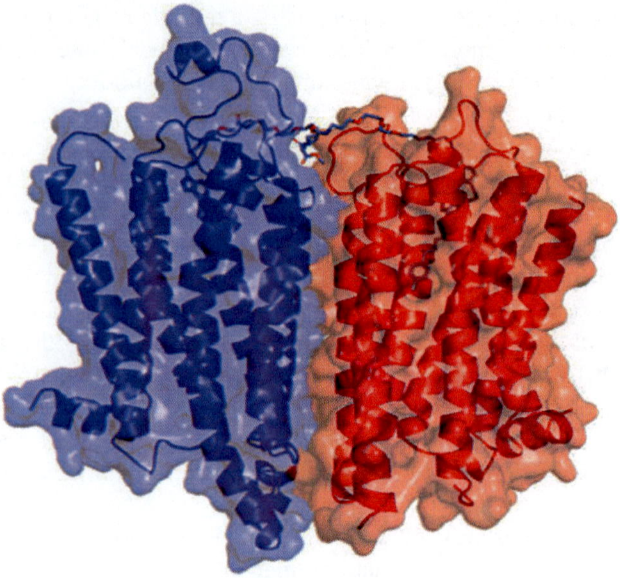

Fig. 11 Heterodimer model [85] of the adenosine A_{2A} (blue) and dopamine D_2 (red) receptor with bivalent ligand **32** (stick model with purple carbons, blue nitrogens, and red oxygens). Model with ligand bound downloaded from supporting information in Pulido et al. [67]

based on the results of bimolecular fluorescence complementation (BiFC) experiments obtained with specific TM peptides mimicking TM receptor domains. This model [85] was then very recently used by Pulido et al. [67] to inform the rational design of bivalent ligands targeted to the $A_{2A}R–D_2R$ heterodimer (Fig. 11, showing model [85] with ligand **32**). The optimal linker length of the bivalent ligands was calculated by adjusting different linker moieties to the van der Waals surface of the heterodimer model. Also, in 2018, the same research team reported the development of a molecular modelling tool for the design of bivalent ligands targeted to GPCR dimers or proteolysis targeting chimeras (PROTACs) [60]. This tool links the chosen ligands via the shortest pathway along the receptors van der Waals surface and then scores the solutions providing prioritisation for the design of new bivalent ligands.

The use of computational dimer models informed by pharmacological data (such as those utilised by the Royo group) as well as tools that consider the TM interfaces, distances between orthosteric binding sites and mode of interaction of the ligands might allow for refinement and more precise rational design of bivalent ligands. Such tools are not only valid for AR-containing dimers, but for all other GPCRs capable of dimerisation.

3 Dual-Acting Ligands

GPCR ligands can also be developed as 'dual-acting' (Fig. 12) where each 'part' of the dual-acting ligand can carry out different binding/functions but not simultaneously within one ligand across multi-receptor binding (as per Fig. 2). Dual-acting

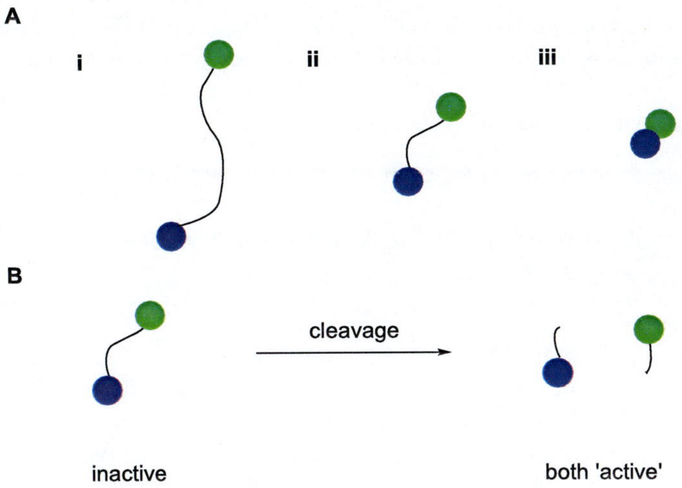

Fig. 12 (**a**) Dual-acting ligands can be (i) 'independent' covalently linked compounds joined by a medium-long linker, (ii) 'fused' compounds with a shorter linker or (iii) or truly 'hybrid' dual-acting ligands within a single small molecule. (**b**) Dual-acting prodrugs (also termed codrugs or mutual prodrugs) are inactive (or inactive for dual-activity) and become 'dual-acting' upon release of two active ligands

ligands can be two 'independent' covalently linked compounds joined by a medium-long linker (Fig. 12a, i) or fused compounds with a shorter linker (Fig. 12a, ii). It is worth noting that 'dual action' does not necessarily mean two 'ligands' that formerly bind to different targets, as the second 'functionality' can be something like a free radical scavenger or other antioxidant mechanisms. Bivalent ligands designed as simultaneously binding ligands (Fig. 2), but that are experimentally determined unlikely to participate in simultaneous multi-receptor binding are actually better defined as 'dual-acting' ligands (Fig. 12a, i).

A fully integrated dual-acting or multi-acting hybrid ligand (Fig. 12a, iii) is where a 'single small molecule' ligand has inherent 'dual' activity for two different receptors with high ligand efficiency at each receptor. Ligands can be rationally designed as fully integrated dual-acting; however, in reality, this 'polypharmacology' is often inherent to some extent in all ligands, including so-called 'selective' ligands designed for a single intended target and is often realised upon wider experimental studies. Purposeful computational approaches to discover hybrid GPCR ligands are increasingly being undertaken [87]. Truly hybrid dual-acting ligands with lower molecular weight have the advantage of better drug-like properties compared to higher molecular weight bivalent ligands. This dual-acting/polypharmacology and associated superior drug-like properties has particular appeal for treating CNS conditions where drug access through the blood-brain barrier is a requirement [88].

Dual-acting prodrugs (Fig. 12b) are when two compounds are linked together to form an inactive compound that upon cleavage/activation releases two 'active' constituents. Also termed ‘codrugs’ or ‘mutual prodrugs’, these are commonly designed to release two ligands that are synergistic and target receptors found in the same immediate vicinity. Codrugs find utility, especially where ligands for each receptor are structurally very different, meaning a ‘dual-acting’ hybrid ligand approach (as per Fig. 12a, iii) is more challenging.

3.1 Dual-Acting Ligands for Neurodegenerative Diseases/Conditions

Monoamine oxidase B (MAO-B) inhibitors decrease oxidative deamination of amine neurotransmitters (such as dopamine) and are commonly administered for the treatment of Parkinson’s disease either as monotherapy or in combination with levodopa (L-DOPA) [89–92]. Dual-acting drugs targeting the $A_{2A}R$ and MAO-B in the treatment of PD may provide therapeutics with neuroprotective potential with improved adverse effect profiles compared to the administration of two separate drugs [93].

One of the first reports of a dual-targeted AR ligand was in 2002, when Chen et al. [94] reported that 8-(3-chlorostyryl)caffeine (CSC) (**36**) (Fig. 13a) was a potent (nanomolar) MAO-B inhibitor in addition to its already well-known $A_{2A}R$ antagonist properties. This dual action was thought to contribute to CSC (**36**) and other caffeine derivative’s neuroprotective properties, in particular in models of PD. Following the discovery of the MAO-B inhibitor properties of CSC (**36**), dual-acting hybrid ligands that block both MAO-B and $A_{2A}R$ were purposefully investigated via exploration of SAR of C-8 substituted caffeinyl analogues (Fig. 13b). A study by Pretorius et al. [95] concluded that (E,E)-8-(4-phenylbutadien-1-yl)caffeines such as **37** are promising lead dual-acting compounds for MAO-B and $A_{2A}R$, although findings did indicate that most substitutions increased one activity but decreased the other. In 2014 [96], triple-action ligands such as **38** with the ability to antagonise A_1R and $A_{2A}R$ and inhibit MAO-B were reported as potential multi-target drugs for the treatment of neurodegenerative disorders. The investigation of tricyclic xanthine derivatives in this study was also driven by the improved water solubility this scaffold offered over previously reported dual $A_{2A}R$ and MAO-B ligands such as CSC (**36**). In subsequent years, more comprehensive SAR studies on these and related compound classes were carried out by the same research team [97], and for example yielded ligand **39** that showed optimised tri-activity for MAO-B inhibition and A_1R and $A_{2A}R$ antagonism, and ligand **40** with optimal dual-activity for $A_{2A}R$ and MAO-B [97].

Extending the dual/triple multi-targeting approach to another GPCR for the potential treatment of PD – Hagenow et al. [99] designed triple action $A_{2A}R$, A_1R and histamine H_3 receptor (H_3R) antagonists by merging a known H_3R scaffold (a cyclic aminopropyloxy-linked phenyl derivative) with a non-xanthine arylindenopyrimidine-based dual $A_{2A}R$ and A_1R antagonist, e.g. ST-2001 (**41**),

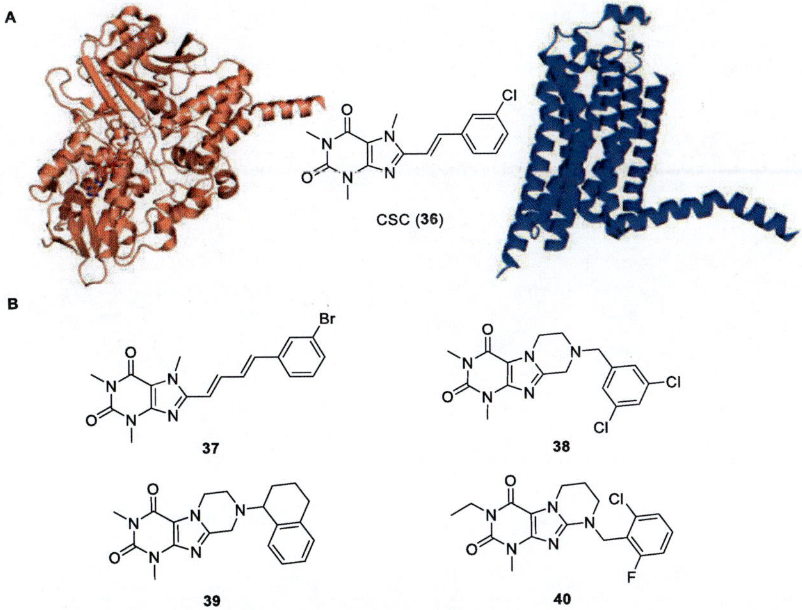

Fig. 13 Dual-acting AR ligands containing a xanthine scaffold that are dual acting at AR and MAO-B. (**a**) One of the first ligands of this class developed was CSC (**36**) that blocked MAO-B (brown; PDB ID: 2BK5, with cofactor) [98] and $A_{2A}R$ (blue, PDB ID: 3eml) [43] (MAO-B and $A_{2A}R$ are not shown with CSC bound) (**b**) Further xanthine derivatives developed as dual-acting blockers of MAO-B and ARs

Fig. 14). The binding affinities of a series of ligands were evaluated in independent radioligand competition binding studies using recombinant human ARs and H_3R. The lead ligand, ST-2001 (**41**), showed low nanomolar affinity for $A_{2A}R$ ($K_i = 7.25 \pm 1.53$ nM), A_1R ($K_i = 11.5 \pm 1.6$ nM) and H_3R (nanomolar K_i), although some binding at $A_{2B}AR$ ($K_i = 36.9 \pm 8.3$ nM) and A_3R ($K_i = 150 \pm 10$ nM) was also measured. Hagenow et al. evaluated two compounds including ST-2001 (**41**) in in vivo in rat and showed that ST-2001 (**41**) improved L-DOPA-induced dyskinesia in their rat model.

Kalash et al. [100] carried out a computer-aided design of tri-targeted ligands for A_1R, $A_{2A}R$ and phosphodiesterase 10A (PDE10A) which could potentially modulate intracellular cAMP levels to prevent neurodegenerative effects. Recently, it has been suggested that PDE10A also plays a role in neurodegenerative diseases such as PD [101], Huntington's disease [102], and schizophrenia [103]. Inhibition of PDE10A results in elevated intracellular cAMP concentrations, and ligands that are dual-acting at an AR and PDE10A could allow for fine-tuning of cAMP levels. Based on computational leads, a series of 2-aminopyridine-3-carbonitriles were synthesised and evaluated biological activity. Of the 25 compounds synthesised and tested, ligands **42** and **43** (Fig. 14) showed the best multi-target profile against

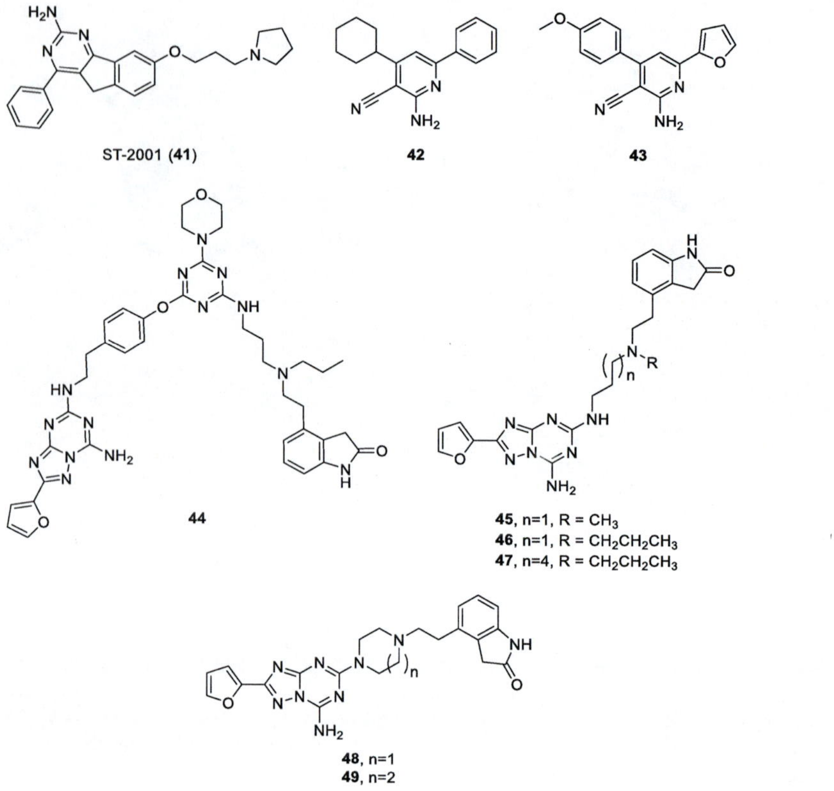

Fig. 14 Dual or tri-acting AR ligands of interest for the potential treatment of neurogenerative diseases, in particular PD

A_1R, $A_{2A}R$ and PDE10A, in particular **43** (A_1R $K_i = 34.2 \pm 2$ nM, $A_{2A}R$ $K_i = 41 \pm 2$ nM, PDE10A $IC_{50} = 3.2 \pm 0.4$ μM).

With the motivation to develop 'drug-like' ligands that are efficacious for the treatment of neurodegenerative disorders such as PD, Jörg et al. [66] reported dual-acting $A_{2A}R$ and D_2R ligands (Fig. 14, **44**–**49**) in the same publication as other 'true' bivalent ligands (Fig. 8). All dual-acting and bivalent ligands reported in Jörg et al. [66] were centred on a ZM241385-based $A_{2A}R$ antagonist and a ropinirole-based D_2R agonist, with dual-acting ligands either linked by longer linkers (**44**), shorter linkers (**45**–**47**) or a 'hybrid' approach with almost completely merged ligands (**48** and **49**). Functional assays were used to evaluate the $A_{2A}R$ inhibitory potency and D_2R functional potency of ligands **44**–**49**, along with monovalent control ligands (not shown). Ligand **44** had a fourfold decrease in $A_{2A}R$ inhibitory potency but equipotent D_2R activity compared to the requisite parent $A_{2A}R$ antagonist ZM241385 and D_2R agonist ropinirole, respectively. This confirmed **44** was indeed dual-acting, although whether this was via a more classical simultaneous heterobivalent ligand mechanism was not further explored. However, based on

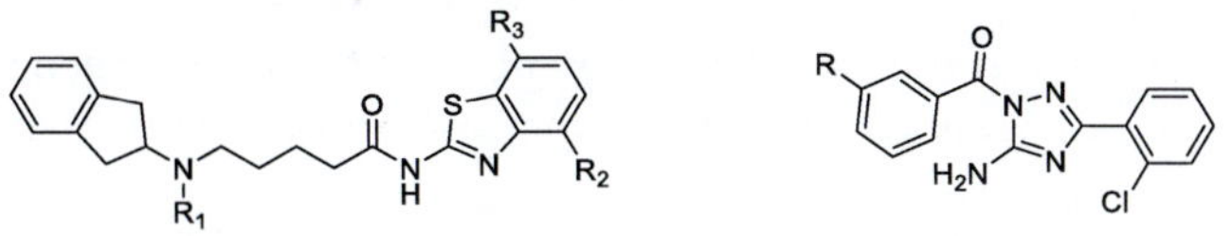

50, R_1 = CH_3, R_2 = CH_3, R_3 = H
51, R_1 = $CH_2CH_2CH_3$, R_2, R_3 = OCH_3

52, R = OMe
53, R = Cl

Fig. 15 Recently reported dual-acting ligands developed by Kampen et al. [104] and Wang et al. [105] for potential PD applications

linker length of other $A_{2A}R$–D_2R heterobivalent ligands (for example as reported by Pulido et al., [67] Fig. 9), one might predict that **44** is more likely to be dual-acting than simultaneous binding. Shorter-linked **45–47** were approximately 1.5 orders of magnitude less potent at $A_{2A}R$ than ZM241385 but had equivalent to eightfold more potency at D_2R compared to ropinirole. These ligands (**45–47**) were the most promising dual-acting ligands reported by Jörg et al. [66] when considering dual-action activity alongside 'drug-like' properties for CNS applications. Hybrid ligands **48** and **49** did not exhibit any detectable D_2R activity, however, showed inhibitory potency at $A_{2A}R$ on par with **45–47** (1.5 orders of magnitude less than ZM241385).

Kampen et al. [104] recently utilised a structure-based modelling approach in an attempt to develop dual-acting bitopic ligands such as **50** and **51** (Fig. 15) (refer to Sect. 4 for a definition and discussion on bitopic ligands) with the ability to modulate both the $A_{2A}R$ and D_2R, again for a potential PD application. Based on structural studies and analysis of both the orthosteric and allosteric ligand binding sites of each receptor, D_2R orthosteric agonist *N*-methyl-2-aminoindane was selected as a starting point for ligand design. Molecular docking calculations suggested that this scaffold could be expanded into a secondary 'allosteric' pocket in the D_2R by linking a building block to the amino moiety. A virtual chemical library with potential dual-target ligands was designed, and a docking screen was carried out to identify the most promising bitopic dual-acting compounds, which would be able to interact with both the orthosteric and allosteric sites of $A_{2A}R$ and D_2R. Ten compounds (including **50**) were selected for synthesis based on predicted interactions with key $A_{2A}R$ residue Asn253$^{6.55}$ (Ballesteros–Weinstein numbering in superscript) in the orthosteric binding site of the $A_{2A}R$, positioning of the aminoindane moiety in the extracellular space, and ease of synthesis. Docking to D_2R showed that the ligands could interact with Asp114$^{3.32}$ in the orthosteric binding site and that the linker allowed the building block to reach into the secondary binding pockets. Three of the synthesised ligands were found to have an affinity for both $A_{2A}R$ and D_2R, and the most promising of these three (**50**) was selected as starting point for hit-to-lead optimisation.

Structure-guided design of second-generation analogues (such as **51**) was performed to improve $A_{2A}R$ and D_2R pharmacology, leading to another series of analogues that were synthesised and biologically tested. Competitive orthosteric radioligand binding assays at the $A_{2A}R$ and D_2R, as well as A_1R, D_3R, dopamine

D_4 receptor (D_4R), and histamine H1 receptor (H_1R) assays were carried out to assess the selectivity of these dual-target ligands. The compounds generally showed high affinity for the D_3R, which was expected considering the high sequence similarity to the D_2R. Compared to the binding at $A_{2A}R$ and D_2R, comparable or substantially weaker ligand binding affinities were measured for the A_1R, D_4R and H_1R. Notably, compound **51** showed high affinity for the $A_{2A}AR$ and D_2R orthosteric sites, but displayed weak or no significant binding to the A_1AR, D_4R and H_1R ($K_i > 10$ μM) orthosteric sites. While ligands such as **51** were originally been designed as bitopic ligands that could interact with both the orthosteric and allosteric binding sites of $A_{2A}R$ and D_2R, the true binding mode was not validated. As such, ligands such as **51** can be classified as 'dual-acting' rather than dual-acting and bitopic.

In 2021, Wang et al. [105] reported a series of dual $A_1R/A_{2A}R$ antagonists in an effort to develop tools that could help solve in both motor and non-motor symptoms in the treatment of PD. This report utilised a multistage approach that sequentially integrated deep learning, pharmacophore models, and molecular docking methods to evaluate 1,178,506 unique compounds. Ultimately, 19 hits were selected from 43 potential compounds, based on visual inspection of ligand-protein interactions between the docked ligands with both the A_1R and $A_{2A}R$ receptors. The biological activity of 19 selected compounds towards A_1R and $A_{2A}R$ was evaluated by cAMP functional assays and radioligand binding assays using [^{3}H]DPCPX at A_1R and [^{3}H] ZM241385 at $A_{2A}R$. Two compounds (**52** and **53**, Fig. 15) containing a 1,2,4-triazole scaffolds possessed the most potent binding affinity and antagonistic activity for A_1R and $A_{2A}R$ at the nanomolar level (A_1R p$Ki = 7.19$ nm, $pIC_{50} = 6.78$ (**52**) and p$Ki = 7.49$ nm, $pIC_{50} = 6.72$ (**53**), $A_{2A}R$ p$Ki = 7.29$ nm, $pIC_{50} = 6.44$ nm (**52**) and p$Ki = 7.16$ nm, $pIC_{50} = 6.31$ nm (**53**)) were identified. It was suggested that **52** and **53** could be potential lead compounds for optimisation towards the development of agents to treat PD.

3.2 *Dual-Acting Ligands for Anticancer, Cardioprotective, Anti-Diabetic and Other Applications*

In 2017, Yu et al. [106] reported the effect of well-known AR agonist IB-MECA (**54**) (Fig. 16) on adipogenesis in human bone marrow mesenchymal stem cells and considered if IB-MECA held potential as an antidiabetic. In deconvoluting IB-MECA's mode of action it was found that it also functioned as a peroxisome proliferator-activated receptor (PPAR) γ partial agonist and PPARδ antagonist. Therefore, IB-MECA and related A_3R ligands may indeed possess multi-binding activity that could be beneficial in the treatment of anti-inflammatory conditions in various clinical settings.

In the same year, Gregg et al. [107] developed dual-acting ligands that contained adenosine (**1**) (Fig. 1) linked to an antioxidant moiety as potentially beneficial agents

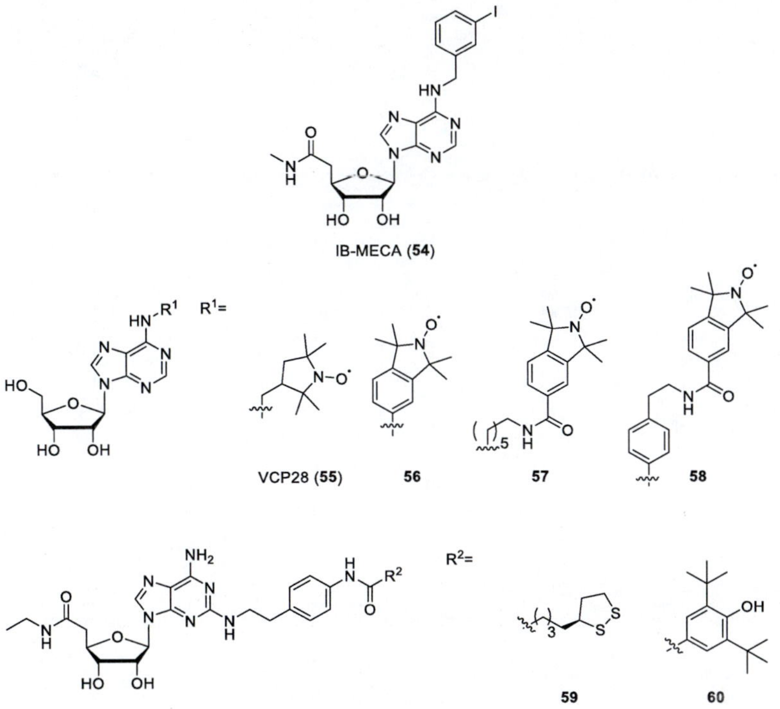

Fig. 16 Ligands designed as dual-acting at one AR subtype and another non-GPCR target, based on an adenosine-like scaffold

to produce cardioprotective effects (**55–58**, Fig. 16). Several ligands were made with a known antioxidant group incorporated directly in the adenosine N^6-position (**55**, **56**), linked via an alkyl linker (**57**), or linked via an alkyl aryl linker (**58**). Competition bindings assays using agonist radioligands were carried out at A_1R and A_3R. Although **58** had the highest A_1R affinity ($K_i = 7 \pm 1$ nM) of all the dual compounds tested, it also had good affinity for A_3R ($K_i = 23 \pm 10$ nM). The directly linked **55** had the best overall combination of A_1R affinity ($K_i = 50 \pm 10$ nM) and selectivity over A_3R ($K_i = 21{,}200 \pm 8{,}300$ nM). In a cardiomyocyte model of ischaemia, **55** demonstrated cardioprotective properties that were reduced but not abolished in the presence of an A_1R antagonist. This indicated there could be a direct benefit from the antioxidant moiety of VCP28 (**55**).

The same research groups [108] later reported several dual-acting antioxidant $A_{2A}R$ agonists based on a novel series of adenosine-5′-N-alkylcarboxamides and N^6-(2,2-diphenylethyl)-adenosine-5′-*N*-alkylcarboxamides, which were linked via the 2-position to an antioxidant moiety (**59, 60**, Fig. 16). The two most potent and selective $A_{2A}R$ agonists **59** and **60**, according to a cAMP accumulation assay, demonstrated strong protective effects relative to a monovalent control ligand

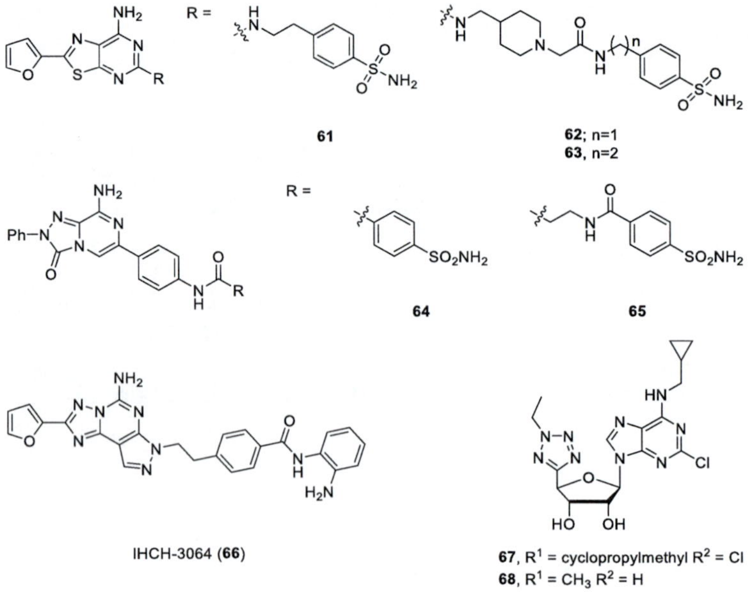

Fig. 17 Dual-acting ligands designed to target at least one AR subtype for anticancer or analgesic applications

lacking the antioxidant moiety (ligand not shown) in an ischemia model using rat atrial cardiomyocytes.

There have been several studies that set out to develop dual-acting ligands for anticancer applications targeting the $A_{2A}R$ and a secondary non-GPCR. In 2020, Varano et al. [109] aimed to develop ligands with the dual function of $A_{2A}R$ antagonism and inhibition of ectonucleotidase CD73, which is a metallophosphatase that catalyses the conversion of adenosine monophosphate (AMP) to adenosine. The CD73 inhibitory hallmark benzenesulfonamide scaffold was merged with a thiazolo [5,4-*d*]pyrimidine $A_{2A}R$ scaffold to give 12 novel ligands, exemplified by **61–63** (Fig. 17). To evaluate AR selectivity, the affinity of the ligands for A_1R, $A_{2A}R$ and A_3R were evaluated independently using radioligand competition binding assays with cells expressing the requisite receptor, and ligand activity for $A_{2B}R$ was determined in a functional cAMP-based assay. All ligands evaluated had weak to no CD73 inhibition, however, **61**, **62** and **63** demonstrated a dual $A_{2A}R/A_{2B}R$ inverse agonist/antagonist profile (e.g. **61**, CD73 activity with 0.1 mM $= 84 \pm 21\%$, A_1R $K_i = 1{,}326 \pm 256$ nM, $A_{2A}R$ $K_i = 5.73 \pm 0.48$ nM, $A_{2A}R$ $IC_{50} = 5.20 \pm 0.46$ nM (inverse agonist), $A_{2B}R$ $IC_{50} = 34 \pm 3$ nM, A_3R $K_i = 1{,}874 \pm 158$ nM). Ligands **61–63** remain a good starting point to develop more efficacious dual-acting ligands for $A_{2A}R$ and CD73 or to further explore the dual $A_{2A}R/A_{2B}R$ activity.

In 2020, Ceni et al. [110] reported the development of multi-target adenosine $A_{2A}R$ antagonists and carbonic anhydrase (CA) IX and XII inhibitors as new potential antitumour agents. These derivatives (e.g. **64** and **65**, Fig. 17) were based on 8-amino-2,6-diphenyl-1,2,4-triazolo[4,3-*a*]pyrazin-3-one, a potent scaffold previously reported as an $A_{2A}R$ antagonist, with substituents added to confer activity towards CA IX and/or XII. A range of triazolopyraizines were screened to determine $A_{2A}R$ affinity and selectivity over other ARs, as well as testing the inhibitory activity against tumour-associated CA IX and XII isoforms. Many of the triazolopyrazines had nanomolar $A_{2A}R$ affinity and high selectivity vs. the other ARs. Overall, compounds **64** and **65** showed the best multi-target profile with good $A_{2A}R$ affinity (**64**, $K_i = 6.4 \pm 1.5$ nM; **65**, $K_i = 108 \pm 25$ nM) and CA IX/XII inhibition (**64**, CA IX K_i = 0.466 μM, CA XII K_i = 0.006 μM; **65**, CA IX K_i = 5.0 nM, CA XII K_i = 27.0 nM). Compounds **64** and **65** also behaved as $A_{2A}R$ antagonists, counteracting NECA-stimulated cAMP production in CHO cells (**64**, $IC_{50} = 90.8 \pm 26$ nM; **65**, $IC_{50} = 926 \pm 111$ nM). Dual-acting $A_{2A}R$ antagonist/CA IX and XII isoform inhibitors such as **64** and **65** could be lead candidates for further investigation as anticancer agents.

Ligands with a dual ability to antagonise $A_{2A}R$ and inhibit histone deacetylase (HDAC) were developed as potential tumour immunotherapeutic agents by Yan et al. [111]. Approximately 40 compounds were designed and synthesised (e.g. IHCH-3064 (**66**), Fig. 17), all containing either a ZM241385- or SCH-58261-based $A_{2A}R$ scaffold linked via different short linkers to hydroxamic acid or 2-amino benzamide-based zinc binding groups, a signature group for HDAC inhibitors. Competition radioligand binding assays were carried out at $A_{2A}R$ and the ability of the new ligands to inhibit HDAC1 and HDAC6 was evaluated. Lead compounds were also tested in an in vitro antiproliferative assay against murine colon cancer cell lines MC38 and CT26 and in an MC38 tumour growth trial in mice. The best compound, IHCH-3064, has high affinity for $A_{2A}R$ ($K_i = 2.2 \pm 0.4$ nM), selective HDAC1 inhibition ($IC_{50} = 80.2 \pm 13.1$ nM), good in vitro antiproliferative activity and in vivo tumour growth inhibition.

There have also been several reports of intentional development of dual-acting ligands across one or more AR subtypes. As an example, in 2017 Petrelli et al. [112] reported the development of 5′-C-ethyltetrazolyladenosine derivatives (such as **67**, **68**, Fig. 17) as potent dual-acting A_1R agonists and A_3R antagonists as potential analgesic compounds. This was based on an earlier report detailing dual-acting A_1R agonists and A_3R antagonists that would be potentially useful for the treatment of glaucoma and epilepsy [113]. The 5′-C-ethyltetrazolyladenosine derivatives were expanded by modifying substituents at the N^6 position of the adenosine-based moiety and compounds such as **67** and **68** (Fig. 17) were identified as a potent dual-acting compound [112]. Compound **67** was found to be an A_1R and A_3R agonist in rat ARs, while in human ARs it was found to be an A_1R agonist and A_3R antagonist, highlighting the importance of species differences for both affinity and efficacy, especially at A_3R. The analgesic activity of dual-acting compounds was examined using the formalin test. Formalin tests were performed on three selected dual-acting compounds, of which the most potent analgesic effect was observed with **68,** which at 0.5 mg/kg reduced both the first and the second phases of formalin test. A combination of **68** (0.3 mg/kg) with A_3R agonist 2-chloro-N^6-(3-iodobenzyl)-5′-N-methylcarboxamidoadenosine (a chloro derivative of

IB-MECA (**54**), Fig. 16), (1 mg/kg) reduced completely the antinocifensive behaviour in both the first and the second phases of formalin test, demonstrating that a combination of an A_1R agonist and an A_3R agonist shows a highly potent analgesic activity.

There are increasingly more reports of purely computational rationally designed ‘polypharmacology’ and dual-acting ligands. In 2021, Liu et al. [114] expanded their previously reported *DrugEx* algorithm [115] to generate drug-like molecules towards multiple targets (A_1R and $A_{2A}R$) while avoiding off-target interactions (in this case, the human Ether-à-go-go-Related Gene, hERG). The *DrugEx* algorithm involves combined evolutionary algorithms (including mutation and crossover operations) with Pareto ranking algorithms in order to generate a large library of novel compounds, which could potentially be dual-acting ligands. Many of the novel molecules generated were similar to known ligands, and a large percentage of generated SMILES were valid and represented desired molecules without many duplications. This purely computational approach allows for rapid design and screening of potential hit-to-lead compounds towards selected multiple targets.

It must be emphasised that many if not all compounds developed as ‘selective’ AR ligands have some activity at other AR subtypes, especially at high concentrations. In addition, without comprehensive data across all AR subtypes for every ligand reported, the number of dual-acting ligands across several AR subtypes is undoubtedly large.

3.3 *Dual-Acting Prodrugs*

In 2012, Dalpiaz et al. [116] reported compound **69** (Fig. 18) which was designed as a dual-targeted prodrug for $A_{2A}R$ and D_2R. As outlined in Sect. 3.1, synergistic effects between $A_{2A}R$ antagonists and D_2R agonists hold promise for the treatment of PD [62–64]. Specifically, $A_{2A}R$ antagonists have been used to amplify the therapeutic effects of L-dopa and reduce motor complications resulting from its long-term use. Prodrug **69** was designed to release dopamine (**71**), which is initially joined via a short succinic linker to a novel $A_{2A}R$ antagonist derivative related to ZM241385. Compound **69** was designed to be hydrolysed at one of the amide bonds, which would release **70** and dopamine (**71**), or the parent $A_{2A}R$ antagonist derivative **72** and **73**, with **73** being further broken down to give dopamine (**71**).

Competitive binding assays were carried out to evaluate **69** and the potential hydrolysis products **70** and **72**. Compound **69** showed a very high affinity for $A_{2A}R$ ($K_i = 2.07 \pm 0.23$ nM in rat striatum, $K_i = 7.32 \pm 0.65$ nM in $hA_{2A}CHO$ cells) while **70** ($A_{2A}R$ $K_i = 24 \pm 2$ nM in rat striatum, $K_i = 35 \pm 3$ nM in $hA_{2A}CHO$ cells) and **72** ($A_{2A}R$ $K_i = 72 \pm 5$ nM in rat striatum, $K_i = 50 \pm 2$ nM in $hA_{2A}CHO$ cells) also showed affinity for $A_{2A}R$, albeit slightly reduced. Interestingly, this meant that ‘prodrug’ **69** had a better affinity for $A_{2A}R$ than either of the hydrolysed products **70** and **72**. Neither intact **69** or **70** showed affinity for A_1R and A_3R, nor any affinity or agonist activity at the D_2R.

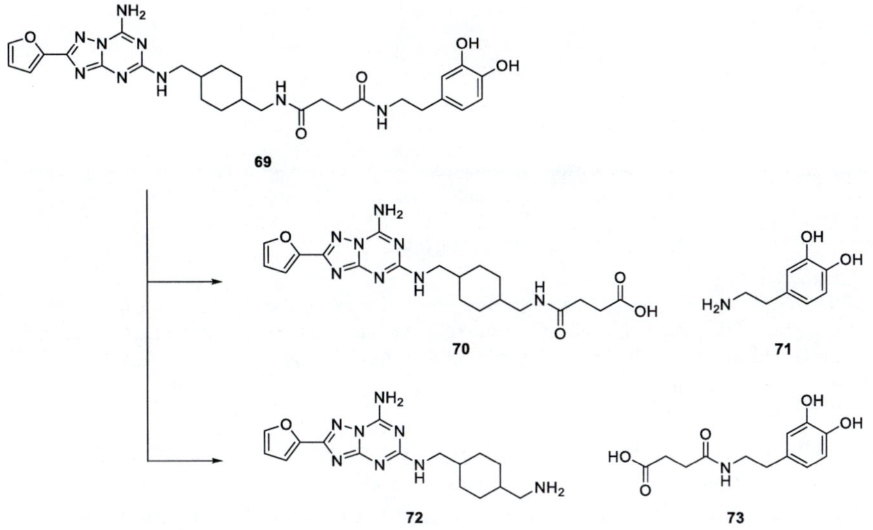

Fig. 18 Dual-acting prodrug **69** developed by Dalpiaz et al. [116], designed to release the endogenous agonist dopamine (**71**) and an $A_{2A}R$ antagonist (**70**, **72**)

The hydrolysis products of compound **69** were analysed in water, phosphate buffer, human whole blood, and rat brain homogenates (all at 37°C). In human blood, **69** was hydrolyzed (half-life $= 2.73 \pm 0.23$ h) to give predominantly compounds **72** and **73**, with smaller levels of **70** and **71** detected. Whereas in rat brain homogenates, **69** was hydrolyzed (half-life approx. 8 h) predominantly to give **70** and **71**, with **72** and **73** not detected within 8 h incubation of **69**. One would normally think of a prodrug as being fully 'inactive' before cleavage to give the active component(s). However, as measured by Dalpiaz et al., [116] 'prodrug' (**69**) does act an as $A_{2A}R$ antagonist even with the succinic-linked dopamine derivative attached. Compound **69** can still be considered a prodrug in the sense it is a dual-acting prodrug – i.e. compound **69** does not have dual activity, only 'mono' activity. Although not a 'prodrug' in the traditional sense, compounds such as **69** could offer a way to advantageously deliver dopamine via a prodrug therapy.

4 Bitopic Ligands

A bitopic ligand is a type of bivalent ligand that binds concomitantly to an orthosteric and allosteric site on the same receptor (Fig. 19), in contrast to targeting two orthosteric sites in the case of more 'typical' bivalent ligands. These so-called hybrid molecules are designed to include orthosteric and/or allosteric moieties connected by a linker or spacer unit. These moieties are typically derived from monovalent ligands, which have established affinities for their respective binding

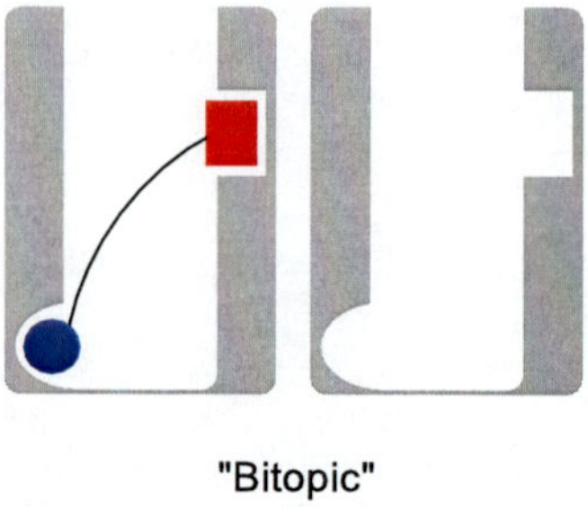

Fig. 19 Bitopic ligands consist of an orthosteric and allosteric moiety and are able to concomitantly interact at the orthosteric and an allosteric site of a receptor monomer. Orthosteric binding moieties are represented by blue circles and allosteric binding moieties by red squares. Adapted from Lane et al. and Jörg et al. [40, 41]

sites. The linker on the other hand must be structured not only to provide the appropriate amount of space and flexibility to allow the individual moieties to bind to their respective sites but also to avoid any unfavourable interactions with the receptor which might lower the overall affinity of the ligand. Although there is some evidence that molecules are able to reach the active site of a GPCR by crossing the transmembrane domains via lipid bilayers [117, 118], access to the binding sites via the extracellular space might be energetically preferred [119, 120].

As for the case of bivalent ligands discussed in Sect. 2 and Fig. 3a, one must consider several other binding possibilities (Fig. 20). A bitopic ligand might be able to interact with an orthosteric and allosteric site of the different monomers within a receptor dimer. The so-called 'flip-flop' mode of binding (Fig. 20) can also occur with bitopic ligands. Alternatively, if only one pharmacophore of the ligand is engaged, two ligands may simultaneously bind, one at each site, imparting something analogous to traditional cooperativity of co-administered ligands (as discussed in Fig. 3a).

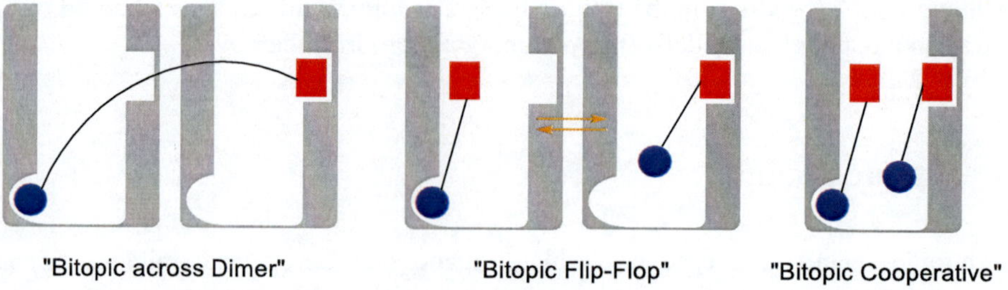

Fig. 20 Depiction of other possible binding modes for bitopic ligands. The flip-flop mode depicts a scenario where the ligand is only able to bind one site at a time and switches between them. Cooperative binding involves two molecules engaged with the receptor at the same time; in case of the bitopic ligand, this is analogous to traditional allosteric modulation. Adapted from Lane et al. and Jörg et al. [40, 41]

While the binding modes of a bitopic ligand can be difficult to differentiate experimentally, one might expect a bitopic binding mode to result in increased affinity over the individual, monovalent congeners. However, as Christopoulos and co-workers explain, that is not always the case [40, 42]. A complex relationship between enthalpy and entropy can result in similar or even reduced affinity, and increased subtype selectivity and signalling bias are other potential signs of bitopic binding [40, 42]. Antony et al. designed a hybrid molecule targeting the muscarinic acetylcholine M_2 receptor by combining a high-affinity orthosteric ligand with a subtype-selective allosteric modulator, resulting in a ligand that not only possessed the properties of the individual congeners but also resulted in previously unseen signalling bias, indicating that the bitopic binding mode was engendering a unique receptor conformation [121]. Langmead et al. investigated the binding mode of the muscarinic acetylcholine M_1 receptor ligand GSK1034702, previously thought to be an allosteric modulator. Through a series of binding, kinetic, and functional assays, they determined the ligand to feature a bitopic binding mode, justifying its lack of subtype selectivity when compared to other, purely allosteric ligands [122]. Similarly, the validation of the binding mode of bivalent ligands is not trivial, as discussed in Sect. 2.

4.1 Bitopic A_1R Ligands

LUF6258 (**74**, Fig. 21) is an example of a bitopic ligand consisting of separate moieties designed to interact with different binding sites on the same receptor [123]. Targeting the A_1R, it consists of the orthosteric endogenous ligand adenosine, a spacer, and an allosteric 2-amino-3-benzoylthiophene (2A3BT) positive allosteric modulator (PAM). The linker stems from the N^6-position of adenosine, is composed of a phenolic ether and 9-carbon alkyl chain, and connects to the thiophene via a tetrahydropyridine moiety, which forms a bicycle at the 4- and 5-positions of the thiophene. Analogues with alkyl chain lengths of 2–10 carbons were also synthesised and it was found that in order to achieve bitopic binding the chain length should be greater than 6 carbons. Homology modelling and docking studies indicated that the allosteric binding component might be pointing into the extracellular space or interacting with extracellular loop 2 (ECL2) [123].

An additional hybrid ligand predicted to bind bitopically to A_1R is VCP746 (**75**) (Fig. 21), developed by Valant et al. in 2014 [124]. Also utilising adenosine as the orthosteric binding moiety and an N^6-linker attachment point, it includes the 2A3BT-based derivative VCP171 (**76**) (Fig. 21) as the allosteric component [125]. The spacer consists of a 6-carbon alkyl chain and benzamide moiety. This ligand showed significant signalling bias towards cAMP inhibition and away from ERK1/2 phosphorylation, and testing in native A_1R models, including rat H9c2(2-1) embryonic cardiomyoblasts and neonatal cardiomyocytes, showed effective cardioprotection, such as protection from cell death in simulated ischemic conditions, while avoiding any slowing of heart rate. While orthosteric A_1R agonists, such

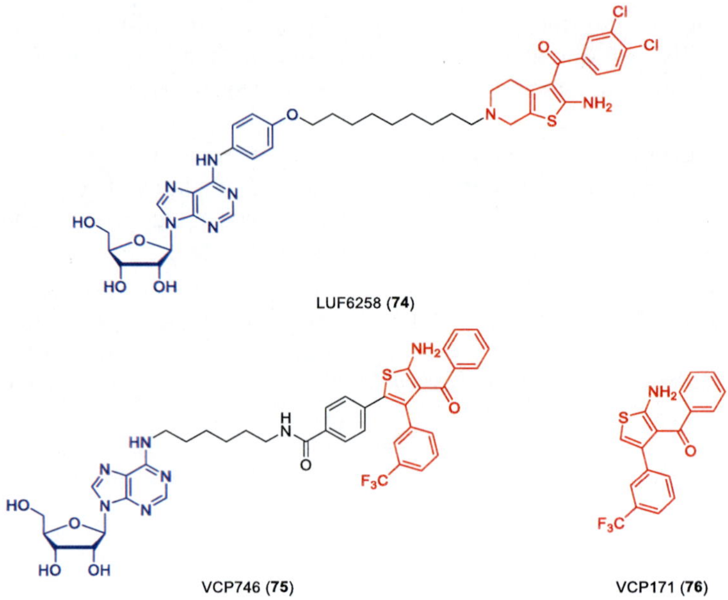

Fig. 21 LUF6258 (**74**), a bitopic A_1R biased agonist developed by Ijzerman and co-workers in 2010 [123], and VCP746 (**75**), developed by Valant et al. in 2014, incorporating allosteric ligand VCL171 (**76**) [124]. Both LUF6258 and VCP746 are hybrid structures, with adenosine as the orthosteric component and allosteric ligands based on the well-known 2-amino-3-benzoyl structure of traditional A_1R PAMs connected via an alkyl spacer

as R-PIA and CPA, also protected against cell death in the same tests, they did produce a significant reduction in heart rate, typical of non-biased A_1R agonists [124]. In 2016, a study by Baltos et al. found that VCP746 (**75**) was also biased away from calcium mobilisation [126].

In a follow-up publication in 2018, Aurelio et al. conducted an SAR study of VCP746 (**75**), again employing a series of pharmacological assays to probe potential signalling bias [127]. The authors synthesised a number of analogues, deleting or altering functional groups at all three of the compound's distinct moieties; the orthosteric and allosteric moieties, as well as the linker. Modifications throughout the molecule had a significant impact on signalling bias, indicating that each moiety plays an important role in the ability of the ligand to effectively bind the receptor.

In particular, their findings highlight the importance of the allosteric binding group in producing the biased profile possessed by VCP746 (**75**). It is important to note that these findings do not align with the conclusions of SAR studies performed on the allosteric moiety alone when studied as an A_1R PAM. In those studies, removal of either the amino or ketone substituents from the core thiophene caused severe reductions in allosteric enhancement whereas these deletions were well tolerated in the bitopic ligand VCP746 (**75**). Additionally, in past studies of VCP171 (**76**), removal of the 4-(3-trifluoromethylphenyl) substituent caused a drop in allosteric enhancement whereas, again, this was well tolerated in VCP746 (**75**). These findings would seem to indicate that the allosteric binding group of the bitopic ligand is not binding to the receptor in the same pose and/or location as the 2A3BT PAMs. This is supported by a computational study performed in 2021 by Reynolds and co-workers, in which they propose the existence of multiple A_1R allosteric binding sites and different binding locations for the allosteric binding moiety of VCP746 (**75**) and traditional 2A3BT PAMs [128].

4.2 *Bitopic $A_{2A}R$, $A_{2B}R$ and A_3R Ligands*

To our knowledge, rationally designed bitopic ligands of the $A_{2A}R$, $A_{2B}R$ and A_3R have not been published to date. Although it is worth pointing out that VCP746 (**75**) is also a high-affinity agonist at the $A_{2B}R$, and upon the deletion of the allosteric moiety a significant drop in binding affinity and functional activity (calcium, IP1, cAMP and pERK1/2) was observed, therefore it can be speculated that **75** is also acting as a bitopic ligand at the $A_{2B}R$ [129].

Additionally, the binding and functional properties of fluorescent AR probes have shown to exhibit unique pharmacological profiles distinctly different from the monovalent parent ligand, and it is reasoned that some of these changes might be due to interactions with allosteric sites by either the spacer and/or fluorophore [130, 131]. For instance, Gao et al. [130] reported a detailed study characterising the binding and functional antagonism profiles of commonly used $A_{2A}R$ pharmacological tools, investigating numerous xanthine and non-xanthine-based congeners and fluorescent probes. They suggest that the slope of the Schild analysis model can be used to indicate if the putative bitopic $A_{2A}R$ pharmacological tools is due to allosteric or competitive antagonism. For example in their study [130], XAC (**4**) (Fig. 1) and XAC630 (**77**) (Fig. 22) had comparable K_i values displacing antagonist [^{3}H]ZM241385 and agonist [^{3}H]CGS21680 in binding assays, yet the slopes calculated using Schild analysis using was significantly different (1.06 ± 0.06 and 0.47 ± 0.04, respectively), which led to the interpretation that **77** (Fig. 22) was an allosteric antagonist rather than competitive antagonist.

Furthermore, Sun et al. [132] reported the crystal structure of $A_{2A}R$ in complex with compound **78** (Fig. 22) (PDBID: 5UIG), a compound displaying high affinity for the $A_{2A}R$ and NMDA NR_2 receptor subtype. The obtained structure of the $A_{2A}R$ revealed unprecedented insights compared to previously published structures with other $A_{2A}R$ antagonists [43, 133–136], which includes the discovery of an exosite interacting with the ortho-methoxyphenyl group. It is proposed that additional interaction of compound **78** with tyrosine residues Tyr9$^{1.35}$ and Tyr271$^{7.36}$ are resulting in a unique conformation that contributes to the high affinity of **78** compared to structurally related analogues [132]. Based on insights from this structural study, compound **78** is interacting with the orthosteric site and an exosite at the $A_{2A}R$, and can consequently be classified as a bitopic ligand.

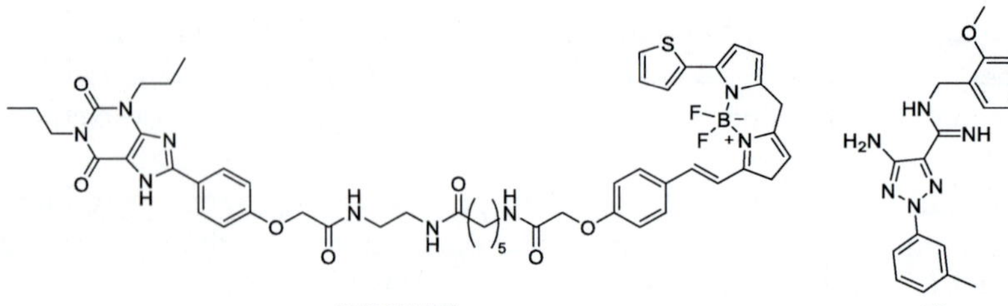

Fig. 22 Chemical structures of fluorescent probe XAC630 (**77**) and bitopic ligand **78**, which was discovered via structural studies

These results suggest that pharmacological tools based on orthosteric ligands may bind differently to a receptor compared to the parent ligand and thus exhibit unique pharmacological profiles. Subtle differences in binding profiles might remain undiscovered initially, but advances in structural biology and pharmacological profiling methods provide novel insights that will lead to the reclassification of such ligands as bitopic ARs ligands. Therefore, the pharmacological profile of novel tool compounds warrants careful validation to ensure the conclusion drawn from such experiments does not result in the false interpretation of the collected data.

5 Conclusions

Advances in GPCR pharmacology and chemical biology have provided novel insights into the complexity of the structure and function of this important drug target. Traditionally, GPCR ligands were designed to either activate or inhibit the signal response by targeting the orthosteric site of a receptor monomer, whereas more recently novel approaches to target allosteric sites as well as homo- and heteromeric complexes have diversified the field.

Purinergic receptors display GPCR features that are universal, yet incompletely understood. Therefore, the purinergic receptor family is a useful model system to study GPCR biology from a chemical, biological and structural standpoint, particularly with respect to functionally distinct GPCR complexes (homo- and hetero-oligomers). The $A_{2A}R$ has been extensively studied, which might be partially attributed to the ligand-bound X-ray structures that were obtained as early as 2008 [43]. Understanding the binding mode of the orthosteric ligands has supported the rational design of bivalent ligands for this target. Additionally, evidence of $A_{2A}R$ heterodimers with the D_2R, which are both implicated in Parkinson's disease, has further increased the interest in the development of bivalent ligands for these targets. Many of the heterobivalent, bitopic and dual-acting A_1R ligands reported in this chapter pre-date the multiple AR structures solved in recent years [70, 71, 132, 133, 135, 137]. However, it can be speculated that structural information will be used to fine-tune the reported ligands while also assisting researchers with the development of new bivalent probes and drug candidates.

When designing bivalent ligands there are some unique challenges that need to be considered. For instance, the high molecular weight of many of these ligands is a direct result of combining two active moieties, which generally results in compounds with poor physico-chemical properties, and blood-brain-barrier permeability. It, therefore, is not surprising that none of the bivalent ligands reported in this chapter have advanced into clinical trials. However, the lower molecular weight dual-acting ligands and prodrugs hold promise for drug discovery projects to be able to overcome some of the limitations of larger bivalent compounds.

In summary, bifunctional ligands have provided us with useful tools to study purinergic receptors. Despite the progress that has been made due to the many contributions of numerous research groups in this research area, there is a necessity for novel bifunctional ligands. On one hand, innovative bivalent tools will allow us to get a better insight into the complex mechanisms underlying purinergic receptor function, and on the other hand, we need to ensure innovation towards the development of novel purinergic therapeutics, thus overcoming the limitations of current drug candidates.

Compliance with Ethical Standards **Conflict of Interest:** The authors have no relevant financial or non-financial interests to disclose.

Funding: C. Payne is supported by a University of Otago Doctoral Scholarship. J.K. Awalt was supported by a Monash International Postgraduate Research Scholarship. M. Jörg holds a Newcastle/Monash University Academic Track (NUMAcT) Fellowship funded by Research England (ref. 131911). L.T. May holds a National Heart Foundation Future Leader Fellowship (101857). The authors received no financial support for the research, authorship and/or publication of this article.

Ethical Approval: This chapter does not contain any studies with human participants or animals performed by the authors.

References

1. Hiller C, Kühhorn J, Gmeiner P (2013) Class A G-protein-coupled receptor (GPCR) dimers and bivalent ligands. J Med Chem 56(17):6542–6559. https://doi.org/10.1021/jm4004335
2. George SR, O'Dowd BF, Lee SP (2002) G-protein-coupled receptor oligomerization and its potential for drug discovery. Nat Rev Drug Discov 1(10):808–820. https://doi.org/10.1038/nrd913
3. Ciruela F, Casadó V, Mallol J et al (1995) Immunological identification of A1 adenosine receptors in brain cortex. J Neurosci Res 42(6):818–828. https://doi.org/10.1002/jnr.490420610
4. Gracia E, Moreno E, Cortes A et al (2013) Homodimerization of adenosine A(1) receptors in brain cortex explains the biphasic effects of caffeine. Neuropharmacology 71:56–69. https://doi.org/10.1016/j.neuropharm.2013.03.005
5. Briddon SJ, Gandía J, Amaral OB et al (2008) Plasma membrane diffusion of G protein-coupled receptor oligomers. Biochim Biophys Acta 1783(12):2262–2268. https://doi.org/10.1016/j.bbamcr.2008.07.006
6. Hill SJ, May LT, Kellam B et al (2014) Allosteric interactions at adenosine A(1) and A(3) receptors: new insights into the role of small molecules and receptor dimerization. Br J Pharmacol 171(5):1102–1113. https://doi.org/10.1111/bph.12345

7. Chandrasekera PC, Wan TC, Gizewski ET et al (2013) Adenosine A1 receptors heterodimerize with beta1- and beta2-adrenergic receptors creating novel receptor complexes with altered G protein coupling and signaling. Cell Signal 25(4):736–742. https://doi.org/10.1016/j.cellsig.2012.12.022
8. Gines S, Hillion J, Torvinen M et al (2000) Dopamine D1 and adenosine A1 receptors form functionally interacting heteromeric complexes. Proc Natl Acad Sci U S A 97(15):8606–8611. https://doi.org/10.1073/pnas.150241097
9. Cortés A, Casadó-Anguera V, Moreno E et al (2019) Chapter three – the heterotetrameric structure of the adenosine A1-dopamine D1 receptor complex: pharmacological implication for restless legs syndrome. In: Clemens S, Ghorayeb I (eds) Advances in pharmacology, vol 84. Academic Press, pp 37–78
10. Kamikubo Y, Tabata T, Sakairi H et al (2015) Complex formation and functional interaction between adenosine A1 receptor and type-1 metabotropic glutamate receptor. J Pharmacol Sci 128(3):125–130. https://doi.org/10.1016/j.jphs.2015.06.002
11. Ciruela F, Escriche M, Burgueno J et al (2001) Metabotropic glutamate 1alpha and adenosine A1 receptors assemble into functionally interacting complexes. J Biol Chem 276(21): 18345–18351. https://doi.org/10.1074/jbc.M006960200
12. Yoshioka K, Hosoda R, Kuroda Y et al (2002) Hetero-oligomerization of adenosine A1 receptors with P2Y1 receptors in rat brains. FEBS Lett 531(2):299–303. https://doi.org/10.1016/s0014-5793(02)03540-8
13. Hinz S, Navarro G, Borroto-Escuela D et al (2018) Adenosine A2A receptor ligand recognition and signaling is blocked by A2B receptors. Oncotarget 9(17):13593–13611. https://doi.org/10.18632/oncotarget.24423
14. Carriba P, Ortiz O, Patkar K et al (2007) Striatal adenosine A2A and cannabinoid CB1 receptors form functional heteromeric complexes that mediate the motor effects of cannabinoids. Neuropsychopharmacology 32(11):2249–2259. https://doi.org/10.1038/sj.npp.1301375
15. Hillion J, Canals M, Torvinen M et al (2002) Coaggregation, cointernalization, and codesensitization of adenosine A2A receptors and dopamine D2 receptors. J Biol Chem 277(20):18091–18097. https://doi.org/10.1074/jbc.M107731200
16. Canals M, Marcellino D, Fanelli F et al (2003) Adenosine A2A-dopamine D2 receptor-receptor heteromerization: qualitative and quantitative assessment by fluorescence and bioluminescence energy transfer. J Biol Chem 278(47):46741–46749. https://doi.org/10.1074/jbc.M306451200
17. Kamiya T, Saitoh O, Yoshioka K et al (2003) Oligomerization of adenosine A2A and dopamine D2 receptors in living cells. Biochem Biophys Res Commun 306(2):544–549. https://doi.org/10.1016/s0006-291x(03)00991-4
18. Ferré S, Karcz-Kubicha M, Hope BT et al (2002) Synergistic interaction between adenosine A2A and glutamate mGlu5 receptors: implications for striatal neuronal function. Proc Natl Acad Sci U S A 99(18):11940–11945. https://doi.org/10.1073/pnas.172393799
19. May LT, Bridge LJ, Stoddart LA et al (2011) Allosteric interactions across native adenosine-A3 receptor homodimers: quantification using single-cell ligand-binding kinetics. FASEB J 25(10):3465–3476. https://doi.org/10.1096/fj.11-186296
20. Rahman A (2009) The role of adenosine in Alzheimer's disease. Curr Neuropharmacol 7(3): 207–216. https://doi.org/10.2174/157015909789152119
21. Huang ZL, Urade Y, Hayaishi O (2011) The role of adenosine in the regulation of sleep. Curr Top Med Chem 11(8):1047–1057. https://doi.org/10.2174/156802611795347654
22. Boison D, Singer P, Shen HY et al (2012) Adenosine hypothesis of schizophrenia – opportunities for pharmacotherapy. Neuropharmacology 62(3):1527–1543. https://doi.org/10.1016/j.neuropharm.2011.01.048
23. Gessi S, Merighi S, Varani K et al (2011) Chapter 2 – adenosine receptors in health and disease. In: Jacobson KA, Linden J (eds) Advances in pharmacology, vol 61. Academic Press, pp 41–75

24. Vecchio EA, Baltos JA, Nguyen ATN et al (2018) New paradigms in adenosine receptor pharmacology: allostery, oligomerization and biased agonism. Br J Pharmacol 175(21): 4036–4046. https://doi.org/10.1111/bph.14337
25. Nguyen KDQ, Vigers M, Sefah E et al (2021) Homo-oligomerization of the human adenosine A2A receptor is driven by the intrinsically disordered C-terminus. Elife 10:e66662. https://doi.org/10.7554/eLife.66662
26. Navarro G, Borroto-Escuela DO, Fuxe K et al (2018) Adenosine receptors as a paradigm to identify dimer/oligomers of G-protein-coupled receptors and as targets in Parkinson's disease and schizophrenia. In: Borea PA, Varani K, Gessi S, Merighi S, Vincenzi F (eds) The adenosine receptors. Springer, pp 239–258
27. Erez M, Takemori AE, Portoghese PS (1982) Narcotic antagonistic potency of bivalent ligands which contain beta-naltrexamine. Evidence for bridging between proximal recognition sites. J Med Chem 25(7):847–849. https://doi.org/10.1021/jm00349a016
28. Müller CE, Jacobson KA (2011) Xanthines as adenosine receptor antagonists. Handb Exp Pharmacol 200:151–199. https://doi.org/10.1007/978-3-642-13443-2_6
29. Jacobson KA (2013) Structure-based approaches to ligands for G-protein-coupled adenosine and P2Y receptors, from small molecules to nanoconjugates. J Med Chem 56(10):3749–3767. https://doi.org/10.1021/jm400422s
30. Jacobson KA, Kirk KL, Padgett WL et al (1985) Functionalized congeners of adenosine: preparation of analogues with high affinity for A1-adenosine receptors. J Med Chem 28(9): 1341–1346. https://doi.org/10.1021/jm00147a039
31. Jacobson KA, Ukena D, Padgett W et al (1987) Molecular probes for extracellular adenosine receptors. Biochem Pharmacol 36(10):1697–1707. https://doi.org/10.1016/0006-2952(87)90056-6
32. Ukena D, Daly JW, Kirk KL et al (1986) Functionalized congeners of 1,3-dipropyl-8-phenylxanthine: potent antagonists for adenosine receptors that modulate membrane adenylate cyclase in pheochromocytoma cells, platelets and fat cells. Life Sci 38(9):797–807. https://doi.org/10.1016/0024-3205(86)90596-5
33. Portoghese PS, Larson DL, Sayre LM et al (1986) Opioid agonist and antagonist bivalent ligands. The relationship between spacer length and selectivity at multiple opioid receptors. J Med Chem 29(10):1855–1861. https://doi.org/10.1021/jm00160a010
34. Karellas P, McNaughton M, Baker SP et al (2008) Synthesis of bivalent beta2-adrenergic and adenosine A1 receptor ligands. J Med Chem 51(19):6128–6137. https://doi.org/10.1021/jm800613s
35. Arnatt CK, Zhang Y (2014) Bivalent ligands targeting chemokine receptor dimerization: molecular design and functional studies. Curr Top Med Chem 14(13):1606–1618. https://doi.org/10.2174/1568026614666140827144752
36. Cappelli A, Manini M, Paolino M et al (2011) Bivalent ligands for the serotonin 5-HT3 receptor. ACS Med Chem Lett 2(8):571–576. https://doi.org/10.1021/ml2000388
37. Portoghese PS, Larson DL, Yim CB et al (1985) Stereostructure-activity relationship of opioid agonist and antagonist bivalent ligands. Evidence for bridging between vicinal opioid receptors. J Med Chem 28(9):1140–1141. https://doi.org/10.1021/jm00147a002
38. Fernandes SM, Lee YS, Gillies RJ et al (2014) Synthesis and evaluation of bivalent ligands for binding to the human melanocortin-4 receptor. Biorg Med Chem 22(22):6360–6365. https://doi.org/10.1016/j.bmc.2014.09.055
39. Huang B, St Onge CM, Ma H et al (2021) Design of bivalent ligands targeting putative GPCR dimers. Drug Discov Today 26(1):189–199. https://doi.org/10.1016/j.drudis.2020.10.006
40. Lane JR, Sexton PM, Christopoulos A (2013) Bridging the gap: bitopic ligands of G-protein-coupled receptors. Trends Pharmacol Sci 34(1):59–66. https://doi.org/10.1016/j.tips.2012.10.003
41. Jörg M, Shonberg J, Mak FS et al (2013) Novel adenosine A(2A) receptor ligands: a synthetic, functional and computational investigation of selected literature adenosine A(2A) receptor

antagonists for extending into extracellular space. Bioorg Med Chem Lett 23(11):3427–3433. https://doi.org/10.1016/j.bmcl.2013.03.070

42. Valant C, Robert Lane J, Sexton PM et al (2012) The best of both worlds? Bitopic orthosteric/allosteric ligands of g protein-coupled receptors. Annu Rev Pharmacol Toxicol 52(1):153–178. https://doi.org/10.1146/annurev-pharmtox-010611-134514
43. Jaakola VP, Griffith MT, Hanson MA et al (2008) The 2.6 angstrom crystal structure of a human A2A adenosine receptor bound to an antagonist. Science 322(5905):1211–1217. https://doi.org/10.1126/science.1164772
44. Bonger KM, van den Berg RJ, Heitman LH et al (2007) Synthesis and evaluation of homo-bivalent GnRHR ligands. Bioorg Med Chem 15(14):4841–4856. https://doi.org/10.1016/j.bmc.2007.04.065
45. Zhang Y, Gilliam A, Maitra R et al (2010) Synthesis and biological evaluation of bivalent ligands for the cannabinoid 1 receptor. J Med Chem 53(19):7048–7060. https://doi.org/10.1021/jm1006676
46. Xu Y, Duggineni S, Espitia S et al (2013) A synthetic bivalent ligand of CXCR4 inhibits HIV infection. Biochem Biophys Res Commun 435(4):646–650. https://doi.org/10.1016/j.bbrc.2013.05.038
47. Kühhorn J, Hübner H, Gmeiner P (2011) Bivalent dopamine D2 receptor ligands: synthesis and binding properties. J Med Chem 54(13):4896–4903. https://doi.org/10.1021/jm2004859
48. Gogoi S, Biswas S, Modi G et al (2012) Novel bivalent ligands for D2/D3 dopamine receptors: significant co-operative gain in D2 affinity and potency. ACS Med Chem Lett 3(12):991–996. https://doi.org/10.1021/ml3002117
49. Shonberg J, Lane JR, Scammells PJ et al (2013) Synthesis, functional and binding profile of (R)-apomorphine based homobivalent ligands targeting the dopamine D2 receptor. Med Chem Commun 4(9):1290–1296. https://doi.org/10.1039/C3MD00154G
50. Pulido D, Casadó-Anguera V, Pérez-Benito L et al (2018) Design of a true bivalent ligand with picomolar binding affinity for a G protein-coupled receptor homodimer. J Med Chem 61(20):9335–9346. https://doi.org/10.1021/acs.jmedchem.8b01249
51. Ullmann T, Gienger M, Budzinski J et al (2021) Homobivalent dopamine D2 receptor ligands modulate the dynamic equilibrium of D2 monomers and homo- and heterodimers. ACS Chem Biol 16(2):371–379. https://doi.org/10.1021/acschembio.0c00895
52. Müller CE, Jacobson KA (2011) Recent developments in adenosine receptor ligands and their potential as novel drugs. Biochim Biophys Acta 1808(5):1290–1308. https://doi.org/10.1016/j.bbamem.2010.12.017
53. Kiesman WF, Elzein E, Zablocki J (2009) A1 adenosine receptor antagonists, agonists, and allosteric enhancers. In: Wilson CN, Mustafa SJ (eds) Adenosine receptors in health and disease. Springer, Berlin, pp 25–58
54. Barlow N, Baker SP, Scammells PJ (2013) Effect of linker length and composition on heterobivalent ligand-mediated receptor cross-talk between the A1 adenosine and beta2 adrenergic receptors. ChemMedChem 8(12):2036–2046. https://doi.org/10.1002/cmdc.201300286
55. Shen J, Zhang L, Song WL et al (2013) Design, synthesis and biological evaluation of bivalent ligands against A(1)-D(1) receptor heteromers. Acta Pharmacol Sin 34(3):441–452. https://doi.org/10.1038/aps.2012.151
56. Müller CE, Schobert U, Hipp J et al (1997) Configurationally stable analogs of styrylxanthines as A2A adenosine receptor antagonist. Eur J Med Chem 32(9):709–719. https://doi.org/10.1016/s0223-5234(97)88913-1
57. González MP, Terán C, Teijeira M (2008) Search for new antagonist ligands for adenosine receptors from QSAR point of view. How close are we? Med Res Rev 28(3):329–371. https://doi.org/10.1002/med.20108
58. Neumeyer JL, Kula NS, Bergman J et al (2003) Receptor affinities of dopamine D1 receptor-selective novel phenylbenzazepines. Eur J Pharmacol 474(2–3):137–140. https://doi.org/10.1016/s0014-2999(03)02008-9

59. Soriano A, Ventura R, Molero A et al (2009) Adenosine A2A receptor-antagonist/dopamine D2 receptor-agonist bivalent ligands as pharmacological tools to detect A2A-D2 receptor heteromers. J Med Chem 52(18):5590–5602. https://doi.org/10.1021/jm900298c
60. Pérez-Benito L, Henry A, Matsoukas MT et al (2018) The size matters? A computational tool to design bivalent ligands. Bioinformatics 34(22):3857–3863. https://doi.org/10.1093/bioinformatics/bty422
61. Ferré S, von Euler G, Johansson B et al (1991) Stimulation of high-affinity adenosine A2 receptors decreases the affinity of dopamine D2 receptors in rat striatal membranes. Proc Natl Acad Sci U S A 88(16):7238–7241. https://doi.org/10.1073/pnas.88.16.7238
62. Fuxe K, Ferré S, Canals M et al (2005) Adenosine A2A and dopamine D2 heteromeric receptor complexes and their function. J Mol Neurosci 26(2–3):209–220. https://doi.org/10.1385/JMN:26:2-3:209
63. Cieślak M, Komoszyński M, Wojtczak A (2008) Adenosine A(2A) receptors in Parkinson's disease treatment. Purinergic Signal 4(4):305–312. https://doi.org/10.1007/s11302-008-9100-8
64. Pinna A, Serra M, Marongiu J et al (2020) Pharmacological interactions between adenosine A2A receptor antagonists and different neurotransmitter systems. Parkinsonism Relat Disord 80(Suppl 1):S37–S44. https://doi.org/10.1016/j.parkreldis.2020.10.023
65. Valle-León M, Callado LF, Aso E et al (2021) Decreased striatal adenosine A2A-dopamine D2 receptor heteromerization in schizophrenia. Neuropsychopharmacology 46(3):665–672. https://doi.org/10.1038/s41386-020-00872-9
66. Jörg M, May LT, Mak FS et al (2015) Synthesis and pharmacological evaluation of dual acting ligands targeting the adenosine A2A and dopamine D2 receptors for the potential treatment of Parkinson's disease. J Med Chem 58(2):718–738. https://doi.org/10.1021/jm501254d
67. Pulido D, Casadó-Anguera V, Gómez-Autet M et al (2022) Heterobivalent ligand for the adenosine A2A-dopamine D2 receptor heteromer. J Med Chem 65(1):616–632. https://doi.org/10.1021/acs.jmedchem.1c01763
68. Pulido D, Albericio F, Royo M (2014) Controlling multivalency and multimodality: up to pentamodal dendritic platforms based on diethylenetriaminepentaacetic acid cores. Org Lett 16(5):1318–1321. https://doi.org/10.1021/ol500022n
69. Jacobson KA, Xie R, Young L et al (2000) A novel pharmacological approach to treating cardiac ischemia. Binary conjugates of A1 and A3 adenosine receptor agonists. J Biol Chem 275(39):30272–30279. https://doi.org/10.1074/jbc.M001520200
70. Glukhova A, Thal DM, Nguyen AT et al (2017) Structure of the adenosine A1 receptor reveals the basis for subtype selectivity. Cell 168(5):867–877 e813. https://doi.org/10.1016/j.cell.2017.01.042
71. Cheng RKY, Segala E, Robertson N et al (2017) Structures of human A1 and A2A adenosine receptors with xanthines reveal determinants of selectivity. Structure 25(8):1275–1285.e1274. https://doi.org/10.1016/j.str.2017.06.012
72. Jespers W, Oliveira A, Prieto-Diaz R et al (2017) Structure-based design of potent and selective ligands at the four adenosine receptors. Molecules 22(11). https://doi.org/10.3390/molecules22111945
73. Sherbiny FF, Schiedel AC, Maaß A et al (2009) Homology modelling of the human adenosine A2B receptor based on X-ray structures of bovine rhodopsin, the β2-adrenergic receptor and the human adenosine A2A receptor. J Comput Aided Mol Des 23(11):807. https://doi.org/10.1007/s10822-009-9299-7
74. Deb PK, Chandrasekaran B, Mailavaram R et al (2019) Molecular modeling approaches for the discovery of adenosine A2B receptor antagonists: current status and future perspectives. Drug Discov Today 24(9):1854–1864. https://doi.org/10.1016/j.drudis.2019.05.011
75. Ezzat HG, Bayoumi AH, Sherbiny FF et al (2021) Design, synthesis, and molecular docking studies of new [1,2,4]triazolo[4,3-a]quinoxaline derivatives as potential A2B receptor antagonists. Mol Divers 25(1):291–306. https://doi.org/10.1007/s11030-020-10070-w

76. Claff T, Klapschinski TA, Tiruttani Subhramanyam UK et al (2022) Single stabilizing point mutation enables high-resolution co-crystal structures of the adenosine A2A receptor with preladenant conjugates. Angew Chem Int Ed 61(22):e202115545. https://doi.org/10.1002/anie.202115545
77. Lee Y, Hou X, Lee JH et al (2021) Subtle chemical changes cross the boundary between agonist and antagonist: new A3 adenosine receptor homology models and structural network analysis can predict this boundary. J Med Chem 64(17):12525–12536. https://doi.org/10.1021/acs.jmedchem.1c00239
78. Almerico AM, Tutone M, Pantano L et al (2013) A3 adenosine receptor: homology modeling and 3D-QSAR studies. J Mol Graph Model 42:60–72. https://doi.org/10.1016/j.jmgm.2013.03.001
79. Ranganathan A, Stoddart LA, Hill SJ et al (2015) Fragment-based discovery of subtype-selective adenosine receptor ligands from homology models. J Med Chem 58(24):9578–9590. https://doi.org/10.1021/acs.jmedchem.5b01120
80. Yang X, van Veldhoven JPD, Offringa J et al (2019) Development of covalent ligands for G protein-coupled receptors: a case for the human adenosine A3 receptor. J Med Chem 62(7): 3539–3552. https://doi.org/10.1021/acs.jmedchem.8b02026
81. Salmaso V, Jacobson KA (2020) In silico drug design for purinergic GPCRs: overview on molecular dynamics applied to adenosine and P2Y receptors. Biomol Ther 10(6). https://doi.org/10.3390/biom10060812
82. Kim SK, Jacobson KA (2006) Computational prediction of homodimerization of the A3 adenosine receptor. J Mol Graph Model 25(4):549–561. https://doi.org/10.1016/j.jmgm.2006.03.003
83. Hübner H, Schellhorn T, Gienger M et al (2016) Structure-guided development of heterodimer-selective GPCR ligands. Nat Commun 7(1):12298. https://doi.org/10.1038/ncomms12298
84. Baltoumas FA, Theodoropoulou MC, Hamodrakas SJ (2016) Molecular dynamics simulations and structure-based network analysis reveal structural and functional aspects of G-protein coupled receptor dimer interactions. J Comput Aided Mol Des 30(6):489–512. https://doi.org/10.1007/s10822-016-9919-y
85. Navarro G, Cordomí A, Casadó-Anguera V et al (2018) Evidence for functional pre-coupled complexes of receptor heteromers and adenylyl cyclase. Nat Commun 9(1):1242. https://doi.org/10.1038/s41467-018-03522-3
86. Huang J, Chen S, Zhang JJ et al (2013) Crystal structure of oligomeric beta1-adrenergic G protein-coupled receptors in ligand-free basal state. Nat Struct Mol Biol 20(4):419–425. https://doi.org/10.1038/nsmb.2504
87. Vass M, Kooistra AJ, Yang D et al (2018) Chemical diversity in the G protein-coupled receptor superfamily. Trends Pharmacol Sci 39(5):494–512. https://doi.org/10.1016/j.tips.2018.02.004
88. Albertini C, Salerno A, de Sena Murteira Pinheiro, P. et al (2021) From combinations to multitarget-directed ligands: a continuum in Alzheimer's disease polypharmacology. Med Res Rev 41(5):2606–2633. https://doi.org/10.1002/med.21699
89. Henchcliffe C, Schumacher HC, Burgut FT (2005) Recent advances in Parkinson's disease therapy: use of monoamine oxidase inhibitors. Expert Rev Neurother 5(6):811–821. https://doi.org/10.1586/14737175.5.6.811
90. Fernandez HH, Chen JJ (2007) Monoamine oxidase-B inhibition in the treatment of Parkinson's disease. Pharmacotherapy 27(12P2):174S–185S. https://doi.org/10.1592/phco.27.12part2.174S
91. Caslake R, Macleod A, Ives N et al (2009) Monoamine oxidase B inhibitors versus other dopaminergic agents in early Parkinson's disease. Cochrane Database Syst Rev 4:Cd006661. https://doi.org/10.1002/14651858.CD006661.pub2

92. Jörg M, Scammells PJ, Capuano B (2014) The dopamine D2 and adenosine A2A receptors: past, present and future trends for the treatment of Parkinson's disease. Curr Med Chem 21(27):3188–3210. https://doi.org/10.2174/1389200215666140217110716
93. Stößel A, Schlenk M, Hinz S et al (2013) Dual targeting of adenosine A(2A) receptors and monoamine oxidase B by 4H-3,1-benzothiazin-4-ones. J Med Chem 56(11):4580–4596. https://doi.org/10.1021/jm400336x
94. Chen JF, Steyn S, Staal R et al (2002) 8-(3-Chlorostyryl)caffeine may attenuate MPTP neurotoxicity through dual actions of monoamine oxidase inhibition and A2A receptor antagonism. J Biol Chem 277(39):36040–36044. https://doi.org/10.1074/jbc.M206830200
95. Pretorius J, Malan SF, Castagnoli Jr N et al (2008) Dual inhibition of monoamine oxidase B and antagonism of the adenosine A(2A) receptor by (E,E)-8-(4-phenylbutadien-1-yl)caffeine analogues. Bioorg Med Chem 16(18):8676–8684. https://doi.org/10.1016/j.bmc.2008.07.088
96. Brunschweiger A, Koch P, Schlenk M et al (2014) 8-Benzyltetrahydropyrazino[2,1-f] purinediones: water-soluble tricyclic xanthine derivatives as multitarget drugs for neurodegenerative diseases. ChemMedChem 9(8):1704–1724. https://doi.org/10.1002/cmdc.201402082
97. Brunschweiger A, Koch P, Schlenk M et al (2016) 8-substituted 1,3-dimethyltetrahydropyrazino[2,1-f]purinediones: water-soluble adenosine receptor antagonists and monoamine oxidase B inhibitors. Bioorg Med Chem 24(21):5462–5480. https://doi.org/10.1016/j.bmc.2016.09.003
98. Hubálek F, Binda C, Khalil A et al (2005) Demonstration of isoleucine 199 as a structural determinant for the selective inhibition of human monoamine oxidase B by specific reversible inhibitors. J Biol Chem 280(16):15761–15766. https://doi.org/10.1074/jbc.M500949200
99. Hagenow S, Affini A, Pioli EY et al (2021) Adenosine A2AR/A1R antagonists enabling additional H3R antagonism for the treatment of Parkinson's disease. J Med Chem 64(12): 8246–8262. https://doi.org/10.1021/acs.jmedchem.0c00914
100. Kalash L, Val C, Azuaje J et al (2017) Computer-aided design of multi-target ligands at A1R, A2AR and PDE10A, key proteins in neurodegenerative diseases. J Cheminform 9(1):67. https://doi.org/10.1186/s13321-017-0249-4
101. Niccolini F, Foltynie T, Reis Marques T et al (2015) Loss of phosphodiesterase 10A expression is associated with progression and severity in Parkinson's disease. Brain 138(Pt 10): 3003–3015. https://doi.org/10.1093/brain/awv219
102. Giampà C, Patassini S, Borreca A et al (2009) Phosphodiesterase 10 inhibition reduces striatal excitotoxicity in the quinolinic acid model of Huntington's disease. Neurobiol Dis 34(3): 450–456. https://doi.org/10.1016/j.nbd.2009.02.014
103. Schmidt CJ, Chapin DS, Cianfrogna J et al (2008) Preclinical characterization of selective phosphodiesterase 10A inhibitors: a new therapeutic approach to the treatment of schizophrenia. J Pharmacol Exp Ther 325(2):681. https://doi.org/10.1124/jpet.107.132910
104. Kampen S, Duy Vo D, Zhang X et al (2021) Structure-guided design of G-protein-coupled receptor polypharmacology. Angew Chem Int Ed Engl 60(33):18022–18030. https://doi.org/10.1002/anie.202101478
105. Wang M, Hou S, Wei Y et al (2021) Discovery of novel dual adenosine A1/A2A receptor antagonists using deep learning, pharmacophore modeling and molecular docking. PLoS Comput Biol 17(3):e1008821. https://doi.org/10.1371/journal.pcbi.1008821
106. Yu J, Ahn S, Kim HJ et al (2017) Polypharmacology of N(6)-(3-Iodobenzyl)adenosine-5'-N-methyluronamide (IB-MECA) and related A3 adenosine receptor ligands: peroxisome proliferator activated receptor (PPAR) gamma partial agonist and PPARdelta antagonist activity suggests their antidiabetic potential. J Med Chem 60(17):7459–7475. https://doi.org/10.1021/acs.jmedchem.7b00805
107. Gregg A, Bottle SE, Devine SM et al (2007) Dual acting antioxidant A1 adenosine receptor agonists. Bioorg Med Chem Lett 17(19):5437–5441. https://doi.org/10.1016/j.bmcl.2007.07.035

108. Hausler NE, Devine SM, McRobb FM et al (2012) Synthesis and pharmacological evaluation of dual acting antioxidant A(2A) adenosine receptor agonists. J Med Chem 55(7):3521–3534. https://doi.org/10.1021/jm300206u
109. Varano F, Catarzi D, Vincenzi F et al (2020) Structural investigation on thiazolo[5,4-d] pyrimidines to obtain dual-acting blockers of CD73 and adenosine A2A receptor as potential antitumor agents. Bioorg Med Chem Lett 30(9):127067. https://doi.org/10.1016/j.bmcl.2020.127067
110. Ceni C, Catarzi D, Varano F et al (2020) Discovery of first-in-class multi-target adenosine A2A receptor antagonists-carbonic anhydrase IX and XII inhibitors. 8-Amino-6-aryl-2-phenyl-1,2,4-triazolo [4,3-a]pyrazin-3-one derivatives as new potential antitumor agents. Eur J Med Chem 201:112478. https://doi.org/10.1016/j.ejmech.2020.112478
111. Yan W, Ling L, Wu Y et al (2021) Structure-based design of dual-acting compounds targeting adenosine A2A receptor and histone deacetylase as novel tumor immunotherapeutic agents. J Med Chem 64(22):16573–16597. https://doi.org/10.1021/acs.jmedchem.1c01155
112. Petrelli R, Scortichini M, Kachler S et al (2017) Exploring the role of N(6)-substituents in potent dual acting 5'-C-ethyltetrazolyladenosine derivatives: synthesis, binding, functional assays, and antinociceptive effects in mice nabla. J Med Chem 60(10):4327–4341. https://doi.org/10.1021/acs.jmedchem.7b00291
113. Petrelli R, Torquati I, Kachler S et al (2015) 5'-C-ethyl-tetrazolyl-N(6)-substituted adenosine and 2-chloro-adenosine derivatives as highly potent dual acting A1 adenosine receptor agonists and A3 adenosine receptor antagonists. J Med Chem 58(5):2560–2566. https://doi.org/10.1021/acs.jmedchem.5b00074
114. Liu X, Ye K, van Vlijmen HWT et al (2021) DrugEx v2: de novo design of drug molecules by pareto-based multi-objective reinforcement learning in polypharmacology. J Cheminform 13(1):85. https://doi.org/10.1186/s13321-021-00561-9
115. Liu X, Ye K, van Vlijmen HWT et al (2019) An exploration strategy improves the diversity of de novo ligands using deep reinforcement learning: a case for the adenosine A2A receptor. J Cheminform 11(1):35. https://doi.org/10.1186/s13321-019-0355-6
116. Dalpiaz A, Cacciari B, Vicentini CB et al (2012) A novel conjugated agent between dopamine and an A2A adenosine receptor antagonist as a potential anti-Parkinson multitarget approach. Mol Pharm 9(3):591–604. https://doi.org/10.1021/mp200489d
117. Szlenk CT, Gc JB, Natesan S (2019) Does the lipid bilayer orchestrate access and binding of ligands to transmembrane orthosteric/allosteric sites of G protein-coupled receptors? Mol Pharmacol 96(5):527–541. https://doi.org/10.1124/mol.118.115113
118. Sykes DA, Stoddart LA, Kilpatrick LE et al (2019) Binding kinetics of ligands acting at GPCRs. Mol Cell Endocrinol 485:9–19. https://doi.org/10.1016/j.mce.2019.01.018
119. Jörg M, Kaczor AA, Mak FS et al (2014) Investigation of novel ropinirole analogues: synthesis, pharmacological evaluation and computational analysis of dopamine D2 receptor functionalized congeners and homobivalent ligands. Med Chem Commun 5(7):891–898. https://doi.org/10.1039/c4md00066h
120. Kaczor AA, Jörg M, Capuano B (2016) The dopamine D2 receptor dimer and its interaction with homobivalent antagonists: homology modeling, docking and molecular dynamics. J Mol Model 22(9):203. https://doi.org/10.1007/s00894-016-3065-2
121. Antony J, Kellershohn K, Mohr-Andra M et al (2009) Dualsteric GPCR targeting: a novel route to binding and signaling pathway selectivity. FASEB J 23(2):442–450. https://doi.org/10.1096/fj.08-114751
122. Bradley SJ, Molloy C, Bundgaard C et al (2018) Bitopic binding mode of an M1 muscarinic acetylcholine receptor agonist associated with adverse clinical trial outcomes. Mol Pharmacol 93(6):645–656. https://doi.org/10.1124/mol.118.111872
123. Narlawar R, Lane JR, Doddareddy M et al (2010) Hybrid ortho/allosteric ligands for the adenosine A(1) receptor. J Med Chem 53(8):3028–3037. https://doi.org/10.1021/jm901252a

124. Valant C, May LT, Aurelio L et al (2014) Separation of on-target efficacy from adverse effects through rational design of a bitopic adenosine receptor agonist. Proc Natl Acad Sci U S A 111(12):4614–4619. https://doi.org/10.1073/pnas.1320962111
125. Aurelio L, Valant C, Flynn BL et al (2009) Allosteric modulators of the adenosine A1 receptor: synthesis and pharmacological evaluation of 4-substituted 2-amino-3-benzoylthiophenes. J Med Chem 52(14):4543–4547. https://doi.org/10.1021/jm9002582
126. Baltos JA, Gregory KJ, White PJ et al (2016) Quantification of adenosine A(1) receptor biased agonism: implications for drug discovery. Biochem Pharmacol 99:101–112. https://doi.org/10.1016/j.bcp.2015.11.013
127. Aurelio L, Baltos JA, Ford L et al (2018) A structure-activity relationship study of bitopic N(6)-substituted adenosine derivatives as biased adenosine A1 receptor agonists. J Med Chem 61(5):2087–2103. https://doi.org/10.1021/acs.jmedchem.8b00047
128. Deganutti G, Barkan K, Ladds G et al (2021) Multisite model of allosterism for the adenosine A1 receptor. J Chem Inf Model 61(4):2001–2015. https://doi.org/10.1021/acs.jcim.0c01331
129. Vecchio EA, Chuo CH, Baltos JA et al (2016) The hybrid molecule, VCP746, is a potent adenosine A2B receptor agonist that stimulates anti-fibrotic signalling. Biochem Pharmacol 117:46–56. https://doi.org/10.1016/j.bcp.2016.08.007
130. Gao ZG, Toti KS, Campbell R et al (2020) Allosteric antagonism of the A2A adenosine receptor by a series of bitopic ligands. Cell 9(5). https://doi.org/10.3390/cells9051200
131. Baker JG, Middleton R, Adams L et al (2010) Influence of fluorophore and linker composition on the pharmacology of fluorescent adenosine A1 receptor ligands. Br J Pharmacol 159(4): 772–786. https://doi.org/10.1111/j.1476-5381.2009.00488.x
132. Sun B, Bachhawat P, Chu ML et al (2017) Crystal structure of the adenosine A2A receptor bound to an antagonist reveals a potential allosteric pocket. Proc Natl Acad Sci U S A 114(8): 2066–2071. https://doi.org/10.1073/pnas.1621423114
133. Lebon G, Warne T, Edwards PC et al (2011) Agonist-bound adenosine A2A receptor structures reveal common features of GPCR activation. Nature 474(7352):521–525. https://doi.org/10.1038/nature10136
134. Xu F, Wu H, Katritch V et al (2011) Structure of an agonist-bound human A2A adenosine receptor. Science 332(6027):322–327. https://doi.org/10.1126/science.1202793
135. Dore AS, Robertson N, Errey JC et al (2011) Structure of the adenosine A(2A) receptor in complex with ZM241385 and the xanthines XAC and caffeine. Structure 19(9):1283–1293. https://doi.org/10.1016/j.str.2011.06.014
136. Lebon G, Edwards PC, Leslie AG et al (2015) Molecular determinants of CGS21680 binding to the human adenosine A2A receptor. Mol Pharmacol 87(6):907–915. https://doi.org/10.1124/mol.114.097360
137. Amelia T, van Veldhoven JPD, Falsini M et al (2021) Crystal structure and subsequent ligand design of a nonriboside partial agonist bound to the adenosine A2A receptor. J Med Chem 64(7):3827–3842. https://doi.org/10.1021/acs.jmedchem.0c01856

Top Med Chem (2023) 41: 223–274
https://doi.org/10.1007/7355_2022_156

Published online: 10 December 2022

Allosteric Modulators of Adenosine Receptors

Elisabetta Barresi, Chiara Giacomelli, Claudia Martini, Federico Da Settimo, Maria Letizia Trincavelli, and Sabrina Taliani

Contents

Abstract Allosterism occurs when a ligand binds to a receptor on a binding site distinct from the orthosteric one to which endogenous ligand normally binds. Among the G protein-coupled receptors (GPCRs), the adenosine receptors (ARs), classified as A_1, A_{2A}, A_{2B}, and A_3 are the first known to be allosterically regulated. The first part of this chapter has focused on the mechanism behind the allosteric modulation on GPCRs, and specifically on ARs. To date, three main mechanisms have been recognized, namely: (1) the allosteric transition, due to the transition from the inactive to the active state; (2) the receptor dimerization; and (3) the ligand of exogenous or endogenous ligands. In the second part, an overview on the medicinal

Elisabetta Barresi and Chiara Giacomelli contributed equally to this work.

E. Barresi, C. Giacomelli, C. Martini, F. Da Settimo, M. L. Trincavelli, and S. Taliani (✉)
Department of Pharmacy, University of Pisa, Pisa, Italy
e-mail: elisabetta.barresi@unipi.it; chiara.giacomelli@unipi.it; claudia.martini@unipi.it; federico.dasettimo@unipi.it; letizia.trincavelli@unipi.it; sabrina.taliani@unipi.it

chemistry of adenosine allosteric modulators, in terms of chemical structure and structure–activity relationships, has been reported, highlighting the most outstanding advances in the development of AR modulators endowed with therapeutic potential.

Keywords Adenosine receptor (AR), Allosteric mechanism, Allosteric modulator, Bias allosteric modulator (BAM), G protein-coupled receptors (GPCRs)

1 Introduction

G protein-coupled receptors (GPCRs) represent the largest family of membrane proteins encoded by the human genome [1]. GPCRs are implicated in several physiological processes and their dysregulation causes the initiation and progression of several pathologies. Thus, GPCRs have been a pivotal target in drug discovery with more than 700 drugs targeting 134 receptors approved by the US Food and Drug Administration (FDA) representing around the 35% of all approved drugs [2]. All the GPCRs share a common seven hydrophobic transmembrane (7TMD; TM1-TM7) domain, with a higher variability reported for both the three extracellular loops (EL1-EL3) of the extracellular domain (ECD) and the three intracellular loops (IL1-IL3) of the intracellular domain (ICD) [3].

GPCRs mediate the signaling of a wide variety of extracellular stimuli including small molecules, ions, lipids, proteins, hormones, and metabolites through the binding on the extracellular ligand-binding pockets highlighting the reason for their high variability. Of note, the EL2 is involved not only in ligand recognition and receptor subtype selectivity, but it also plays an important role in receptor activation and ligand efficacy [4, 5]. GPCRs transmit the external signals to the intracellular side of the receptors promoting the coupling with a heterotrimeric G-proteins. The scenario is more complex considering the role of regulatory processes, such as β-arrestin recruitment, receptor phosphorylation, and desensitization, that modulate the GPCR signal transduction. Furthermore, β-arrestins per se act as transducers activating alternative signaling pathways [6]. The importance of this alternative signaling has prompted [3] to propose the change of GPCR in favor of the term 7 transmembrane receptors (7TMR). Conversely, the ECD is linked with the G-protein-coupling region by evolutionarily conserved sequences such as the CWxP motif, the PIF motif; the sodium binding site, the NPxxY motif in TM7, and the E/DRY motif at the border between TM3 and IL2 [7]. During the last decades, the research on the structure of GPCRs has grown mainly owing to breakthroughs in computational biology, protein engineering, lipid-based crystallography, X-ray diffraction, and cryo-electron microscopy (cryo-EM) that prompt new insights on the binding site of endogenous and exogenous molecules and the features related to the GPCR activation stages [8].

The great importance of this family of receptors and their physiological role in eukaryotic cells have been acknowledged by the scientific community with the award of the 2012 Nobel Prize in chemistry to Robert J. Lefkowitz and Brian K. Kobilka "for studies of G protein-coupled receptors." This award shows how the study of these receptors has contributed and is still helping the development of new drugs interacting with this family of proteins.

For decades, the development of ligands in traditional GPCR drug discovery has focused on targeting the endogenous agonist-binding (orthosteric) site of the receptor. However, the orthosteric sites of different receptor isoforms often display high homology limiting the drug selectivity. Accordingly, the interest in the discovery and development of molecules targeting topographically distinct allosteric sites on GPCRs is growing [9]. Ligands that bind to these allosteric sites are called "allosteric modulators" and offer higher selectivity of action and responsiveness in a spatially and temporally advantageous manner [10].

2 Evolution of the Allosteric Models

An allosteric effect was first described in 1904 by Christian Bohr demonstrating the ability of carbon dioxide to affect the binding affinity of oxygen to hemoglobin [11]. However, the term "allosteric" (from the Greek meaning "other site") was first used in 1961 by Jacques Monod and Francois Jacob [12] describing the mechanism of a ligand to inhibit an enzymatic activity as "the inhibitor is not a steric analogue of the substrate."

The simplest mechanism of the allosteric modulator of GPCRs is the coupling with the G-protein itself. All the GPCRs present at least two binding sites, one for the endogenous ligand and one for the G-protein. This represents the simplest model for a positive allosteric interaction [13, 14]. Indeed, endogenous agonist binding modifies the receptor conformation increasing the affinity to the G-protein, thus favoring the coupling process [15]. Conversely, the binding of GTP to the G-protein alters the G-protein structure causing a negatively cooperative effect on agonist binding, thus promoting the un-coupling of the activated G-protein from the receptor.

The interest in the development of agonist, antagonist, and/or allosteric ligands of GPCR has prompted the development of mathematical models to describe the affinity and efficacy parameters due to the ligand–receptor interaction (Fig. 1). Generally, if one considers a 1:1 binding interaction, the affinity (Kd) of a ligand is defined as the ratio between the rate of association and dissociation constant (k_{on}/k_{off}) (Fig. 1a). This simple model was refined in the original ternary complex model in which the binding of the ligand allows the coupling with the G-protein complex resulting in the activation, as described by De Lean et al. [16] (Fig. 1b). The demonstration that GPCR could couple to G-protein also in the absence of endogenous agonist highlights the plausibility of a constitutive GPCR activity [17]. This evidence leads to the development of the two-state model of receptor activation [18] as well as the necessity to improve the De Lean model resulting in the two-state model (Fig. 1c).

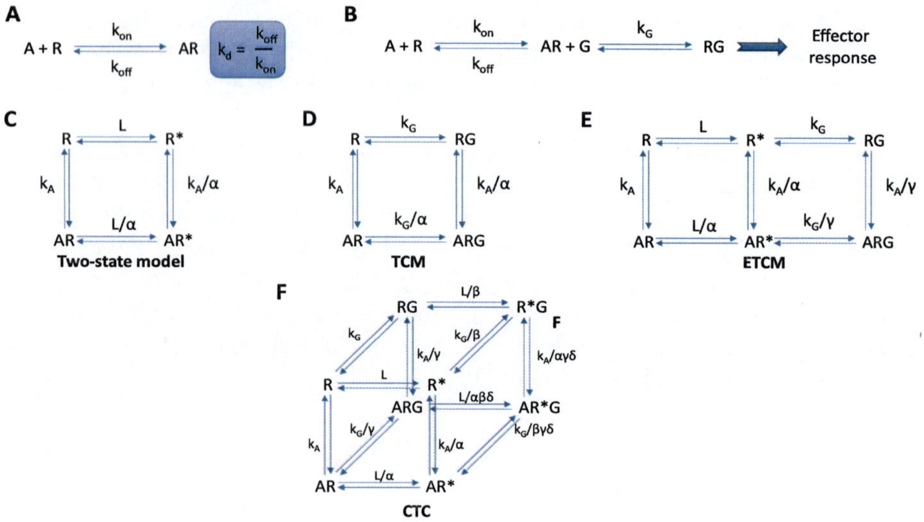

Fig. 1 Mathematical models to describe ligand-receptor binding parameters. (**a**) The affinity of a ligand (A) for a receptor (R). (**b**) Original ternary complex model considering the coupling with the G-protein. (**c**) The two-state model of receptor activation. (**d**) The ternary complex model (TCM). (**e**) The extended ternary complex model (ETCM). (**f**) The cubic ternary complex model (CTC)

The receptors exist in an equilibrium between an inactive (R) and active (R*) state defined by the isomerization constant (I). In this model, the agonist shifts the equilibrium preferentially to the active state, whereas the inverse agonist shifts the equilibrium preferentially to the inactive state. The "efficacy" of a ligand is determined by its preference for binding the active over the inactive form of the receptor (α). Values of α greater than one describe a shift toward the active state; conversely, values lower than one stabilize the inactive state. Similarly, the model can also describe the effects that GPCR ligands and G-proteins had on each other in the so-called ternary complex model (Fig. 1d). This represents the simplest probabilistic models that describe the actions of agonists, antagonists, and inverse agonists [18–20].

The "extended ternary complex model" (ETCM Fig. 1e) combines the two-state model and the TCM model to consider both the effect of G-protein coupling and the isomerization of the receptor [21]. However, also the ETCM is thermodynamically incomplete, and Weiss et al. [22] develop the "cubic ternary complex model" (CTC, Fig. 1f).

The main limitation of these models derives from the limitation of the two-state model. Most of the evidence is based on the GPCR high-resolution structures that are only snapshots of a dynamic system. The solved structures are the most stable conformations obtained using detergents, fusion proteins, or mutations, limiting the dynamics of the receptor. These limitations have been in part overcome by the use of NMR spectroscopy and molecular dynamics simulations that allow us to

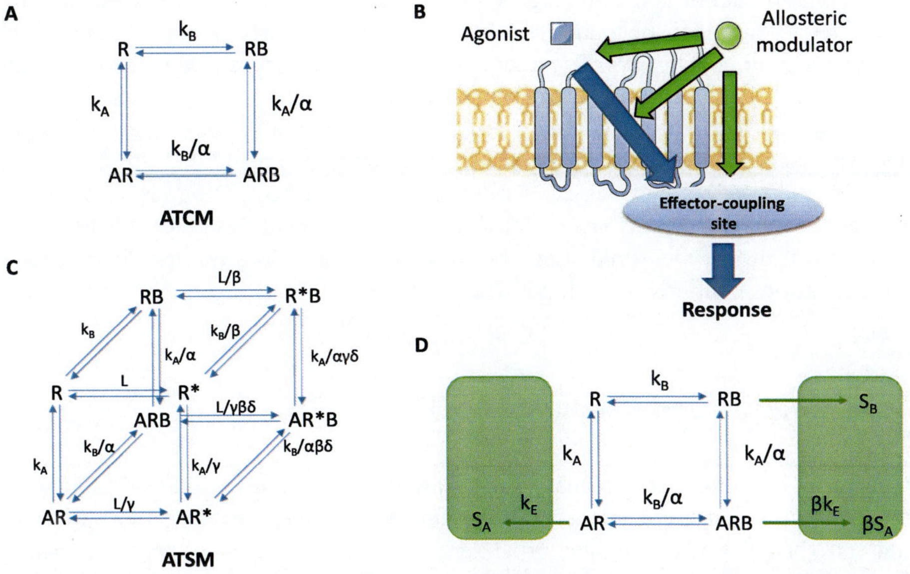

Fig. 2 (**a**) The allosteric ternary complex model (ATCM). (**b**) Effects of orthosteric and allosteric ligands of GPCR: range of possible activities. (**c**) The allosteric two-state model (ATSM). (**d**) The schematic combination of the simple ATCM with an operational model of agonism

disclose the GPCR plasticity and demonstrate the existence of several active and inactive conformations [23].

Based on its nature, an allosteric modulator can bind the GPCR modifying the affinity of the orthosteric ligand. Thus, the TCM has been applied to describe an allosteric interaction of two ligands between distinct binding sites on a receptor. This is commonly referred to as the "allosteric ternary complex model" (ATCM) (Fig. 2a) [24–26]. In this model, the binding cooperativity (α) defines the ratio of the dissociation constants of the free receptor to that of the occupied receptor and represents the modulation of the affinity of each other. α values greater than 1 represent an enhancement of affinity (positive cooperativity), α values minor than 1 a reduction of affinity (negative cooperativity), and a value of 1 denotes no effect on binding affinity at equilibrium (neutral cooperativity).

The ATCM model fails to describe the effects of allosteric modulators on orthosteric ligand efficacy (Fig. 2b) [27–31]. For example, the allosteric modulator, CGP7930 (**1**), enhances both the potency and the maximal effect of $GABA_B$ receptor agonist [28], without affecting the binding of the orthosteric antagonist [^{3}H] CGP62349.

To model the effects on both orthosteric binding parameters, the ATCM was extended by Hall [32] as the "allosteric two-state model" (ATSM) (Fig. 2c). However, the complexity of this model limited its application to the experimental data, and an alternative approach has been based on the combination of the ATCM with an operational model of agonism (Fig. 2d) [33]. The model also takes into account the effect (E), which can be quantified by measuring the modulation of one of the receptor signaling pathways. Together with the α parameters related to the modulator on orthosteric agonist affinity, additional parameters, such as the operational efficacy (τ_B) and β, are introduced. β values represent the modulation degree of orthosteric agonist efficacy. $\alpha\beta > 1$ shows a positive modulation, values of $0 < \alpha\beta < 1$ show a negative modulation, whereas $\alpha\beta = 1$ defines a neutral effect. The τ_B is related to the K_E and the receptor density ($\tau_B = [R]/K_E$). The limitation of this model resides in the dependence on the measured signal-pathway and that it does not take into account the strength of coupling to each pathway.

3 Allosteric Modulation of GPCR

Generally, the allosteric modulators are classified into "positive allosteric modulators" (PAMs), which enhance orthosteric agonist activity, "negative allosteric modulators" (NAMs), which reduce orthosteric agonist activity, and "neutral allosteric ligands" (NALs). NALs have no effects on orthosteric agonist at the equilibrium but bind to the allosteric site competing with the PAM or NAMs decreasing their activity [34]. PAMs can increase agonist binding (α >1 and/or agonist efficacy (β >1), NAMs can decrease agonist binding (α <1) and/or agonist efficacy (β <1).

Some PAMs or NAMs may exhibit a mixed mode of action acting as both allosteric and orthosteric ligands. These ligands are known as ago-allosteric modulators (ago-PAM or ago-NAM). Several examples have been reported for GPCR, such as A_1 adenosine receptor ago-PAM [35], $GABA_B$ receptor ago-PAM [36], the M_1 receptor ago-PAM [37], or glucagon-like peptide-1 receptor ago-PAM [38]. Allosteric modulators can also enrich the active-state GPCR conformations in their own right, expressing the property of agonism in the absence of bound orthosteric ligand [39].

Recently, hybrid ligands also called bitopic ligands have been reported. These are characterized by the presence of two pharmacophores, one that binds the orthosteric site connected with a linker to a second scaffold able to bind the allosteric site [40]. Several ligands have been developed in the last two decades such as the VCP746 (**2**) A_1 bitopic ligand [41]; the VU0364572 (**3**) and VU0357017 (**4**) M_1 ligands [42, 43]. Bitopic ligands targeting dimeric GPCR receptors have been also developed [44]. Of interest, it has been recently reported a ligand able to bind the A_{2A}-D_2 receptor heterodimer that enhances dimer stability [45]. Despite the bitopic ligands can improve the specificity and the site direct activity, they generally have a large structure that decreases their drug-like physicochemical properties.

3.1 Methodological Aspect for Allosteric Ligand Characterization

One of the unique characteristics associated with the allosteric modulators is the so-called probe dependence. The magnitude and the effects on different signaling pathways can change based on the nature of the orthosteric ligand used as a probe in the experimental evaluation [46]. An example of this property has been reported for the A_{2B} adenosine receptor (AR) PAMs that differently modify the potency of A_{2B} AR agonists (NECA, BAY 60-6583, and adenosine) [47].

The investigations on the allosteric properties of a new ligand are tricky. The concept that agonist kinetics can be linked to its efficacy was first proposed more than 60 years ago. Despite affinity is an equilibrium parameter, it is defined by the ratio between two kinetic parameters, the association and dissociation constants, highlighting the need to apply kinetic evaluation to assess the allosteric properties of new ligands. Allosteric modulators can modify only the association or dissociation constant or both highlighting their allosteric nature. However, the binding site of the allosteric ligand can be the same as the orthosteric or a spatially distinct site. The association constant per se can vary in both cases hampering its use as a parameter to describe the exact mechanism of action of new modulators. Conversely, the dissociation rate can be modified only if a ligand binds a distinct site causing a conformational change. In fact, most of the literature on PAMs and NAMs reports dissociation kinetic studies (Fig. 3).

Despite the use of kinetic experiments to characterize allosteric modulators being widely accepted, some limitations need to be highlighted. The alteration of dissociation kinetic can support the allosteric nature of the ligand; conversely, the lack of alteration is not resolutive to discard an allosteric effect. For example, the allosteric modulator can exert its effects predominantly on orthosteric agonist efficacy rather than affinity.

Of note, kinetic experiments have been also applied to assess cooperative interactions in dimers and oligomers of different types of GPCR such as adenosine and dopamine [49, 50]. However, the advance in fluorescence microscopy approach in live cells can prompt the research to better define the allosteric mechanism on different GPCRs.

3.2 GPCR Bias Allosteric Modulators (BAMs)

GPCRs transmit an extracellular input to the intracellular space. Specifically, the signal derived from the binding of an orthosteric ligand is translated to an intracellular transducer that is typically around 40 Å apart. GPCRs couple to different signaling heterotrimeric G-protein complexes (e.g., Gs, Gq, or Gi), are phosphorylated by different isoforms of G protein-coupled receptor kinases (GRKs), and bind one or more β-arrestins. It is accepted that orthosteric agonists can promote a specific

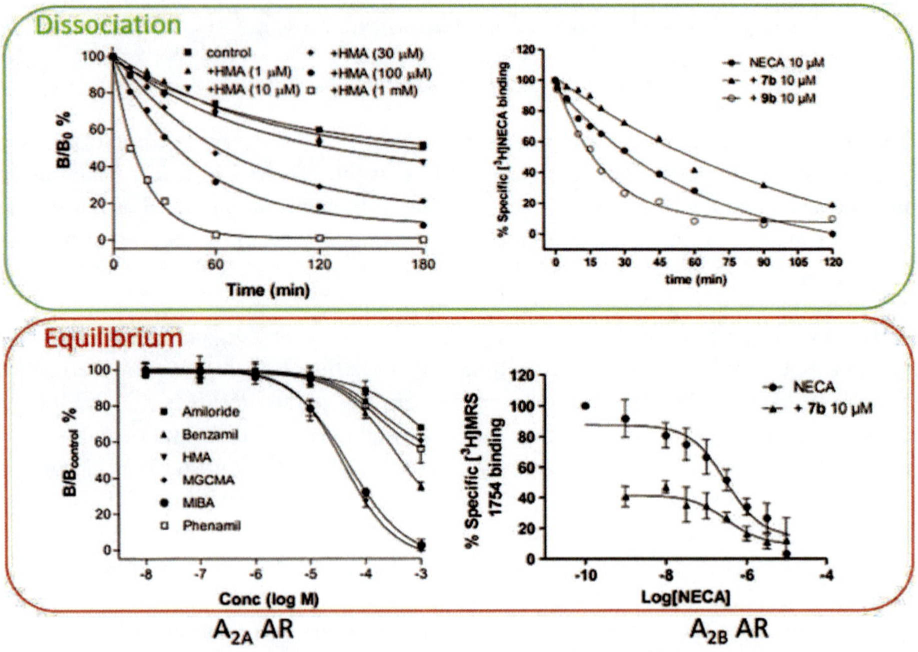

Fig. 3 Difference in the effects of allosteric modulator in kinetic or equilibrium conditions. Panel up left: 5-(*N*,*N*-hexamethylene)amiloride (HMA, **5**) is an NAM of the A_{2A} AR. After the pre-equilibration of [^{3}H]ZM241385, the dissociation was started by the addition of 1 mM CGS15943 (**6**) mixed with various concentrations of HMA. Panel down left: Concentration dependence modulation of tested compounds on specific [^{3}H]ZM241385 binding to A_{2A} AR. Images adapted with permission from Ref. [48]. Panel up right: After the pre-equilibration of [^{3}H]NECA, the dissociation was initiated by the addition of 10 μM NECA in the absence or presence of the tested compound. **7b** (compound **65b**, vide infra) and **9b** (compound **67b**, vide infra) are PAM and NAM of A_{2B} AR, respectively. Panel down right: Effects of 7b on the binding of the non-selective agonist NECA on [^{3}H]MRS 1754. Image adapted with permission from Ref. [47]

conformational state of the GPCR leading to a selectivity toward a specific; this phenomenon is called "functional selectivity" or "biased agonism" [51].

GPCR bound to a G-protein was crystallized only in 2011 (β2AR complex with Gs) [52], then several other solved structures of GPCR-G protein have been reported. Similarly, the crystal structure of a GPCR in complex with the β-arrestins has been also reported (β1AR–βarr1) [53]. It is well known that arrestins preferentially bind active phosphorylated GPCRs, so receptor phosphorylation by GRKs is a prerequisite. Despite the pivotal role of GRKs, less is known regarding the structural features of their interaction with GPCR. Independently of the molecular features, the phosphorylation of the GPCR C-terminal in specific conformation recognizes the C-tail of arrestin ("hanging conformation"), then the N- and C-terminal domains of arrestin rotate to interact with the intracellular core of the GPCR ("core conformation") [54, 55]. Thus, the pattern of GPCR phosphorylation addressed as the "phosphorylation barcode" [55] guides the fate of the engaged signaling pathways. Different ligands interacting with the GPCR produced diverse conformational changes modulating the initiation of different pathways and this is the basis of the bias agonism. Of note, the fully engaged arrestin-GPCR complex excludes the binding of the G-protein, producing a bias toward the β-arrestin-signaling. Similarly, the arrestin hanging conformation allows a simultaneous binding with the G-protein to the GPCR. It has been recently proposed a "megaplex" in which the G-protein binds the GPCR core region and simultaneously the β-arrestin binds to the phosphorylated C-terminal tail demonstrating the possibility of a sustained G-protein signaling after the internalization [56]. Despite this evidence further highlighting possible mechanisms of bias agonism, the role of the transducer has to be better determined in future studies.

What is known on bias agonism can be translated to the allosteric modulators that can engage allosteric binding promoting a pathway-specific selection; these molecules have been called "bias allosteric modulators" (BAMs) [14, 57]. A BAM lacks intrinsic activity and can affect or not the binding of the orthosteric ligand. It can direct the selectivity of an agonist without bias agonism features (Gs vs. β-arrestins; Fig. 4a). For example, a BAM can enhance selectively the Gs signaling showing a PAM activity (Fig. 4b), or it can reduce the β-arrestin signaling showing an NAM activity (Fig. 4c) or both (Fig. 4d). The ability of a BAM to exert a bias agonistic activity leads to the discovery of the so-called bias ago-allosteric modulators (ago-BAMs). They shift the agonist signaling in favor of a specific transducer, and alone were able to act as a bias agonist (Fig. 4e).

The new insight on BAM drug discovery has been recently reviewed by Slosky et al. [58]. BAMs represent a new frontier of drug discovery and development allowing the mitigation of the side effects of conventional orthosteric molecules. However, several challenges regarding their mechanism of action and their experimental evaluation have to be faced in the future years.

3.3 Advantages and Disadvantages of GPCR Allosteric Modulators

Despite one-third of approved drugs targeting GPCR, only a few of them act as allosteric modulators. Allosteric modulators of ligand-gated ion channels are well known, e.g. the benzodiazepines diazepam (Valium®), which enhances the GABA effect on $GABA_A$ receptor. However, only a few GPCR allosteric modulators have been approved: cinacalcet (Mimpara®), a PAM of the calcium-sensing receptor for the treatment of hyperparathyroidism; maraviroc (Selzentry®), an NAM of the chemokine receptor CXCR5 as a virus entry inhibitor in HIV therapy; and plerixafor (AMD3100), an NAM of the chemokine receptor CXCR4 for the mobilization of the stem cells [59].

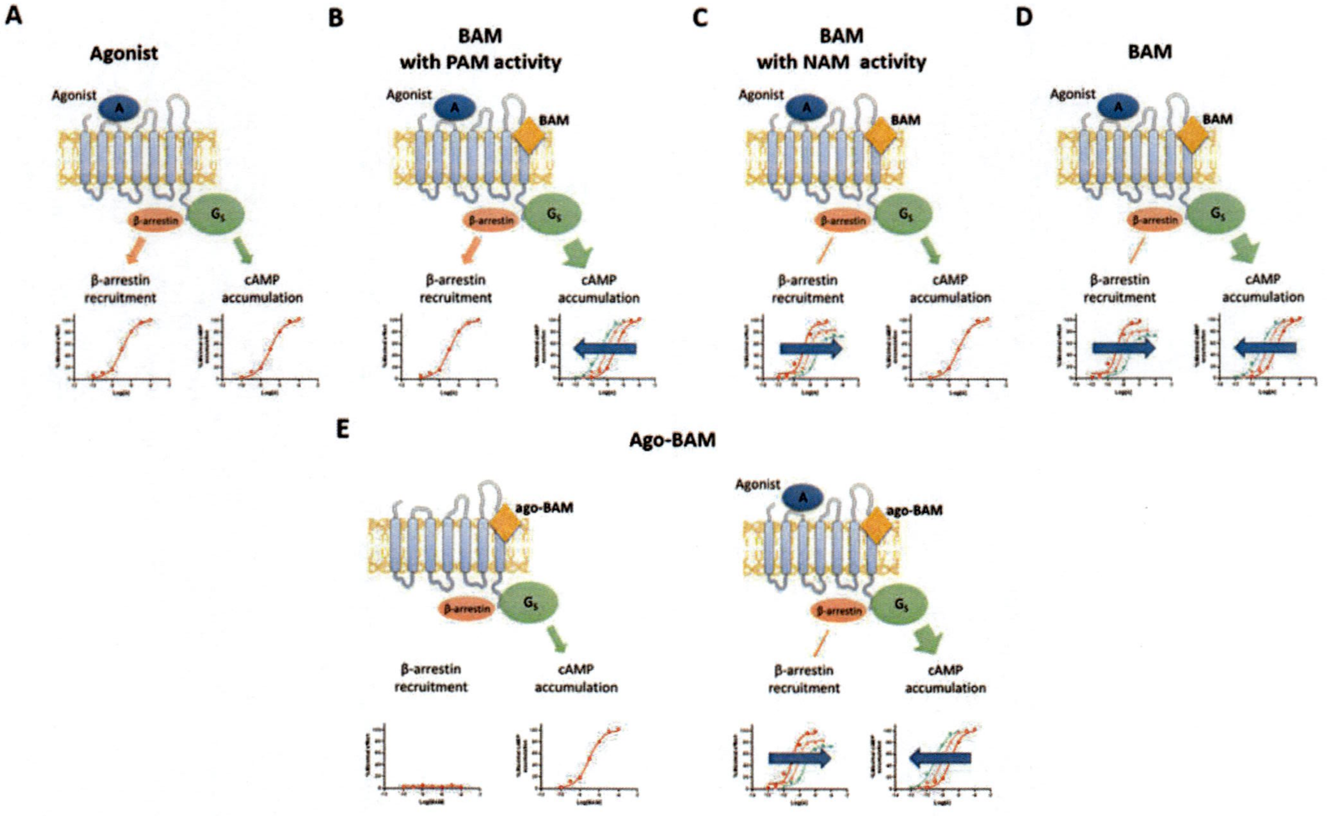

Fig. 4 A schematization of the different effects of a BAM and an ago-BAM on the signaling of a GPCRs coupled receptor

The use of allosteric modulators in favor of orthosteric ones has several advantages. Parameters such as affinity and residence times well describe the magnitude of action of orthosteric ligand. Thus, the therapeutic effects of such ligands are mainly related to the concentration on the receptor site independently of the fluctuations of the endogenous ligand levels. As a consequence, these ligands generate on-target side effects, promoting activation of the receptor in different compartments in which the concentration of the endogenous ligand is low. Furthermore, the prolonged activation of the GPCR causes their desensitization and receptor up-/down-regulation modifying their response to the drug. Orthosteric ligands have often also off-target side effects due to the high degree of conservation of the orthosteric binding site between the receptor subtypes.

Allosteric modulators may overcome some of such limitations. Allosteric modulators act only in the presence of the endogenous agonist with a high site- and event- selectivity. If an orthosteric ligand induces or inhibits a receptor response, the allosteric modulator makes a "fine-tuning" of the response. They preserve better the spatiotemporal "rhythms" of the endogenous ligand limiting long-term disruption of the signaling pathway. Overdose situations are limited by the "ceiling effect" of allosteric modulators due to their dependency on the presence of endogenous ligand; however, this advantage is missed in ago-allosteric modulators or ago-BAMs. Since allosteric sites are less conserved than orthosteric ones, the allosteric modulators generally are more selective reducing the off-target effects. Similarly, the outbreak of BAMs further highlights the possibility of selectivity target therapy-relevant signaling pathways reducing the side effects. Last but not least, allosteric modulators can improve the targeting of undruggable GPCR. In fact, the peculiar structure of these receptor binding sites or the high site occupancy from the endogenous ligand makes these receptors a not trivial target of orthosteric ligands [58].

Despite their advantages, allosteric modulators have also disadvantages [60]. First of all, the less knowledge of the allosteric sites hampered the advance in drug discovery. Furthermore, slight structural modifications of allosteric modulators may often change the nature of the allosteric effects (from PAM to NAM, or vice versa) leading to a non-tractable structure–activity relationship (SAR). Often, these modulators are poorly soluble in water limiting their progression in clinical trials. Finally, since the effects of allosteric modulators depend on the endogenous agonist concentration, their effects may vary during the disease progression [58, 61].

4 Mechanism of Adenosine Receptor (ARs) Allostery

Burnstock in 1976 first used the term "purinergic receptors." Adenosine receptors (ARs) belong to the P1 family and to date four receptor subtypes (A_1, A_{2A}, A_{2B}, and A_3) have been reported. All these receptors are activated by adenosine that is a ubiquitous purine nucleotide involved in several physiological and pathological functions [62]. There is a large sequence identity between the human A_1 and

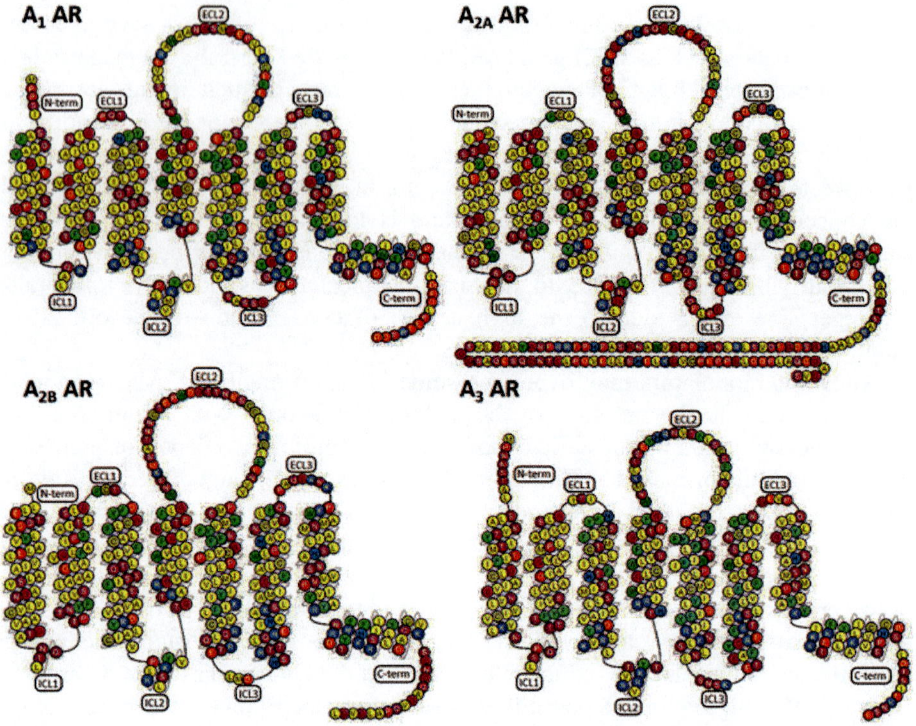

Fig. 5 Snake view of the human A_1, A_{2A}, A_{2B}, and A_3 AR sequences. The polarity of the residues is highlighted by different color. The images are generated by GPCRDB (https://gpcrdb.org/) [63]

A_3AR (≈49%) and the human A_{2A} and A_{2B}ARs (≈59%) whereas A_3 is more variable (Fig. 5).

Adenosine concentration is in the nanomolar range in physiological conditions and is able to activate only the so-called high-affinity subtypes (A_1 and A_{2A}AR). In hypoxic, ischemic, or inflammatory conditions, the adenosine intracellular levels arise to micromolar concentrations activating the so-called low-affinity subtypes (A_{2B} and A_3AR) [64, 65]. Of note, the receptor density may profoundly affect the magnitude of adenosine activity, as well as the bias agonism toward a specific intracellular signaling pathway [65].

The A_1AR and A_3AR subtypes are coupled with the G_i proteins inhibiting the adenylate cyclase, both increase the activity of PLC and calcium mobilization [66]. Conversely, A_{2A}AR and A_{2B}AR subtypes are coupled with a G_s protein increasing the activity of adenylyl cyclase [66]. The A_{2B}AR activates the PLC pathways through the coupling of $G_{q/11}$ proteins and MAP kinase [67, 68]. The A_{2A}AR also induces formation of inositol phosphates to raise intracellular calcium and activate protein kinase C via $G_{\alpha 15}$ and $G_{\alpha 16}$ proteins [69].

To date (May 2022), 61 solved structures of the A_{2A} AR and 5 structures of the A_1 AR have been recorded in the Protein Data Bank (PDB) [70]. Structures are

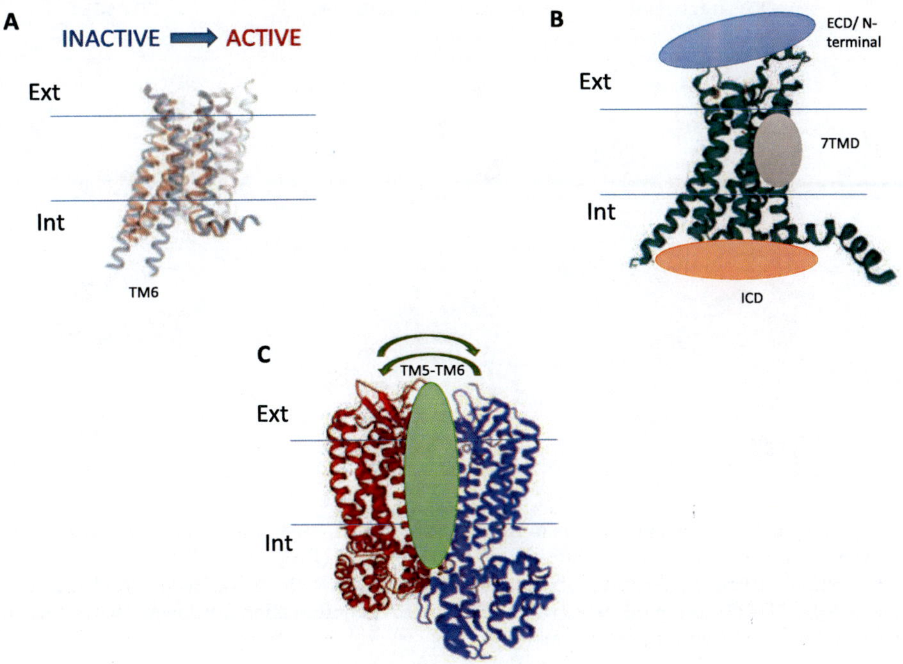

Fig. 6 Three main mechanisms of AR (GPCR) allosteric modulation. (**a**) Binding with an agonist causes a conformational change favoring the G-protein binding. Blue: (5NLX, PDB) inactive in complex with ZM-241385. Red: (6GDG, PDB) active in complex with NECA and coupled with NECA. Image generated with GPCDB (https://gpcrdb.org/) [63, 71]. (**b**) Site of interaction of exogenous and endogenous ligands. The extracellular domain (ECD), the seven transmembrane domain (7TMD), and intracellular domain (ICD) are evidenced on the crystal structure of human A_{2A} AR in complex with the agonist adenosine (2YDO, PDB) (3OE0, PDB). Image generated with PDB [70]. (**c**) Site of modulation in receptor–receptor interaction on the crystal structure of the dimeric CXCR4 chemokine receptor in complex with the antagonist CVX15 (3OE0, PDB). Image generated with PDB [70]

reported in the inactive state based on the complex with different antagonist and in the active state based on the complex with agonist. Of note, both subtypes are solved also in the complex with the respective G-protein highlighted the full active state of both receptors. Unfortunately, no structure derived by X-ray crystallography or cryo-electron microscopy (cryo-EM) for A_{2B}AR and A_3AR has been reported yet.

The availability of crystal structure of both active and inactive state of AR has opened the way to a better understanding of the mechanism of AR allosteric modulation. Three main mechanisms of allosteric modulation can be summarized in general for GPCR and specifically for the AR (Fig. 6): (1) the “allosteric transition,” due to the transition from the inactive to the active state; (2) the receptor dimerization; and (3) the binding of exogenous or endogenous ligands.

The solved structures of GPCR are only snapshots of a highly dynamic system that rarely can resemble the plethora of different conformations adopted by the

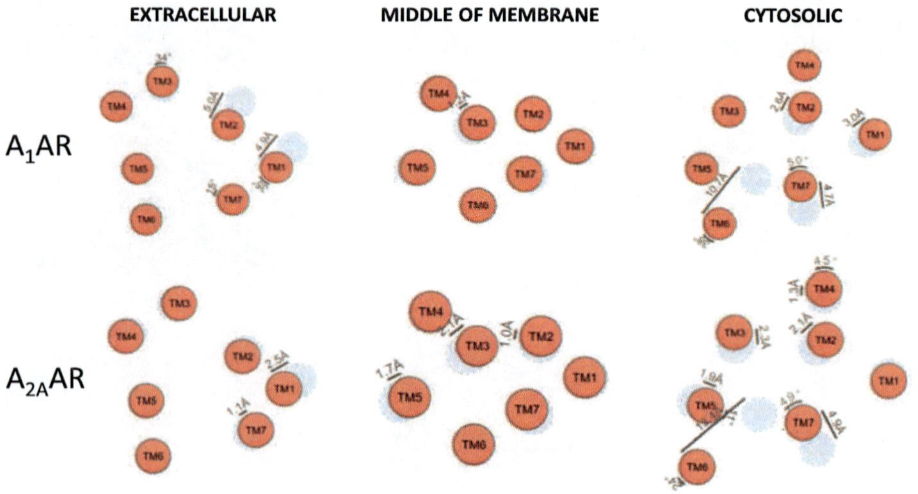

Fig. 7 The TM1-7 segment movement of the extracellular, middle of membrane, and cytosolic domains. For the A_1AR the modification from the active (Red, PDB: 7LD4) to the inactive (Blue, PDB: 5UEN) conformation is shown. For the $A_{2A}AR$ the modification from the active (Red, PDB: 6GDG) to the inactive (Blue, PDB: 5NLX) conformation is shown. The images are generated by GPCRDB (https://gpcrdb.org/) [63]

receptors. Interestingly, most of the GPCR crystal structures are in the inactive state. The structures of ARs (A_1 and A_{2A} AR) bound to both agonist and antagonist have been solved highlight different conformation in the inactive and active state. However, among the 66 crystal structures, only five are in complex with the Gi or Gs protein without structures in complex with the β-arrestin complex (https://gpcrdb.org/). The binding with the orthosteric ligands or the transducer is the first example of allostery in AR.

Several authors have deeply investigated the molecular features involved in AR activation [72]. The main movements interested in A_1AR and $A_{2A}R$ activation are the large outward movement of the TM6 and inward movement of TM7 (Fig. 7) that facilitate the opening of the pocket for the G-protein binding. Several "microswitches" have been implicated in AR activation such as the "salt bridge," the "rotamer toggle switch," and the "tyrosine toggle switch" that slightly differ between the two receptor subtypes. In the $A_{2A}R$, the tyrosine residue $Y288^{7.53}$ changes the conformation modifying the interaction of the TM2 and TM6. The Ballesteros–Weinstein numbering is used in which the "288" indicates the residue number, the "7" indicates the TM domain, and "53" indicates the residue in the TM helix. The "salt bridge" or "ion lock" in the $A_{2A}R$ occurs between the $D101^{3.50}$ and the $E228^{6.30}$. Another important residue in the $A_{2A}R$ activation is $D52^{2.50}$ that plays a crucial role in the allosteric transmission of the ligand binding and the activation of the signaling pathway [73]. Eddy et al. by the means of NMR analysis demonstrate that this residue mediates the dynamic interaction between the two faces of the GPCR coordinating the allosteric switch of $A_{2A}R$ activation.

Most of the solved structure, all for the AR, are in the presence of a ligand in the orthosteric or allosteric sites. The solved structure of the unbound receptor "apoGPCR" can improve the knowledge on the modification of the extracellular side of the GPCR. Regarding the AR, the binding of a ligand mainly involves the EL2 that also represents the less conserved domain among the different subtypes. Interestingly, modifications of the EL2 conformation stabilize the agonist-bound receptor highlighting its involvement in the allosteric modulation of extracellular ligand binding [72].

4.1 AR Homo- and Hetero-Dimerization as an Allosteric Mechanism

The ability of GPCR to establish receptor–receptor interaction is known since the 1980s [74]. The evolution of biochemical and biophysical techniques improves the knowledge of the molecular features of receptor–receptor interactions (RRIs) [75]. Notably, agonist binding promotes an outward movement of TM5 and TM6 (~ 14 Å), and several homomers or heteromers interact via the TM5/6 highlighting the physical interaction as a pivotal mechanism of allosteric modulation to modify the functional response of such complex.

A growing body of evidence demonstrates the ability of AR to form homodimers, heterodimers, and heterotetramers with each other and also with other receptors. In the G protein-coupled receptor heterodimer network (GPCR-HetNet, http://www.gpcr-hetnet.com/) [76], 11 interactions for the A_1AR and 12 interactions for the $A_{2A}R$ are reported. The tools only report experimental evidence of physical interactions; no data are available for the $A_{2B}AR$, and to date, only the interaction with the A_1AR is reported for the A_3AR subtypes. Recently, Franco et al. [77] reviewed the data on homodimers and heteromers formed by all the AR subtypes highlighting the functional relevance and the structural features of these interactions.

The prototypical AR heteromer is the $A_{2A}R$-D_2R, the binding of the $A_{2A}AR$ agonist on the dimer causes a negative allosteric modulation of D_2R signaling [78]. Interestingly, a reciprocal interaction also exists, as D_2R activation negatively modulates the $A_{2A}AR$ functional responsiveness [79]. Based on these interactions, the $A_{2A}AR$–D_2R heteromer has been considered a therapeutic target for schizophrenia, drug addiction, and neurodegenerative diseases such as Parkinson's disease. The use of $A_{2A}AR$ antagonists reduces the negative allosteric modulation of the D_2R protomer contributing to the antiparkinsonian effects [80, 81]. Of note, the $A_{2A}AR$ antagonist istradefylline is approved in the treatment of Parkinson's disease, however, evidence of the effects of $A_{2A}AR$ agonists and antagonists on motor function in dopamine D_2 receptor-deficient mice has been reported [82, 83] questioning the importance of the dimerization process.

4.2 Allosteric Modulation of AR by Sodium and Cholesterol

A pivotal example of an endogenous allosteric modulator is sodium ion (Na^+). The first high-resolution structure of a GPCR bind to the sodium was the $A_{2A}AR$ [84]. The effects of sodium on GPCR have been widely investigated leading to a large body of evidence on its negative effects on receptor response [85]. The $A_{2A}AR$ crystal structure is in complex with the antagonist 4-(2-[7-amino-2-(2-furyl[1,2,4]-triazolo[2,3-a[1,3,5]triazin-5-yl-aminoethyl)phenol ZM241385 in an inactive state. The sodium ion binding site is composed of a series of several GPCR-conserved residues, specifically the $D52^{2.50}$, the $S91^{3.39}$, the $W246^{6.48}$, the $N280^{7.45,}$ and the $N284^{7.49}$ [84]. Based on the solved structure, several other computational and biochemical analyses have been carried out revealing an allosteric control of sodium that stabilizes the inactive conformation of the receptor [86]. The sodium is detectable exclusively in the presence of antagonists demonstrating its ability to shift the conformational equilibrium toward the inactive state of the receptors [87]. The cavity of the allosteric sodium binding site is reduced in the active state of the receptors hampering the sodium entry from the extracellular space. Recently, microsecond-scale molecular dynamics (MD) simulations on 18 GPCRs have demonstrated that sodium interacts with its allosteric binding site from the extracellular space [88]. Conversely, sodium can leave the binding pocket not only by the extracellular space but also by translocating into the intracellular space [89]. Mechanistic aspects of sodium dissociation have been investigated. Upon agonist binding, the shift of TM3, TM6, and TM7 leads to the collapse of the sodium-binding site due to the H bond formed between the $N284^{7.49}$ and $D52^{2.50}$. The main feature promoting this conformational change is the protonation state of $D52^{2.50}$ changing from the deprotonated (inactive receptor) to protonated state (active receptor) and this change is facilitated by the $S91^{3.39}$ in A_1 and $A_{2A}ARs$ [72].

In this scenario, the extracellular concentration of sodium has a pivotal role in controlling the conformational landscape of the AR. The sodium binding acts as a negative allosteric modulator modifying the equilibrium of the conformational states increasing the inactive-like receptor population and decreasing the affinity toward agonists. Thus, ligands able to interact with the sodium-binding site can affect the receptor response. Of note, the volume of this cavity is limited (150–250 Å) hampering the development of ligands able to act at this site [90]. To date, few compounds have been discovered acting on ion binding sites, among which the most studied is amiloride (**7**) and its derivatives acting as $A_{2A}AR$ NAMs [91]. Recently, Lu et al. reported a new ligand (Fg754, **8**, Fig. 18) able to stabilize the inactive conformation of the $A_{2A}AR$ through direct interaction with the $D52^{2.50}$ residue [92].

Another endogenous ligand proposed as AR allosteric modulator is cholesterol. However, whether cholesterol acts through specific interactions or indirectly via modifications of the cell membrane properties is still unclear. Based on the high-resolution crystal structures of β2-adrenergic receptors bound to cholesterol a cleft between TM2 and TM3 was identified as the “cholesterol consensus motif” (CCM),

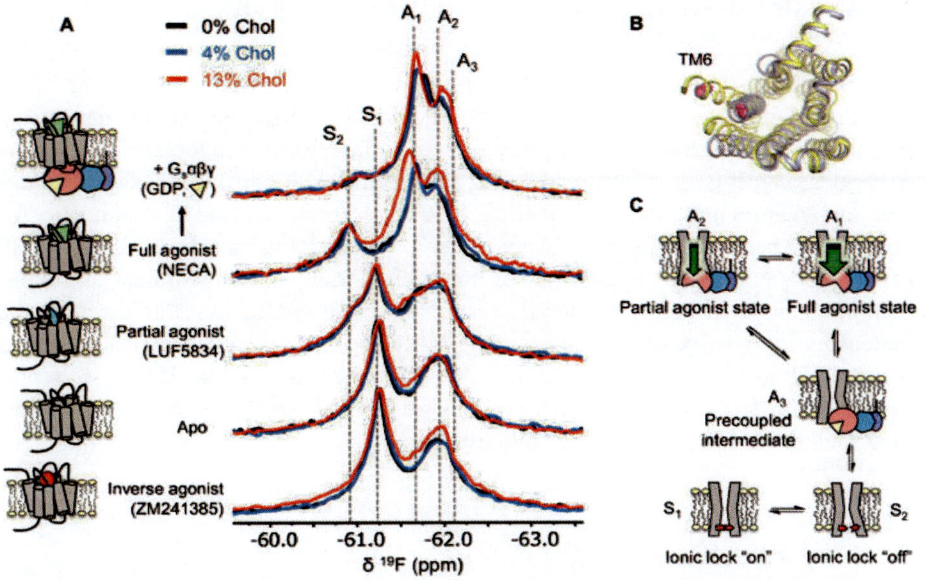

Fig. 8 Effect of cholesterol on conformer population of A_{2A}AR. (**a**) ^{19}F NMR spectra of A_{2A}AR in the presence of different cholesterol concentrations (0, 4, and 13%). Two inactive states (S1-2) and three active states (A1-3) are shown (gray dashed lines). (**b**) Intracellular view of an inactive (gray, PDB: 4EIY) and an active (yellow, PDB: 5G53) crystal structure of A_{2A}AR. (**c**) Cartoon representations of the key functional states of A_{2A}AR are indicated in (**a**). The figure is reproduced from Huang et al. [96]

and it has a high prevalence (around 44%) across the class A GPCRs [93]. The A_{2A}AR has both this CCM motif and the CRAC (cholesterol recognition/interaction amino acid consensus) motif; however, the role of cholesterol in modulating the receptor function is debated due to the opposite effects evidenced by its depletion [94, 95]. Recently, Huang et al. [96] try to respond to this question by the means of a computation approach and NMR analysis (Fig. 8).

The authors demonstrate that cholesterol acts as a weak PAM of A_{2A}AR thanks to the stabilization of the receptor population in the active-state conformers (A3) and a stronger coupling to the G-protein (A1 and A2). However, the cholesterol interactions have a short-life indicating a lack of high-affinity binding sites or direct allosteric modulation. Of note, the authors demonstrate that a similar effect is obtained by the modification of the physical properties of the membrane supposing that the allosteric effect is a result of indirect effects through changes in bilayer fluidity and thickness [96].

Despite the huge efforts in the study of cholesterol's role on A_{2A}AR activity, less is known regarding its role in the modulation of A_1AR, A_{2B}AR, and A_3AR.

5 Allosteric Modulators of Adenosine Receptors

In recent years, a great interest in the development of AR ligands able to modulate specifically their activity limiting the on-target and off-target side effects has arisen. Adenosine itself has been approved as non-selective agonist in therapy (Adenocard®). It is in widespread use in the treatment of paroxysmal supraventricular tachycardia due to its activation of A_1 receptors and as a diagnostic for myocardial perfusion imaging (Adenoscan®, Astellas Pharma, Inc.). In addition, adenosine is under investigation in several clinical trials for the treatment of inflammation, neuropathic and perioperative pain, and cardioprotection. Interestingly, the increasing knowledge of the mechanism of AR allosteric modulation has prompted the research in the development of new AR allosteric modulators. Herein, we report the main allosteric modulators directed toward the different AR subtypes as well as their main possible application in therapy.

5.1 *A_1 AR Allosteric Modulators*

The first A_1AR allosteric modulators were discovered in 1990 [97, 98] by the research group of Bruns, who described a class of 2-amino-3-benzoylthiophenes (2A3BTs) able to selectively increase the binding affinity and slow down the dissociation of the agonist radioligand [^{3}H]-N^6-cyclohexyladenosine ([^{3}H]CHA) from the A_1 ARs, without altering the dissociation rate of antagonists [99, 100]. Additional studies demonstrated the ability of 2A3BTs to act as PAM at A_1AR, both in binding and functional assays. In this series, structure–activity relationship (SAR) studies evidenced the 2-amino function and the 3-carbonyl group as key structural elements, leading the authors to hypothesize the active conformation is the one possessing an intramolecular hydrogen bond between the amino nitrogen and carbonyl oxygen. In addition, the replacement of the thiophene core with benzene led to equally potent compounds, while the introduction of a nitrogen-containing ring determined a decreased activity. Some derivatives belonging to this class showed an additional antagonist activity toward the A_1AR at high concentrations. Further studies evidenced that the presence of an alkyl chain at 4-position, as well as lipophilic substituents on the 3-benzoyl ring, favored allosteric activity with respect to antagonist one, while bulky substitution at 5-position combined with hydrogen or small alkyl substitution at 4-position favored the antagonist one.

A more detailed biological evaluation was performed on the four most active compounds, namely PD81,723 (**9**), PD71,605 (**10**), PD78,416 (**11**), and PD117,975 (**12**). Among them, PD81,723 (**9**) showed the best potency and the most favorable ratio of allosteric enhancement to antagonism (Fig. 9), soon becoming the prototypical positive allosteric modulator (PAM) [98, 100].

Van der Klein and co-workers developed a new series of 2-amino-3-benzoylthiophenes by cyclizing the 4,5-dimethyl groups as in PD71,605 (**10**) and

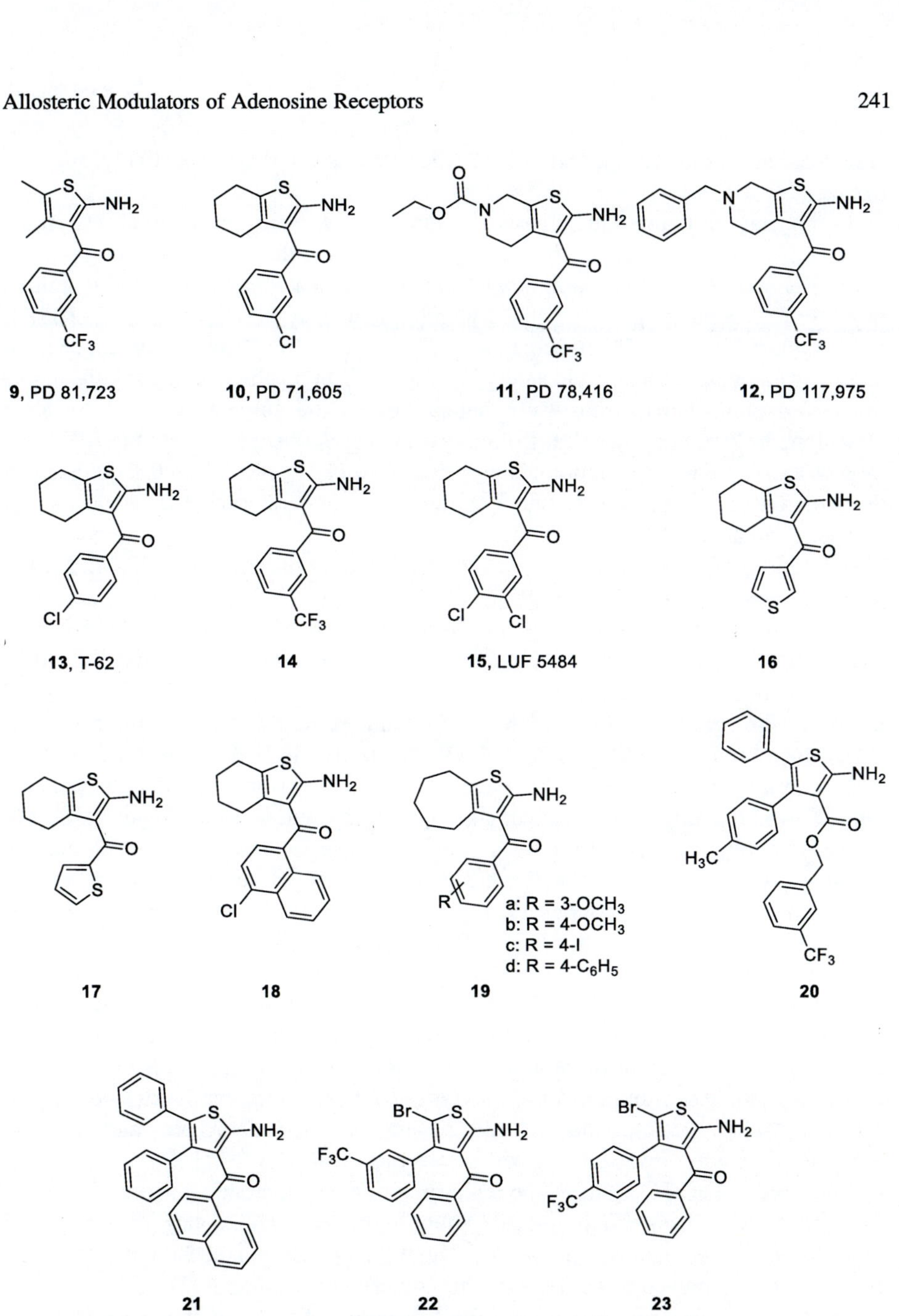

Fig. 9 2-Amino-3-substituted thiophene derivatives as A_1 allosteric modulators

introducing different substitutions at the 3-benzoyl ring [101, 102]. Among these compounds, derivatives **13–15** (Fig. 9) proved to be more potent enhancers with respect to PD81,723 (**9**), with comparable antagonistic activity. Biological assays carried out on rat-brain cortex membranes revealed a 123%, 122%, and 151%

enhancer activity for compounds **13–15** (measured at 10 μM), respectively, superior to that of **9** (100%).

In this class of compounds, lipophilic substituents, mainly halogen atoms, in the *meta* and/or *para* position of the benzoyl ring led to enhancer activity compounds with respect to **9**. Notably, compound **13** resulted efficacious when tested in animal models of neuropathic pain associated with hyperalgesia and allodynia [103–105]. It was evaluated by King Pharmaceuticals and progressed to Phase II clinical trials to assess its analgesic efficacy and safety in subjects with postherpetic neuralgia and its associated pain. Those trials were halted after some subjects showed transient elevations in liver enzymes (https://clinicaltrials.gov/ct2/show/NCT00809679).

In 2006, this class of compounds was successfully exploited with the synthesis of the tritiated derivative of **13**, thus becoming the first radiolabeled adenosine A_1 allosteric enhancer [106].

An additional exhaustive survey of the structural requirements of PD81,723 was also performed by the research group of Baraldi [107–110]. The most interesting results were obtained by compounds in which the 3-phenyl was replaced with either 2- or 3-thienyl moiety or the 1-naphthoyl ring. In detail, thienyl derivatives **16** and **17** exhibited similar enhancing activity compared to PD81,723 (**9**), while naphthoyl **18** increased the agonist binding to hA_1 CHO (Chinese hamster ovary), brain, and rat cortex membrane by 149%, 27%, and 43%, respectively [109, 110]. Further structural modifications of PD81,723 (**9**) were performed by Tranberg and collaborators, who showed that allosteric enhancer (AE) activity increased according to the size of the C4-C5 polymethylene bridge; also, for this subseries, lipophilic substituents, in the *meta* and/or *para* position of the 3-benzoyl ring lead to enhancer activity compounds, such as compounds **19**, Fig. 9 [111]. All these studies emphasized the presence of two regions in the A_1AR allosteric binding site: the first able to accommodate the 2-amino-3-aroyl thiophene scaffold, the second able to host the alkyl substituents at C-4 and C-5 positions of the thiophene. Few after, Scammells et al. demonstrated that the presence of two bulky lipophilic groups at C4 and C5 of the thiophene core led to more potent allosteric enhancer compounds compared to PD81,723 (**9**). Accordingly, 4,5-diphenyl substituted compounds **20** and **21** and 5-bromo-4-phenyl compounds **22** and **23** (Fig. 9) showed a three- and five-fold increase of activity, respectively, with respect to PD81,723 (**9**) [112].

A few years later [113], the same research group demonstrated that the removal of 3-benzoyl group from 2-amino-5,6,7,8-tetrahydro-4*H*-cyclohepta[*b*]thiophene did not alter allosteric enhancer properties. Notably, **24** maintained the potency of the parents **19** and showed to be more potent and efficacious than **9** (Fig. 10).

Further investigations concerning the 5-position of the 2-aminothiophene scaffold were also performed [114]. In the 3-benzoyl substituted series, 5-unsubstituted compound **25** proved to be the most efficacious (EC_{50} 15.8 μM, AE score 77%), while in the second series of 3-ethyl ester substituted compounds, the 5-(4-chlorophenyl) analog **26** was the most potent and efficacious one (EC_{50} 6.6 μM, AE score 57%), (Fig. 10). Specifically, AE score points out the ability of the candidate AE to stabilize the agonist–A_1AR–G-protein ternary complex, ranging from 0 to 100%. A score of 100% means no dissociation and a score of zero means

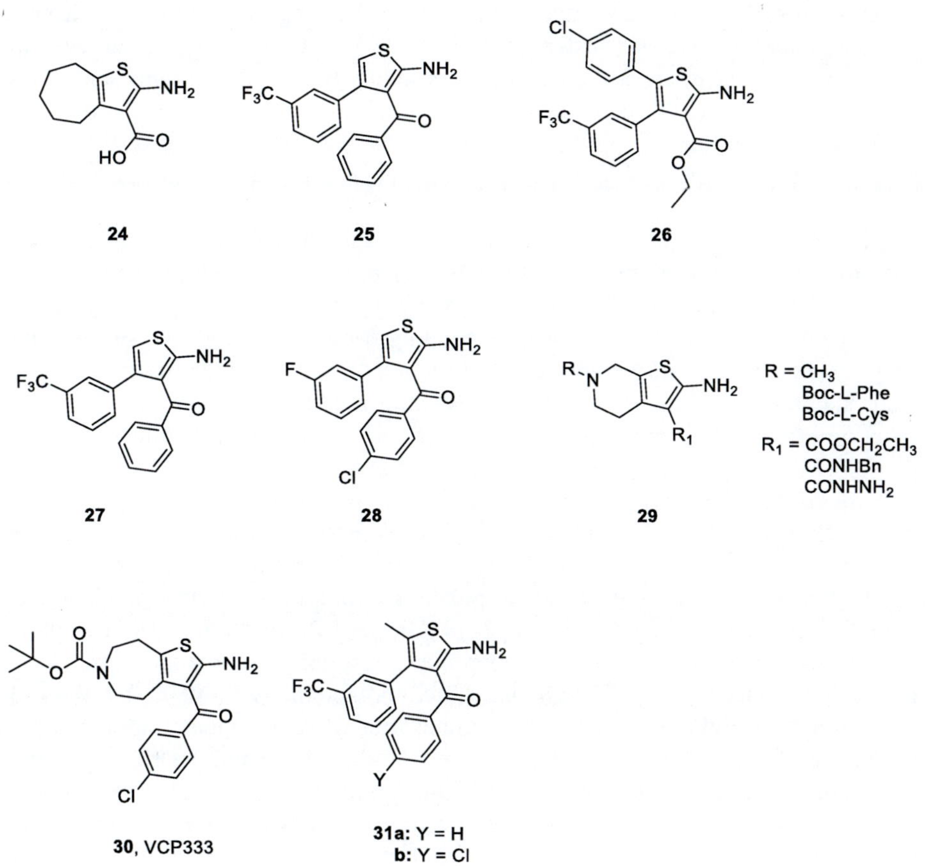

Fig. 10 2-Amino-3-substituted thiophene derivatives as A_1 allosteric modulators

complete dissociation. In another study, it was evidenced that the replacement of the ethoxycarbonyl moiety at the 3-position with a hydrazide or an amide one led to a drastic reduction of AE activity [115].

Furthermore, the authors evidenced that 5-unsubstituted compounds maintained the ability to allosterically modulate the activity of the orthosteric agonist N^6-R-phenylisopropyladenosine (R-PIA) using a functional assay of A_1AR-mediated phosphorylation of ERK1/2 (pERK1/2) in intact CHO cells [116]. Among them, compounds **27–28** showed to be the most interesting ones (Fig. 10). However, these compounds showed also an intrinsic agonistic activity, resulting able to activate the receptor in absence of R-PIA. In the same year, the authors used PD78,416 (**11**) and PD117,975 (**12**) as lead compounds to develop a new series of 2-amino-4,5,6,7-tetrahydrothieno[2,3-*c*]pyridines **29** by introducing different substituents at 3- and 6-positions [117]. The potency of these compounds as allosteric modulators of A_1AR was investigated by evaluating their ability to stabilize the ternary complex [^{125}I]-aminobenzyladenosine ([^{125}I]ABA, orthosteric agonist)–

A_1AR–G-protein in an in vitro dissociation kinetic binding assay. Results evidenced that the greatest degree of allosteric stabilization of such complex was achieved by compounds that featured a 3-ethyl ester combined with *N*-Boc amino acid at the 6-position or a 6-methyl combined with an amide or hydrazide at the 3-position. Unfortunately, further biological experiments revealed some degree of orthosteric antagonism properties of this class, even more marked with respect to the allosteric one.

More recently, an analog of **13** (T-62) with a 7-membered heterocyclic ring fused with the thiophene, namely VCP333 (**30**), was reported in a tentative to improve functional recovery in mouse hearts after an ischemic injury (Fig. 10) [118]; on the other hand, starting from **25**, the corresponding 5-methyl analogs **31a-b**, differing only for the presence of a chloro atom at the *para*-position of the benzoyl moiety, were identified as biased allosteric agonists and modulators of the A_1AR agonist function in two different functional assays of ERK1/2 phosphorylation and inhibition of cAMP production in CHO cells (Fig. 10) [46].

Further investigations concerning the 4-position of the 2-aminothiophene scaffold were also performed by the research group of Romagnoli through the synthesis of a series of 2-amino-3-(4-chlorobenzoyl)thiophene derivatives tethered to different alkyl or aryl piperazine moieties at 4-position through a methylene linker (compounds with general formula **32**, Fig. 11) [119, 120] Results evidenced the crucial role played by the presence of a lipophilic substitution at *N*-4 position of the piperazine. Indeed, *N*-cyclohexyl-substituted compound was more active with respect to *N*-methyl one; in addition, aromatization of the cyclohexyl moiety led to an increased activity. On the contrary, the introduction of one (pyridine) or two (pyrimidine) nitrogen atoms on the phenyl ring led to reduced activity. The synthesis of several compounds with one or more electron-withdrawing or electron-releasing substituents, such as F, Cl, CN, CF_3, CH_3, 3,4-methylenedioxy, in all five positions of the phenyl ring linked to the pyrazine moiety, evidenced a beneficial role played by such substitution. Among them, the 4-chloro (**32a**), 4-trifluoromethyl (**32b**), 3,4-difluoro (**32c**), 3-chloro-4-fluoro (**32d**), and 4-trifluoromethoxy (**32e**) derivatives resulted the most active ones, both in binding (saturation and competition) and functional assays (Fig. 11).

Starting from these results and with the aim to further refine the SAR around this class of derivatives, the same authors prepared a series of 2-amino-3-(4-chlorobenzoyl)-4-[(arylpiperazinyl)methyl]-thiophenes, characterized by the presence of several groups at the 5-position (Fig. 11); in particular, a methyl, an ethyl, a bromine, or different *para*-substituted phenyl ring were introduced at such position combined with the presence of different substituents on the phenyl of the 4-arylpiperazine moiety, selected as the best-performing substituents in terms of AE activity [121]. In general, 5-aryl substituted derivatives showed superior activity as A_1AE with respect to methyl, ethyl, and bromine counterparts, with **32f-h** being the most active within the series (Fig. 11).

In addition, these compounds showed the ability to induce: (1) a shift of the A_1ARs toward the active state, as suggested by the increase of 2-chloro-N^6-cyclopentyladenosine (CCPA) affinity and (2) an increase in B_{max} shift to hA_1ARs

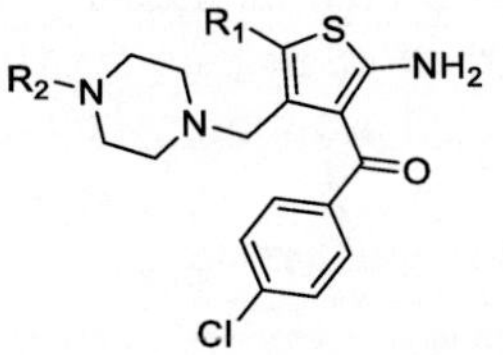

a: R_1 = H; R_2 = 4-Cl-C_6H_4
b: R_1 = H; R_2 = 4-CF_3-C_6H_4
c: R_1 = H; R_2 = 3,4-diF-C_6H_3
d: R_1 = H; R_2 = 3-Cl-4-F-C_6H_3
e: R_1 = H; R_2 = 4-OCF_3-C_6H_4
f: R_1 = 4-F-C_6H_4; R_2 = C_6H_5
g: R_1 = 4-Cl-C_6H_4; R_2 = 4-CF_3-C_6H_4
h: R_1 = 4-Cl-C_6H_4; R_2 = 3-Cl-C_6H_4

32

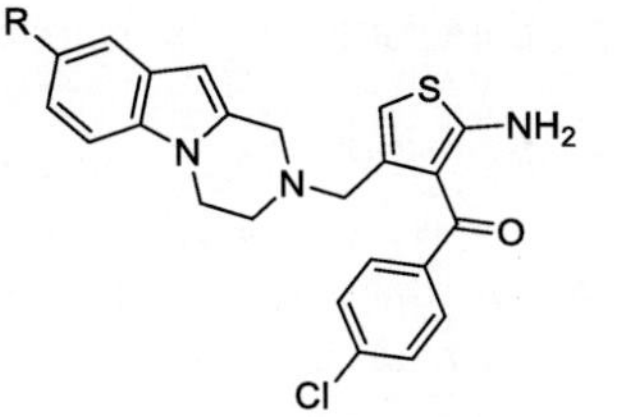

33 **a**: R = H
b: R = F
c: R = Cl

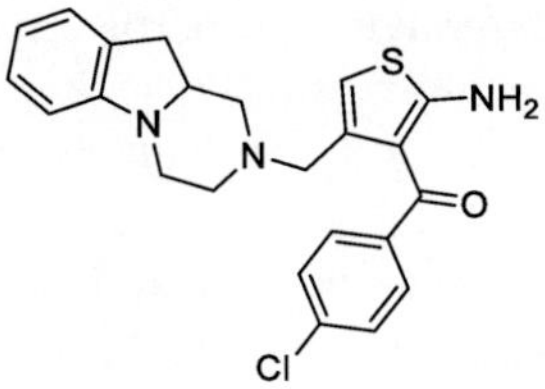

34

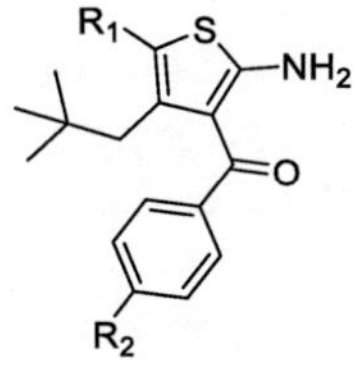

35

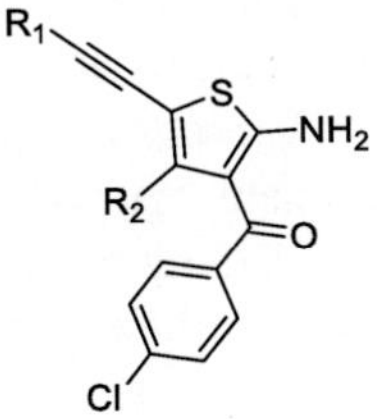

36

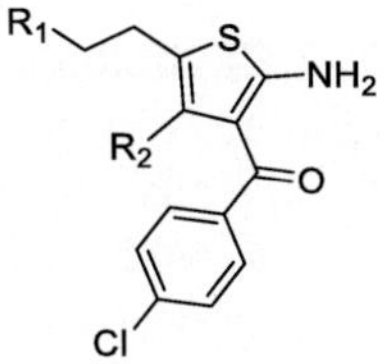

37

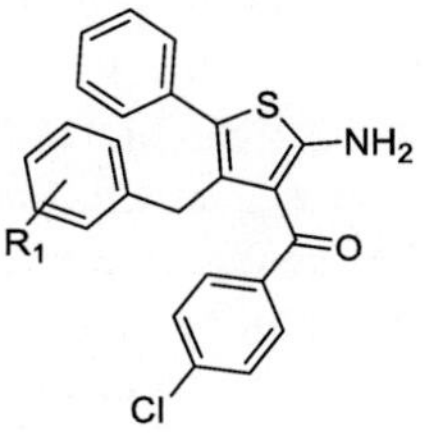

38

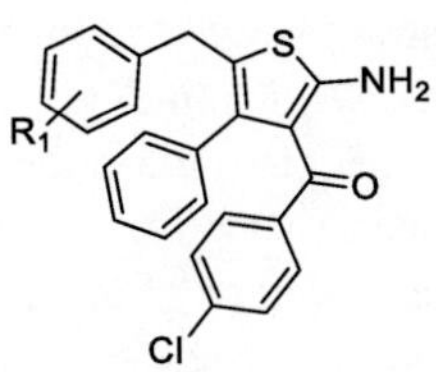

39

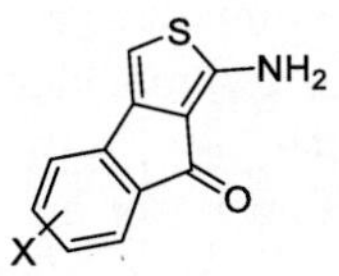

40

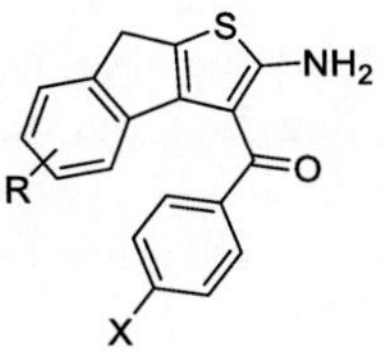

41

a: R = H; X = H
b: R = H; X = Cl
c: R = 7-CF_3; X = H
d: R = 7-CF_3; X = Cl
e: R = 5-CF_3; X = H
f: R = 5-CF_3; X = Cl

Fig. 11 2-Amino-3-substituted thiophene derivatives as A_1 allosteric modulators

derived from [^{3}H]CCPA saturation binding assays, suggesting the ability of such derivatives to mediate a shift from the ground state (R) to the activated one (R*) of the A_1ARs.

Further competition binding studies performed on human A_1, A_{2A}, and A_3 ARs evidenced that such compounds, behind their enhancing activity, were not able to displace the radioligands [^{3}H]-8-cyclopentyl-1,3-dipropylxanthine ([H^3]DPCPX), [^{3}H]ZM241385 and [^{3}H]-*N*-[2-(2-furanyl)-8-propyl-8*H*-pyrazolo[4,3-*e*][1,2,4] triazolo[1,5-*c*]pyrimidin-5-yl]-*N'*-(4-methoxyphenyl)urea ([^{3}H]MRE3008F20) from the orthosteric site.

Subsequently, compound **32f** was tested for its anti-nociceptive properties in different mouse model of acute and chronic pain, showing analgesic effects in formalin and writhing tests comparable with that of morphine. In addition, **32f** displayed anti-allodynic properties in a neuropathic pain model without locomotor or cataleptic side effects [122].

In a following study, the authors decided to replace the arylpiperazine moiety with different chemical functionalities [123]. The most interesting results were obtained by compounds characterized by a tetrahydropyrazino[1,2-*a*]indole nucleus with general structures **33** and **34**, conceived as conformationally constrained analogs of compounds **32**, in which the *ortho* position of the phenyl ring was linked to the 2-carbon of the piperazine by a methylene chain (Fig. 11). As observed for compounds **32**, when tested for their ability to modify the agonist-binding parameters, these derivatives resulted able to mediate an increase in Bmax shift to A_1AR of more than sevenfold and enhance the CCPA affinity.

With the aim to study the role of different alkyl chains at 4-position, a series of 2-amino-3-(4-chlorobenzoyl)-4-alkylthiophenes was prepared and biologically evaluated as A_1AE. From this preliminary evaluation, the neopentyl derivatives emerged as the most active ones. Subsequently, a series of 2-amino-3-(4-chlorobenzoyl)-4-neopentylthiophenes substituted with variable modifications (bromine, aryl, heteroaryl, aryl/heteroarylacetylene) at the 5-position, as well as in the benzoyl ring (compounds **35**, Fig. 11) was synthesized [124].

All compounds, with only a few exceptions, proved to be superior with respect to PD81,723 (**9**) in a functional assay. In particular, from this study, it has emerged that: (1) the introduction of a heterocyclic ring, potentially able to establish hydrogen bound at 5-position of the thiophene ring resulted in compounds with reduced (pyridine-2-yl, pyridine-3-yl, pyridine-4-yl), maintained (furan-2-yl, furan-3-yl), or slightly increased (thiophen-2-yl, thiophen-3-yl) activity with respect to the 5-phenyl one; (2) the insertion of electron-withdrawing or electron-releasing substituents on different position of the 5-phenyl ring does not cause a consistent pattern of effects. Indeed, compounds substituted with opposite electronic effect moieties showed comparable AE activity.

In a following study [125], the authors investigated the effects derived from the insertion at 5-position of the 2-amino-3-(*para*-chlorobenzoyl)thiophene nucleus of a (hetero)aryl or alkyl moiety by means of a two-carbon rigid or a flexible linker combined with different substituents at 4-position (hydrogen, methyl, phenyl, neopentyl, *N*,*N*-dicyclohexylamine, or *N*-arylpiperazine by a methylene linker, compounds **36** and **37**, Fig. 11). In general, the 4-neopentyl derivatives resulted in the most active A_1AEs. In addition, within this subseries, the presence of an acetylene spacer at 5-position determined more beneficial effects with respect to that produced by compounds substituted with a more flexible ethyl linker.

In the following year, two other series of 2-amino-3-(*p*-chlorobenzoyl)thiophenes were investigated [126]. In the first series, a variously substituted benzyl ring was introduced at 4-position, while maintaining an unsubstituted 5-phenyl ring (compounds **38**, Fig. 11). The second series consisted of the corresponding regioisomeric 4-aryl-5-benzylthiophene analogs **39**, Fig. 11. Overall, the position of the benzyl and the phenyl moieties seems to be crucial for AE activity, with the 4-benzyl-5-aryl analogs **38** proved to be superior relative to regioisomers **39**.

Two series of compounds, conceived as conformationally constrained analogs of 2A3BTs, were also developed by the same authors, Fig. 11 [127]. In detail, in series **40**, the 3-benzoyl ring was linked to the 4-position of the thiophene moiety, while in series **41**, the 4-phenyl ring was tethered to the 5-position of thiophene through a methylene linker. The most interesting results were obtained with compounds **41**, being **40** inactive or antagonist. In particular, **41b**, **41d**, and **41f** behaved as AE, being able to potentiate the A_1AR-mediated stimulation of ERK1/2 phosphorylation in the presence of an EC_{50} concentration of R-PIA.

The lack of allosteric activity of the other compounds from series **41** (**41a**, **41c**, **41e**), although from a structural point of view they only differ for a chloro atom at the *para*-position of the 3-benzoyl ring, evidenced a subtle balance between allosteric enhancement and antagonism or lack of activity.

Besides 2A3BTs, other alternative classes of A_1AE have been developed by different research groups.

Among them, the 2-aminothiazole class was firstly reported by the research group of Chordia [128]. Three compounds belonging to this class, namely **42–44** (Fig. 12),

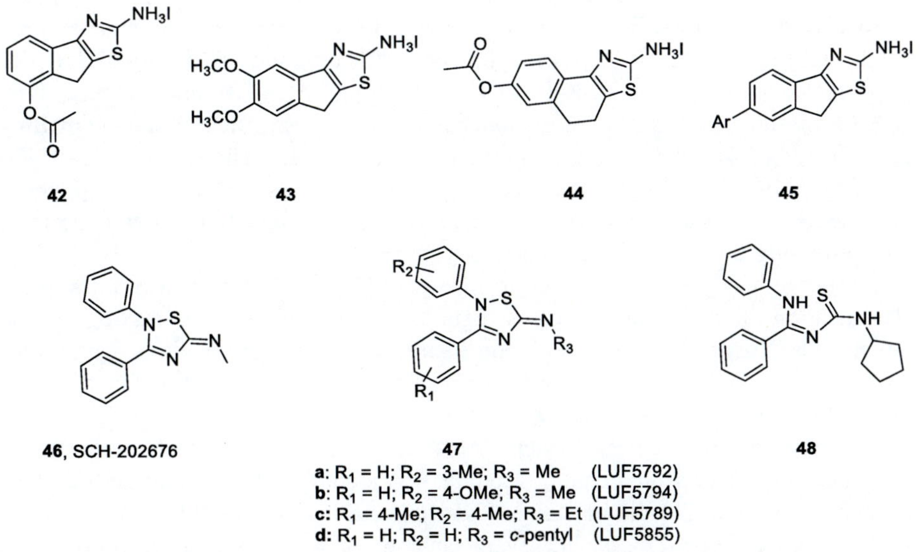

Fig. 12 2-Aminothiazoles and [1,2,4]thiadiazole analogs as A_1 allosteric modulators

synthesized as hydroiodides, showed the ability to reduce the agonist dissociation rate from A_1AR with EC_{50} of 0.3, 1.2, and 3.8 μM, respectively.

However, soon after, results obtained by Ijzerman et al., on either identical or similar compounds, resulted in contrast with that obtained by Chordia [129]. Indeed, no compounds, tested as free bases, proved to affect [^{3}H]CCPA binding to A_1AR in CHO cells or as well as its association or dissociation kinetics.

Despite these mixed results, Chordia et al. pursued their study on 2-aminothiazoles and synthesized a series of 6-arylindeno[1,2-*d*]thiazoles **45** (Fig. 12) [130]. The most active compounds, tested for their ability to retard dissociation of the agonist-receptor-G-protein complex, resulted in those featuring a 4-methoxy, 4-phenyloxy, 3,4-dimethoxy, and 4-dimethylamino substituents on the pendant phenyl ring, which showed EC_{50} of 2.2, 1.5, 1.0, and 0.9 μM, respectively. However, tested as free bases, such compounds showed decreased activity, in accordance with that previously reported by Ijzerman; in addition, activity increases to a constant value after 4 h. Actually, the reason behind this change of activity over time remained an unresolved issue.

Also, the [1,2,4]thiadiazole scaffold was exploited in order to develop potent AE of A_1AR. Starting from SCH-202676 **46**, which proved to be a pan allosteric modulator of both agonist and antagonist binding to several GPCRs including [131–133] opioid, muscarinic, adrenergic, dopaminergic receptors, the research group of van den Nieuwendijk synthesized a series of 2,3,5-substituted[1,2,4] thiadiazole analogs and evaluated their modulatory properties on AR [134, 135], the most representative compounds being **47 a-d** (Fig. 12). Derivatives **47 a-c** resulted able to decrease the agonist [^{3}H]CCPA binding to *h*A_1AR at a final concentration of 10 μM, while little effects were observed on the antagonist [^{3}H] DPCPX binding. They were deeply characterized, together with **46** (SCH-202676), all showing to inhibit [^{3}H]CCPA binding in a concentration-dependent manner. Derivative **47a** resulted in the most active with EC_{50} of 0.64 μM, four times more potent with respect to **46** [134]. On the contrary, compound **47d**, tested at 1 μM, proved to significantly inhibit [^{3}H]CCPA binding to A_1AR and increase [^{3}H] DPCPX binding. Further studies by both equilibrium and kinetic radioligand binding experiments of **47d** together with **46** were also performed. The EC_{50} value of **47d** was 503 ± 39 nM with respect to the EC_{50} of **46** that was 2,800 ± 200 nM. From these studies, it was also evidenced a nonstoichiometric interaction with the receptor, thus providing the proof of an unusual interaction.

Further bioanalytical analysis performed with HPLC and HPLC-MS allowed the classification of these compounds as highly reactive sulfhydryl modifying agents, rather than allosteric modulators. Upon incubation, they were converted into thiourea (see **48** as an example of thiourea formed by **47d**) by thiol group of cysteines in the cell membranes [135].

In the search for A_1AR allosteric modulators, Scammells et al. performed a pharmacophore-based screening, leading to identify ethyl 5-amino-3-(4-*tert*-butylphenyl)-4-oxo-3,4-dihydrothieno[3,4-d]pyridazine-1-carboxylate **49** [136] as lead compound (Fig. 13). Then, several analogs of **49** have been synthesized by introducing different substituents on the phenyl ring, in the place of the 4-*t*-butyl

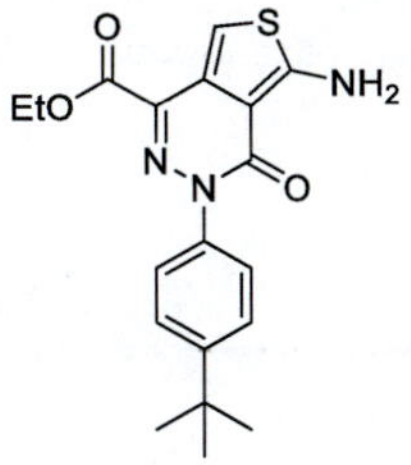

Fig. 13 5-Amino-4-oxo-3-phenyl-3,4-dihydrothieno[3,4-d]pyridazine-1-carboxylate derivative

group. Unfortunately, behind their ability to stabilize agonist-receptor-G-protein ternary complex, this class of compounds behaved preferentially as orthosteric antagonists/inverse agonists, as observed in both equilibrium binding and two different functional assays. These findings allowed the authors to speculate that the allosteric site on the A_1AR is in close proximity to the orthosteric one.

5.1.1 Bitopic Allosteric Modulators of the A_1AR

The concept of bivalent ligand has been deeply exploited by chemist researchers to generate novel GPCR ligands. A bivalent ligand is a hybrid molecule constituted by two pharmacophores covalently tethered by a linker of different nature and lengths [40].

Ijzerman et al. with the aim to locate the allosteric site on the A_1AR, prepared a series of bivalent ligands featuring both orthosteric and allosteric pharmacophores tethered by an alkyl linker of variable length [137]. The bivalent ligand N^6-[2-amino-3-(3,4-dichlorobenzoyl)-4,5,6,7-tetrahydrothieno[2,3-*c*]-pyridin-6-yl-9-nonyloxy-4-phenyl]-adenosine **50** (LUF6258, Fig. 14) with a 9-carbon atom spacer did not show significant changes in affinity or potency in the presence of **9**, indicating that this ligand bridged both sites on A_1AR.

Also, the research group of Christopoulos designed a bitopic ligand for the A_1 AR by tethering **27** and endogenous agonist adenosine linked by a 4-benzamidohexyl chains, compound **51** (Fig. 15) [41]. Such a compound is able to protect against ischemia in native A_1AR-expressing cardiomyoblasts and cardiomyocytes without effects on rat atrial heart rate. In addition, a more deeply pharmacological evaluation evidence significant bias away from calcium mobilization with respect to ERK1/2 phosphorylation, Akt phosphorylation, and cAMP accumulation inhibition relative to the reference agonist NECA.

Shortly after, starting from **51** [41], three series of compounds featuring structural modification to the orthosteric pharmacophore (**52**), the linker (**53**), or the allosteric pharmacophore (**54**) were synthesized and pharmacologically evaluated, Fig. 15 [138].

Fig. 14 Structure of A_1AR bitopic allosteric modulator

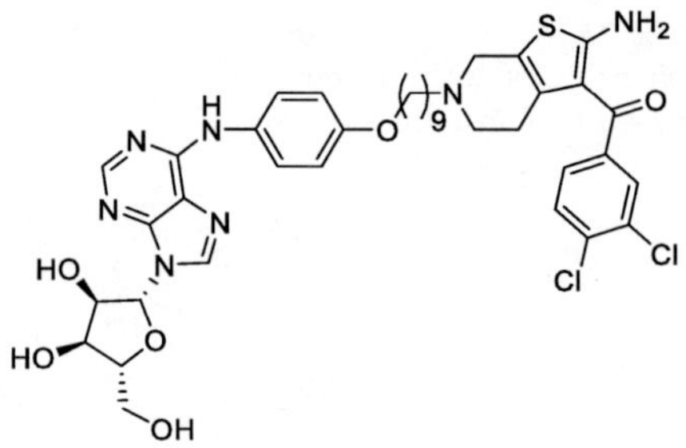

50

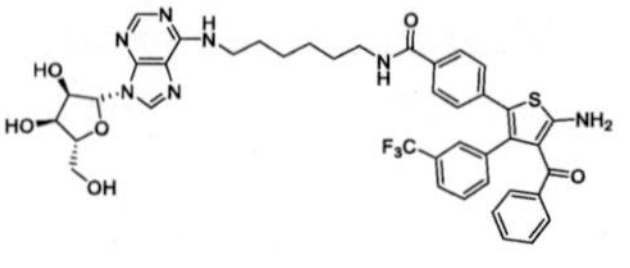

51

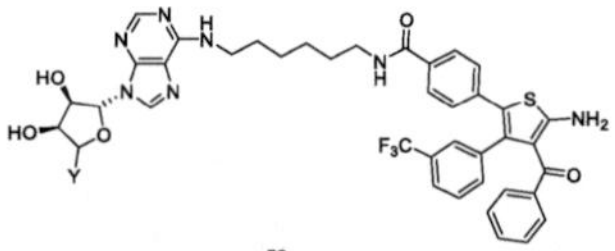

52

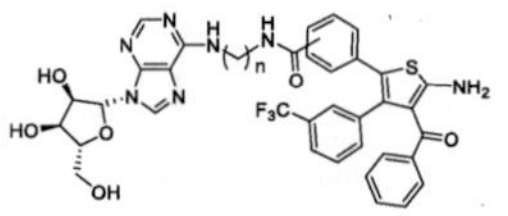

53

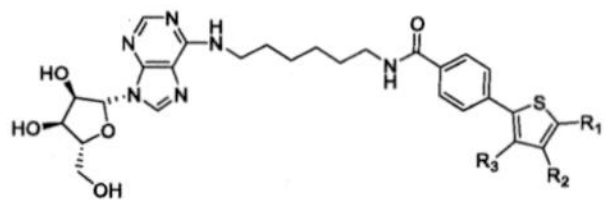

54

Fig. 15 Structure of A_1AR bitopic allosteric modulator

Overall, the results evidenced the crucial role played by the allosteric pharmacophore, particularly the 4-(trifluoromethyl)phenyl moiety, in governing bitopic ligand bias at the A_1AR. Of note, removal of the 4-(trifluoromethyl)phenyl moiety (series **54**, $R_1 = NH_2$, $R_2 = COPh$, $R_3 = H$) eliminated the bias, while deletion of all three substituents of the thiophene scaffold (series **54**, $R_1 = R_2 = R_3 = H$) showed a reversal in the bias profile relative to **51**, causing a significant bias toward calcium mobilization. In addition, the incorporation of an additional methylene unit (series **53**, $n = 7$) resulted in a compound with increased bias and A_1AR selectivity relative to **51**. Finally, molecular modeling studies, performed on **51** into an active A_1AR homology model based on an active $A_{2A}AR$ crystal structure, evidenced that the interaction between the allosteric pharmacophore and the linker with the extracellular region may underlie the early activation of the biased agonist profile of such compound.

5.1.2 Biological Activity

Based on the A_1AR distribution in the central terminals of peripheral nociceptive neurons, receptor agonists have been proposed as promising agents for chronic and neuropathic pain. The distribution of the receptors in the CNS has also highlighted the antidepressant and anxiolytic-like effects of the A_1AR agonists. The cardiovascular system is the other pivotal target of the A_1AR agonists; in fact, adenosine is approved as a drug for paroxysmal supraventricular tachycardia. A_1AR activation in the heart has beneficial effects against ischemic cardiac injury, coronary and ventricular dysfunction, and ventricular remodeling. A_1AR agonists have been also proposed for the management of metabolic disorders such as type 2 diabetes, obesity, and hyperlipidemia due to their ability to inhibit lipolysis and insulin release and improve insulin sensitivity [139]. Unfortunately, most of the proposed ligands failed to reach the clinical application due to their on-target and off-target side effects.

The use of A_1AR allosteric modulators aims to overcome orthosteric agonist limitations. Based on their allosteric effects, they have a "cooperative" activity with the endogenous ligand supporting tissue- or context-based selectivity; in fact, they exert their action only in the desired site where adenosine concentration changes as a result of the disease. The most elegant application of A_1AR PAM as an analgesic agent has been recently reported [140]. In response to the sciatic nerve injury, adenosine concentration increases in the rat spinal dorsal horn promoting neuropathic pain [141]. To better understand the effects of allosteric modulation of A_1AR the PAM VCP171 (**55**) has been compared with the MIPS521 (**56**), Fig. 16. Despite its high affinity ($K_B = 3.16$ μM), the in vivo administration of the VCP171 fails to exert a potent analgesic effect. The lack of effects has been related to its low cooperativity with adenosine ($\alpha\beta = 5.8$). Conversely, the MIPS521 (**56**) has a lower affinity ($K_B = 11$ μM) but higher positive cooperativity with adenosine ($\alpha\beta = 65$). Its in vivo administration produces a robust decrease in the synaptic current amplitude in the neuron of the spinal cord from nerve-injured rats. Notably, the effects are present in the nerve-injured but not in the sham surgery controls indicating a site direct effect due to the increase of adenosine concentration. The lower affinity of MIPS521 produces also a low effect on the rat atrial beat rate as an A_1AR agonist side effect. These data highlight that the analgesic efficacy of PAMs is determined by the magnitude of on-target positive cooperativity with adenosine and it is possible only where an elevated adenosine concentration occurs such as in the spinal cord neurons after an injury.

The interesting results on MIPS521 lead to its crystallization in complex with A_1AR-G_{i2} complex [140]. The solved structure of this complex (PDB: 7LD3) represents the first high-resolution structure of an AR receptor with an allosteric

Fig. 16 Structures of VCP171 (**55**) and MIPS521 (**56**)

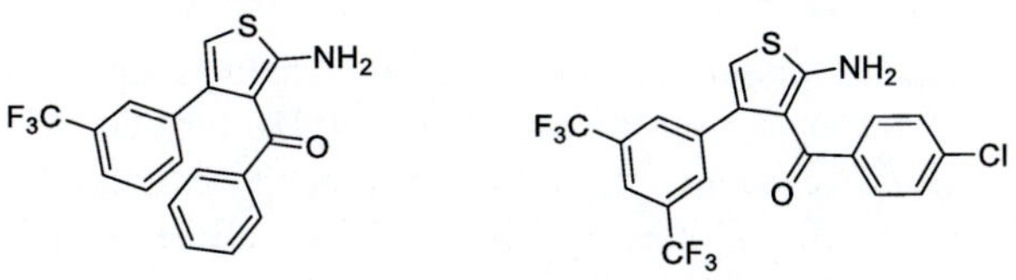

55, VCP171

56,MIPS521

modulator. This structure reveals an unexpected allosteric binding pocket on the A_1AR that has not been observed previously for other GPCRs. The pocket is the outer side of the TM1, TM6, and TM7 and the residues involved in the binding are L242$^{6.43}$, L245$^{6.46}$, S246$^{6.47}$, L276$^{7.41}$, M283$^{7.48}$, F275$^{7.40}$, V22$^{1.45}$, L18$^{1.41}$, and I19$^{1.42}$. The other face of the binding pocket is exposed to the lipid interface. The biochemical analysis combined with Gaussian accelerated molecular dynamics (GaMD) analysis reveals that the high cooperativity of MIPS521 is probably related to the effects on TM6 and TM7 that stabilize the A_1AR-G_{i2} complex. The cryo-EM structure of a PAM prompts the knowledge of the molecular mechanism involved in allosteric AR activation facilitating the rational design of novel PAMs.

5.2 A_{2A} Allosteric Modulators

Scattered examples of A_{2A} allosteric modulators have been reported to date in literature [142]. Among them, 4-methyl-7-([methylamino]carbonyl)oxy)-2*H*-1-benzopyran-2-one) **57** (Fig. 17) showed the ability to increase A_{2A}AR agonist activity in the rat striatum [143]. Thiazole derivative SCH202676 (**46**) (Fig. 12), already mentioned in Sect. 5.1, is a pan allosteric modulator [135]. Actually, it showed to inhibit radioligand binding to human A_1, A_{2A}, and A_3 with IC_{50} values ranging from 0.5 to 0.8 μM, altering both agonist and antagonist binding. In detail, tested at 10 μM, **46** enhanced the dissociation rate of the antagonist [^{3}H]-ZM241385 without effects that of the agonist [^{3}H]-2-[*p*-(2-carboxyethyl)phenylethyl-amino]-5′-*N*-ethylcarboxamido adenosine ([^{3}H]-CGS21680) at A_{2A}AR. At the same time, it decreased the dissociation rate of the antagonist [^{3}H]DPCPX at A_1ARs and enhanced that of the agonist [^{125}I]-N^6-(4-amino-3-iodobenzyl)-5′-*N*-methylcarbamoyladenosine ([^{125}I]I-AB-MECA) at A_3 ARs, without effects on the dissociation of both A_1AR agonist [^{3}H]R-PIA and the A_3AR antagonist [^{3}H]8-ethyl-4-methyl-2-phenyl-(8*R*)-4,5,7,8-tetrahydro-1*H*-imidazo[2,1-i]-purin-5-one ([^{3}H] PSB-11) [132].

Sodium ions behaved as A_{2A}AR allosteric modulators by decreasing the dissociation rate of either agonists [^{3}H]CGS21680, NECA and N^6-[2-(3,5-dimethoxyphenyl)-2-(2-methylphenyl)-ethyl]adenosine (DPMA) or antagonist [^{3}H]ZM241385 in a concentration-dependent manner [48]. Following site-directed mutagenesis studies evidenced the involvement of two definite and closely linked amino acids, located at the first and seventh transmembrane domain, namely Glu13 and His278, in sodium ions modulation [48].

The potassium-sparing diuretic amiloride **7** and its analogs behaved as A_{2A}AR allosteric modulators too (Fig. 16). Among them, 5-(*N*,*N*-hexamethylene)amiloride (**5**, HMA), 5-(*N*-methyl-*N*-isobutyl)amiloride (**58**, MIBA), 5-(*N*-methyl-*N*-guanidinocarbonyl-methyl)amiloride (**59**, MCGMA), phenamil **60,** and benzamil **61** were found to be more effective allosteric inhibitors than amiloride at rat A_{2A}Rs in increasing the dissociation rate of the antagonist [^{3}H]ZM241385 [48, 144]. Furthermore, all these compounds did not affect the dissociation rate of

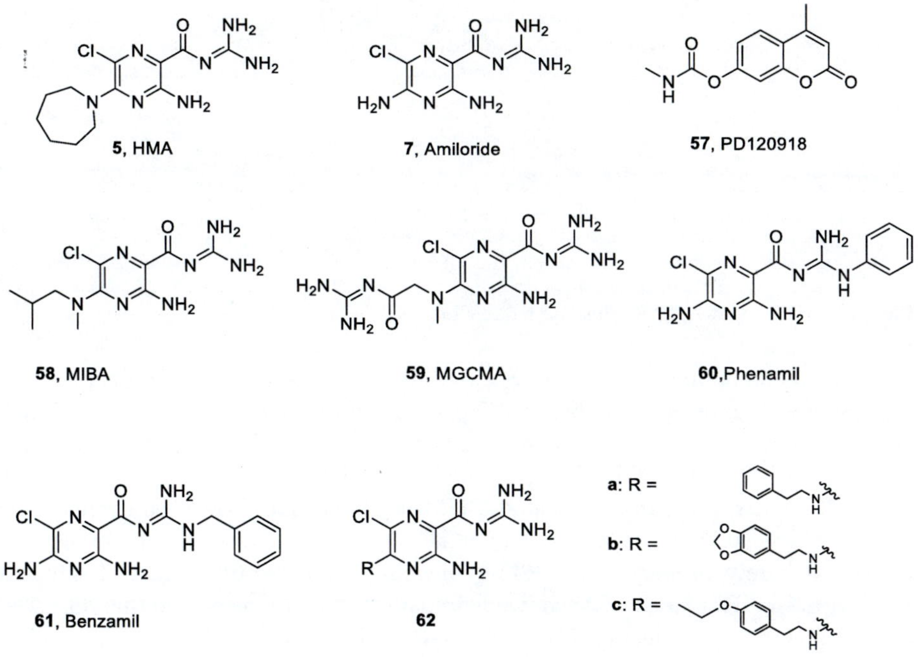

Fig. 17 Structures of PD120918 (**57**), amiloride (**7**) and amiloride analogs

the agonist [^{3}H]CGS21680. Further studies evidenced the capacity of such compounds to affect also other AR agonists and antagonists binding. In particular, they all proved to increase the dissociation rate of antagonists [^{3}H]DPCPX and [^{3}H]PSB-11 from A_1 and A_3 ARs, respectively and to decrease that of A_3 agonist [^{125}I]I-AB-MECA.

Docking studies on amiloride **7** and HMA (**5**) into the high-resolution X-ray crystal structure of $hA_{2A}AR$ showed the occupancy of the sodium ion binding site by such compounds [84, 87]. Similar to sodium ions, the positively charged guanidinium moiety of **7** and its analogs established strong interaction with the carboxylate of Asp52. In addition, the Trp246 was forced to rotate away in the presence of **7** or **5**. These findings were confirmed by mutation studies; indeed, the activity of both compounds was decreased by D52A mutation, whereas it was increased by W246A mutation.

The higher activity showed by 5′-substituted amilorides (**5, 58–61**) with respect to amiloride **7** strongly spurred the research group of Ijzerman to further exploit such position to develop more potent amiloride derivatives (compounds **62**, Fig. 17) [91].

Firstly, all compounds were tested for binding to the wild-type (WT) and W246A $hA_{2A}AR$. The most interesting compound was the 5-phenylethyl derivative (**62a**), thus representing the starting point for the following series of amiloride derivatives featuring substituents on the phenyl ring, Fig. 17.

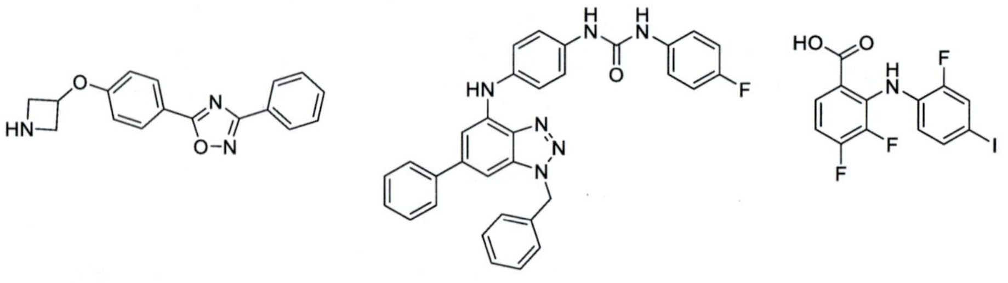

Fig. 18 Structures of $A_{2A}AR$ allosteric modulators

All novel compounds, including **5**, exhibited higher potency at W246A $hA_{2A}AR$ compared to WT receptor, indicating Trp246 as an obstacle in the binding of such amiloride derivatives. The most potent amiloride derivatives of series **62** were selected and tested for their effects on [^{3}H]ZM241385 dissociation from both WT and W246A $hA_{2A}AR$. In such a study, HMA (**5**) and **62c** showed to decrease the residence time (RT) of [^{3}H]ZM241385 of 2.2- and 1.5-fold, respectively. In the case of **62b**, it barely influenced the RT, evidencing a competitive effect of **62b**. At W246A $hA_{2A}AR$, **62a** and **62c** showed similar effects with respect to that observed in WT receptors, producing 1.9- and 2.1-fold decreases in RT; otherwise, **62b** significantly decreased the RT by 2.3-fold, suggesting a noncompetitive mechanism for such compound in the absence of the bulky tryptophan side chain. Overall, these findings emphasized the key role played by Trp246 in governing the allosteric behavior of this class of compounds.

Very recently [92], the research group of Shui developed an affinity mass spectrometry (MS) fragment screening, leading to the discovery of a negative allosteric modulator (NAM) of $A_{2A}AR$ characterized by an azetidine moiety, namely Fg754 (**8**, Fig. 18), able to engage contacts within the allosteric sodium ion binding site. Subsequent molecular dynamics (MD) simulations, SAR, and nuclear magnetic resonance (NMR) analyses revealed a unique binding mechanism and conformation-related functional effect of **8**, that is considerably distinct from HMA (**5**) [92].

The research group of Giorgi et al. contributed to this field by developing a class of purine- and 8-azapurine-N^6-1,3-diphenylurea derivatives, Fig. 18 [145]. Compound **63** possessed the best modulatory property on $A_{2A}AR$. Tested at 10 μM in the equilibrium and kinetic dissociation studies, **63** showed to increase the binding of both radioligands [^{3}H]ZM241385 and [^{3}H]CGS21680 and to decrease the dissociation rate of [^{3}H]ZM241385 from the $A_{2A}AR$. Finally, in functional assays performed on rat aortic rings, **63** enhanced the vasorelaxant effect induced by the agonist CGS21680, by increasing both its efficacy and potency.

Finally, the AEA061, whose structure was not unveiled, has been reported to act as $A_{2A}AR$ PAM, by increasing adenosine potency as well as efficacy, thereby enhancing the functional responsiveness of the receptor [146].

5.2.1 Biological Activity

The $A_{2A}AR$ as well as the A_1AR is an attractive drug target for several pathological conditions. Despite several clinical trials on $A_{2A}AR$ agonists and antagonists have been reported, only three drugs have been approved for clinical use: adenosine for myocardial perfusion imaging, the short-life agonist Regadenoson for myocardial perfusion imaging, and the receptor antagonist Istradefylline for the treatment of Parkinson's disease. The other drug is the xanthine antagonist istradefylline recently approved by the US FDA for the treatment of Parkinson's disease. As for the A_1AR subtype, the application of an allosteric modulator can improve the positive effects of receptor stimulation by reducing the side effects. The anti-inflammatory activity of adenosine is mainly mediated by the activation of the $A_{2A}AR$ subtype. Consistently, the AEA061 counteracts the lipopolysaccharide (LPS)-induced inflammation in mouse monocytes and splenocytes decreasing the production and release of pro-inflammatory cytokines such as tumor necrosis factor-alpha (TNF-α), macrophage inflammatory protein-1α, -1β, -2, interleukin-1α, and keratinocyte chemokine [146]. Interestingly, the AEA061 potentiates also the inosine-mediated $A_{2A}R$ activation promoting the anti-inflammatory activity of inosine in splenic monocytes [147]. In addition, oral administration of AEA061 caused a reduction in ear swelling, skin thickness, erythema, scale formation, and inflammatory cytokine expression in mice with imiquimod-induced psoriasis-like dermatitis [148].

The $A_{2A}AR$ is expressed in the CNS and as reported above, its interaction with the D_2R leads to the development of drugs for neurodegenerative disorders. Interestingly, the $A_{2A}AR$ is expressed in the neurons of Nucleus Accumbens (NAc) [149] that represents a new sleep-regulating area through the integration of motivational stimuli. Despite the $A_{2A}AR$ activation has a sleep-inducing effects [150], the use of orthosteric agonists is limited by the cardiovascular side effects. Interestingly, the administration of the $A_{2A}R$ PAM-1 (**64**) in mice promotes slow-wave sleep (SWS) without the modification of the body temperature, blood pressure, or heart rate [151], Fig. 18.

5.3 A_{2B} AR Allosteric Modulators

The first class of allosteric modulators of A_{2B} AR was serendipitously discovered in 2013 by Taliani et al. [152]. In detail, since 2001 the same research group has disclosed a class of AR antagonists featuring the [1,2,4]triazino[4,3-*a*]benzimidazole scaffold (TBI) [153–155], previously exploited for the identification of potent BzR ligands [153, 156] as geometrically constrained analogs of the indol-3-ylglyoxylamides [157–164]. Variable substitution patterns at position 3 (R) and 10 (R′) of the tricyclic system allowed the identification of some potent and selective A_1 AR and A_{2B} AR antagonists [153–155].

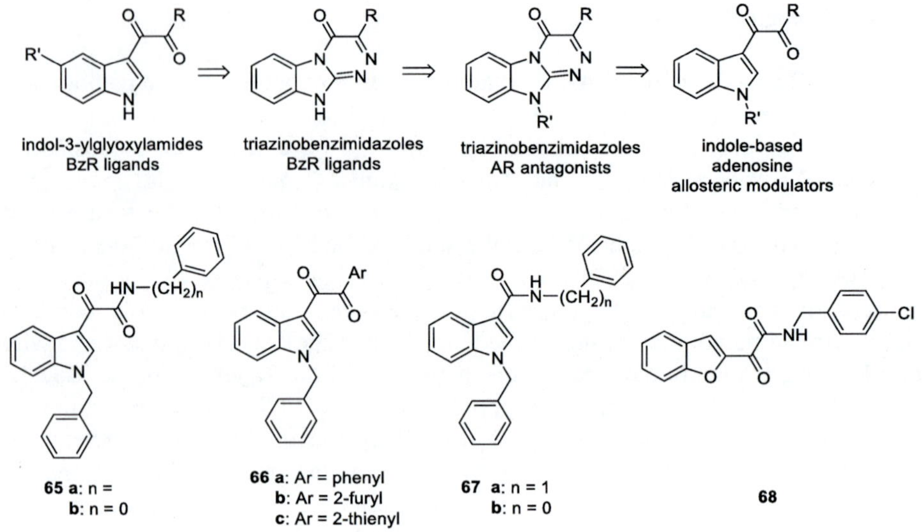

Fig. 19 Structures of allosteric modulators of A_{2B} AR

Continuing the searches for novel lead compounds binding to ARs, a series of 1-benzyl-3-ketoindoles (**65a-b**, **66a-c**, **67a-b**, Fig. 19), designed as open-chain analogs of the TBI, was investigated [47, 152]. The pharmacological profile of these compounds was investigated and validated in a deep study by means of binding and functional assays performed on CHO cells expressing A_1, A_{2A}, A_{2B}, and A_3 receptors [165]. Concerning A_1, A_{2A}, and A_3 receptors, only **66a** and **66b** displayed submicromolar affinities for the A_1 AR (Ki values of 161 nM and 343 nM, respectively), whereas the remaining five indoles showed no appreciable affinity for the three AR subtypes. Functional experiments revealed that **66a** and **66b** acted as A_1 AR antagonists. When tested for their ability to modify the cAMP levels in CHO expressing A_{2B} AR, no compounds increase the cAMP levels at the concentration of 10 μM, clearly showing a lack of A_{2B} AR agonist activity. However, when the experiments were repeated in the presence of NECA, **65a,b** and **66a** potentiated its agonist effects, suggesting that these three compounds act as PAMs of the A_{2B} AR. Conversely, **66b,c** and **67a,b** potently counteracted the NECA-mediated increase in cAMP, indicating that they behave as A_{2B} AR NAMs. In particular, PAMs increase the efficacy of A_{2B} AR agonists (NECA, BAY 60-6583, adenosine) without producing any effect on potency, while NAMs reduce both potency and the efficacy of agonists.

A plausible hypothesis is that PAMs **65a,b** and **66a** affect specific conformational states of A_{2B} AR to improve the functional coupling to the intracellular signaling system without altering the conformation of the orthosteric site, while **66b,c** and **67a, b** shifted the conformational state toward the inactive one and modified the conformation of the orthosteric site [47].

The most interesting feature of this class is that small differences in the decoration around the indole scaffold determine a large difference in the efficacy of compounds, passing from positive to negative modulators. However, only preliminary SARs were delineated because of the limited number of compounds. In particular, **65** and **67** behave as PAMs or NAMs according to the nature of the linker between the indole nucleus and thc side chain, namely glyoxylamide for **65** and carboxamide for **67**. For **66**, the interaction with the protein strictly depends on the nature of the pendant aromatic ring: phenyl-substituted **66a** is a PAM while **66b** and **66c**, featuring a furyl or thienyl ring, respectively, are NAMs.

Very recently [166], a novel series of **65–67** chemically related compounds was developed and investigated as bone mineralization agents (Fig. 19). This library was constituted by novel indole-based derivatives bearing various substitutions at 1- and/or 3-positions and by derivatives characterized by different aromatic heterocycles in place of indole.

Notably, SAR study in terms of matrix mineralization stimulation activity in MSCs (either in the presence or in the absence of the agonist BAY60-6583) evidenced that the indole ring and N^1-arylalkyl group do not represent key pharmacophoric elements to display A_{2B} AR positive allosteric modulation. Compound **68** (Fig. 19), characterized by a particular chemical structure with respect to the reference indoles **65–67**, was identified as the most active compound, so represents a lead structure for further future development in this field.

5.3.1 Biological Activity

To date no A_{2B}AR ligands are in the clinical application; however, several pre-clinical data have been reported. The A_{2B}AR agonist and antagonist have been proposed for acute lung injury, ischemia, vascular leakage, metabolic disorders, and bone defects [167]. The recent biological applications of the A_{2B}AR ligands have been recently reviewed by Barresi et al. [167].

Regarding the use of allosteric modulators, due to the recent discovery of the first PAMs, few data have been reported in the literature. The A_{2B} AR PAM **65b (7b**, Fig. 3) has been investigated as a therapeutic tool for the treatment of bone-related diseases (e.g., osteoporosis, rheumatoid arthritis, osteogenesis imperfecta, multiple myeloma, fracture malunion) [165]. Results evidenced the ability of **65b** (**7b**, Fig. 3) to potentiate the effects of A_{2B} AR agonists (adenosine, NECA, and BAY 60-6583) in promoting MSC differentiation to osteoblasts in vitro by increasing the expression of osteoblast-related genes (runx2 and osterix) and osteoblast marker proteins (phosphatase alkaline and osteocalcin) combined with stimulation of osteoblast mineralization.

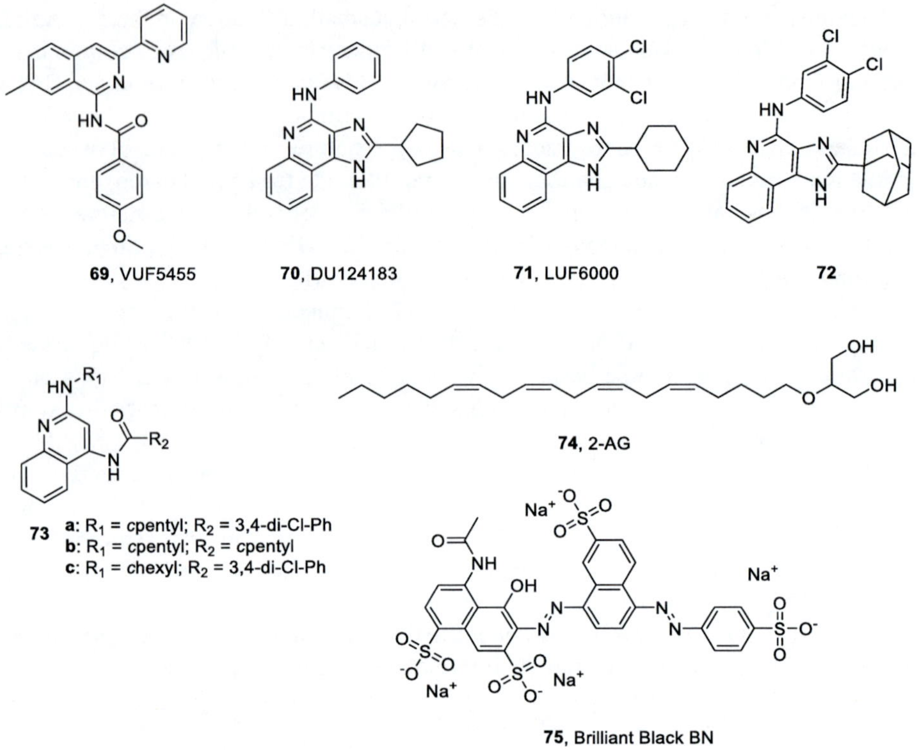

Fig. 20 Structures of A_3AR allosteric modulators

5.4 A_3 AR Allosteric Modulators

The first A_3AR allosteric modulators were discovered in 2001 [168] by the research group of Jacobson, who described a class of 3-(2-pyridinyl)isoquinoline derivatives (Fig. 20) able to selectively slow down the dissociation of the agonist radioligand $[^{125}I]$I-AB-MECA from the A_3ARs in a concentration-dependent manner, without altering the dissociation of the antagonist $[^3H]$PBS-11. Such compounds were also able to significantly increase forskolin-induced cAMP production mediated by the same agonist. In addition, they did not affect the dissociation of both agonist $[^3H]$R-PIA from A_1ARs and the agonist $[^3H]$CGS21680 from $A_{2A}ARs$.

Previously described as A_3AR antagonists by IJzerman and collaborators [169], the majority of compounds displayed also competitive binding properties, with K_i values in the submicromolar/nanomolar range. Among all compounds, compound **69**, also called VUF5455 (Fig. 20) showed higher activity as allosteric modulator (EC_{50} = 10 μM) together with the weakest orthosteric antagonist potency (K_i = 1,680 nM). Therefore, compound **69** was highlighted for the best potency and the most favorable degree of allosteric modulation versus receptor antagonism, becoming the reference lead for the design of pure allosteric enhancer of A_3AR. In addition, this class of compounds showed to be selective allosteric modulator for A_3AR with respect to A_1 and A_{2A} AR subtypes.

An additional class of A_3AR allosteric modulators are the 1*H*-imidazo-[4,5-*c*] quinolin-4-amines [170]. Analogous to the former class, they were first described as non-xanthine antagonists, mainly on A_1AR, by the research group of Van Galen in 1991 [171]. All these compounds resulted able to selectively decrease the dissociation of the agonist [^{125}I]I-AB-MECA from the A_3ARs, without altering that of the antagonist [^{3}H]PBS-11 and the agonist [^{3}H]R-PIA from rat brain A_1ARs and that of [^{3}H]CGS21680 from rat striatal $A_{2A}ARs$. Nevertheless, all these compounds showed also orthosteric effects with K_i values ranging from 120 nM to 101 μM at the same receptor subtype; in addition, some members belonging to this series also bound competitively at the A_1AR. Overall, the most promising lead compound within the current series of derivatives, the 2-cyclopentyl-1*H*-imidazo[4,5-*c*]quinoline DU124183 (**70**) (Fig. 20) showed not only the most favorable ratio of allosteric enhancement to antagonism but also the ability to enhance the maximum efficacy of A_3AR agonist, although decreasing its potency, in the cAMP assays with cloned A_3AR. DU124183 (**70**) displayed also moderate affinity as orthosteric antagonist to *h*A_3AR with K_i of 820 nM.

In a following study, the same authors used **70** as a lead compound to generate a series of analogs featuring modifications at the 4-amino and 2-positions, partly applying the Topliss approach [172]. Compound **71**, also named LUF6000 (Fig. 20), characterized by a 2-cyclohexyl and a 4-(3,4-dichlorophenyl) emerged as the most active compound, being able to enhance agonist efficacy in a functional assay by 45% and reduce dissociation rate without altering agonist potency. This is most probably due to its reduced ability to interact with the orthosteric binding site. In fact, the majority of allosteric modulators belonging to this series are weak inhibitors of orthosteric ligand binding. Clear divergence between the structural requirements for allosteric activity and the inhibition of orthosteric radioligand binding has been observed. In particular, the most favorable groups for allosteric enhancement are a phenyl ring or monocyclic phenyl ring at the 4-amino position and a medium-sized cycloalkyl substituents (cyclopentyl and cyclohexyl) at 2-position. LUF6000 (**71**) also increases the efficacy of inosine, which serves as a second, even though weak, A_3AR endogenous agonist [173].

Further exploration of steric and electronic effects around the 4-arylamino and 2-cycloalkyl moieties of the imidazo[4,5-*c*]quinoline scaffold is then reported by Kim et al. [174], Fig. 20. Results evidenced that only a restricted number of substituents at 2- and 4-positions allow to maintain allosteric activity. In particular, concerning the 4-phenylamino moiety, the best substitution patterns are obtained with the presence of two chloro atoms at different positions (2,4, 3,4, or 3,5), as in **71**, while for the 2-position the insertion of a multiple methylene bridge leads to more active allosteric modulators, without effects on the orthosteric radioligand binding. In addition, the hydrophobic substituents at 2-position are strictly required because the introduction of a nitrogen atom at such level results detrimental to allosteric activity. Of note, the highest ability to enhance the maximum efficacy of agonist Cl-IB-MECA is observed for compound **72**, with 210% of the activity with respect to control in the [^{35}S]GTPγS binding assay.

In the same year [175], IJzerman et al., searching for a new chemical template for the development of selective A_3AR allosteric modulators devoid of orthosteric effects, formally opened the imidazole ring of LUF6000 (**71**) and obtained a series of 2,4-disubstituted quinolines (general formula **73**) endowed with interesting allosteric enhancement activity (Fig. 20). Most compounds, tested at 10 μM, modestly inhibited orthosteric radioligand binding to A_1 and A_3AR. In particular, compounds featuring a cyclopentyl moiety (**73a-b**) decreased the ^{125}I-IB-MECA dissociation rate from A_3AR, without affecting that of A_1AR. Among them, the highest allosteric enhancement was obtained by those featured a (substituted)phenyl ring at R_2, while the introduction at such position of a benzyl, cyclopentyl, or indazole resulted detrimental. In particular, the most potent allosteric compound within this subseries resulted in **73a**, characterized by a 3,4-dichlorophenyl ring at R_2 position. The replacement of the cyclopentyl ring with the cyclobutyl one at R_1 position led to less active compounds; otherwise, the introduction of a cyclohexyl moiety proved to be beneficial, leading to the most potent enhancer of the series, namely **73c** (LUF6096). In addition, compound **73c** showed a decreased, almost negligible, orthosteric effect on A_1 and A_3AR. In a functional assay, **73c** increased the intrinsic efficacy of Cl-IB-MECA, and, differently from **71**, also the potency of Cl-IB-MECA.

The already mentioned amiloride **7** and its analogs behaved also as allosteric inhibitors of antagonist binding at A_3AR, Fig. 17 [144], with HMA (**5**) being the most potent one.

Given that cannabinoid receptors and A_3 AR are co-localized, i.e. in glia and immune cells such as mast cells, the research group of Lane et al. investigated the effects of several endogenous as well as synthetic cannabinoid ligands on A_3 AR [176]. Of note, 2-arachidonylglycerol (2-AG (**74**), Fig. 20), proved to be the most potent compound, resulting able to inhibit agonist [^{125}I]I-AB-MECA binding at the hA_3AR and enhance its dissociation rate, suggesting such compound as NAM of hA_3AR. In a functional assay, **74** decreased both the potency of an agonist at the hA_3AR and the basal signaling of this receptor. As the hA_3AR is expressed in astrocytes and microglia, these results may be noteworthy in cerebral ischemia, a pathological condition characterized by high levels of 2-AG (**74**).

Brilliant Black BN (**75**, Fig. 20) is a food dye used in calcium assays at concentrations ranging from 100 to 500 μM to quench extracellular fluorescence and to avoid washing cells before stimulation [177]. To verify its pharmacological inertness, its influence on ligand-receptor complexes was evaluated. Results evidenced the ability of **75** to decrease the affinity of antagonists at both A_1AR and A_3AR with an allosteric mechanism of action. In addition, **75** did not affect the potency of NECA in stimulating a calcium response through A_1AR and A_3AR, whereas it decreased the antagonism mediated by 1 μM xanthine amine congener (XAC) and its fluorescent derivative, XAC-X-BY630, in a dose-dependent manner at both A_1AR and A_3AR. Overall, these findings evidenced that **75** is not an inert substance and as such its use as a food dye or pharmaceutical excipient should be carefully evaluated.

5.4.1 Biological Activity

A_3AR as well as $A_{2A}AR$ offers a promising therapeutic target for inflammatory diseases, such as rheumatoid arthritis and psoriasis [178]. Its importance in treating cancer is also increasing [179] hope. Similarly, to the $A_{2B}AR$, the biological application of the A_3AR allosteric modulators is limited. LUF6000 (**71**) has been reported to exhibit anti-inflammatory effects in three different experimental animal models of rat adjuvant-induced arthritis, monoiodoacetate-induced osteoarthritis, and concanavalin A-induced liver inflammation in mice, suggesting its potential use in the treatment of different autoimmune inflammatory diseases [180]. The molecule is termed also CF602 and is on a translational path for the treatment of erectile dysfunction [181]. LUF6096 (**73c**) protects the heart from the ischemic injury in in vivo model of canine cardiac ischemia [182].

6 Final Remarks

ARs have been studied for several decades due to their involvement in pathological processes. Despite the huge amount of agonists and antagonists synthesized so far, to date, only few molecules have reached clinical application. The highly conserved orthosteric binding site among the different receptor subtypes has hampered the development of new drugs. In fact, the classical agonists and antagonists often produce on- and off-target side effects and they do not discriminate specific signaling pathways.

With the rapid breakthroughs in structural biology, several allosteric modulators of ARs have been revealed. The allosteric modulators act only when the endogenous ligand is present producing more physiological and site-specific effects. Despite the advantages of allosteric modulators with respect to the orthosteric ones, the development of such ligands is hampered by the lack of structural information on the allosteric binding site. Of note, the first high-resolution crystal structure of a PAM bound to an ARs has been reported only a few years ago. The increased knowledge of the AR allosteric binding site can direct and accelerate the drug discovery thanks also to the advances in the NMR spectroscopy, the molecular dynamic simulations, and the imaging technique.

The frontiers of the allostery have been recently moved to the discovery of BAMs. The pathological processes are often driven by the specific activation of intracellular signaling. BAMs bind to the allosteric site as the non-bias counterpart, however, they perform a fine-tuning of a specific receptor signaling. The development of unrecognized BAMs for ARs can prompt their targeting for novel therapeutic applications and increase the knowledge of the involvement of specific signaling in pathological mechanisms.

Despite the fields of allosteric modulation, un-bias and bias, are growing, several challenges have to be faced. The activity of these ligands is affected by the in vivo environment and for BAMs also by the specific effector ratio in the site of action. The possibility to recapitulate such conditions in vitro to improve drug screening is challenging. Furthermore, allosteric sites are less conserved across species reducing the possibility to set up efficient animal models. Finally, the last discovery on A_1 allosteric modulator [140] reveals the importance to perform a more detailed biological characterization of the new allosteric modulators. The affinity cannot be exhaustive of their therapeutic potential. In this respect, probe dependence and cooperativity with the agonist are two features of pivotal importance to consider.

In conclusion, despite the critical questions that remain, the allostery may represent an attractive field in AR drug discovery in the future years.

Compliance with Ethical Standards **Conflict of Interest**: "The authors have no conflicts of interest to declare."

Funding: The Authors received no financial support.

Ethical Approval: "The chapter is a review and no humans or animals were involved in the preparation of the chapter."

References

1. Schiöth HB, Lagerström MC (2008) Structural diversity of G protein-coupled receptors and significance for drug discovery. Nat Rev Drug Discov 7:339–357. https://doi.org/10.1038/nrd2518
2. Sriram K, Insel PA (2018) G protein-coupled receptors as targets for approved drugs: how many targets and how many drugs? Mol Pharmacol 93:251–258. https://doi.org/10.1124/mol.117.111062
3. Kobilka BK (2007) G protein coupled receptor structure and activation. Biochim Biophys Acta Biomembr 1768:794–807. https://doi.org/10.1016/j.bbamem.2006.10.021
4. Avlani VA, Gregory KJ, Morton CJ, Parker MW, Sexton PM, Christopoulos A (2007) Critical role for the second extracellular loop in the binding of both orthosteric and allosteric G protein-coupled receptor ligands. J Biol Chem 282:25677–25686. https://doi.org/10.1074/jbc.M702311200
5. Seibt BF, Schiedel AC, Thimm D, Hinz S, Sherbiny FF, Müller CE (2013) The second extracellular loop of GPCRs determines subtype-selectivity and controls efficacy as evidenced by loop exchange study at A2 adenosine receptors. Biochem Pharmacol 85:1317–1329. https://doi.org/10.1016/j.bcp.2013.03.00
6. Lefkowitz RJ, Shenoy SK (2022) Transduction of receptor signals by β-arrestins. Science 308:512–517. https://doi.org/10.1126/science.1109237
7. Zhou Q, Yang D, Wu M, Guo Y, Guo W, Zhong L et al (2019) Common activation mechanism of class a GPCRs. Elife 8:e50279. https://doi.org/10.7554/eLife.50279
8. Tesmer JJG (2016) Hitchhiking on the heptahelical highway: structure and function of 7TM receptor complexes. Nat Rev Mol Cell Biol 17:439–450. https://doi.org/10.1038/nrm.2016.36
9. Wold EA, Zhou J (2018) GPCR allosteric modulators: mechanistic advantages and therapeutic applications. Curr Top Med Chem 18:2002–2006. https://doi.org/10.2174/1568026619999190101151837

10. Keov P, Sexton PM, Christopoulos A (2011) Allosteric modulation of G protein-coupled receptors: a pharmacological perspective. Neuropharmacology 60:24–35. https://doi.org/10.1016/j.neuropharm.2010.07.010
11. Bohr C, Hasselbalch K, Krogh A (1904) Ueber einen in biologischer Beziehung wichtigen Einfluss, den die Kohlensäurespannung des Blutes auf dessen Sauerstoffbindung übtl. Skand Arch Physiol 16:402–412
12. Monod J, Jacob F (1961) General conclusions: teleonomic mechanisms in cellular metabolism, growth, and differentiation. Cold Spring Harb Symp Quant Biol 26:389–401. https://doi.org/10.1101/sqb.1961.026.01.048
13. Ehlert FJ (1985) The relationship between muscarinic receptor occupancy and adenylate cyclase inhibition in the rabbit myocardium. Mol Pharmacol 28:410–421
14. Kenakin TP, Kenakin T (2012) Biased signalling and allosteric machines: new vistas and challenges for drug discovery. Br J Pharmacol 165:1659–1669. https://doi.org/10.1111/j.1476-5381.2011.01749.x
15. Christopoulos A (2002) Allosteric binding sites on cell-surface receptors: novel targets for drug discovery. Nat Rev Drug Discov 1:198–210. https://doi.org/10.1038/nrd746
16. De Lean A, Stadel JM, Lefkowitz RJ (1985) A ternary complex model explains the agonist-specific binding properties of the adenylate cyclase-coupled beta-adrenergic receptor. J Biol Chem 255:7108–7117
17. Costa T, Herz A (1989) Antagonists with negative intrinsic activity at δ opioid receptors coupled to GTP-binding proteins. Proc Natl Acad Sci U S A 86:7321–7325. https://doi.org/10.1073/pnas.86.19.7321
18. Leff P (1995) The two-state model of receptor activation. Trends Pharmacol Sci 16:89–97. https://doi.org/10.1016/s0165-6147(00)88989-0
19. Kenakin T (1996) The classification of seven transmembrane receptors in recombinant expression systems. Pharmacol Rev 48:413–463
20. Lefkowitz RJ, Cotecchia S, Samama P, Costa T (1993) Constitutive activity of receptors coupled to guanine nucleotide regulatory proteins. Trends Pharmacol Sci 14:303–307. https://doi.org/10.1016/0165-6147(93)90048-O
21. Samama P, Cotecchia S, Costa T, Lefkowitz RJ (1993) A mutation-induced activated state of the beta 2-adrenergic receptor. Extending the ternary complex model. J Biol Chem 268:4625–4636
22. Weiss JM, Morgan PH, Lutz MW, Kenakin TP (1996) The cubic ternary complex receptor-occupancy model III. Resurrecting efficacy. J Theor Biol 181:381–397. https://doi.org/10.1006/jtbi.1996.0139
23. Ye L, Van Eps N, Zimmer M, Ernst OP, Scott Prosser R (2016) Activation of the A2A adenosine G-protein-coupled receptor by conformational selection. Nature 533:265–268. https://doi.org/10.1038/nature17668
24. May LT, Leach K, Sexton PM, Christopoulos A (2007) Allosteric modulation of G protein-coupled receptors. Annu Rev Pharmacol Toxicol 47:1–51. https://doi.org/10.1146/annurev.pharmtox.47.120505.105159
25. Ehlert FJ (1988) Estimation of the affinities of allosteric ligands using radioligand binding and pharmacological null methods. Mol Pharmacol 33:187–194
26. Stockton JM, Birdsall NJ, Burgen AS, Hulme EC (1983) Modification of the binding properties of muscarinic receptors by gallamine. Mol Pharmacol 23:551–557
27. Litschig S, Gasparini F, Rueegg D, Stoehr N, Flor PJ, Vranesic I et al (1999) CPCCOEt, a noncompetitive metabotropic glutamate receptor 1 antagonist, inhibits receptor signaling without affecting glutamate binding. Mol Pharmacol 55:453–461
28. Urwyler S, Mosbacher J, Lingenhoehl K, Heid J, Hofstetter K, Froestl W et al (2001) Positive allosteric modulation of native and recombinant γ-aminobutyric acid B receptors by 2,6-di-tert-butyl-4-(3-hydroxy-2,2-dimethyl-propyl)-phenol (CGP7930) and its aldehyde analog CGP13501. Mol Pharmacol 60:963–971

29. Zahn K, Eckstein N, Tränkle C, Sadée W, Mohr K (2002) Allosteric modulation of muscarinic receptor signaling: alcuronium-induced conversion of pilocarpine from an agonist into an antagonist. J Pharmacol Exp Ther 301:720–728. https://doi.org/10.1124/jpet.301.2.720
30. Watson C, Jenkinson S, Kazmierski W, Kenakin T (2005) The CCR5 receptor-based mechanism of action of 873140, a potent allosteric noncompetitive HIV entry inhibitor. Mol Pharmacol 67:1268–1282. https://doi.org/10.1124/mol.104.008565
31. Price MR, Baillie GL, Thomas A, Stevenson LA, Easson M, Goodwin R et al (2005) Allosteric modulation of the cannabinoid CB1 receptor. Mol Pharmacol 68:1484–1495. https://doi.org/10.1124/mol.105.016162
32. Hall DA (2000) Modeling the functional effects of allosteric modulators at pharmacological receptors: an extension of the two-state model of receptor activation. Mol Pharmacol 58:1412–1423. https://doi.org/10.1124/mol.58.6.1412
33. Leach K, Sexton PM, Christopoulos A (2007) Allosteric GPCR modulators: taking advantage of permissive receptor pharmacology. Trends Pharmacol Sci 28:382–389. https://doi.org/10.1016/j.tips.2007.06.004
34. Christopoulos A, Changeux JP, Catterall WA, Fabbro D, Burris TP, Cidlowski JA et al (2014) International Union of Basic and Clinical Pharmacology. XC. Multisite pharmacology: recommendations for the nomenclature of receptor allosterism and allosteric ligands. Pharmacol Rev 66:918–947. https://doi.org/10.1124/pr.114.008862
35. Baraldi PG, Iaconinoto MA, Moorman AR, Carrion MD, Lopez Cara C, Preti D et al (2007) Allosteric enhancers for A1 adenosine receptor. Mini Rev Med Chem 7:559–569. https://doi.org/10.2174/138955707780859459
36. Binet V, Brajon C, Le Corre L, Acher F, Pin JP, Prézeau L (2004) The heptahelical domain of GABAB2 is activated directly by CGP7930, a positive allosteric modulator of the GABAB receptor. J Biol Chem 279:29085–29091. https://doi.org/10.1074/jbc.M400930200
37. Langmead CJ, Fry VAH, Forbes IT, Branch CL, Christopoulos A, Wood MD et al (2006) Probing the molecular mechanism of interaction between 4-n-Butyl-1-[4-(2-methylphenyl)-4-oxo-1-butyl]-piperidine (AC-42) and the muscarinic M1 receptor: direct pharmacological evidence that AC-42 is an allosteric agonist. Mol Pharmacol 69:236–246. https://doi.org/10.1124/mol.105.017814
38. Cong Z, Chen LN, Ma H, Zhou Q, Zou X, Ye C et al (2021) Molecular insights into ago-allosteric modulation of the human glucagon-like peptide-1 receptor. Nat Commun 12:1–11. https://doi.org/10.1038/s41467-021-24058-z
39. Cao AM, Quast RB, Fatemi F, Rondard P, Pin JP, Margeat E (2021) Allosteric modulators enhance agonist efficacy by increasing the residence time of a GPCR in the active state. Nat Commun 12:1–13. https://doi.org/10.1038/s41467-021-25620-5
40. Lane JR, Sexton PM, Christopoulos A (2013) Bridging the gap: bitopic ligands of G-protein-coupled receptors. Trends Pharmacol Sci 34:59–66. https://doi.org/10.1016/j.tips.2012.10.003
41. Valant C, May LT, Aurelio L, Chuo CH, White PJ, Baltos JA et al (2014) Separation of on-target efficacy from adverse effects through rational design of a bitopic adenosine receptor agonist. Proc Natl Acad Sci U S A 111:4614–4619. https://doi.org/10.1073/pnas.1320962111
42. Digby GJ, Utley TJ, Lamsal A, Sevel C, Sheffler DJ, Lebois EP et al (2012) Chemical modification of the M(1) agonist VU0364572 reveals molecular switches in pharmacology and a bitopic binding mode. ACS Chem Nerosci 3:1025–1036. https://doi.org/10.1021/cn300103e
43. Melancon BJ, Gogliotti RD, Tarr JC, Saleh SA, Chauder BA, Lebois EP et al (2012) Continued optimization of the MLPCN probe ML071 into highly potent agonists of the hM1 muscarinic acetylcholine receptor. Bioorg Med Chem Lett 22:3467–3472. https://doi.org/10.1016/j.bmcl.2012.03.088
44. Arnatt C, Zhang Y (2014) Bivalent ligands targeting chemokine receptor dimerization: molecular design and functional studies. Curr Top Med Chem 14:1606–1618. https://doi.org/10.2174/1568026614666140827144752

45. Pulido D, Casadó-Anguera V, Gómez-Autet M, Llopart N, Moreno E, Casajuana-Martin N et al (2022) Heterobivalent ligand for the adenosine A2A-dopamine D2 receptor heteromer. J Med Chem 65:616–632. https://doi.org/10.1021/acs.jmedchem.1c01763
46. Valant C, Aurelio L, Devine SM, Ashton TD, White JM, Sexton PM et al (2012) Synthesis and characterization of novel 2-amino-3-benzoylthiophene derivatives as biased allosteric agonists and modulators of the adenosine A 1 receptor. J Med Chem 55:2367–2375. https://doi.org/10.1021/jm201600e
47. Trincavelli ML, Giacomelli C, Daniele S, Taliani S, Cosimelli B, Laneri S et al (2014) Allosteric modulators of human A2B adenosine receptor. Biochim Biophys Acta 1840: 1194–1203. https://doi.org/10.1016/j.bbagen.2013.12.021
48. Gao ZG, Ijzerman AP (2000) Allosteric modulation of A2A adenosine receptors by amiloride analogues and sodium ions. Biochem Pharmacol 60:669–676. https://doi.org/10.1016/s0006-2952(00)00360-9
49. Kara E, Lin H, Strange PG (2010) Co-operativity in agonist binding at the D2 dopamine receptor: evidence from agonist dissociation kinetics. J Neurochem 112:1442–1453. https://doi.org/10.1111/j.1471-4159.2009.06554
50. May LT, Bridge LJ, Stoddart LA, Briddon SJ, Hill SJ (2011) Allosteric interactions across native adenosine-A3 receptor homodimers: quantification using single-cell ligand-binding kinetics. FASEB J 25:3465–3476. https://doi.org/10.1096/fj.11-186296
51. Kenakin T (2011) Functional selectivity and biased receptor signaling. J Pharmacol Exp Ther 336:296–302. https://doi.org/10.1124/jpet.110.173948
52. Rasmussen SGF, Devree BT, Zou Y, Kruse AC, Chung KY, Kobilka TS et al (2011) Crystal structure of the β2 adrenergic receptor–Gs protein complex. Nature 477:549–555. https://doi.org/10.1038/nature10361
53. Lee Y, Warne T, Nehmé R, Pandey S, Dwivedi-Agnihotri H, Chaturvedi M et al (2020) Molecular basis of β-arrestin coupling to formoterol-bound β1-adrenoceptor. Nature 583:862–866. https://doi.org/10.1038/s41586-020-2419-1
54. Kim M, Vishnivetskiy SA, Van Eps N, Alexander NS, Cleghorn WM, Zhan X et al (2012) Conformation of receptor-bound visual arrestin. Proc Natl Acad Sci U S A 109:18407–18412. https://doi.org/10.1073/pnas.1216304109
55. Zhou XE, He Y, de Waal PW, Gao X, Kang Y, Van Eps N et al (2017) Identification of phosphorylation codes for arrestin recruitment by G protein-coupled receptors. Cell 170:457–469. https://doi.org/10.1016/j.cell.2017.07.002.e13
56. Thomsen ARB, Plouffe B, Cahill TJ, Shukla AK, Tarrasch JT, Dosey AM et al (2016) GPCR-G protein-β-arrestin super-complex mediates sustained G protein signaling. Cell 166:907–919. https://doi.org/10.1016/j.cell.2016.07.004
57. Müller CE, Schiedel AC, Baqi Y (2012) Allosteric modulators of rhodopsin-like G protein-coupled receptors: opportunities in drug development. Pharmacol Ther 135:292–315. https://doi.org/10.1016/j.pharmthera.2012.06.002
58. Slosky LM, Caron MG, Barak LS (2021) Biased allosteric modulators: new Frontiers in GPCR drug discovery. Trends Pharmacol Sci 42:283–299. https://doi.org/10.1016/j.tips.2020.12.005
59. Scholten DJ, Canals M, Maussang D, Roumen L, Smit MJ, Wijtmans M et al (2012) Pharmacological modulation of chemokine receptor function. Br J Pharmacol 165:1617–1643. https://doi.org/10.1111/j.1476-5381.2011.01551.x
60. Raddatz R, Schaffhauser H, Marino MJ (2007) Allosteric approaches to the targeting of G-protein-coupled receptors for novel drug discovery: a critical assessment. Biochem Pharmacol 74:383–391. https://doi.org/10.1016/j.bcp.2007.05.007
61. Urwyler S (2011) Allosteric modulation of family C G-protein-coupled receptors: from molecular insights to therapeutic perspectives. Pharmacol Rev 63:59–126. https://doi.org/10.1124/pr.109.002501
62. Klotz KN, Hessling J, Hegler J, Owman C, Kull B, Fredholm BB et al (1998) Comparative pharmacology of human adenosine receptor subtypes – characterization of stably transfected receptors in CHO cells. Naunyn Schmiedebergs Arch Pharmacol 357:1–9. https://doi.org/10.1007/pl00005131

63. Kooistra AJ, Munk C, Hauser AS, Gloriam DE (2021) An online GPCR structure analysis platform. Nat Struct Mol Biol 28:875–878. https://doi.org/10.1038/s41594-021-00675-6
64. Beukers MW, Den Dulk H, Van Tilburg EW, Brouwer J, Ijzerman AP (2000) Why are A2B receptors low-affinity adenosine receptors? Mutation of Asn273 to Tyr increases affinity of human A2B receptor for 2-(1-hexynyl)adenosine. Mol Pharmacol 58:1349–1356. https://doi.org/10.1124/mol.58.6.1349
65. Fredholm BB, Irenius E, Kull B, Schulte G (2001) Comparison of the potency of adenosine as an agonist at human adenosine receptors expressed in Chinese hamster ovary cells. Biochem Pharmacol 61:443–448. https://doi.org/10.1016/s0006-2952(00)00570-0
66. Fredholm BB, IJzerman AP, Jacobson KA, Linden J, Müller CE (2011) International union of basic and clinical pharmacology. LXXXI. Nomenclature and classification of adenosine receptors – an update. Pharmacol Rev 63:1–34. https://doi.org/10.1124/pr.110.003285
67. Giacomelli C, Daniele S, Romei C, Tavanti L, Neri T, Piano I et al (2018) The A2B adenosine receptor modulates the epithelial- mesenchymal transition through the balance of cAMP/PKA and MAPK/ERK pathway activation in human epithelial lung cells. Front Pharmacol 9:54. https://doi.org/10.3389/fphar.2018.00054
68. Linden J, Thai T, Figler H, Jin X, Robeva AS (1999) Characterization of human A2B adenosine receptors: radioligand binding, western blotting, and coupling to Gqin human embryonic kidney 293 cells and HMC-1 mast cells. Mol Pharmacol 56:705–713
69. Offermanns S, Simon MI (1995) Gα15 and Gα16 couple a wide variety of receptors to phospholipase C. J Biol Chem 270:15175–15180. https://doi.org/10.1074/jbc.270.25.15175
70. Berman HM, Westbrook J, Feng Z, Gilliland G, Bhat TN, Weissig H et al (2000) The protein data bank. Nucleic Acids Res 28:235–242. https://doi.org/10.1093/nar/28.1.235
71. Pándy-Szekeres G, Esguerra M, Hauser AS, Caroli J, Munk C, Pilger S et al (2022) The G protein database, GproteinDb. Nucleic Acids Res 50:D518–D525. https://doi.org/10.1093/nar/gkab852
72. Jespers W, Schiedel AC, Heitman LH, Cooke RM, Kleene L, van Westen GJP et al (2018) Structural mapping of adenosine receptor mutations: ligand binding and signaling mechanisms. Trends Pharmacol Sci 39:75–89. https://doi.org/10.1016/j.tips.2017.11.001
73. Eddy MT, Lee MY, Gao ZG, White KL, Didenko T, Horst R et al (2018) Allosteric coupling of drug binding and intracellular signaling in the A2A adenosine receptor. Cell 172:68–80. https://doi.org/10.1016/j.cell.2017.12.004
74. Fuxe K, Agnati LF, Benfenati F, Celani M, Zini I, Zoli M et al (1983) Evidence for the existence of receptor–receptor interactions in the central nervous system. Studies on the regulation of monoamine receptors by neuropeptides. J Neural Transm Suppl 18:165–179
75. Guidolin D, Marcoli M, Tortorella C, Maura G, Agnati LF (2018) G protein-coupled receptor-receptor interactions give integrative dynamics to intercellular communication. Rev Neurosci 29:703–726. https://doi.org/10.1515/revneuro-2017-0087
76. Borroto-Escuela DO, Brito I, Romero-Fernandez W, Di Palma M, Oflijan J, Skieterska K et al (2014) The G protein-coupled receptor heterodimer network (GPCR-HetNet) and its hub components. Int J Mol Sci 15:8570–8590. https://doi.org/10.3390/ijms15058570
77. Franco R, Cordomí A, Llinas del Torrent C, Lillo A, Serrano-Marín J, Navarro G et al (2021) Structure and function of adenosine receptor heteromers. Cell Mol Life Sci 78:3957–3968. https://doi.org/10.1007/s00018-021-03761-6
78. Fuxe K, Ferré S, Zoli M, Agnati LF (1998) Integrated events in central dopamine transmission as analyzed at multiple levels. Evidence for intramembrane adenosine A2A/dopamine D2 and adenosine A1/dopamine D1 receptor interactions in the basal ganglia. Brain Res Rev 26:258–273. https://doi.org/10.1016/s0165-0173(97)00049-0
79. Trincavelli ML, Daniele S, Orlandini E, Navarro G, Casadó V, Giacomelli C et al (2012) A new D2 dopamine receptor agonist allosterically modulates A2A adenosine receptor signalling by interacting with the A2A/D2 receptor heteromer. Cell Signal 24:951–960. https://doi.org/10.1016/j.cellsig.2011.12.018

80. Fuxe K, Guidolin D, Agnati LF, Borroto-Escuela DO (2015) Dopamine heteroreceptor complexes as therapeutic targets in Parkinson's disease. Expert Opin Ther Targets 19:377–398. https://doi.org/10.1517/14728222.2014.981529
81. Borroto-Escuela DO, Fuxe K (2019) Adenosine heteroreceptor complexes in the basal ganglia are implicated in Parkinson's disease and its treatment. J Neural Transm 126:455–471. https://doi.org/10.1007/s00702-019-01969-2
82. Jenner P, Mori A, Aradi SD, Hauser RA (2021) Istradefylline – a first generation adenosine A2A antagonist for the treatment of Parkinson's disease. Expert Rev Neurother 21:317–333. https://doi.org/10.1080/14737175.2021.1880896
83. Chen JF, Moratalla R, Impagnatiello F, Grandy DK, Cuellar B, Rubinstein M et al (2001) The role of the D2 dopamine receptor (D2R) in A2A adenosine receptor (A2AR)-mediated behavioral and cellular responses as revealed by A2A and D2 receptor knockout mice. Proc Natl Acad Sci U S A 98:1970–1975. https://doi.org/10.1073/pnas.98.4.1970
84. Liu W, Chun E, Thompson AA, Chubukov P, Xu F, Katritch V et al (2012) Structural basis for allosteric regulation of GPCRS by sodium ions. Science 337:232–236. https://doi.org/10.1126/science.1219218
85. Katritch V, Fenalti G, Abola EE, Roth BL, Cherezov V, Stevens RC (2014) Allosteric sodium in class A GPCR signaling. Trends Biochem Sci 39:233–244. https://doi.org/10.1016/j.tibs.2014.03.002
86. Massink A, Gutiérrez-De-Terán H, Lenselink EB, Zacarías NVO, Xia L, Heitman LH et al (2015) Sodium ion binding pocket mutations and adenosine A2A receptor function. Mol Pharmacol 87:305–313. https://doi.org/10.1124/mol.114.095737
87. Gutiérrez-De-Terán H, Massink A, Rodríguez D, Liu W, Han GW, Joseph JS et al (2013) The role of a sodium ion binding site in the allosteric modulation of the A2A adenosine G protein-coupled receptor. Structure 21:2175–2285. https://doi.org/10.1016/j.str.2013.09.020
88. Selvam B, Shamsi Z, Shukla D (2018) Universality of the sodium ion binding mechanism in class A G-protein-coupled receptors. Angew Chem Int Ed 57:3048–3053. https://doi.org/10.1002/anie.201708889
89. Vickery ON, Carvalheda CA, Zaidi SA, Pisliakov AV, Katritch V, Zachariae U (2018) Intracellular transfer of Na^+ in an active-state G-protein-coupled receptor. Structure 26:171–180. https://doi.org/10.1016/j.str.2017.11.013
90. Zarzycka B, Zaidi SA, Roth BL, Katritch V (2019) Harnessing ion-binding sites for GPCR pharmacology. Pharmacol Rev 71:571–595. https://doi.org/10.1124/pr.119.017863
91. Massink A, Louvel J, Adlere I, Van Veen C, Huisman BJH, Dijksteel GS et al (2016) 5′-substituted amiloride derivatives as allosteric modulators binding in the sodium ion pocket of the adenosine A2A receptor. J Med Chem 59:4769–4777. https://doi.org/10.1021/acs.jmedchem.6b00142
92. Lu Y, Liu H, Yang D, Zhong L, Xin Y, Zhao S et al (2021) Affinity mass spectrometry-based fragment screening identified a new negative allosteric modulator of the adenosine A2A receptor targeting the sodium ion pocket. ACS Chem Biol 16:991–1002. https://doi.org/10.1021/acschembio.0c00899
93. Hanson MA, Cherezov V, Griffith MT, Roth CB, Jaakola VP, Chien EYT et al (2008) A specific cholesterol binding site is established by the 2.8 Å structure of the human β2-adrenergic receptor. Structure 16:897–905. https://doi.org/10.1016/j.str.2008.05.001
94. McGraw C, Yang L, Levental I, Lyman E, Robinson AS (2019) Membrane cholesterol depletion reduces downstream signaling activity of the adenosine A2A receptor. Biochim Biophys Acta Biomembr 1861:760–767. https://doi.org/10.1016/j.bbamem.2019.01.001
95. Guixà-González R, Albasanz JL, Rodriguez-Espigares I, Pastor M, Sanz F, Martí-Solano M et al (2017) Membrane cholesterol access into a G-protein-coupled receptor. Nat Commun 28:1–12. https://doi.org/10.1038/ncomms14505
96. Huang SK, Almurad O, Pejana RJ, Morrison ZA, Pandey A, Picard LP et al (2022) Allosteric modulation of the adenosine A2A receptor by cholesterol. Elife 11:e73901. https://doi.org/10.7554/eLife.73901

97. Bruns RF, Fergus JH, Coughenour LL, Courtland GG, Pugsley TA, Dodd JH et al (1990) Structure-activity relationships for enhancement of adenosine A1 receptor binding by 2-amino-3-benzoylthiophenes. Mol Pharmacol 38:950–958
98. Bruns RF, Fergus JH (1990) Allosteric enhancement of adenosine A1 receptor binding and function by 2-amino-3-benzoylthiophenes. Mol Pharmacol 38:939–949
99. Musser B, Mudumbi RV, Liu J, Olson RD, Vestal RE (1999) Adenosine A1 receptor-dependent and -independent effects of the allosteric enhancer PD 81,723. J Pharmacol Exp Ther 288:446–454
100. Romagnoli R, Baraldi PG, Moorman AR, Borea PA, Varani K (2015) Current status of A1 adenosine receptor allosteric enhancers. Future Med Chem 7:1247–1259. https://doi.org/10.4155/fmc.15.65
101. Van Der Klein PAM, Kourounakis AP, IJzerman AP (1999) Allosteric modulation of the adenosine A1 receptor. Synthesis and biological evaluation of novel 2-amino-3-benzoylthiophenes as allosteric enhancers of agonist binding. J Med Chem 42:3629–3635. https://doi.org/10.1021/jm991051d
102. Kourounakis AP, Van Der Klein PAM, Ijzerman AP (2000) Elucidation of structure-activity relationships of 2-amino-3-benzoylthiophenes: study of their allosteric enhancing vs. antagonistic activity on adenosine A 1 receptors. Drug Dev Res 49:227–237. https://doi.org/10.1002/1098-2299
103. Childers SR, Li X, Xiao R, Eisenach JC (2005) Allosteric modulation of adenosine A1 receptor coupling to G-proteins in brain. J Neurochem 93:715–723. https://doi.org/10.1111/j.1471-4159.2005.03044.x
104. Li X, Conklin D, Ma W, Zhu X, Eisenach JC (2002) Spinal noradrenergic activation mediates allodynia reduction from an allosteric adenosine modulator in a rat model of neuropathic pain. Pain 97:117–125. https://doi.org/10.1016/s0304-3959(02)00011-8
105. Li X, Conklin D, Pan HL, Eisenach JC (2003) Allosteric adenosine receptor modulation reduces hypersensitivity following peripheral inflammation by a central mechanism. J Pharmacol Exp Ther 305:950–955. https://doi.org/10.1124/jpet.102.047951
106. Baraldi PG, Pavani MG, Leung E, Moorman AR, Varani K, Vincenzi F et al (2006) Synthesis and biological characterization of [3H] (2-amino-4,5,6,7-tetrahydrobenzo[b]thiophen-3-yl)-(4-chlorophenyl)-methanone, the first radiolabelled adenosine A1 allosteric enhancer. Bioorg Med Chem Lett 16:1402–1404. https://doi.org/10.1016/j.bmcl.2005.11.03
107. Baraldi PG, Zaid AN, Lampronti I, Fruttarolo F, Pavani MG, Tabrizi MA et al (2000) Synthesis and biological effects of a new series of 2-amino-3-benzoylthiophenes as allosteric enhancers of A1-adenosine receptor. Bioorg Med Chem Lett 10:1953–1957. https://doi.org/10.1016/s0960-894x(00)00379-6
108. Baraldi PG, Romagnoli R, Pavani MG, Del Carmen NM, Tabrizi MA, Shryock JC et al (2003) Synthesis and biological effects of novel 2-amino-3-naphthoylthiophenes as allosteric enhancers of the A1, adenosine receptor. J Med Chem 46:794–809. https://doi.org/10.1021/jm0210212
109. Romagnoli R, Baraldi PG, Moorman AR, Iaconinoto MA, Carrion MD, Cara CL et al (2006) Microwave-assisted synthesis of thieno[2,3-c]pyridine derivatives as a new series of allosteric enhancers at the adenosine A1 receptor. Bioorg Med Chem Lett 16:5530–5533. https://doi.org/10.1016/j.bmcl.2006.08.041
110. Baraldi PG, Pavani MG, Shryock JC, Moorman AR, Iannotta V, Borea PA et al (2004) Synthesis of 2-amino-3-heteroaroylthiophenes and evaluation of their activity as potential allosteric enhancers at the human A1 receptor. Eur J Med Chem 39:855–865. https://doi.org/10.1016/j.ejmech.2004.06.009
111. Tranberg CE, Zickgraf A, Giunta BN, Luetjens H, Figler H, Murphree LJ et al (2002) 2-Amino-3-aroyl-4,5-alkylthiophenes: agonist allosteric enhancers at human a1 adenosine receptors. J Med Chem 45:382–389. https://doi.org/10.1021/jm010081p
112. Lütjens H, Zickgraf A, Figler H, Linden J, Olsson RA, Scammells PJ (2003) 2-Amino-3-benzoylthiophene allosteric enhancers of A1, adenosine agonist binding: new 3, 4-, and 5-modifications. J Med Chem 46:1870–1877. https://doi.org/10.1021/jm020295m

113. Nikolakopoulos G, Figler H, Linden J, Scammells PJ (2006) 2-Aminothiophene-3-carboxylates and carboxamides as adenosine A1 receptor allosteric enhancers. Bioorg Med Chem 14: 2358–2365. https://doi.org/10.1016/j.bmc.2005.11.018
114. Aurelio L, Figler H, Flynn BL, Linden J, Scammells PJ (2008) 5-substituted 2-aminothiophenes as A1 adenosine receptor allosteric enhancers. Bioorg Med Chem 16: 1319–1327. https://doi.org/10.1016/j.bmc.2007.10.065
115. Aurelio L, Christopoulos A, Flynn BL, Scammells PJ, Sexton PM, Valant C (2011) The synthesis and biological evaluation of 2-amino-4,5,6,7,8,9-hexahydrocycloocta[b]thiophenes as allosteric modulators of the A1 adenosine receptor. Bioorg Med Chem Lett 21:3704–3707. https://doi.org/10.1016/j.bmcl.2011.04.080
116. Aurelio L, Valant C, Flynn BL, Sexton PM, Christopoulos A, Scammells PJ (2009) Allosteric modulators of the adenosine A1 receptor: synthesis and pharmacological evaluation of 4-substituted 2-amino-3-benzoylthiophenes. J Med Chem 52:4543–4547. https://doi.org/10.1021/jm9002582
117. Aurelio L, Valant C, Figler H, Flynn BL, Linden J, Sexton PM et al (2009) 3- and 6-substituted 2-amino-4,5,6,7-tetrahydrothieno[2,3-c]pyridines as A1 adenosine receptor allosteric modulators and antagonists. Bioorg Med Chem 17:7353–7361. https://doi.org/10.1016/j.bmc.2009.08.024
118. Butcher A, Scammells PJ, White PJ, Devine SM, Rose'meyer RB (2013) An allosteric modulator of the adenosine A1 receptor improves cardiac function following ischaemia in murine isolated hearts. Pharmaceuticals (Basel) 6:546–556. https://doi.org/10.3390/ph6040546
119. Romagnoli R, Baraldi PG, Carrion MD, Cara CL, Cruz-Lopez O, Iaconinoto MA et al (2008) Synthesis and biological evaluation of 2-amino-3-(4-chlorobenzoyl)-4-[N-(substituted) piperazin-1-yl]thiophenes as potent allosteric enhancers of the A1 adenosine receptor. J Med Chem 51:5875–5879. https://doi.org/10.1021/jm800586p
120. Romagnoli R, Baraldi PG, Carrion MD, Cara CL, Cruz-Lopez O, Salvador MK et al (2012) Structure–activity relationships of 2-amino-3-aroyl-4-[(4-arylpiperazin-1-yl)methyl]thiophenes. Part 2: probing the influence of diverse substituents at the phenyl of the arylpiperazine moiety on allosteric enhancer activity at the A1 adenosine receptor. Bioorg Med Chem 20: 996–1007. https://doi.org/10.1016/j.bmc.2011.11.044
121. Romagnoli R, Baraldi PG, Carrion MD, Cara CL, Cruz-Lopez O, Salvador MK et al (2012) Synthesis and biological evaluation of 2-amino-3-(4-chlorobenzoyl)-4-[(4- arylpiperazin-1-yl) methyl]-5-substituted-thiophenes. Effect of the 5-modification on allosteric enhancer activity at the A1 adenosine receptor. J Med Chem 55:7719–7735. https://doi.org/10.1021/jm3007504
122. Vincenzi F, Targa M, Romagnoli R, Merighi S, Gessi S, Baraldi PG et al (2014) TRR469, a potent A1 adenosine receptor allosteric modulator, exhibits anti-nociceptive properties in acute and neuropathic pain models in mice. Neuropharmacology 81:6–14. https://doi.org/10.1016/j.neuropharm.2014.01.028
123. Romagnoli R, Baraldi PG, Carrion MD, Cara CL, Salvador MK, Preti D et al (2013) Synthesis and biological effects of novel 2-amino-3-(4-chlorobenzoyl)-4-substituted thiophenes as allosteric enhancers of the A1 adenosine receptor. Eur J Med Chem 67:409–427. https://doi.org/10.1016/j.ejmech.2013.07.002
124. Romagnoli R, Baraldi PG, Carrion MD, Cruz-Lopez O, Cara CL, Saponaro G et al (2014) Synthesis and biological evaluation of novel 2-amino-3-aroyl-4-neopentyl-5-substituted thiophene derivatives as allosteric enhancers of the A1 adenosine receptor. Bioorg Med Chem 22: 148–166. https://doi.org/10.1016/j.bmc.2013.11.043
125. Romagnoli R, Baraldi PG, Ijzerman AP, Massink A, Cruz-Lopez O, Lopez-Cara LC et al (2014) Synthesis and biological evaluation of novel allosteric enhancers of the A1adenosine receptor based on 2-amino-3-(4′-chlorobenzoyl)-4-substituted-5-arylethynyl thiophene. J Med Chem 57:7673–7686. https://doi.org/10.1021/jm5008853
126. Romagnoli R, Baraldi PG, Lopez-Cara C, Cruz-Lopez O, Moorman AR, Massink A et al (2015) Synthesis and biological evaluation of a new series of 2-amino-3-aroyl thiophene

derivatives as agonist allosteric modulators of the A1 adenosine receptor. A position-dependent effect study. Eur J Med Chem 101:185–204. https://doi.org/10.1016/j.ejmech.2015.06.041

127. Aurelio L, Valant C, Flynn BL, Sexton PM, White JM, Christopoulos A et al (2010) Effects of conformational restriction of 2-amino-3-benzoylthiophenes on A1 adenosine receptor modulation. J Med Chem 53:6550–6559. https://doi.org/10.1021/jm1008538
128. Chordia MD, Murphree LJ, Macdonald TL, Linden J, Olsson RA (2002) 2-Aminothiazoles: a new class of agonist allosteric enhancers of A1 adenosine receptors. Bioorg Med Chem Lett 12:1563–1566. https://doi.org/10.1016/s0960-894x(02)00236-6
129. Göblyös A, Santiago SN, Pietra D, Mulder-Krieger T, Künzel JVFD, Brussee J et al (2005) Synthesis and biological evaluation of 2-aminothiazoles and their amide derivatives on human adenosine receptors. Lack of effect of 2-aminothiazoles as allosteric enhancers. Bioorg Med Chem 13:2079–2087. https://doi.org/10.1016/j.bmc.2005.01.006
130. Chordia MD, Zigler M, Murphree LJ, Figler H, Macdonald TL, Olsson RA et al (2005) 6-Aryl-8H-indeno[1,2-d]thiazol-2-ylamines: A1 adenosine receptor agonist allosteric enhancers having improved potency. J Med Chem 48:5131–5139. https://doi.org/10.1021/jm049132j
131. Fawzi AB, Macdonald D, Benbow LL, Smith-Torhan A, Zhang H, Weig BC et al (2001) SCH-202676: an allosteric modulator of both agonist and antagonist binding to G protein-coupled receptors. Mol Pharmacol 59:30–37. https://doi.org/10.1124/mol.59.1.30
132. Gao ZG, Gross AS, Jacobson KA (2004) Effects of the allosteric modulator SCH-202676 on adenosine and P2Y receptors. Life Sci 74:3173–3180. https://doi.org/10.1016/j.lfs.2003.11.014
133. Lanzafame A, Christopoulos A (2004) Investigation of the interaction of a putative allosteric modulator, N-(2,3-diphenyl-1,2,4-thiadiazole-5-(2H)-ylidene) methanamine hydrobromide (SCH-202676), with M1 muscarinic acetylcholine receptors. J Pharmacol Exp Ther 308: 830–837. https://doi.org/10.1124/jpet.103.060590
134. Van Den Nieuwendijk AMCH, Pietra D, Heitman L, Göblyös A, IJzerman AP (2004) Synthesis and biological evaluation of 2,3,5-substituted [1,2,4]thiadiazoles as allosteric modulators of adenosine receptors. J Med Chem 47:663–672. https://doi.org/10.1021/jm030863d
135. Göblyös A, De Vries H, Brussee J, Ijzerman AP (2005) Synthesis and biological evaluation of a new series of 2,3,5-substituted [1,2,4]-thiadiazoles as modulators of adenosine A1 receptors and their molecular mechanism of action. J Med Chem 48:1145–1151. https://doi.org/10.1021/jm049337s
136. Ferguson GN, Valant C, Horne J, Figler H, Flynn BL, Linden J et al (2008) 2-Aminothienopyridazines as novel adenosine A1 receptor allosteric modulators and antagonists. J Med Chem 51:6165–6172. https://doi.org/10.1021/jm800557d
137. Narlawar R, Lane JR, Doddareddy M, Lin J, Brussee J, Ijzerman AP (2010) Hybrid ortho/allosteric ligands for the adenosine A1 receptor. J Med Chem 53:3028–3037. https://doi.org/10.1021/jm901252a
138. Aurelio L, Baltos JA, Ford L, Nguyen ATN, Jörg M, Devine SM et al (2018) A structure-activity relationship study of bitopic N 6 -substituted adenosine derivatives as biased adenosine a 1 receptor agonists. J Med Chem 61:2087–2103. https://doi.org/10.1021/acs.jmedchem.8b00047
139. Vincenzi F, Pasquini S, Battistello E, Merighi S, Gessi S, Borea PA et al (2020) A1 adenosine receptor partial agonists and allosteric modulators: advancing toward the clinic? Front Pharmacol 11:2117. https://doi.org/10.3389/fphar.2020.625134
140. Draper-Joyce CJ, Bhola R, Wang J, Bhattarai A, Nguyen ATN, Cowie-Kent I et al (2021) Positive allosteric mechanisms of adenosine A1 receptor-mediated analgesia. Nature 597:571–576. https://doi.org/10.1038/s41586-021-03897-2
141. Imlach WL, Bhola RF, May LT, Christopoulos A, Christie MJ (2015) A positive allosteric modulator of the adenosine A1 receptor selectively inhibits primary afferent synaptic transmission in a neuropathic pain model. Mol Pharmacol 88:460–468. https://doi.org/10.1038/s41586-021-03897-2

142. La Motta C, Sartini S, Morelli M, Taliani S, Da Settimo F (2010) Allosteric modulators for adenosine receptors: an alternative to the orthosteric ligands. Curr Top Med Chem 10:976–992. https://doi.org/10.2174/156802610791293136
143. Ribeiro JA, Sebastião AM, De Mendonça A (2002) Adenosine receptors in the nervous system: pathophysiological implications. Prog Neurobiol 68:377–392. https://doi.org/10.1016/s0301-0082(02)00155-7
144. Gao ZG, Melman N, Erdmann A, Kim SG, Müller CE, IJzerman AP et al (2003) Differential allosteric modulation by amiloride analogues of agonist and antagonist binding at A1 and A3 adenosine receptors. Biochem Pharmacol 65:525–534. https://doi.org/10.1016/s0006-2952(02)01556-3
145. Giorgi I, Biagi G, Bianucci AM, Borghini A, Livi O, Leonardi M et al (2008) N^6-1,3-Diphenylurea derivatives of 2-phenyl-9-benzyladenines and 8-azaadenines: synthesis and biological evaluation as allosteric modulators of A2A adenosine receptors. Eur J Med Chem 43:1639–1647. https://doi.org/10.1016/j.ejmech.2007.10.021
146. Welihinda AA, Amento EP (2014) Positive allosteric modulation of the adenosine A2a receptor attenuates inflammation. J Inflamm 11:1–14. https://doi.org/10.1186/s12950-014-0037-0
147. Welihinda AA, Kaur M, Raveendran KS, Amento EP (2018) Enhancement of inosine-mediated A2AR signaling through positive allosteric modulation. Cell Signal 42:227–235. https://doi.org/10.1016/j.cellsig.2017.11.002
148. Welihinda A, Ravikumar P, Kaur M, Mechanic J, Yadav S, Kang GJ et al (2022) Positive allosteric modulation of A2AR alters immune cell responses and ameliorates psoriasis-like dermatitis in mice. J Invest Dermatol 142:624–632. https://doi.org/10.1016/j.jid.2021.07.174
149. Lazarus M, Shen HY, Cherasse Y, Qu WM, Huang ZL, Bass CE et al (2011) Arousal effect of caffeine depends on adenosine A2A receptors in the shell of the nucleus accumbens. J Neurosci 31:10067–10075. https://doi.org/10.1523/JNEUROSCI.6730-10.2011
150. Scammell TE, Gerashchenko DY, Mochizuki T, McCarthy MT, Estabrooke IV, Sears CA et al (2001) An adenosine A2a agonist increases sleep and induces Fos in ventrolateral preoptic neurons. Neuroscience 107:653–963. https://doi.org/10.1016/s0306-4522(01)00383-9
151. Korkutata M, Saitoh T, Cherasse Y, Ioka S, Duo F, Qin R et al (2019) Enhancing endogenous adenosine A2A receptor signaling induces slow-wave sleep without affecting body temperature and cardiovascular function. Neuropharmacology 144:122–132. https://doi.org/10.1016/j.neuropharm.2018.10.022
152. Taliani S, Trincavelli ML, Cosimelli B, Laneri S, Severi E, Barresi E et al (2013) Modulation of A2B adenosine receptor by 1-benzyl-3-ketoindole derivatives. Eur J Med Chem 69:331–337. https://doi.org/10.1016/j.ejmech.2013.09.001
153. Da Settimo F, Primofiore G, Taliani S, Marini AM, La Motta C, Novellino E et al (2001) 3-Aryl[1,2,4]triazino[4,3-α]benzimidazol-4(10H)-ones: a new class of selective A1 adenosine receptor antagonists. J Med Chem 44:316–327. https://doi.org/10.1021/jm001054m
154. Da Settimo F, Primofiore G, Taliani S, La Motta C, Novellino E, Greco G et al (2004) A1 adenosine receptor antagonists, 3-aryl[1,2,4]triazino[4,3-a]benzimidazol-4-(10H)-ones (ATBIs) and N-alkyl and N-acyl-(7-substituted-2-phenylimidazo[1,2-a][1,3,5]triazin-4-yl) amines (ITAs): different recognition of bovine and human binding sites. Drug Dev Res 63: 1–7. https://doi.org/10.1002/ddr.10366
155. Taliani S, Pugliesi I, Barresi E, Simorini F, Salerno S, La Motta C et al (2012) 3-Aryl-[1,2,4]triazino[4,3-a]benzimidazol-4(10 H)-one: a novel template for the design of highly selective A2B adenosine receptor antagonists. J Med Chem 55:1490–1499. https://doi.org/10.1021/jm201177b
156. Primofiore G, Da Settimo F, Taliani S, Marini AM, Simorini F, Novellino E et al (2003) Geometrically constrained analogues of N-benzylindolylglyoxylylamides: [1, 2, 4]triazino[4, 3-a]benzimidazol-4(10H)-one derivatives as potential new ligands at the benzodiazepine receptor. Arch Pharm 336:413–421. https://doi.org/10.1002/ardp.200300788

157. Da Settimo A, Primofiore G, Da Settimo F, Marini AM, Novellino E, Greco G et al (1998) N'-phenylindol-3-ylglyoxylohydrazide derivatives: synthesis, structure- activity relationships, molecular modeling studies, and pharmacological action on brain benzodiazepine receptors. J Med Chem 41:3821–3830. https://doi.org/10.1021/jm9800301
158. Da Settimo A, Primofiore G, Da Settimo F, Marini AM, Novellino E, Greco G et al (1996) Synthesis, structure-activity relationships, and molecular modeling studies of N-(indol-3-ylglyoxylyl)benzylamine derivatives acting at the benzodiazepine receptor. J Med Chem 39: 5083–5091. https://doi.org/10.1021/jm960240i
159. Primofiore G, Da Settimo F, Taliani S, Marini AM, Novellino E, Greco G et al (2001) Novel N-(arylalkyl)indol-3-ylglyoxylylamides targeted as ligands of the benzodiazepine receptor: synthesis, biological evaluation, and molecular modeling analysis of the structure-activity relationships. J Med Chem 44:2286–2297. https://doi.org/10.1021/jm010827j
160. Primofiore G, Taliani S, Da Settimo F, Marini AM, La Motta C, Simorini F et al (2007) Novel N-substituted indol-3-ylglyoxylamides probing the LDi and L1/L2 lipophilic regions of the benzodiazepine receptor site in search for subtype-selective ligands. J Med Chem 50:1627–1634. https://doi.org/10.1021/jm0607707
161. Salerno S, Da Settimo F, Taliani S, Simorini F, La Motta C, Fornaciari G et al (2012) Medicinal chemistry of indolylglyoxylamide GABAA/BzR high affinity ligands: identification of novel anxiolytic/non sedative agents. Curr Top Med Chem 12:286–311. https://doi.org/10.2174/156802612799078847
162. Taliani S, Cosimelli B, Da Settimo F, Marini AM, La Motta C, Simorini F et al (2009) Identification of anxiolytic/nonsedative agents among indol-3- ylglyoxylamides acting as functionally selective agonists at the γ-aminobutyric acid-A (GABAA) α2 benzodiazepine receptor. J Med Chem 52:3723–3734. https://doi.org/10.1021/jm9001154
163. Cosimelli B, Cosconati S, Marinelli L, Novellino E, Greco G (2012) From the pharmacophore to the homology model of the benzodiazepine receptor: the Indolylglyoxylamides affair. Curr Top Med Chem 12:321–332. https://doi.org/10.2174/156802612799078702
164. Primofiore G, Da Settimo F, Marini AM, Taliani S, La Motta C, Simorini F et al (2006) Refinement of the benzodiazepine receptor site topology by structure-activity relationships of new N-(heteroarylmethyl)indol-3-ylglyoxylamides. J Med Chem 49:2489–2495. https://doi.org/10.1021/jm0511841
165. Trincavelli ML, Daniele S, Giacomelli C, Taliani S, Da Settimo F, Cosimelli B et al (2014) Osteoblast differentiation and survival: a role for A2B adenosine receptor allosteric modulators. Biochim Biophys Acta 1843:2957–2966. https://doi.org/10.1016/j.bbamcr.2014.09.013
166. Barresi E, Giacomelli C, Marchetti L, Baglini E, Salerno S, Greco G et al (2021) Novel positive allosteric modulators of A2B adenosine receptor acting as bone mineralisation promoters. J Enzyme Inhib Med Chem 36:286–294. https://doi.org/10.1080/14756366.2020.1862103
167. Barresi E, Martini C, Da Settimo F, Greco G, Taliani S, Giacomelli C et al (2021) Allosterism vs. orthosterism: recent findings and future perspectives on A2B AR physiopathological implications. Front Pharmacol 12:384. https://doi.org/10.3389/fphar.2021.652121
168. Gao ZG, Van Muijlwijk-Koezen JE, Chen A, Müller CE, Ijzerman AP, Jacobson KA (2001) Allosteric modulation of A3 adenosine receptors by a series of 3-(2-pyridinyl)isoquinoline derivatives. Mol Pharmacol 60:1057–1063
169. Van Muijlwijk-Koezen JE, Timmerman H, Link R, Van der Goot H, Ijzerman AP (1998) A novel class of adenosine A3 receptor ligands. 1. 3-(2- pyridinyl)isoquinoline derivatives. J Med Chem 41:3987–3993. https://doi.org/10.1021/jm980036q
170. Gao ZG, Kim SG, Soltysiak KA, Melman N, Ijzerman AP, Jacobson KA (2002) Selective allosteric enhancement of agonist binding and function at human A3 adenosine receptors by a series of imidazoquinoline derivatives. Mol Pharmacol 62:81–89. https://doi.org/10.1124/mol.62.1.81

171. Ijzerman AP, Soudijn W, van Galen PJM, Nissen P, van Wijngaarden I (1991) 1H-Imidazo [4,5-c]quinolin-4-amines: novel non-xanthine adenosine antagonists. J Med Chem 34:1202–1206. https://doi.org/10.1021/jm00107a046
172. Göblyös A, Gao ZG, Brussee J, Connestari R, Santiago SN, Ye K et al (2006) Structure-activity relationships of new 1H-imidazo[4,5-c]quinolin-4-amine derivatives as allosteric enhancers of the A3 adenosine receptor. J Med Chem 49:3354–3361. https://doi.org/10.1021/jm060086s
173. Gao ZG, Ye K, Göblyös A, IJzerman AP, Jacobson KA (2008) Flexible modulation of agonist efficacy at the human A3 adenosine receptor by the imidazoquinoline allosteric enhancer LUF6000. BMC Pharmacol 8:1–11. https://doi.org/10.1186/1471-2210-8-20
174. Yoonkyung K, De Castro S, Gao ZG, IJzerman AP, Jacobson KA (2009) Novel 2-and 4-substituted 1H-Imidazo[4,5-c]quinolin-4-amine derivatives as allosteric modulators of the A 3 adenosine receptor. J Med Chem 52:2098–2108. https://doi.org/10.1021/jm801659w
175. Heitman LH, Göblyös A, Zweemer AM, Bakker R, Mulder-Krieger T, Van Veldhoven JPD et al (2009) A series of 2,4-disubstituted quinolines as a new class of allosteric enhancers of the adenosine A3 receptor. J Med Chem 52:926–931. https://doi.org/10.1021/jm8014052
176. Lane JR, Beukers MW, Mulder-Krieger T, IJzerman AP (2010) The endocannabinoid 2-arachidonylglycerol is a negative allosteric modulator of the human A3 adenosine receptor. Biochem Pharmacol 79:48–56. https://doi.org/10.1016/j.bcp.2009.07.024
177. May LT, Briddon SJ, Hill SJ (2010) Antagonist selective modulation of adenosine A1 and A3 receptor pharmacology by the food dye brilliant black BN: evidence for allosteric interactions. Mol Pharmacol 77:678–686. https://doi.org/10.1124/mol.109.063065
178. Fishman P, Cohen S (2016) The A3 adenosine receptor (A3AR): therapeutic target and predictive biological marker in rheumatoid arthritis. Clin Rheumatol 35:2359–2362. https://doi.org/10.1007/s10067-016-3202-4
179. Mazziotta C, Rotondo JC, Lanzillotti C, Campione G, Martini F, Tognon M (2021) Cancer biology and molecular genetics of A3 adenosine receptor. Oncogene 41:301–308. https://doi.org/10.1038/s41388-021-02090-z
180. Cohen S, Barer F, Bar-Yehuda S, Ijzerman AP, Jacobson KA, Fishman P (2014) A3 adenosine receptor allosteric modulator induces an anti-inflammatory effect: in vivo studies and molecular mechanism of action. Mediators Inflamm 204:708746. https://doi.org/10.1155/2014/708746
181. Cohen S, Fishman P, Tikva P (2016) MP89-04 CF602 improves erectile dysfunction in diabetic rats. J Urol 195(4S). https://doi.org/10.1016/j.juro.2016.02.2465
182. Du L, Gao ZG, Nithipatikom K, IJzerman AP, Van Veldhoven JPD, Jacobson KA et al (2012) Protection from myocardial ischemia/reperfusion injury by a positive allosteric modulator of the A3 adenosine receptor. J Pharmacol Exp Ther 340:210–217. https://doi.org/10.1124/jpet.111.187559

Top Med Chem (2023) 41: 275–316
https://doi.org/10.1007/7355_2023_164

Published online: 19 August 2023

In Silico Insights Toward the Exploration of Adenosine Receptors Ligand Recognition

Davide Bassani and Stefano Moro

Contents

Abstract Adenosine receptors have been studied for many years now, their structures have been resolved and many of the determinants for their interaction patterns have been highlighted. Nevertheless, they still represent extremely fascinating but challenging biological targets, and much effort has been put by both academic and industrial drug discovery groups into finding novel and proficient molecular candidates for the regulations of these purinergic species. Computational methods have been vastly implemented for the research of adenosine receptors modulators, allowing to speed up and optimize the development of potent and selective compounds. In this paper, we describe the main computer-based methods that have been used and that are still exploited in this field, reporting also many of the cases in

D. Bassani and S. Moro (✉)
Department of Pharmaceutical and Pharmacological Sciences, Molecular Modeling Section (MMS), University of Padova, Padova, Italy
e-mail: stefano.moro@unipd.it

which their application brought to light novel and promising molecular species with great therapeutic potential. While the research on adenosine receptors is still very open, the evolution and amelioration of computational approaches is expected to fill the gaps that we still miss about these beautiful targets, in a way that we can better communicate with them, finally providing valid and efficacious therapeutic solutions through their regulation.

Keywords Adenosine, Adenosine receptors, CADD, Computational chemistry, Drug design, Drug discovery, GPCRs, MD simulations, Molecular docking, Molecular dynamics, Molecular modeling, Purine

1 The Drug Discovery Process

Pharmaceutical research represents one of the most costly and risky fields in the industrial scenario. It is enough to think that a single New Molecular Entity (NME) may require more than 14 years and billions of dollars of investment to pass from the early target identification phase up to reaching the market [1, 2]. Moreover, as the years pass, these numbers become more and more important, and the advent of biological drugs, gene therapy, and monoclonal antibodies contributed to further increasing the costs of the research.

According to the US Food and Drug Administration (FDA), the entire process defined as "drug discovery" can be divided into five main phases [3].

The first one is named "discovery and development" and comprises the steps from target identification up to lead optimization. After having selected a target with a functional link to a disease area and, more specifically, a defined disorder, one of the classical approaches is to perform one or more screenings on that specific target. With the advancement of technology, at the present day it is possible to evaluate compound pools in the order of magnitude of hundreds of thousands, or even millions, in a matter of few days [4]. This, of course, is very dependent on the type of assay and on the resources available for the research. Once the early screening campaigns are completed, hopefully, some compounds were found to be able to bind to the target and exert a functional response. In this context, a molecule is named a "hit" when the activity on the desired endpoint has been assessed, but it is still presenting some pharmacokinetics (PK) or pharmacodynamics (PD) issues. Hits represent very valuable starting points for design purposes, both if the structure of the target is known and, even more, when it is not known [5]. Indeed, lacking structural data about the biological entity of interest, the information guiding the optimization processes will be derived mainly from the ligands that present some on-target activity. The following steps are referred to as "hit-to-lead optimization" (in many contexts, abbreviated as "H2L"). In this section, several aspects of the developed molecules are tackled, starting from aiming to improve potency and selectivity on the desired target, up to the evaluation and prediction of the PK/PD profile, of course taking into account the ADME properties and the compound-related toxicity [6]. With this process, the hits are ameliorated on several aspects, progressively shifting to a "lead" profile. Leads are few molecules (usually never more than 5–10 in a classical drug discovery campaign) with high potency (usually in the nanomolar range or even below, depending on the target) and reasonable selectivity, with an ADME-Tox profile that is coherent with the in vivo studies that will follow [7–9]. Indeed, the second main phase in the whole process is called "pre-clinical research" and comprises all the in vitro and in vivo assays that are necessary for addressing the efficacy and safety of the molecular entities tested. Among this pool of compounds, only one usually can make it to the next phase, taking the title of "candidate" [10, 11]. This last molecule enters then the "clinical research", by far the riskiest and most expensive in the whole discovery process [12, 13]. Clinical trials are divided into three main sections I, II, and III, each one with a defined endpoint and involving an increasing amount of patients. Only if the efficacy and safety of the compounds are assessed in all the trial steps, and only if their therapeutical value is demonstrated to be higher than what already exists on the market for that disease or disorder, then the marketing approval can be required to the regulatory agencies (e.g., EMEA for Europe or FDA for the USA) [14, 15]. This passage constitutes the fourth main phase of drug discovery, while the fifth and last one is named "post-market safety monitoring," and its focus is on pharmacovigilance and the feedback about the drug and its application.

2 The Benefits of Computation and Automation

As already mentioned, the technological advances and the rampaging improvements in hardware and software development gave a huge boost to the industrial field, and the pharmaceutical scenario was of course one of the sectors in which the benefits were most appreciated [16–18]. The automation of synthetic and assay-related processes not only improved their speed and efficiency but also was extremely beneficial in terms of reproducibility and rigor [19]. One of the main examples is the "High-Throughput Screening" (HTS) technology, with which is nowadays possible to screen millions of compounds at a rate that was simply unimaginable some years ago [20].

Even if these improvements are relevant, it is important to remember that the chemical space in which scientists search for novel potential small molecule drugs is exaggeratedly big. Indeed, it was estimated that the "drug-like" chemical space would comprise 10^{60}compounds [21], which is a number greater than the seconds of the existence of the Universe itself. Such vastness cannot be investigated in an experimental setup, for both time and resource-related reasons. With this initial idea, the possibility of exploiting computers to perform such a search in a "virtual" way instead that in the lab started to spread in the pharmaceutical world. Nowadays, with the most advanced HTVS ("High-Throughput Virtual Screening") pipelines, the evaluation of billions of molecules/day is achievable, of course depending on the kind of approach implemented for such a screening [22].

Nevertheless, the great advantages of computer-based methods in drug discovery are not limited to a simple speed-up in the process of screening. Indeed, combining structural information of the target with the power of graphical vision is today routine to transfer the process of drug design directly into the site of action, allowing scientists to explore the ligand modifications that would more suitably improve its therapeutic profile in a visual and user-friendly fashion [23–25]. In addition, even if the structural information of the target is lacking, computational methods have been developed to predict the three-dimensional organization of biological entities, based mainly on sequence data [26, 27]. If the biological target is very new, or if no information is available to get a reasonable prediction, then scientists may apply different approaches to get the most out of simple ligand activity data. This family of methods, usually referred to as "cheminformatics" [28], was very exploited in the past, and now their application has slightly decreased with the exponential rise of the number of biological structure available in the open-source databases. One relevant example is represented by the Protein Data Bank (usually referred to as "PDB") [29], in which the number of entries deposited raised from 13,583 in the year 2000, to 203,084 in 2023 (last update: 4thApril 2023) [30] (Fig. 1).

Even if their popularity has decreased in the latest decades, some of the cheminformatics methods are still applied, and many of them have encountered a sort of "evolution," being included in the more actual Machine Learning (ML) approaches. This last field has shown its great potential in recent years, being a game-changer in topics such as image recognition, data prediction, and automatic problem-solving [31]. Not only but this family of methods is also now widely used for generative purposes, being able to "rationally" create novel plans,

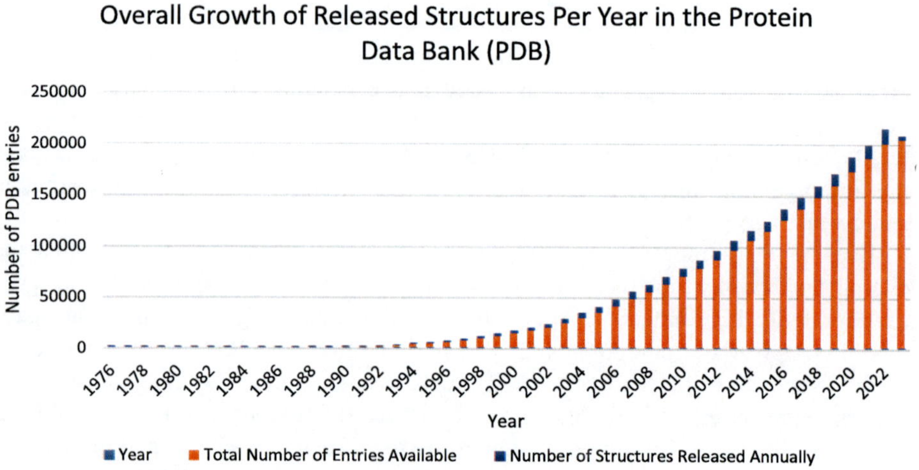

Fig. 1 Plot representing the increase in the number of publicly available structures in the Protein Data Bank from 1976 up to now (last update: 4th April 2023)

figures, and, very interestingly in the medicinal chemistry scenario, novel chemical structures [32]. All the cited techniques and methods, which fall in the great family of "Computer-Aided Drug Design" (CADD) [33–35], will be covered in detail in the following paragraphs, together with their application in the field of adenosine receptor research.

3 The Main Branches in Computer-Aided Drug Design

As already cited, many are the methods and techniques available to CADD scientists to generate better compounds in a fast and efficient way. The main question is related to which approach to choose in the various design scenarios. The primary discriminant for this decision is usually related to the availability of structural data for the biological entity investigated.

If this information is present, then the path chosen is usually characterized by strong exploitation of methods falling into the category of "Structure-Based Drug Design" (SBDD) [36]. These techniques are based on the modeling of the ligands together with the target binding site, or the target as a whole. In the case in which no structural data is available to the CADD scientist, then there are mainly two paths available.

The first is the prediction of the structure of the target based on sequence data. One way to execute this calculation is the classical "homology modeling" approach, which relies on the dogmatic fact that, whenever two proteins share 30% or more of their sequence [37], their three-dimensional organization will roughly be the same. Many different programs have been developed for this purpose, among which MODELLER [38], SWISS-MODEL [39], Phyre2 [40], ROBETTA [41], and MOE [42] *homology modeler* applications have gained much popularity. The second way for protein structure prediction is to exploit Machine Learning approaches to connect sequence data with tertiary structure information. In this field, the absolute game-changing event was the release of AlphaFold [43], from the DeepMind company. This specific platform uses deep-learning methods for protein structure prediction, and it has been demonstrated to perform incredibly well both in the 2018 Critical Assessment of Structure Prediction [44] (CASP, 1st version of the software) and in the same competition 2 years later (in the AlphaFold 2 version). Even if the revolution brought by this application is astonishing, many are gaps that still have to be filled. For example, the algorithm was up to now exploited to investigate mainly the human and mammalian protein worlds, and for this reason, many very interesting biological targets coming from other organisms still are lacking in the AlphaFold database. Second, the program predicts just one of the states of the targets, and this aspect is very relevant for proteins such as GPCRs, ion channels, and chaperones, in which the difference between the various activation states is critical for the modeling. Whichever strategy is chosen for the generation of the model, it is always a good practice to validate it, and this is usually performed with some replicas of Molecular Dynamics simulations. These experiments allow for assessing the coherency of the model and eventually allow us to investigate some gaps in the prediction of certain domains' structures [45].

In the case of the adenosine receptors, A_1R, $A_{2A}R$, and $A_{2B}R$ have different structures deposited in the PDB database at the present date, both in the apo- and holo-forms (5 for A_1R, 69 for $A_{2A}R$, and 2 for $A_{2B}R$, updated 4th April 2023), while A_3R still lacks an experimentally resolved structure. Recent work from Margiotta and collaborators demonstrated how, even if higher sequence similarity exists between $A_{2A}R$ and A_3R, a valid alternative is represented by the exploitation of A_1R as the template for homology modeling [46]. Of course, nowadays the inactive form of the A_3R structure is available in the AlphaFold database, with a good confidence score for the three-dimensional structure prediction in all its domains.

The other main scenario is the one in which no structural information about the target is available, and also the three-dimensional predictions are not possible or not enough reliable. In this case, the methods available to the CADD scientist will have to more heavily rely on the ligands' structural information. This family of approaches takes the name of "Ligand-Based Drug Design" (LBDD) [47], and its older members constitute the early representatives of rational drug design back in the middle of the twentieth century. Among these, much popularity was gained from the analysis suggested by Hansch, Hammet, and Taft [48], as well as the Free-Wilson method [49]. Pharmacophore-based approaches are also an important component of LBDD [50] and now are mainly exploited in their three-dimensional version (3D-pharmacophores), in which much importance is given to the small molecule conformer generation and sampling [51]. Also in this case, Machine Learning methods have become increasingly more important in the field, allowing finding structural patterns in huge compound datasets, and being also able to propose chemical modifications and, more recently, even novel active entities [52] (Fig. 2).

4 Structural Biology Elements in Adenosine Receptor Research

As mentioned, the availability of structural information about the biological entities investigated is determinant for computational scientists. In the experimental setup, the classical ways to obtain such data rely on two main techniques: X-ray crystallography and Nuclear Magnetic Resonance (NMR) [53, 54]. The first is probably one of the most historically implemented, with the vast majority of the available PDB proteins being resolved with this technique. It is important to always remember that the pros and cons of these methods are compensating for each other. Indeed, X-ray crystallography can be used on big molecular entities and offers a real "snapshot" of the molecular scenario. The first drawback of the technique is not being able to resolve hydrogen atoms (which are in many cases fundamental for the chemical interaction analysis) and the second, and probably major, is related to the fact that, offering just a "photography" of the biological entity (or complex,

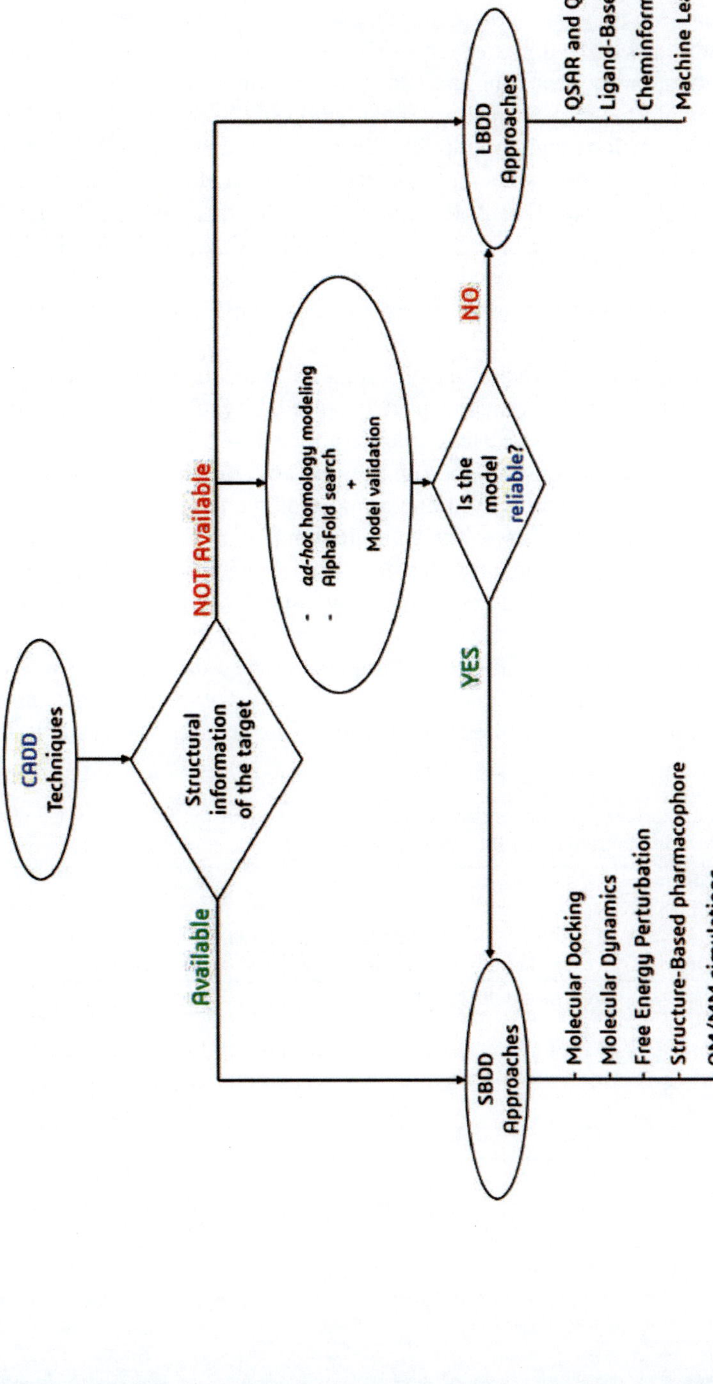

Fig. 2 Schematic representation of the main computational design methods available in drug discovery. A crucial aspect is determined by the availability of structural information about the target investigated. The acronyms are the following: *QM/MM* Quantum Mechanics-Molecular Mechanics, *QSAR* Quantitative Structure-Activity Relationship, *QSPR* Quantitative Structure-Property Relationship, *AI* Artificial Intelligence

depending on the cases), it cannot be suitable to realistically describe the physiological scenario, in which all the elements are moving. The NMR, on the other side, can give information about hydrogen atoms and analyzes the molecular environment in a dynamic context, resolving the cons of X-ray. Nevertheless, NMR can be proficiently used only on relatively small (usually below 50 kDa of molecular weight), water-soluble entities, and for this reason, complex systems such as biological membranes are challenging to analyze. Concerning this aspect, a very important methodology gained huge popularity in the very latest years, called Cryogenic Electron Microscopy (or cryo-EM). This method, which is different from the other two, being based on microscopy, relies on a fine-controlled electron beam, and it is exploited to get structural information of biological systems that can also be high in dimensions and complexity. Cryo-EM has been largely exploited in recent times, and also adenosine receptor research gained benefit from it [55–58]. Indeed, three out of the five A_1R structures, 2 out of the 69 $A_{2A}R$, and both the two $A_{2B}R$ PDB entries have been obtained with this technique (updated 4th April 2023) (Table 1).

Membrane systems were reported to be most probably the research field that gained the most out of this approach, but of course, also this technology has some disadvantages. The first is related to the very high costs for the infrastructure and its maintenance, and the more technical aspect regards the resolution performance of the method, which in many cases reaches values above 3.5–4.0 Å, obliging the cryo-EM scientist to not precisely assign the right amino acid to long chains [59].

When the information coming from one or more of such approaches is available, then which one to prioritize is up to the computational scientist, who will take into account both the quality of the provided information and the specific project in which the protein is involved. For example, if searching for a competitive antagonist targeting the inactive form of a receptor, it will be more proficient to take as a reference a high-resolution protein-ligand complex between the inactive form of the investigated target and an already known antagonist, if possible. Otherwise, also the simple inactive form can be exploited if no complex is available. The final choice of

Table 1 Table summarizing the number and experimental source of the four adenosine receptor isoforms available at the present date (4th April 2023)

Human receptor	Available experimental structures in the PDB	X-ray crystallography	NMR spectroscopy	Cryo-EM	Comments
A_1R	5	2	0	3	–
$A_{2A}R$	69	65	0	2	2 structures were resolved by microcrystal electron diffraction (MicroED)
$A_{2B}R$	2	0	0	2	–
A_3R	0	0	0	0	A good-confidence AlphaFold model is available

the experimental data to prioritize, or to even combine, is then crucial for the outcomes of the computational studies.

5 Structure-Based Drug Design (SBDD)

With the exponential raise of the availability of structural data, structure-based methods gained increasing more and more popularity in computational drug design [60–62]. Indeed, the information about the target of interest can have a great impact in both the screening and optimization phases, allowing the drug designers to operate on the protein–ligand interaction problem like artists on a canvas. Among the many structure-based approaches, the most popular is for sure represented by molecular docking and molecular dynamics (MD), which will be discussed in detail in the following sections. The possibility to operate with a visual representation of the binding site made also routines for the generation of proper 3D-pharmacophores that take into account the hindrance and the properties of the chemical environment in which the ligand should place. These methods are heavily implemented after molecular docking, to filter the conformations generated by the algorithm in terms of "suitability" with the pharmacophore hypothesis. Moreover, machine learning-based techniques have been exploited for pose prediction [63], MD evolution forecasting [64], pocket mapping [65], and interaction energy prediction [66]. Regarding this last point, Free-Energy methods have been increasingly applied in recent times [67], allowing getting a very precise determination of the binding free energy between a series of co-generic ligands and a defined target (Fig. 3).

Being present a good amount of structural information on the adenosine receptors (apart from A_3R, which nevertheless can be treated as a model with a decent degree of confidence), it is not surprising that the vast majority of studies for compounds influencing their activity rely strongly on structure-based methods. In the following subchapters, we will give an overview of the main SBDD approaches, as well as their implementation in adenosine receptor research.

5.1 *Molecular Docking*

By far the most applied SBDD method, molecular docking is a computational approach for the prediction of the best conformation with which a molecule binds to another to form a stable complex. The technique takes its name from the very first software created for this purpose, which was called "DOCK," and was published by Kuntz in 1982 [68].

Molecular docking programs are formed by two main engines: the conformational search algorithm and the scoring function. The first of these has the purpose of extensively sampling the conformational space of the ligand, while the scoring function assigns a defined value (or "score") to each of these conformations, to rank them and prioritize the best ones [69].

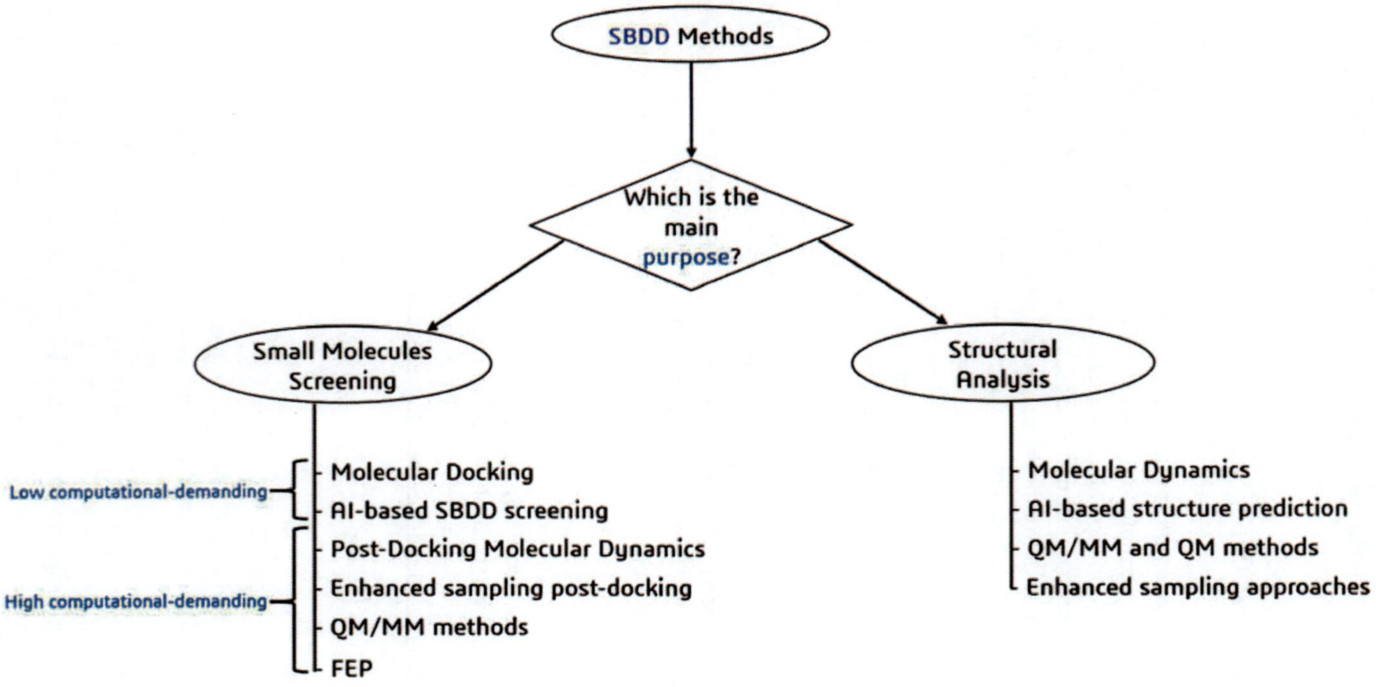

Fig. 3 Schematic representation of the main SBDD approaches, grouped based on their main design purpose. When screening small molecules against a biological target, another relevant factor to take into account is the computational power available, which allows to roughly discriminate the techniques based on the number of molecules/day screened with the same computational infrastructure. The acronyms are the following: *AI* Artificial Intelligence, *QM/MM* Quantum Mechanics-Molecular Mechanics, *FEP* Free-Energy Perturbation

Scoring functions can be classified into three main branches: knowledge-based, force field-based, and empirical. The first ones rely on statistical analyses of the most observed interactions between a specific ligand's atom type and a particular protein's atom type, ranking the poses according to the ones that are more similar to what is statistically retrievable in the high-quality X-ray databases. Force field-based functions simply exploit force fields to evaluate the energy of a system. Finally, empirical scoring methods are made of different terms, each representing different intermolecular interactions and weighted by using experimental values for its specific kind of interaction. The closest the calculated values to the experimental ones, the highest the score, and the closest the ligand pose to the top of the raking for the algorithm [70].

Moreover, also the docking method per se can be categorized based on the level of "freedom" that is given to the ligand-target system. The less computationally expensive method is referred to as "rigid docking," and in this case, both ligand and target are kept rigid [71]. Of course, this requires a pre-generation of small molecule conformation for the screening, and this task can be easily executed with different software nowadays (from the easily code-implementable methods based on the RDkit cheminformatics package [72] to the golden standard OMEGA from the OpenEye company [73]), which are usually relying on force fields for the creation and enumeration of conformers. Going a little up in the computational cost, we encounter the "flexible-ligand docking" methodologies, in which the ligand can explore different conformational changes within the binding site, which is kept always rigid [74]. In the case in which also the binding region of the target is given the freedom to move, we enter the "induced-fit docking" scenario [75]. This last technique was much investigated by several groups, both in academia and industry, being considered a sort of "link" between molecular docking and molecular dynamics methods. Indeed, with this approach, small shifts in the side chains in the binding site are allowed, allowing the ligand to explore some small regions that would be excluded if the target structure would stay rigid. The possibility to implement implicit solvation and some correction terms for water molecules and protein flexibility made induced-fit approaches appealing [76], but the level of approximation that is intrinsic to the technique cannot be overcome unless shifting to a level of physicochemical theory that is so high that the versatility and speed advantages of molecular docking would be completely lost. Finally, "ensemble docking" is a method in which the flexible-ligand docking is executed against an ensemble of conformations of the target, usually coming from molecular dynamics. In this case, also larger changes in the binding site can be sampled, being the time for molecular dynamics for ensemble generation arbitrary. In this way, the full flexibility of the protein can be taken into account [77]. Of course, the application of this approach intrinsically requires a much higher computational cost than the previous ones.

Nowadays, much different software for molecular docking has been developed and released, and they are usually classified into the type of scoring function that they rely on. If the algorithm is genetic, very relevant examples are constituted by programs like GOLD [78] (which stands for "Genetic Optimization for Ligand

Docking," developed and commercialized by the Cambridge Crystallographic Data Center) and AutoDock [79] (developed by the Scripps Research Institute). One main example of systematic algorithm-based software is Glide [80] (distributed with a commercial license by Schrödinger), while programs like PLANTS [81] (from the Tübingen University) are based on an Ant-Colony Optimization (ACO) algorithm.

This process is heavily dependent on the programs used and, while search algorithms perform reasonably well in the majority of the scenarios, the scoring functions can give very different outcomes if different docking engines are compared to one another [82, 83]. Indeed, in many scenarios, scoring functions have been demonstrated to be unreliable in prioritizing the best compounds in a docking experiment [84–86], and this raised the need for a computational medicinal chemist to find other ways to rank the compounds after docking. Some very simple filtering methods are just based on a certain threshold (which could be arbitrary or based on previous knowledge of the case study examined) for a specific three-dimensional property of the pose [87, 88]. Specifically, the elimination of all those conformations presenting clashes with the receptor or unfavorable electrostatic interaction with the binding site was reported to be part of the workflow in different studies [89, 90]. Of course, even if this family of approaches is very simple and fast, they cannot be considered exhausting in their efficiency and allow only to eliminate some unfavorable molecules, without touching the order in which the scoring function places the non-filtered ones. The broad family of techniques that are routinely used for more specifically addressing this issue can be referred to as "post-docking" approaches, and they will be deeply analyzed in the following paragraphs.

Other than the scoring functions, there are still other issues related to molecular docking. First, relying on molecular mechanics (MM) [91], it cannot take into consideration the polarizability effects on the different parts of both protein and ligand. Indeed, some recent implementations (mainly based on polarizable force fields) have introduced some terms for getting a more "realistic" representation of the analyzed problem, but of course, always keeping a relevant level of approximation [92–94].

Then, the calculations are executed in a vacuum, without taking into account the effect of the solvent [95]. The water molecules network around the ligand and close to the binding site has been demonstrated to be extremely crucial for molecular modeling [96], and even more when dealing with GPCRs [97, 98]. Indeed, the correct placement of water molecules within the binding site can completely change the prioritized ligands and their top-ranked conformations, and for this reason, it requires a deep and proper investigation based on the target examined [99, 100]. The placement is usually executed taking into account the previous knowledge of the target, the availability of experimental data, and eventually the results of some high-quality calculations (e.g., Free-Energy Methods) [101].

Another intuitive issue is related to the fact that working in MM, any bond can be broken or created during the experiment. The most used programs for covalent docking tendentially use some smart artifacts to give a good scientific meaning to the results that they provide [102–104], but in many cases, CADD scientists still prefer to model such interactions by hand (using their knowledge and eventually combining some ML methods) or with more rigorous QM/MM (quantum mechanics/molecular mechanics) approaches [105, 106]. This latest technique is the golden standard for covalent bond modeling [107], but of course, its implementation requires exponentially higher computational power than classical docking, and for this reason, is not usually applied in large virtual screening campaigns.

Even with these well-known drawbacks, the versatility, speed, and reliability (with the right precautions) of molecular docking still make it the most used structure-based method in computational drug design. Indeed, with a good hardware infrastructure, it is nowadays possible to screen several hundred million small molecules per day [108, 109], making then the choice of the desired post-docking method the bottleneck of the HTVS campaign.

In the field of Adenosine Receptors, where many experimentally resolved structures are available, molecular docking has played a piloting role in many discovery projects. Indeed, by exploiting both docking and molecular dynamics simulations, Inamdar et al. were able to develop a series of potent benzamides and furamides, with different degrees of selectivity on the four adenosine receptor subtypes [110]. In this scenario, molecular modeling was crucial in rationalizing the behavior of the investigated ligands. Molecular docking was also implemented to evaluate and confirm the binding mode of a series of adenosine derivatives displaying a low-nanomolar affinity for $A_{2A}R$ and A_3R, synthesized and characterized by Rodriguez et al. [111]. In another recent study, Federico et al. explored the modifications granting selectivity toward A_3R for a series of [1,2,4]triazolo[1,5-c] pyrimidines, using molecular docking as an explorative tool for the design and rationalization [112] (Fig. 4).

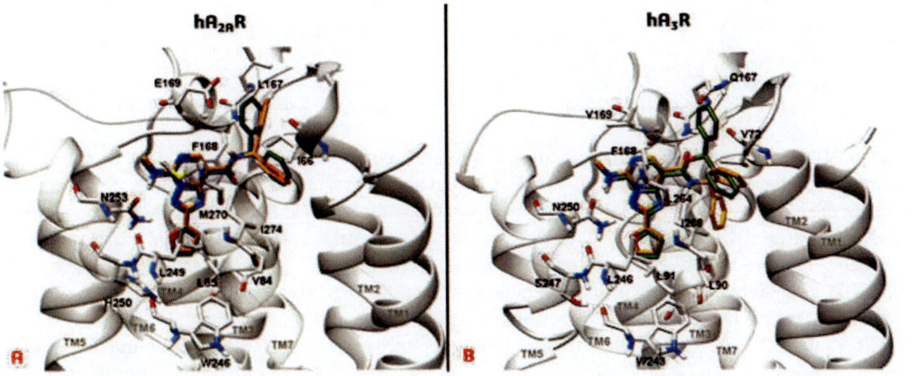

Fig. 4 Example of an application of molecular docking in adenosine receptor small molecule discovery. Panel (**a**) Representation of the predicted structure of the complex between $hA_{2A}R$ (gray) and compounds 5 (green, $K_iA_{2A}R = 197$ nM) and 21 (orange, $K_iA_{2A}R = 61.6$ nM) taken from the cited article from Federico et al. [112] for adenosine receptor antagonist discovery. Panel (**b**) Representation of the predicted structure of the complex between hA_3R (gray) and compounds 5 (green, $K_iA_3R = 1.99$ nM) and 21 (orange, $K_iA_3R = 5.08$ nM) from the same study

Wang et al. also exploited a computational pipeline including molecular docking, pharmacophore modeling, and deep learning to prioritize some potent compounds against A_1R and $A_{2A}R$ [113]. Glide was used as the docking tool by Sarkar et al. in a workflow that led to the identification of benzothiazolylquinoline conjugates as novel A_3R antagonists for cancer treatment [114]. Preti et al. exploited Vina for the placement of novel A_1R nanomolar agonists in the receptor's pocket, facilitating the rationalization of their selectivity profile.

From a mechanistic perspective, Margiotta et al. exploited molecular docking (and, more specifically, the *DockBench* software [115]) to investigate the impact of considering or not the allosteric sodium ion in the lower 7TM region of the adenosine A_{2A} receptor in a virtual screening setup [116]. This study was then extended by Bassani et al. 4 years later, in an experiment considering all the available complexes between a GPCR and an orthosteric antagonist available in the PDB at that time [117] (Fig. 5).

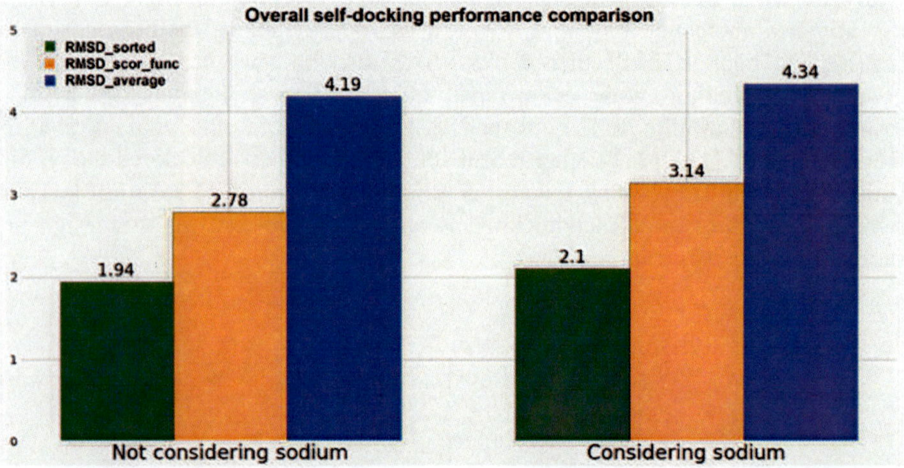

Fig. 5 Plot represents the comparison of the overall performance of the different docking algorithms implemented in the cited study by Bassani et al. [117] when the sodium ion is not considered during the calculation (on the left) and when both the sodium ion and the crystal water molecules at 4 Å or nearer to it are included (on the right). The metrics used for the comparison are the "RMSD_average" (which represents the mean RMSD of all the poses obtained), the "RMSD_scor_func" (which is the average value of the RMSDs obtained by the poses which were top-ranked by the scoring functions in each docking run), and the "RMSD_sorted" (the mean value of the RMSDs obtained from the poses with the lowest RMSD value in each docking calculation). As it can be noticed, molecular docking is not able to proficiently take into consideration the presence of sodium ion in the 7TM region of GPCRs, as no relevant differences in performance are predictable when the ion and its surrounding water molecules are taken into account or not

5.2 Molecular Dynamics

Molecular Dynamics (MD) represents the main computational technique used to investigate the evolution of a chemical and/or biological system with time. Its main principle is the iterative resolution of Newton's equation of motion to predict the atomic positions during a simulation. MD is heavily used in drug discovery, mainly to investigate conformational changes of relevant biological entities, other than, of course, evaluating how molecules behave one to the others in a dynamic context [118, 119]. Leaving molecules the freedom to move, MD can resemble the physiological environment much more than molecular docking, giving so a more "realistic" perspective of the ligand–target interaction. Practically speaking, the MD simulation is formed by a continuous series of snapshots, in which the atoms change position from the previous to the next, following the mathematical equation of the force field. This series of "photographs," in the end of the experiment, forms the so-called MD-trajectory. This "realism" is also translated in the computational implementation of the solvent, which can be taken into account which much more rigor in respect to simple "high-freedom" molecular docking experiments [120]. Indeed, the solvation model used can be of two main kinds: implicit or explicit [121]. In the first case, water molecules are not introduced in the simulations, but the force field is influenced by a series of equations resembling the dielectric constant and the electrostatic behavior of a desired solvent (water, for what concerns drug discovery). Some relevant examples of such methods are SMD ("Solvation Model based on Density") [122], COSMO (COnductor-like Screening MOdel) [123], and CPCM ("Conductor-like Polarizable Continuum Model") [124]. The second type of method is referred to as "explicit solvation" models, and they are based on the explicit treatment of the single water molecules in the system. Very remarkable examples of this kind are the simple SPC (Simple Point-Charge) model [125], which is now less implemented in comparison with the more advanced TIP3P ("Transferable Intermolecular Potential 3P") and TIP4P ("Transferable Intermolecular Potential 4P") architectures [126]. In this case, the simulation is much more rigorous with respect to the implicit solvation case, of course at the expense of the computational power required, which is much higher [127] (Fig. 6).

Moreover, the analysis of trajectories characterized by explicit solvation allows a deep evaluation of the evolution of water molecules' positions. This is crucial to get information regarding which water molecules can effectively mediate or impair some target–target or target–ligand contacts [130, 131] and, in addition to that, allows to identify the water molecules which tendentially do not change their position with time [132]. In the field of GPCRs research, correctly assigning the position and the energy of solvent molecules has been demonstrated to be extremely relevant in drug discovery campaigns [133, 134], and several tools and applications have so been developed to make such analyses routine and automatized [135, 136]. Some relevant examples of these are WaterMaps (commercialized and developed by Schrödinger) [137], WaterFLAP (distributed by Molecular Discovery) [138], and AquaMMapS (created by the Moro group, University of Padua) [139].

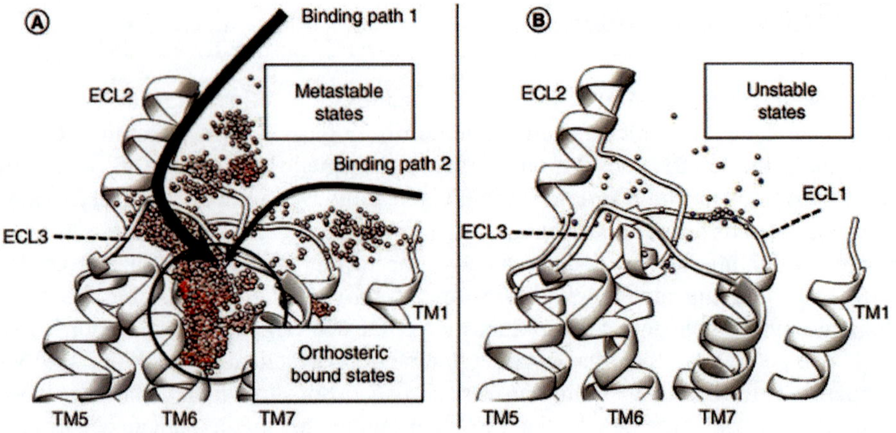

Fig. 6 Example of the application of Molecular Dynamics in GPCR research, taken form a recent study from Deganutti et al. [128]. Specifically, the figure represents the metastable and transition states along the binding pathway of a G-protein-coupled receptor ligand. Panel (**a**) The centroid of a GPCR ligand (shown as a small red sphere) populates different metastable states during the binding process, before reaching the orthosteric site. Energetically favored transitions from different metastable states describe complete binding paths (in this example, binding paths 1 and 2). Panel (**b**) Centroids corresponding to energetically unfavorable states (white spheres) are widely dispersed at the extracellular vestibule, indicating the receptor/ligand desolvation as a putative kinetic bottleneck, as proposed in a paper by Dror et al. [129]

In the pharmaceutical field, molecular dynamics can be used just on the target/s of interest, as well as on the target–ligand complex. In this last case, the analysis of the MD behavior of a complex has been highlighted as a good system for the evaluation of the affinity between the two entities, and this can also be extended to several applications. One of the very first of them is the analysis of the target's dynamic evolution, to extrapolate the most information possible about the chain flexibility, the binding site conformational freedom, and its accessibility with time [140, 141]. A second very relevant usage regards the filtration of docking poses, in the so-called MD-based post-docking methods [142], which will be discussed in the next paragraph. Another important case in which MD is used for discovery purposes is the investigation of the best modification that can be executed on a defined series of compounds to improve their binding free energy. This is the basis for the Free-Energy perturbation methods, which will be also treated in the following section.

As in the case of molecular docking, also MD is not free from defects and drawbacks. First, being based on MM, it is strongly relying on force fields, and so it approximates all chemical bonds to springs and all atoms to spheres with uniform properties. The partial charges are defined before the simulation, and no polarization effect is taken into account. Among the most relevant force fields for MD simulations, AMBER [143], GROMOS (GROningenMOlecular Simulation) [144], OPLS [145], and CHARMM [146] are worthy to be mentioned. It is true that, as previously discussed, some relevant polarizable force fields have been developed and put into practice [93], but in many cases, the benefits introduced by the slightly more rigorous description of the system do not compensate for the higher computational costs of these implementations. With the rights assumptions, MD can be made suitable for the investigation of enzymatic processes or covalent bonding. This is the case of the already cited QM/MM scenario, in which just the protein region involved in the reaction is considered at the QM level of theory, allowing a proper description of the bond's oscillation, breaking, and creation, whilc thc whole remaining part of the system is treated with the MM laws [147]. In this way, the calculation can run much faster than a totally QM-set simulation, but of course, requires special care for the definition of the threshold in space between the QM and MM regions. Another very relevant drawback of MD with respect to docking and other ligand-based methods is related to the speed, the computational power, and the storage required by these approaches. Even if the recent technological advances allowed a tremendous speed-up in the simulations [148], the production rate of MD, expressed in processed molecules/day, is still several orders of magnitude below the one that can be offered by molecular docking or ligand-based methods. Moreover, the storage of docking poses (usually, some kilobytes/pose) is much lighter in respect of MD trajectories (which can weight several gigabytes each), and this is even more true if the simulations are carried out with an explicit solvent model (due to the huge increase in the number of atoms in the system). Clearly, the detail of description of the environment offered by MD is much higher in respect of the cited techniques, but in many cases, this is not enough to justify the costs. This is the reason why MD-based virtual screenings had just limited success up to now [149], confining molecular dynamics to be used mainly as an investigative or filtration instrument. Even if the situation changes heavily between the different companies, it is also possible to assert that MD has a much more limited diffusion in the industrial environment rather than in academia, always for the benefit/costs balanced previously explained [150].

The application of the MD-based methods in the Adenosine Receptor field will be discussed in the following sections, which will be first introduced by the explanation of the single techniques implemented.

5.3 Post-Docking Refinement

As already mentioned, one of the most relevant applications of MD is to be a filtration tool for the outcomes of a molecular docking experiment. Indeed, the ligand-target conformations coming from a docking run can each be considered a complex suitable for an MD simulation. In this scenario, the system needs one or more proper preparation steps (usually involving solvation, charge neutralization through salt addition, solvent relaxation, and energy minimization) to be run in a dynamic setup. After obtaining the trajectories (which should be, depending on the simulation setup, multiple for a single complex, allowing to get a more suitable sampling of the possible dynamic evolution steps [151, 152]) their analysis usually starts from the evaluation of the classical geometrical parameters of RMSD (root-mean-squared deviation) and RMSF (root-mean-squared fluctuation) of atomic positions with time [153–155]. The first indicates the magnitude of the displacement of the ligand from its initial position, while the second informs about the fluctuation from the most populated position during the trajectory. In both cases, high values can be considered typical of a weakly-interacting starting conformation of the ligand, while lower values indicate a molecular conformation that is kept stable during the simulation. The main fact is that these parameters offer a poor description of the dynamic evolution of the target-ligand contact, not taking into account any interaction between the two [156]. More reliable descriptors can be identified in the hydrogen bonds and hydrophobic interaction dynamic analysis [157], but also in this case, even if the information obtained is higher in quality than the previously described geometrical values, it still lacks hints regarding the various parts of the ligand and of the target which are relevant for the interaction. Some Python-based available codes and also some commercial suites (e.g., Desmond, by Schrödinger) offer nowadays some qualitative data in this respect [158]. The implementation of filtration through a structure-based three-dimensional pharmacophore model, which can be created with a knowledge-based approach or in an experimental complex-biased way (e.g., consensus pharmacophore), has been demonstrated to be a rapid and valuable strategy for pose selection [117, 159], but not so extensively for their prioritization. One of the ways to overcome this limitation is to exploit what is called a "protein-ligand interaction fingerprint" (PLIF) [160, 161]. Indeed, each of these maps the type of interaction between a certain ligand atom and a certain amino acid residue in the target, and so offers very relevant information from a relationship perspective between the two entities [162, 163] (Fig. 7).

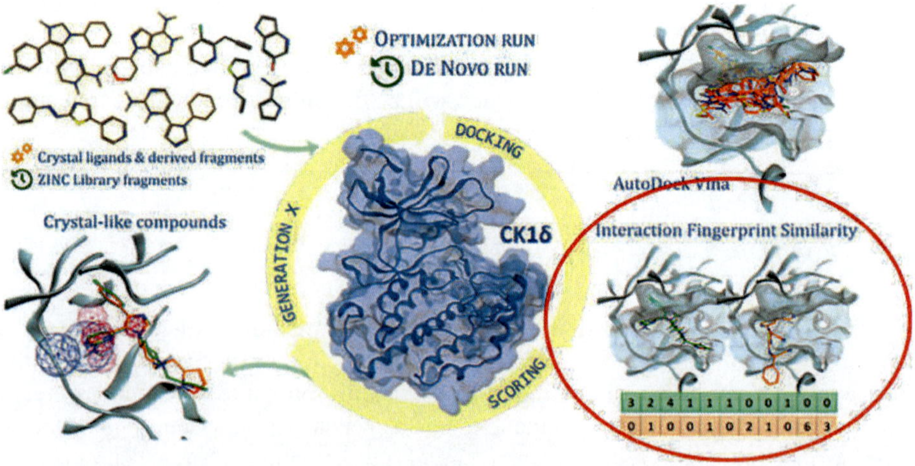

Fig. 7 Schematic example of a practical implementation of Protein–Ligand Interaction Fingerprints (PLIFs) in a recent article by Pavan et al. [163]. In this scenario, the PLIFs were used as a metric to guide the novel molecules generated by the AutoGrow4 [164] de novo approach toward the interactive pattern of the known crystallographic ligands of a desired target. In this specific paper, the protein used as a case study was Casein Kinase 1δ

Mapping in a dynamic fashion the PLIFs of all the ligand's atoms through time, it is possible to get a very detailed overview of what has changed during the trajectory from an interactive point of view. One relevant example of the application of PLIFs as metrics for post-docking purposes has been recently proposed by Pavan et al., in which the authors exploit the newly designed "Thermal Titration Molecular Dynamics" (TTMD) methodology for ligand classification based on residence time [165]. This last approach, which was first validated on crystal structures, and then its potential was also confirmed for post-docking purposes [166], relies on the principle by which, increasing the temperature in an MD simulation, the more stable ligands will tendentially keep their position and their interaction pattern within the binding site for a longer time in respect of weaker ligands, also being less sensitive to the raises in temperature. This method has proven to be effective on kinases and proteases targets, but with the right implementations, its application on GPCRs seems promising (Fig. 8).

Post-docking methods have been implemented in several studies in Adenosine Receptor discovery, for example, Anighoro and Rastelli used an MD and Molecular Mechanics Poisson-Boltzmann Surface Area (MM-PBSA) based post-docking approach (called "BEAR" by the authors, the acronym for "Binding Estimation After Refinement") to enrich a selected pool of molecules of active $A_{2A}R$ ligands [169]. MD and MM-PBSA were also pointed out as the refinement methods after a virtual screening executed with GOLD in a work by Martinelli and Ortore [170]. Catarzi et al. exploited a post-docking rescoring method relying on the enthalpic contribution to the free energy of binding in a study for the development of potent A_3R antagonists [171]. A very similar strategy was used by Lenzi et al. in a workflow for the discovery of pyrazolo[3,4-c]quinoline derivatives as adenosine receptor antagonists. Both MD and FEP were recently implemented by Jespers et al. for the post-docking refinement of several different ligands, taking into account all four adenosine receptor isoforms [172]. Sabbadin et al. presented a protocol based on Membrane Molecular Dynamics as a post-docking strategy for the description of the stability of $A_{2A}R$-ligand complexes, proposing the possibility to extend this method also to other GPCRs [173].

5.4 Enhanced Sampling Approaches

One of the main limitations of MD is based on the very low sampling that the method offers for a defined system. Indeed, also very long, repeated simulations are not able to explore the whole energetic landscape of the entities included in the system [174, 175]. This restriction opened an entire field of research for the generation of methodologies able to introduce some bias in the MD environment, making it suitable for sampling wider and wider parts of the energy landscape, permitting the escape from eventual local minima and the extensive search for the effective most probable states. Because of this feature, these methods take the name of "enhanced sampling" approaches, and their application spread in the scientific

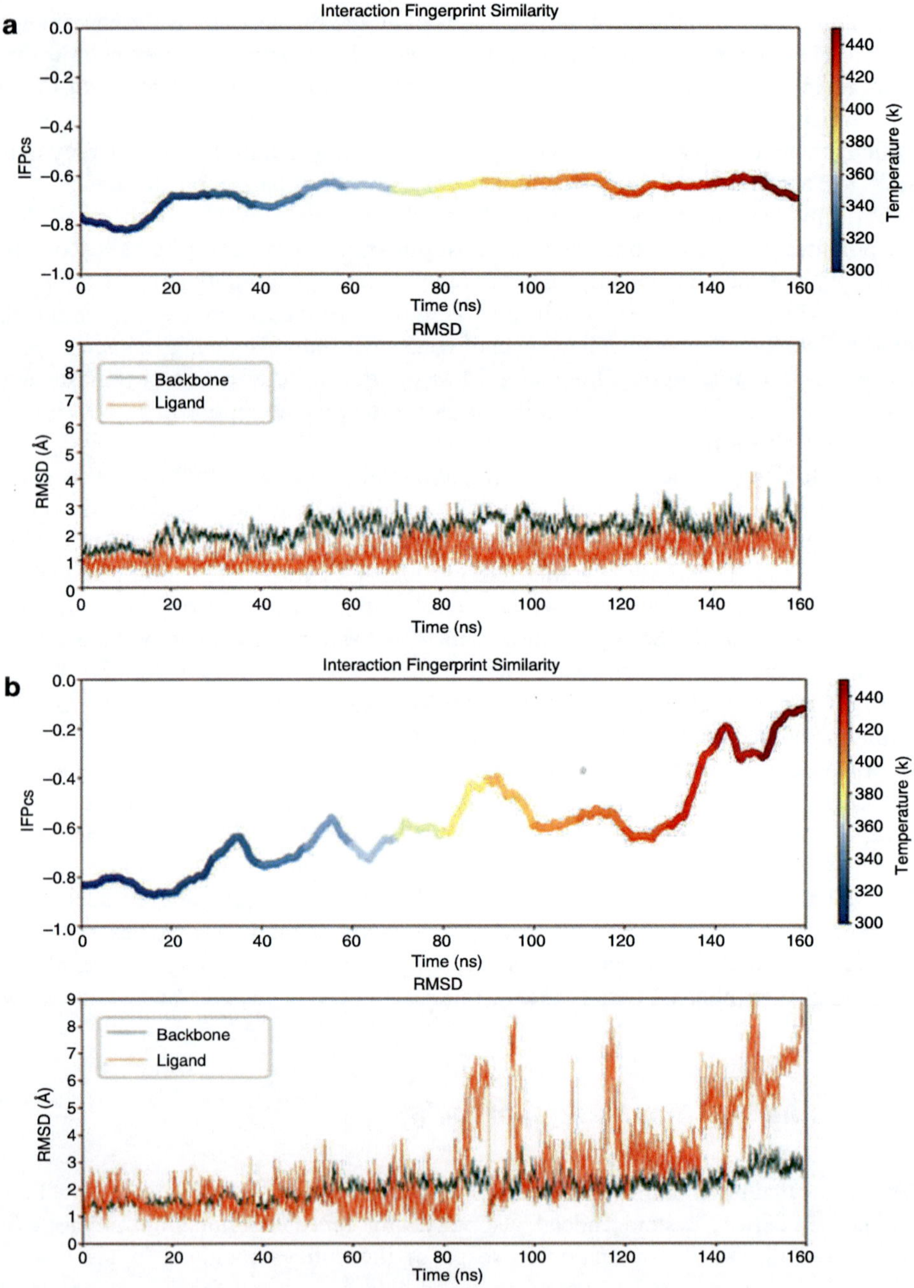

Fig. 8 Example of two TTMD profiles in comparison, coming from the study by Pav et al. [165]. On the upper side of each panel, is it possible to depict the IFP_{CS} score change during the simulation, higher scores represent a loss in the initial PLIFs and so, consequently, a loss of the biding mode. The lower part of each panel shows the RMSDs of the protein backbones (in green) and the ligand (in orange) against the simulation time. These last plots allow assessing that the

community in the latest decades [176]. Among the many techniques, some of the most relevant are steered MD [177], scaled MD [178], replica-exchange MD [179], supervised molecular dynamics (SuMD) [180], and metadynamics approaches [181]. The first is based on the application of a defined directed force on the ligand to favor its unbinding. The Jarzynski equality is then used to reconstruct the free-energy profile of the system. In scaled MD, a "scaling factor" is introduced for regulating the potential energy of the solute's degree within the simulation. Replica exchange is quite different as an approach, because it is based on the swapping of atomic positions between simulations which are run in parallel with one another and carried out at different temperatures. This is possible by exploiting independent Monte Carlo random walks.

An Ensemble-Based Steered MD approach has been implemented by Potterton et al. for the determination of ligand-$A_{2A}R$ residence time of 17 different known $A_{2A}R$ binders [182]. Similarly, Akhunzada et al. exploited steered MD combined with the Bell-Evans model for the estimation of the residence time of the $A_{2A}R$ antagonist ZMA241385 and the agonist NECA [183].

Supervised Molecular Dynamics (SuMD) is a more recent approach, and it is specifically designed for increasing the sampling of ligand approaching and binding events. This method does not introduce a bias in the system but just limits to supervise the distance between the ligand and the binding site during the simulation. Specifically, the entire trajectory is formed by the several so-called SuMD steps, and each of these is itself a short MD simulation (600 ps, by default) in which the ligand-binding site distance is measured at the beginning and the end. If this distance diminishes in the SuMD step, then this is considered productive, is saved, and will concur in forming the final merged trajectory, otherwise, it is discarded, and the ligand's coordinates are restored at the beginning of the step itself [184]. This method has already been extensively applied to several targets, like kinases [185, 186], proteases [187], ribosomal proteins [188], and GPCRs [180, 189] (Fig. 9).

Among the latest, adenosine receptors are one of the drug target families that benefited the most from this enhanced sampling technique [190]. The methodology was implemented by Sabbadin et al. for the description of the ligand-protein recognition path of adenosine to $A_{2A}R$, allowing also to spot a potential meta-binding site [191]. Going deeper into the topic, Deganutti et al. used SuMD to analyze the differences in the recognition mechanism of adenosine and inosine to

Fig. 8 (continued) increase in temperature is not affecting the protein folding in a relevant fashion. Panel (**a**) representation of the results of the TTMD simulation for the ligand PF670462 in the pocket of Casein Kinase 1δ (starting from the bound conformation of the crystal 3UZP [167]). Panel (**b**) depiction of the results of the TTMD experiment for an Epiblastin A brominated derivative bound to Casein Kinase 1δ, coming from the crystal with PDB code 5IH6 [168]. As it can be observed, the nanomolar ligand PF670462 keeps its PLIFs during the simulation, while the micromolar Epiblastin A derivative progressively loses the initial binding mode, assessing the instability of its contacts with the protein

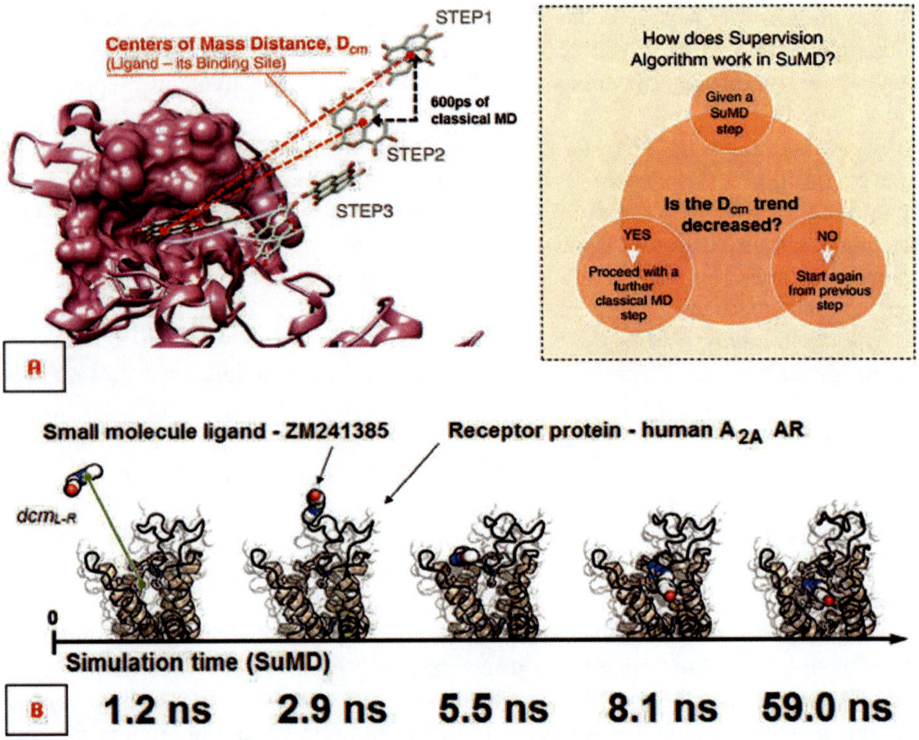

Fig. 9 Panel (**a**) schematic representation of the Supervised Molecular Dynamics (SuMD) algorithm, as presented in the study by Sabbadin et al. [190] As can be seen, the "supervision" algorithm analyzes the ligand-binding site distance at the end of each SuMD step (600 ps of MD). If this distance diminishes, the final coordinates are saved and used as the starting point for the next step, otherwise, the SuMD step is discarded and the coordinates are restored as the initial ones. The supervision algorithm using the ellagic acid recognition process to human Casein Kinase 2 (CK2) is given as an example. Panel (**b**) picture of the evolution of a ligand-GPCR SuMD trajectory with time. The receptor is represented in $A_{2A}R$, while the ligand is the known antagonist ZM241385 (ZMA). The image is taken from the same article by Sabbadin et al., as in the previous panel

$A_{2A}R$, highlighting the possibility that the lower potency of the second ligand could come from a missed hydrogen bond interaction in the binding site [192]. Moreover, De Filippo et al. highlighted how the presence of a meta-binding site with a high affinity for adenosine in the ECL2 of $A_{2B}R$ determinates a several orders of magnitude lowering of potency to this receptor for the endogenous ligand, without affecting the binding of the potent agonists N-ethylcarboxamidoadenosine (NECA) and CGS-21680 [193].

In a recent work by Bolcato et al., the authors exploited supervised MD to rationalize the partial agonist behavior of a famous series of dicyanopyridines on $A_{2A}R$ [189] (Fig. 10).

In a recent paper by Deganutti et al., SuMD was also applied in the attempt to rationalize at the molecular level the mechanism of action of the A_3R positive

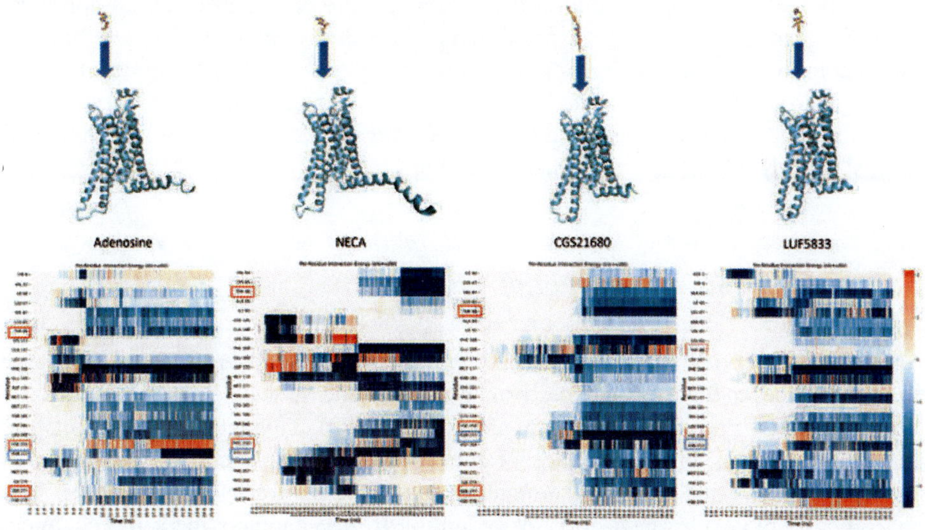

Fig. 10 Example of an application of Supervised Molecular Dynamics (SuMD) to the elucidation of the differences in the ligand-binding paths between full agonists (Adenosine, NECA, and CGS21680) and a non-ribosidic partial agonist (LUF5833) of the adenosine A_{2A} receptor, coming from the cited study from Bolcato et al. [189]. The panels on the upper part of the figure represent the initial configuration of each SuMD simulation, with the ligand placed away from the orthosteric binding site. The plots in the lower part depict the outcomes of the time-based per-residue interaction analysis, in which the summation of electrostatic and van der Waals contributions for each of the 25 most contacted protein residues is reported for each frame of the trajectories produced. In this case study, SuMD highlighted the main differences between the ligand-protein recognition paths of full agonists and LUF5833, which can exert a partial agonism to $A_{2A}R$ even if important residues such as Thr88 and Ser277 are not recruited directly by this specific compound. An explanation of the molecular reasons behind this behavior is reported in the publication

allosteric modulator LUF6000 [194]. The methodology has also proven to be proficient for the identification of the key determinants justifying the selectivity between A_1R and $A_{2A}R$ antagonists in a publication by Bolcato et al. [195]. Coherently to what was already presented for molecular docking, also SuMD was used by Bissaro et al. to examine the influence of the sodium ion in the 7TM region of $A_{2A}R$, and specifically, it was demonstrated that its presence diminishes the stabilization of the endogenous ligand adenosine in different steps of the binding process [196].

Finally, metadynamics approaches are based on the iterative "filling" of the potential energy in the simulation with a series of Gaussian curves. This forces the system to explore different energetic minima and, so it improves the sampling of the so-called rare events (like the binding or conformational shifts). Cao et al. used well-tempered metadynamics combined with free-energy calculations to reproduce the experimentally measured binding affinity for the known ZMA $A_{2A}R$ antagonist, giving reason to the role of lipids and ECL2 for the ligand's recognition process [197]. Moreover, metadynamics was successfully applied by Li et al. for detailing the ligand binding-induced shift between inactive and active states in $A_{2A}R$ [198].

Similarly to SuMD, Deganutti et al. recently developed suMetaD, an approach combining supervised MD and metadynamics for the simulation of ligand binding and unbinding events. This tool provides energetics and solvation/desolvation-related information, which can be crucial in the development and optimization steps for small molecules, especially in the GPCR scenario. The tool was validated with six $A_{2A}R$ inhibitors on the same receptor [199].

5.5 *Free-Energy Perturbation (FEP)*

One of the methods that have acquired the most success in recent times, mainly due to the advances in the hardware performance that made these calculations accessible, is the so-called Free-Energy Perturbation (FEP) [200]. This technique exploits the fact that, by the simple principles of the thermodynamic cycle, it is possible to correctly retrieve the binding free energy for a series of co-generic ligands. The investigated molecules have to be very similar to one another (the same chemical series) because the technique is much more reliable and sensitive if the differences in energies that have to be calculated involve very small changes in the structure [201, 202]. FEP has been demonstrated to correctly predict the binding free energies for various chemotypes on different targets, also with errors below 1 kcal/mol (comparable with the experimental uncertainty) [203]. The huge improvements in the informatics architectures, mostly regarding the Graphics Processing Units (GPUs) field, are expected to make FEP methods more accessible, allowing a wider spread of their application.

In the field of adenosine receptors, FEP has already been implemented for small molecule discovery purposes. Indeed, Matricon et al. exploited an approach based on molecular dynamics and FEP to guide compound optimization in the A_{2A} receptor, also applying the technique for the prospective optimization of non-purine fragments [204]. The same MD/FEP technique was implemented by Matricon et al. in the investigation of ligand-induced water reorganization in $A_{2A}R$, leading to the proficient design of a novel nanomolar-potent agonist [205]. Jespers et al. exploited FEP to build structure-energetics relationships for $A_{2A}R$ activation, which have been used for the pharmacological profiling of different series of $A_{2A}R$ agonists [206]. The same group had already demonstrated the validity of FEP in guiding the SBDD pipelines on the different adenosine receptor subtypes in a previous paper [172]. Moreover, Jespers et al. used FEP in an iterative design cycle, combining it with synthesis and X-ray crystallography, which led to the creation of novel small molecules with micromolar potency and a validated binding mode on $A_{2A}R$ [207]. Wang et al. investigated the binding mode and the selectivity determinants of the compound ISAM-140 to $A_{2B}R$ by implementing rigorous FEP calculations [208]. A series of different ligands active on $A_{2A}R$ were investigated by Deflorian et al. using FEP, demonstrating how water molecules and amino acid ionization states play a crucial role in the success of the computational free-energy experiment [209].

6 Ligand-Based Drug Design (LBDD)

Even if nowadays the structure-based approaches are the most used, mainly thanks to the increase in the experimentally available structures for the vast majority of the protein families, ligand-based methods are still part of several drug discovery pipelines. Indeed, even if it is much rarer to lack structural information about the target in the past, this information may come just from a few experimental entries or models, and it could also be that these techniques are used together with the SBDD ones, allowing to get some "consensus" results in terms of candidate prioritization.

LBDD methods were developed starting in the 60s when scientists tried to find correlations between physicochemical properties and structural moieties in small molecules. Among the researchers, it is important to remember the efforts put in by Hansch, Hammet, and Taft [48]. The goal was then extended to the attempt to correlate structural components to the change in activity to a desired target or disease. This was the initial rising of QSAR (Quantitative Structure-Activity Relationship) modeling [210], which then became concrete in the more advanced methodologies of the Free-Wilson analysis, the pharmacophore search, and cheminformatics.

In their three-dimensional version, pharmacophores are still widely used in the CADD environment [211]. They can both be implemented with an SBDD perspective, as a filtering method for docking poses (as mentioned above) or a pre-virtual screening filter, to preventively select only those molecules able to generate a conformer respecting a defined pharmacophore hypothesis. Other than that, cheminformatics still finds a lot of room in drug design, from the clustering approaches to the t-SNE-based chemical space analysis [212], to the ligand-based scaffold hopping approaches [213].

Looking deeper into the application of ligand-based methods for adenosine receptor research, noticeable is the creation of the bioinformatics and cheminformatics platform called "Adenosiland" by Floris et al. in 2012. This platform is provided with both ligand and structure-based queries for adenosine receptor drug design [214]. In a recent article, Xu et al. developed a consensus pharmacophore hypothesis on known $A_{2A}R$ ligands, validating it on a virtual database of 106 diverse small molecules [215]. Similarly, Bacilieri et al. developed a novel $A_{2A}R$ ligand and structure-based 3D-QSAR model based on 751 known antagonists, validating it on a set of 29 newly synthesized compounds, which showed a potency from the low to the high nanomolar range [216]. Tafi et al. exploited the program Catalyst for the generation of three pharmacophoric hypotheses for human A_3R antagonists, demonstrating their validity on a set of new molecules [217].

7 Machine Learning and Artificial Intelligence in Drug Discovery

7.1 The Machine Learning Method

In the middle of the 60s, the idea that machines could gather transferable knowledge in a human-like fashion started to spread through the scientific community [218]. Since that moment, huge steps have been taken, and machine learning (ML) is now part of everyone's lives. From image recognition, to object classification, to properties forecasting, ML has manifested all its potential [219]. All these advances are of course possible because of the huge quantities of data that have been gathered in the latest decades on most various topics and fields. These data are organized in databases, in which each entry constitutes a "data point," while its characteristics take the name of "features." Each of the points is associated with a "label," which can be varied depending on a case-by-case scenario. Very generally speaking, the ML mechanism consists in a three-phase approach, which is training, testing, and validation. In the first one, the algorithm examines a series of labeled data points, learning all the possible links between the features and the labels, and so trying to extract the necessary knowledge for the prediction of novel entries' labels. This predictive performance is assessed in the test phase, in which the algorithm tries to execute predictions on a set of points of which the label is known, but is not supplied to the model. In this phase, it is possible to assess the ability of the trained ML model before really making it operational on an unlabeled dataset, which is the validation one [220]. The performance of the models can be measured by using different metrics, depending mainly on the kind of problem that is examined. If it is a binary classification one, balanced accuracy, specificity (or "true negative ratio"), and sensitivity (or "true positive ratio"), together with the Matthews Correlation Coefficient (MCC) [221], are for sure the ones that are most closely tackled, while the correlation coefficient (R^2), the mean average percentage error (MAPE), and the root-mean-squared error (RMSE) are the most important in a regression problem [222].

Many are the model architectures that have been developed for ML purposes, from the early and simple Linear Regression and Logistic Regression to the tree-based methods such as Decision Trees, Random Forest, ExtraTrees, and Gradient Boosting, to the most recent deep neural networks (DNN), which are an adaptation of the classical Multilayer Perceptron [31]. Specifically, deep neural networks have been demonstrated to be the best-performing models when dealing with databases of very important dimensions [223] (Fig. 11).

Of course, even if the potential of these methods is astonishing, and even if they have proven in many different scenarios to outperform the golden standard techniques, also ML has several drawbacks. First, relying on great quantities of high-quality data, the algorithm's performance is directly dependent on both these factors [224]. If the data are lacking, then novel predictions will have a higher chance to fall into some data space that was never encompassed by the algorithm, and so in this

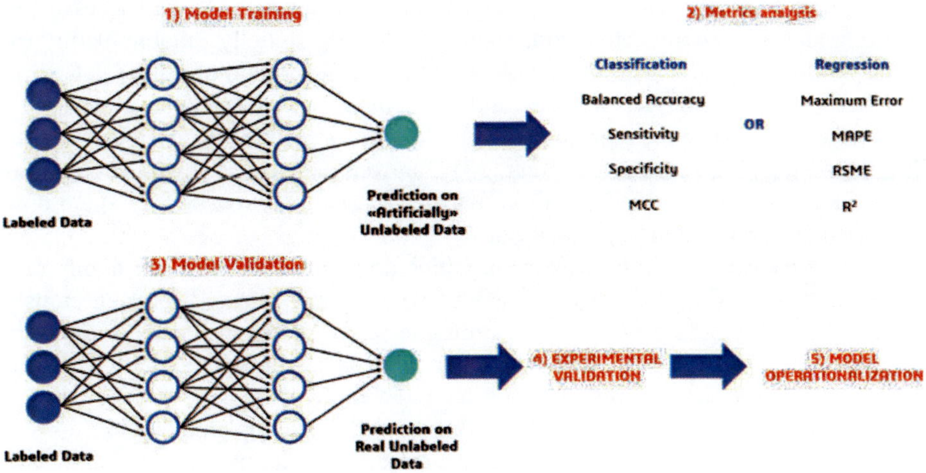

Fig. 11 Schematic representation of the Machine Learning process. First, the model (which is depicted as a Multilayer Perceptron) is trained on labeled data, and its predictive performance on a set of data points of which the label is voluntarily hidden allows extracting the relevant metrics for the model evaluation. Once the model is trained and it is associated with good confidence, it can be used to predict a set of unlabeled points, which are then experimentally validated to see how the model performs in a real-case scenario. If the validation phase is passed with success (usually it is repeated several times, to increase the confidence in the model), the ML tool becomes operational and is used for further predictions

case the probability of failing the prediction raises. On the other hand, if the data are not qualitatively high enough, the knowledge extracted by the model will also be not good enough for performing reliable predictions (as a common phrase in the ML lingo says: "garbage in, garbage out") [225]. The data availability and curation are so crucial for the success of these methods, which indeed show their weakness in the cases of unprecedented problems and poorly investigated research scenarios.

7.2 *Machine Learning in Drug Discovery and Its Implementation in the Adenosine Receptor Field*

With the great raising of data availability in recent decades, ML has spread to all the fields of human knowledge, wherever there is enough data to train a solid predictive algorithm. The drug discovery scenario is for sure one of the environments that benefited the most from these methods, from the toxicity prediction [226] to the ADME properties estimation [227], to the more recent generative approaches for de novo design [228]. Indeed, in the latest decade, many efforts have been successfully put into teaching the ML algorithms to recognize and reproduce SMILES strings, to subsequently generate novel, valid, and reliable small molecule structures. Several discovery studies relying on ML generative algorithms (mainly based on the Long

Short-Term Memory neural network architecture) have proven to be successful in creating and prioritizing active compounds [229, 230]. Initially suitable mainly for ligand-based purposes, with the development in the generation of proper data from the structures of biological entities and from their complexes with ligands, ML methods have expanded enormously also in the structure-based field. Indeed, ML-based docking scoring functions [231], as well as ML-based MD simulations [232] and virtual screenings based on machine learning approaches [233] are now routine in several labs around the world.

As already mentioned, in the case in which an unprecedented target family or a novel case scenario enters into research, ML methods have to be consciously handled to avoid getting just some random guesses out of the generated models. In any case, the knowledge and experience of scientists are always required, and even more in the early phase of investigation on a new topic.

In the case of adenosine receptors, ML methods were implemented in different projects by several research groups. For example, Liu et al. used $A_{2A}R$ as the case study to prove the efficacy of their Recurrent Neural Network (RNN)-based model DrugEx in generating novel and diverse molecules predicted to be active on the receptor [234]. Goßen et al. used a mixed approach combining structural information and a random forest classifier to guide the identification of novel $A_{2A}R$ antagonists, finally leading to the discovery of a nanomolar-potent small molecule ligand [235]. A deep learning-based docking method (called "Deep Docking") was proposed by Tang et al. for the discovery of two novel $A_{2A}R$ antagonists, starting from 1.4 billion compounds, in a low computationally-demanding fashion [236]. With a repurposing strategy, Puhl et al. built and validated a Bayesian machine learning model, which led to the identification of already approved molecules that could show potential activity in the adenosine receptor world [237].

8 Conclusions

Computational approaches revolutionized the field of drug discovery, bringing a total change in how a novel compound is designed, optimized, and progressed in the pipeline. From the very early research, many steps ahead have been executed, both from an efficiency (in terms of time and money) and from a methodological perspective. Further improvements in the upcoming future will be for sure oriented to increasing the power of ML and Artificial Intelligence (AI) approaches, to make them more and more reliable and predictive [238]. Together with that, the rising computational power, also assisted by the advent of quantum computing, is expected to allow a spread in the applicability of quantum mechanical methods, progressively reducing the degree of approximation that the CADD scientists have been obliged to accept up to date because of efficiency and computational resources-related reasons [239]. All these fascinating improvements will have a noticeable impact on the pharmaceutical discovery scenario, and we are confident that, together with a progressively better understanding of human physiology and adenosine signaling, computer-based approaches will guide the generation of enhanced, safer, more selective, and more efficacious therapeutics in the adenosine discovery field.

Compliance with Ethical Standards Conflict of Interest

The author declares that they have no conflict of interest.

Funding The authors received no financial support for the research, authorship, and/or publication of this article.

Ethical Approval This chapter does not contain any studies with human participants or animals performed by the authors.

References

1. Martin L, Hutchens M, Hawkins C (2017) Clinical trial cycle times continue to increase despite industry efforts. Nat Rev Drug Discov 16(3):157–157. https://doi.org/10.1038/nrd.2017.21
2. Simoens S, Huys I (2021) R&D costs of new medicines: a landscape analysis. Front Med 8. https://doi.org/10.3389/fmed.2021.760762
3. FDA. The drug development process. https://www.fda.gov/patients/learn-about-drug-and-device-approvals/drug-development-process
4. Hertzberg RP, Pope AJ (2000) High-throughput screening: new technology for the 21st century. Curr Opin Chem Biol 4(4):445–451. https://doi.org/10.1016/S1367-5931(00)00110-1
5. Keserű GM, Makara GM (2006) Hit discovery and hit-to-lead approaches. Drug Discov Today 11(15–16):741–748. https://doi.org/10.1016/j.drudis.2006.06.016
6. Deprez-Poulain R, Deprez B (2004) Facts, figures and trends in lead generation. Curr Top Med Chem 4(6):569–580. https://doi.org/10.2174/1568026043451168
7. Hevener KE, Pesavento R, Ren J, Lee H, Ratia K, Johnson ME (2018) Hit-to-lead: hit validation and assessment. Methods Enzymol:265–309. https://doi.org/10.1016/bs.mie.2018.09.022
8. Bleicher KH, Böhm H-J, Müller K, Alanine AI (2003) Hit and lead generation: beyond high-throughput screening. Nat Rev Drug Discov 2(5):369–378. https://doi.org/10.1038/nrd1086
9. Hughes J, Rees S, Kalindjian S, Philpott K (2011) Principles of early drug discovery. Br J Pharmacol 162(6):1239–1249. https://doi.org/10.1111/j.1476-5381.2010.01127.x
10. Hefti FF (2008) Requirements for a lead compound to become a clinical candidate. BMC Neurosci 9(S3):S7. https://doi.org/10.1186/1471-2202-9-S3-S7
11. Keserü GM, Makara GM (2009) The influence of lead discovery strategies on the properties of drug candidates. Nat Rev Drug Discov 8(3):203–212. https://doi.org/10.1038/nrd2796
12. (2012) The truly staggering cost of inventing new drugs. [Online]. Available: https://www.forbes.com/sites/matthewherper/2012/02/10/the-truly-staggering-cost-of-inventing-new-drugs/?sh=4f5fd0104a94
13. Congressional Budget Office (2021) Research and development in the pharmaceutical industry. [Online]. Available: https://www.cbo.gov/publication/57126
14. FDA. Development & approval process | drugs. [Online]. Available: https://www.fda.gov/drugs/development-approval-process-drugs
15. EMA. Authorisation of medicines. [Online]. Available: https://www.ema.europa.eu/en/about-us/what-we-do/authorisation-medicines

16. Assenberg R, Wan PT, Geisse S, Mayr LM (2013) Advances in recombinant protein expression for use in pharmaceutical research. Curr Opin Struct Biol 23(3):393–402. https://doi.org/10.1016/j.sbi.2013.03.008
17. D'Atri V, Fekete S, Clarke A, Veuthey J-L, Guillarme D (2019) Recent advances in chromatography for pharmaceutical analysis. Anal Chem 91(1):210–239. https://doi.org/10.1021/acs.analchem.8b05026
18. Denora N, Trapani A, Laquintana V, Lopedota A, Trapani G (2009) Recent advances in medicinal chemistry and pharmaceutical technology-strategies for drug delivery to the brain. Curr Top Med Chem 9(2):182–196. https://doi.org/10.2174/156802609787521571
19. Porta R, Benaglia M, Puglisi A (2016) Flow chemistry: recent developments in the synthesis of pharmaceutical products. Org Process Res Dev 20(1):2–25. https://doi.org/10.1021/acs.oprd.5b00325
20. Hann MM, Oprea TI (2004) Pursuing the leadlikeness concept in pharmaceutical research. Curr Opin Chem Biol 8(3):255–263. https://doi.org/10.1016/j.cbpa.2004.04.003
21. Bohacek RS, McMartin C, Guida WC (1996) The art and practice of structure-based drug design: a molecular modeling perspective. Med Res Rev 16(1):3–50. https://doi.org/10.1002/(SICI)1098-1128(199601)16:1<3::AID-MED1>3.0.CO;2-6
22. Tripathi NM, Bandyopadhyay A (2022) High throughput virtual screening (HTVS) of peptide library: technological advancement in ligand discovery. Eur J Med Chem 243:114766. https://doi.org/10.1016/j.ejmech.2022.114766
23. Pirhadi S, Sunseri J, Koes DR (2016) Open source molecular modeling. J Mol Graph Model 69:127–143. https://doi.org/10.1016/j.jmgm.2016.07.008
24. Leelananda SP, Lindert S (2016) Computational methods in drug discovery. Beilstein J Org Chem 12:2694–2718. https://doi.org/10.3762/bjoc.12.267
25. Pavan M, Bassani D, Sturlese M, Moro S (2022) From the Wuhan-Hu-1 strain to the XD and XE variants: is targeting the SARS-CoV-2 spike protein still a pharmaceutically relevant option against COVID-19? J Enzyme Inhib Med Chem 37(1):1704–1714. https://doi.org/10.1080/14756366.2022.2081847
26. Vyas V, Ukawala R, Chintha C, Ghate M (2012) Homology modeling a fast tool for drug discovery: current perspectives. Indian J Pharm Sci 74(1):1. https://doi.org/10.4103/0250-474X.102537
27. Cavasotto CN, Phatak SS (2009) Homology modeling in drug discovery: current trends and applications. Drug Discov Today 14(13–14). https://doi.org/10.1016/j.drudis.2009.04.006
28. Fourches D, Muratov E, Tropsha A (2010) Trust, but verify: on the importance of chemical structure curation in cheminformatics and QSAR modeling research. J Chem Inf Model 50(7):1189–1204. https://doi.org/10.1021/ci100176x
29. Berman HM (2000) The protein data bank. Nucleic Acids Res 28(1):235–242. https://doi.org/10.1093/nar/28.1.235
30. PDB statistics: overall growth of released structures per year. [Online]. Available: https://www.rcsb.org/stats/growth/growth-released-structures
31. Sarker IH (2021) Machine learning: algorithms, real-world applications and research directions. SN Comput Sci 2(3):160. https://doi.org/10.1007/s42979-021-00592-x
32. Mouchlis VD et al (2021) Advances in de novo drug design: from conventional to machine learning methods. Int J Mol Sci 22(4):1676. https://doi.org/10.3390/ijms22041676
33. Sabe VT et al (2021) Current trends in computer aided drug design and a highlight of drugs discovered via computational techniques: a review. Eur J Med Chem 224:113705. https://doi.org/10.1016/j.ejmech.2021.113705
34. Pecoraro C et al (2023) 1,2,4-Amino-triazine derivatives as pyruvate dehydrogenase kinase inhibitors: synthesis and pharmacological evaluation. Eur J Med Chem 249:115134. https://doi.org/10.1016/j.ejmech.2023.115134
35. Pavan M, Bassani D, Sturlese M, Moro S (2022) Bat coronaviruses related to SARS-CoV-2: what about their 3CL proteases (MPro)? J Enzyme Inhib Med Chem 37(1):1077–1082. https://doi.org/10.1080/14756366.2022.2062336

36. Anderson AC (2003) The process of structure-based drug design. Chem Biol 10(9):787–797. https://doi.org/10.1016/j.chembiol.2003.09.002
37. Baker D, Sali A (2001) Protein structure prediction and structural genomics. Science 294(5540):93–96. https://doi.org/10.1126/science.1065659
38. Webb B, Sali A (2016) Comparative protein structure modeling using modeller. Curr Protoc Bioinformatics 54(1). https://doi.org/10.1002/cpbi.3
39. Waterhouse A et al (2018) SWISS-MODEL: homology modelling of protein structures and complexes. Nucleic Acids Res 46(W1):W296–W303. https://doi.org/10.1093/nar/gky427
40. Kelley LA, Mezulis S, Yates CM, Wass MN, Sternberg MJE (2015) The Phyre2 web portal for protein modeling, prediction and analysis. Nat Protoc 10(6):845–858. https://doi.org/10.1038/nprot.2015.053
41. Song Y et al (2013) High-resolution comparative modeling with RosettaCM. Structure 21(10): 1735–1742. https://doi.org/10.1016/j.str.2013.08.005
42. Chemical Computing Group ULC (2023) Molecular operating environment (MOE) 2022.02
43. Jumper J et al (2021) Highly accurate protein structure prediction with AlphaFold. Nature 596(7873):583–589. https://doi.org/10.1038/s41586-021-03819-2
44. Protein Structure Prediction Center. https://predictioncenter.org/
45. David A, Islam S, Tankhilevich E, Sternberg MJE (2022) The AlphaFold database of protein structures: a biologist's guide. J Mol Biol 434(2):167336. https://doi.org/10.1016/j.jmb.2021.167336
46. Margiotta E, Moro S (2019) A comparison in the use of the crystallographic structure of the human A1 or the A2A adenosine receptors as a template for the construction of a homology model of the A3 subtype. Appl Sci 9(5):821. https://doi.org/10.3390/app9050821
47. Shim J, MacKerell Jr AD (2011) Computational ligand-based rational design: role of conformational sampling and force fields in model development. MedChemComm 2(5):356. https://doi.org/10.1039/c1md00044f
48. Silakari O, Singh PK (2021) QSAR: descriptor calculations, model generation, validation and their application. In: Concepts and experimental protocols of modelling and informatics in drug design. Elsevier, pp 29–63. https://doi.org/10.1016/B978-0-12-820546-4.00002-7
49. Kubinyi H (1988) Free Wilson analysis. Theory, applications and its relationship to Hansch analysis. Quant Struct Relationships 7(3):121–133. https://doi.org/10.1002/qsar.19880070303
50. Yang S-Y (2010) Pharmacophore modeling and applications in drug discovery: challenges and recent advances. Drug Discov Today 15(11–12):444–450. https://doi.org/10.1016/j.drudis.2010.03.013
51. Ebejer J-P, Morris GM, Deane CM (2012) Freely available conformer generation methods: how good are they? J Chem Inf Model 52(5):1146–1158. https://doi.org/10.1021/ci2004658
52. Bhisetti G, Fang C (2022) Artificial intelligence–enabled de novo design of novel compounds that are synthesizable. Methods Mol Biol 2390:409–419. https://doi.org/10.1007/978-1-0716-1787-8_17
53. PDB-101. Methods for determining atomic structures. https://pdb101.rcsb.org/learn/guide-to-understanding-pdb-data/methods-for-determining-structure#:~:text=Several methods are cur rently used,create the final atomic model.
54. Thompson MC, Yeates TO, Rodriguez JA (2020) Advances in methods for atomic resolution macromolecular structure determination. F1000Research 9:667. https://doi.org/10.12688/f1000research.25097.1
55. Benjin X, Ling L (2020) Developments, applications, and prospects of cryo-electron microscopy. Protein Sci 29(4):872–882. https://doi.org/10.1002/pro.3805
56. Carroni M, Saibil HR (2016) Cryo electron microscopy to determine the structure of macromolecular complexes. Methods 95:78–85. https://doi.org/10.1016/j.ymeth.2015.11.023
57. Wang H-W, Wang J-W (2017) How cryo-electron microscopy and X-ray crystallography complement each other. Protein Sci 26(1):32–39. https://doi.org/10.1002/pro.3022
58. Callaway E (2015) The revolution will not be crystallized: a new method sweeps through structural biology. Nature 525(7568):172–174. https://doi.org/10.1038/525172a

59. Renaud J-P et al (2018) Cryo-EM in drug discovery: achievements, limitations and prospects. Nat Rev Drug Discov 17(7):471–492. https://doi.org/10.1038/nrd.2018.77
60. Batool M, Ahmad B, Choi S (2019) A structure-based drug discovery paradigm. Int J Mol Sci 20(11):2783. https://doi.org/10.3390/ijms20112783
61. Bassani D, Ragazzi E, Lapolla A, Sartore G, Moro S (2022) Omicron variant of SARS-CoV-2 virus: in silico evaluation of the possible impact on people affected by diabetes mellitus. Front Endocrinol (Lausanne) 13. https://doi.org/10.3389/fendo.2022.847993
62. Sartore G, Bassani D, Ragazzi E, Traldi P, Lapolla A, Moro S (2021) In silico evaluation of the interaction between ACE2 and SARS-CoV-2 spike protein in a hyperglycemic environment. Sci Rep 11(1):22860. https://doi.org/10.1038/s41598-021-02297-w
63. Terayama K, Iwata H, Araki M, Okuno Y, Tsuda K (2018) Machine learning accelerates MD-based binding pose prediction between ligands and proteins. Bioinformatics 34(5): 770–778. https://doi.org/10.1093/bioinformatics/btx638
64. Doerr S et al (2021) TorchMD: a deep learning framework for molecular simulations. J Chem Theory Comput 17(4):2355–2363. https://doi.org/10.1021/acs.jctc.0c01343
65. Shi W, Singha M, Srivastava G, Pu L, Ramanujam J, Brylinski M (2022) Pocket2Drug: an encoder-decoder deep neural network for the target-based drug design. Front Pharmacol 13. https://doi.org/10.3389/fphar.2022.837715
66. Dong L, Qu X, Zhao Y, Wang B (2021) Prediction of binding free energy of protein–ligand complexes with a hybrid molecular mechanics/generalized born surface area and machine learning method. ACS Omega 6(48):32938–32947. https://doi.org/10.1021/acsomega.1c04996
67. Cournia Z, Allen B, Sherman W (2017) Relative binding free energy calculations in drug discovery: recent advances and practical considerations. J Chem Inf Model 57(12):2911–2937. https://doi.org/10.1021/acs.jcim.7b00564
68. Kuntz ID, Blaney JM, Oatley SJ, Langridge R, Ferrin TE (1982) A geometric approach to macromolecule-ligand interactions. J Mol Biol 161(2):269–288. https://doi.org/10.1016/0022-2836(82)90153-X
69. Pavan M, Bassani D, Bolcato G, Bissaro M, Sturlese M, Moro S (2022) Computational strategies to identify new drug candidates against neuroinflammation. Curr Med Chem 29(27):4756–4775. https://doi.org/10.2174/0929867329666220208095122
70. Li J, Fu A, Zhang L (2019) An overview of scoring functions used for protein–ligand interactions in molecular docking. Interdiscip Sci Comput Life Sci 11(2):320–328. https://doi.org/10.1007/s12539-019-00327-w
71. Alogheli H, Olanders G, Schaal W, Brandt P, Karlén A (2017) Docking of macrocycles: comparing rigid and flexible docking in glide. J Chem Inf Model 57(2):190–202. https://doi.org/10.1021/acs.jcim.6b00443
72. Landrum G (2010) RDKit: Open-source cheminformatics. [Online]. Available: https://www.rdkit.org/
73. Hawkins PCD, Skillman AG, Warren GL, Ellingson BA, Stahl MT (2010) Conformer generation with OMEGA: algorithm and validation using high quality structures from the protein databank and Cambridge structural database. J Chem Inf Model 50(4):572–584. https://doi.org/10.1021/ci100031x
74. Huang S-Y (2018) Comprehensive assessment of flexible-ligand docking algorithms: current effectiveness and challenges. Brief Bioinform 19(5):982–994. https://doi.org/10.1093/bib/bbx030
75. Sherman W, Day T, Jacobson MP, Friesner RA, Farid R (2006) Novel procedure for modeling ligand/receptor induced fit effects. J Med Chem 49(2):534–553. https://doi.org/10.1021/jm050540c
76. Miller EB et al (2021) Reliable and accurate solution to the induced fit docking problem for protein–ligand binding. J Chem Theory Comput 17(4):2630–2639. https://doi.org/10.1021/acs.jctc.1c00136

77. Amaro RE et al (2018) Ensemble docking in drug discovery. Biophys J 114(10):2271–2278. https://doi.org/10.1016/j.bpj.2018.02.038
78. Jones G, Willett P, Glen RC, Leach AR, Taylor R (1997) Development and validation of a genetic algorithm for flexible docking. J Mol Biol 267(3). https://doi.org/10.1006/jmbi.1996.0897
79. Morris GM et al (2009) AutoDock4 and AutoDockTools4: automated docking with selective receptor flexibility. J Comput Chem 30(16):2785–2791. https://doi.org/10.1002/jcc.21256
80. Friesner RA et al (2004) Glide: a new approach for rapid, accurate docking and scoring. 1. Method and assessment of docking accuracy. J Med Chem 47(7). https://doi.org/10.1021/jm0306430
81. Korb O, Stützle T, Exner TE (2006) PLANTS: application of ant colony optimization to structure-based drug design. https://doi.org/10.1007/11839088_22
82. Bassani D, Pavan M, Bolcato G, Sturlese M, Moro S (2022) Re-exploring the ability of common docking programs to correctly reproduce the binding modes of non-covalent inhibitors of SARS-CoV-2 protease Mpro. Pharmaceuticals 15(2):180. https://doi.org/10.3390/ph15020180
83. Bolcato G, Cuzzolin A, Bissaro M, Moro S, Sturlese M (2019) Can we still trust docking results? An extension of the applicability of DockBench on PDBbind database. Int J Mol Sci 20(14):3558. https://doi.org/10.3390/ijms20143558
84. Ramírez D, Caballero J (2018) Is it reliable to take the molecular docking top scoring position as the best solution without considering available structural data? Molecules 23(5):1038. https://doi.org/10.3390/molecules23051038
85. Gabel J, Desaphy J, Rognan D (2014) Beware of machine learning-based scoring functions – on the danger of developing black boxes. J Chem Inf Model 54(10):2807–2815. https://doi.org/10.1021/ci500406k
86. Wang Z et al (2016) Comprehensive evaluation of ten docking programs on a diverse set of protein–ligand complexes: the prediction accuracy of sampling power and scoring power. Phys Chem Chem Phys 18(18):12964–12975. https://doi.org/10.1039/C6CP01555G
87. Ha EJ, Lwin CT, Durrant JD (2020) LigGrep: a tool for filtering docked poses to improve virtual-screening hit rates. J Cheminform 12(1):69. https://doi.org/10.1186/s13321-020-00471-2
88. Gushchina IV, Polenova AM, Suplatov DA, Švedas VK, Nilov DK (2020) vsFilt: a tool to improve virtual screening by structural filtration of docking poses. J Chem Inf Model 60(8): 3692–3696. https://doi.org/10.1021/acs.jcim.0c00303
89. Spinaci A et al (2023) 'Dual Anta-Inhibitors' of the A2A adenosine receptor and casein kinase CK1delta: synthesis, biological evaluation, and molecular modeling studies. Pharmaceuticals 16(2):167. https://doi.org/10.3390/ph16020167
90. Carbone D et al (2023) Discovery of the 3-amino-1,2,4-triazine-based library as selective PDK1 inhibitors with therapeutic potential in highly aggressive pancreatic ductal adenocarcinoma. Int J Mol Sci 24(4):3679. https://doi.org/10.3390/ijms24043679
91. Boeyens J (2001) Molecular mechanics: theoretical basis, rules, scope and limits. Coord Chem Rev 212(1):3–10. https://doi.org/10.1016/S0010-8545(00)00353-2
92. Wang J, Cieplak P, Li J, Hou T, Luo R, Duan Y (2011) Development of polarizable models for molecular mechanical calculations I: parameterization of atomic polarizability. J Phys Chem B 115(12):3091–3099. https://doi.org/10.1021/jp112133g
93. Jing Z et al (2019) Polarizable force fields for biomolecular simulations: recent advances and applications. Annu Rev Biophys 48(1):371–394. https://doi.org/10.1146/annurev-biophys-070317-033349
94. Sighel D et al (2023) Streptogramin a derivatives as mitochondrial translation inhibitors to suppress glioblastoma stem cell growth. Eur J Med Chem 246:114979. https://doi.org/10.1016/j.ejmech.2022.114979

95. Heilmann E et al (2023) SARS-CoV-2 3CL pro mutations selected in a VSV-based system confer resistance to nirmatrelvir, ensitrelvir, and GC376. Sci Transl Med 15(678). https://doi.org/10.1126/scitranslmed.abq7360
96. de Beer S, Vermeulen N, Oostenbrink C (2010) The role of water molecules in computational drug design. Curr Top Med Chem 10(1):55–66. https://doi.org/10.2174/156802610790232288
97. Hu J, Sun X, Kang Z, Cheng J (2023) Computational investigation of functional water molecules in GPCRs bound to G protein or arrestin. J Comput Aided Mol Des 37(2): 91–105. https://doi.org/10.1007/s10822-022-00492-z
98. Rappas M et al (2020) Comparison of orexin 1 and orexin 2 ligand binding modes using X-ray crystallography and computational analysis. J Med Chem 63(4):1528–1543. https://doi.org/10.1021/acs.jmedchem.9b01787
99. Breznik M et al (2023) Prioritizing small sets of molecules for synthesis through in-silico tools: a comparison of common ranking methods. ChemMedChem 18(1). https://doi.org/10.1002/cmdc.202200425
100. Huang N, Shoichet BK (2008) Exploiting ordered waters in molecular docking. J Med Chem 51(16):4862–4865. https://doi.org/10.1021/jm8006239
101. Wahl J, Smieško M (2019) Assessing the predictive power of relative binding free energy calculations for test cases involving displacement of binding site water molecules. J Chem Inf Model 59(2):754–765. https://doi.org/10.1021/acs.jcim.8b00826
102. Song Q, Zeng L, Zheng Q, Liu S (2023) SCARdock: a web server and manually curated resource for discovering covalent ligands. ACS Omega 8(11):10397–10402. https://doi.org/10.1021/acsomega.2c08147
103. Toledo Warshaviak D, Golan G, Borrelli KW, Zhu K, Kalid O (2014) Structure-based virtual screening approach for discovery of covalently bound ligands. J Chem Inf Model 54(7): 1941–1950. https://doi.org/10.1021/ci500175r
104. Kumalo H, Bhakat S, Soliman M (2015) Theory and applications of covalent docking in drug discovery: merits and pitfalls. Molecules 20(2):1984–2000. https://doi.org/10.3390/molecules20021984
105. Groenhof G (2013) Introduction to QM/MM simulations. Methods Mol Biol:43–66. https://doi.org/10.1007/978-1-62703-017-5_3
106. Chaskar P, Zoete V, Röhrig UF (2014) Toward on-the-fly quantum mechanical/molecular mechanical (QM/MM) docking: development and benchmark of a scoring function. J Chem Inf Model 54(11):3137–3152. https://doi.org/10.1021/ci5004152
107. Mihalovits LM, Ferenczy GG, Keserű GM (2022) The role of quantum chemistry in covalent inhibitor design. Int J Quantum Chem 122(8). https://doi.org/10.1002/qua.26768
108. Lyu J et al (2019) Ultra-large library docking for discovering new chemotypes. Nature 566(7743):224–229. https://doi.org/10.1038/s41586-019-0917-9
109. Gentile F et al (2022) Artificial intelligence–enabled virtual screening of ultra-large chemical libraries with deep docking. Nat Protoc 17(3):672–697. https://doi.org/10.1038/s41596-021-00659-2
110. Inamdar GS et al (2013) New insight into adenosine receptors selectivity derived from a novel series of [5-substituted-4-phenyl-1,3-thiazol-2-yl] benzamides and furamides. Eur J Med Chem 63:924–934. https://doi.org/10.1016/j.ejmech.2013.03.020
111. Rodríguez A et al (2015) New selective a 2A agonists and a 3 antagonists for human adenosine receptors: synthesis, biological activity and molecular docking studies. MedChemComm 6(6): 1178–1185. https://doi.org/10.1039/C5MD00086F
112. Federico S et al (2018) [1,2,4]triazolo[1,5-c]pyrimidines as adenosine receptor antagonists: modifications at the 8 position to reach selectivity towards A3 adenosine receptor subtype. Eur J Med Chem 157:837–851. https://doi.org/10.1016/j.ejmech.2018.08.042
113. Wang M, Hou S, Wei Y, Li D, Lin J (2021) Discovery of novel dual adenosine A1/A2A receptor antagonists using deep learning, pharmacophore modeling and molecular docking. PLoS Comput Biol 17(3):e1008821. https://doi.org/10.1371/journal.pcbi.1008821

114. Sarkar B, Maiti S, Jadhav GR, Paira P (2018) Discovery of benzothiazolylquinoline conjugates as novel human a 3 receptor antagonists: biological evaluations and molecular docking studies. R Soc Open Sci 5(2):171622. https://doi.org/10.1098/rsos.171622
115. Cuzzolin A, Sturlese M, Malvacio I, Ciancetta A, Moro S (2015) DockBench: an integrated informatic platform bridging the gap between the robust validation of docking protocols and virtual screening simulations. Molecules 20(6):9977–9993. https://doi.org/10.3390/molecules20069977
116. Margiotta E, Deganutti G, Moro S (2018) Could the presence of sodium ion influence the accuracy and precision of the ligand-posing in the human A2A adenosine receptor orthosteric binding site using a molecular docking approach? Insights from Dockbench. J Comput Aided Mol Des 32(12):1337–1346. https://doi.org/10.1007/s10822-018-0174-2
117. Bassani D, Pavan M, Sturlese M, Moro S (2022) Sodium or not sodium: should its presence affect the accuracy of pose prediction in docking GPCR antagonists? Pharmaceuticals 15(3):346. https://doi.org/10.3390/ph15030346
118. Hollingsworth SA, Dror RO (2018) Molecular dynamics simulation for all. Neuron 99(6):1129–1143. https://doi.org/10.1016/j.neuron.2018.08.011
119. Pavan M, Moro S (2023) Lessons learnt from COVID-19: computational strategies for facing present and future pandemics. Int J Mol Sci 24(5):4401. https://doi.org/10.3390/ijms24054401
120. Bolcato G et al (2021) A computational workflow for the identification of novel fragments acting as inhibitors of the activity of protein kinase CK1δ. Int J Mol Sci 22(18):9741. https://doi.org/10.3390/ijms22189741
121. Case PAKDA, Aktulga HM, Belfon K, Ben-Shalom IY, Berryman JT, Brozell SR, Cerutti DS, Cheatham III TE, Cisneros GA, Cruzeiro VWD, Darden TA, Duke RE, Giambasu G, Gilson MK, Gohlke H, Goetz AW, Harris R, Izadi S, Izmailov SA (2017) Comparison of implicit and explicit solvent models for the calculation of solvation free energy in organic solvents. J Chem Theory Comput 13(3):1034–1043. https://doi.org/10.1021/acs.jctc.7b00169
122. Marenich AV, Cramer CJ, Truhlar DG (2009) Universal solvation model based on solute electron density and on a continuum model of the solvent defined by the bulk dielectric constant and atomic surface tensions. J Phys Chem B 113(18):6378–6396. https://doi.org/10.1021/jp810292n
123. Klamt A, Schüürmann G (1993) COSMO: a new approach to dielectric screening in solvents with explicit expressions for the screening energy and its gradient. J Chem Soc Perkin Trans 2(5):799–805. https://doi.org/10.1039/P29930000799
124. Takano Y, Houk KN (2005) Benchmarking the conductor-like polarizable continuum model (CPCM) for aqueous solvation free energies of neutral and ionic organic molecules. J Chem Theory Comput 1(1):70–77. https://doi.org/10.1021/ct049977a
125. Dyer KM, Perkyns JS, Stell G, Montgomery Pettitt B (2009) Site-renormalised molecular fluid theory: on the utility of a two-site model of water. Mol Phys 107(4–6):423–431. https://doi.org/10.1080/00268970902845313
126. Jorgensen WL, Chandrasekhar J, Madura JD, Impey RW, Klein ML (1983) Comparison of simple potential functions for simulating liquid water. J Chem Phys 79(2):926–935. https://doi.org/10.1063/1.445869
127. Florová P, Sklenovský P, Banáš P, Otyepka M (2010) Explicit water models affect the specific solvation and dynamics of unfolded peptides while the conformational behavior and flexibility of folded peptides remain intact. J Chem Theory Comput 6(11):3569–3579. https://doi.org/10.1021/ct1003687
128. Deganutti G, Moro S, Reynolds CA (2019) Peeking at G-protein-coupled receptors through the molecular dynamics keyhole. Future Med Chem 11(6):599–615. https://doi.org/10.4155/fmc-2018-0393
129. Dror RO et al (2011) Pathway and mechanism of drug binding to G-protein-coupled receptors. Proc Natl Acad Sci 108(32):13118–13123. https://doi.org/10.1073/pnas.1104614108
130. Schaller D, Pach S, Wolber G (2019) PyRod: tracing water molecules in molecular dynamics simulations. J Chem Inf Model 59(6):2818–2829. https://doi.org/10.1021/acs.jcim.9b00281

131. Araya-Secchi R et al (2014) Characterization of a novel water pocket inside the human Cx26 hemichannel structure. Biophys J 107(3):599–612. https://doi.org/10.1016/j.bpj.2014.05.037
132. Bellissent-Funel M-C et al (2016) Water determines the structure and dynamics of proteins. Chem Rev 116(13):7673–7697. https://doi.org/10.1021/acs.chemrev.5b00664
133. Venkatakrishnan AJ et al (2019) Diverse GPCRs exhibit conserved water networks for stabilization and activation. Proc Natl Acad Sci 116(8):3288–3293. https://doi.org/10.1073/pnas.1809251116
134. Ciancetta A, Sabbadin D, Federico S, Spalluto G, Moro S (2015) Advances in computational techniques to study GPCR–ligand recognition. Trends Pharmacol Sci 36(12):878–890. https://doi.org/10.1016/j.tips.2015.08.006
135. Gowers R et al (2016) MDAnalysis: a python package for the rapid analysis of molecular dynamics simulations. pp 98–105. https://doi.org/10.25080/Majora-629e541a-00e
136. Zhang S et al (2021) ProDy 2.0: increased scale and scope after 10 years of protein dynamics modelling with python. Bioinformatics 37(20):3657–3659. https://doi.org/10.1093/bioinformatics/btab187
137. Abel R, Young T, Farid R, Berne BJ, Friesner RA (2008) Role of the active-site solvent in the thermodynamics of factor Xa ligand binding. J Am Chem Soc 130(9):2817–2831. https://doi.org/10.1021/ja0771033
138. Baroni M, Cruciani G, Sciabola S, Perruccio F, Mason JS (2007) A common reference framework for analyzing/comparing proteins and ligands. Fingerprints for ligands and proteins (FLAP): theory and application. J Chem Inf Model 47(2):279–294. https://doi.org/10.1021/ci600253e
139. Cuzzolin A, Deganutti G, Salmaso V, Sturlese M, Moro S (2018) AquaMMapS: an alternative tool to monitor the role of water molecules during protein-ligand association. ChemMedChem 13(6):522–531. https://doi.org/10.1002/cmdc.201700564
140. Shaw DE et al (2021) Anton 3. In: Proceedings of the international conference for high performance computing, networking, storage and analysis, pp 1–11. https://doi.org/10.1145/3458817.3487397
141. Fornasier E et al (2022) A new inactive conformation of SARS-CoV-2 main protease. Acta Crystallogr Sect D Struct Biol 78(3):363–378. https://doi.org/10.1107/S2059798322000948
142. Bolcato G, Bissaro M, Sturlese M, Moro S (2020) Comparing fragment binding poses prediction using HSP90 as a key study: when bound water makes the difference. Molecules 25(20):4651. https://doi.org/10.3390/molecules25204651
143. Case DA (2022) Amber22. University of California, San Francisco
144. Scott WRP et al (1999) The GROMOS biomolecular simulation program package. J Phys Chem A 103(19):3596–3607. https://doi.org/10.1021/jp984217f
145. Jorgensen WL, Tirado-Rives J (1988) The OPLS [optimized potentials for liquid simulations] potential functions for proteins, energy minimizations for crystals of cyclic peptides and crambin. J Am Chem Soc 110(6):1657–1666. https://doi.org/10.1021/ja00214a001
146. Brooks BR et al (2009) CHARMM: the biomolecular simulation program. J Comput Chem 30(10):1545–1614. https://doi.org/10.1002/jcc.21287
147. Lodola A, De Vivo M (2012) The increasing role of QM/MM in drug discovery. Adv Protein Chem Struct Biol 87:337–362. https://doi.org/10.1016/B978-0-12-398312-1.00011-1
148. Harvey MJ, De Fabritiis G (2012) High-throughput molecular dynamics: the powerful new tool for drug discovery. Drug Discov Today 17(19–20):1059–1062. https://doi.org/10.1016/j.drudis.2012.03.017
149. Bissaro M, Bolcato G, Pavan M, Bassani D, Sturlese M, Moro S (2021) Inspecting the mechanism of fragment hits binding on SARS-CoV-2 M pro by using supervised molecular dynamics (SuMD) simulations. ChemMedChem 16(13):2075–2081. https://doi.org/10.1002/cmdc.202100156
150. Durrant JD, McCammon JA (2011) Molecular dynamics simulations and drug discovery. BMC Biol 9(1):71. https://doi.org/10.1186/1741-7007-9-71

151. Knapp B, Ospina L, Deane CM (2018) Avoiding false positive conclusions in molecular simulation: the importance of replicas. J Chem Theory Comput 14(12):6127–6138. https://doi.org/10.1021/acs.jctc.8b00391
152. Kubitzki MB, de Groot BL (2007) Molecular dynamics simulations using temperature-enhanced essential dynamics replica exchange. Biophys J 92(12):4262–4270. https://doi.org/10.1529/biophysj.106.103101
153. Aier I, Varadwaj PK, Raj U (2016) Structural insights into conformational stability of both wild-type and mutant EZH2 receptor. Sci Rep 6(1):34984. https://doi.org/10.1038/srep34984
154. Ivanova L, Tammiku-Taul J, García-Sosa AT, Sidorova Y, Saarma M, Karelson M (2018) Molecular dynamics simulations of the interactions between glial cell line-derived neurotrophic factor family receptor GFRα1 and small-molecule ligands. ACS Omega 3(9): 11407–11414. https://doi.org/10.1021/acsomega.8b01524
155. Martínez L (2015) Automatic identification of mobile and rigid substructures in molecular dynamics simulations and fractional structural fluctuation analysis. PloS One 10(3):e0119264. https://doi.org/10.1371/journal.pone.0119264
156. Knapp B, Frantal S, Cibena M, Schreiner W, Bauer P (2011) Is an intuitive convergence definition of molecular dynamics simulations solely based on the root mean square deviation possible? J Comput Biol 18(8):997–1005. https://doi.org/10.1089/cmb.2010.0237
157. Roe DR, Cheatham TE (2013) PTRAJ and CPPTRAJ: software for processing and analysis of molecular dynamics trajectory data. J Chem Theory Comput 9(7):3084–3095. https://doi.org/10.1021/ct400341p
158. Bowers KJ et al (2006) Molecular dynamics – scalable algorithms for molecular dynamics simulations on commodity clusters. In: Proceedings of the 2006 ACM/IEEE conference on supercomputing – SC '06, p 84. https://doi.org/10.1145/1188455.1188544
159. Cescon E et al (2020) Scaffold repurposing of in-house chemical library toward the identification of new casein kinase 1 δ inhibitors. ACS Med Chem Lett 11(6):1168–1174. https://doi.org/10.1021/acsmedchemlett.0c00028
160. Da C, Kireev D (2014) Structural protein–ligand interaction fingerprints (SPLIF) for structure-based virtual screening: method and benchmark study. J Chem Inf Model 54(9):2555–2561. https://doi.org/10.1021/ci500319f
161. Bouysset C, Fiorucci S (2021) ProLIF: a library to encode molecular interactions as fingerprints. J Cheminform 13(1):72. https://doi.org/10.1186/s13321-021-00548-6
162. Wójcikowski M, Zielenkiewicz P, Siedlecki P (2015) Open drug discovery toolkit (ODDT): a new open-source player in the drug discovery field. J Cheminform 7(1):26. https://doi.org/10.1186/s13321-015-0078-2
163. Pavan M, Menin S, Bassani D, Sturlese M, Moro S (2022) Implementing a scoring function based on interaction fingerprint for Autogrow4: protein kinase CK1δ as a case study. Front Mol Biosci 9. https://doi.org/10.3389/fmolb.2022.909499
164. Spiegel JO, Durrant JD (2020) AutoGrow4: an open-source genetic algorithm for de novo drug design and lead optimization. J Cheminform 12(1):25. https://doi.org/10.1186/s13321-020-00429-4
165. Pavan M, Menin S, Bassani D, Sturlese M, Moro S (2022) Qualitative estimation of protein–ligand complex stability through thermal titration molecular dynamics simulations. J Chem Inf Model. https://doi.org/10.1021/acs.jcim.2c00995
166. Menin S, Pavan M, Salmaso V, Sturlese M, Moro S (2023) Thermal titration molecular dynamics (TTMD): not your usual post-docking refinement. Int J Mol Sci 24(4):3596. https://doi.org/10.3390/ijms24043596
167. Long A, Zhao H, Huang X (2012) Structural basis for the interaction between casein kinase 1 delta and a potent and selective inhibitor. J Med Chem 55(2):956–960. https://doi.org/10.1021/jm201387s
168. Ursu A et al (2016) Epiblastin a induces reprogramming of epiblast stem cells into embryonic stem cells by inhibition of casein kinase 1. Cell Chem Biol 23(4):494–507. https://doi.org/10.1016/j.chembiol.2016.02.015

169. Anighoro A, Rastelli G (2013) Enrichment factor analyses on G-protein coupled receptors with known crystal structure. J Chem Inf Model 53(4):739–743. https://doi.org/10.1021/ci4000745
170. Martinelli A, Ortore G (2013) Molecular modeling of adenosine receptors. Methods Enzymol:37–59. https://doi.org/10.1016/B978-0-12-407865-9.00003-0
171. Catarzi D et al (2013) Pyrazolo[1,5-c]quinazoline derivatives and their simplified analogues as adenosine receptor antagonists: synthesis, structure–affinity relationships and molecular modeling studies. Bioorg Med Chem 21(1):283–294. https://doi.org/10.1016/j.bmc.2012.10.031
172. Jespers W et al (2017) Structure-based design of potent and selective ligands at the four adenosine receptors. Molecules 22(11):1945. https://doi.org/10.3390/molecules22111945
173. Sabbadin D, Ciancetta A, Moro S (2014) Bridging molecular docking to membrane molecular dynamics to investigate GPCR–ligand recognition: the human a 2A adenosine receptor as a key study. J Chem Inf Model 54(1):169–183. https://doi.org/10.1021/ci400532b
174. Lotz SD, Dickson A (2018) Unbiased molecular dynamics of 11 min timescale drug unbinding reveals transition state stabilizing interactions. J Am Chem Soc 140(2):618–628. https://doi.org/10.1021/jacs.7b08572
175. Hartmann C, Banisch R, Sarich M, Badowski T, Schütte C (2013) Characterization of rare events in molecular dynamics. Entropy 16(1):350–376. https://doi.org/10.3390/e16010350
176. Bernardi RC, Melo MCR, Schulten K (2015) Enhanced sampling techniques in molecular dynamics simulations of biological systems. Biochim Biophys Acta Gen Subj 1850(5):872–877. https://doi.org/10.1016/j.bbagen.2014.10.019
177. Patel JS, Berteotti A, Ronsisvalle S, Rocchia W, Cavalli A (2014) Steered molecular dynamics simulations for studying protein–ligand interaction in cyclin-dependent kinase 5. J Chem Inf Model 54(2):470–480. https://doi.org/10.1021/ci4003574
178. Sinko W, Miao Y, de Oliveira CAF, McCammon JA (2013) Population based reweighting of scaled molecular dynamics. J Phys Chem B 117(42):12759–12768. https://doi.org/10.1021/jp401587e
179. Qi R, Wei G, Ma B, Nussinov R (2018) Replica exchange molecular dynamics: a practical application protocol with solutions to common problems and a peptide aggregation and self-assembly example. Methods Mol Biol:101–119. https://doi.org/10.1007/978-1-4939-7811-3_5
180. Sabbadin D, Moro S (2014) Supervised molecular dynamics (SuMD) as a helpful tool to depict GPCR–ligand recognition pathway in a nanosecond time scale. J Chem Inf Model 54(2):372–376. https://doi.org/10.1021/ci400766b
181. Bussi G, Laio A (2020) Using metadynamics to explore complex free-energy landscapes. Nat Rev Phys 2(4):200–212. https://doi.org/10.1038/s42254-020-0153-0
182. Potterton A et al (2019) Ensemble-based steered molecular dynamics predicts relative residence time of a 2A receptor binders. J Chem Theory Comput 15(5):3316–3330. https://doi.org/10.1021/acs.jctc.8b01270
183. Akhunzada MJ, Yoon HJ, Deb I, Braka A, Wu S (2022) Bell-Evans model and steered molecular dynamics in uncovering the dissociation kinetics of ligands targeting G-protein-coupled receptors. Sci Rep 12(1):15972. https://doi.org/10.1038/s41598-022-20065-2
184. Cuzzolin A et al (2016) Deciphering the complexity of ligand-protein recognition pathways using supervised molecular dynamics (SuMD) simulations. J Chem Inf Model 56(4):687–705. https://doi.org/10.1021/acs.jcim.5b00702
185. Panday SK, Sturlese M, Salmaso V, Ghosh I, Moro S (2019) Coupling supervised molecular dynamics (SuMD) with entropy estimations to Shine light on the stability of multiple binding sites. ACS Med Chem Lett 10(4):444–449. https://doi.org/10.1021/acsmedchemlett.8b00490
186. Bissaro M, Federico S, Salmaso V, Sturlese M, Spalluto G, Moro S (2018) Targeting protein kinase CK1δ with Riluzole: could it be one of the possible missing bricks to interpret its effect in the treatment of ALS from a molecular point of view? ChemMedChem 13(24):2601–2605. https://doi.org/10.1002/cmdc.201800632

187. Pavan M, Bolcato G, Bassani D, Sturlese M, Moro S (2021) Supervised molecular dynamics (SuMD) insights into the mechanism of action of SARS-CoV-2 main protease inhibitor PF-07321332. J Enzyme Inhib Med Chem 36(1):1645–1649. https://doi.org/10.1080/14756366.2021.1954919
188. Pavan M, Bassani D, Sturlese M, Moro S (2022) Investigating RNA–protein recognition mechanisms through supervised molecular dynamics (SuMD) simulations. NAR Genomics Bioinforma 4(4). https://doi.org/10.1093/nargab/lqac088
189. Bolcato G, Pavan M, Bassani D, Sturlese M, Moro S (2022) Ribose and non-ribose A2A adenosine receptor agonists: do they share the same receptor recognition mechanism? Biomedicine 10(2):515. https://doi.org/10.3390/biomedicines10020515
190. Sabbadin D, Salmaso V, Sturlese M, Moro S (2018) Supervised molecular dynamics (SuMD) approaches in drug design, pp 287–298. https://doi.org/10.1007/978-1-4939-8630-9_17
191. Sabbadin D, Ciancetta A, Deganutti G, Cuzzolin A, Moro S (2015) Exploring the recognition pathway at the human a 2A adenosine receptor of the endogenous agonist adenosine using supervised molecular dynamics simulations. MedChemCommun 6(6):1081–1085. https://doi.org/10.1039/C5MD00016E
192. Deganutti G, Welihinda A, Moro S (2017) Comparison of the human a 2A adenosine receptor recognition by adenosine and inosine: new insight from supervised molecular dynamics simulations. ChemMedChem 12(16):1319–1326. https://doi.org/10.1002/cmdc.201700200
193. De Filippo E et al (2020) A2A and A2B adenosine receptors: the extracellular loop 2 determines high (A2A) or low affinity (A2B) for adenosine. Biochem Pharmacol 172:113718. https://doi.org/10.1016/j.bcp.2019.113718
194. Deganutti G, Cuzzolin A, Ciancetta A, Moro S (2015) Understanding allosteric interactions in G protein-coupled receptors using supervised molecular dynamics: a prototype study analysing the human A3 adenosine receptor positive allosteric modulator LUF6000. Bioorg Med Chem 23(14):4065–4071. https://doi.org/10.1016/j.bmc.2015.03.039
195. Bolcato G, Bissaro M, Deganutti G, Sturlese M, Moro S (2020) New insights into key determinants for adenosine 1 receptor antagonists selectivity using supervised molecular dynamics simulations. Biomol Ther 10(5):732. https://doi.org/10.3390/biom10050732
196. Bissaro M, Bolcato G, Deganutti G, Sturlese M, Moro S (2019) Revisiting the allosteric regulation of sodium cation on the binding of adenosine at the human A2A adenosine receptor: insights from supervised molecular dynamics (SuMD) simulations. Molecules 24(15):2752. https://doi.org/10.3390/molecules24152752
197. Cao R, Giorgetti A, Bauer A, Neumaier B, Rossetti G, Carloni P (2018) Role of extracellular loops and membrane lipids for ligand recognition in the neuronal adenosine receptor type 2A: an enhanced sampling simulation study. Molecules 23(10):2616. https://doi.org/10.3390/molecules23102616
198. Li J, Jonsson AL, Beuming T, Shelley JC, Voth GA (2013) Ligand-dependent activation and deactivation of the human adenosine a 2A receptor. J Am Chem Soc 135(23):8749–8759. https://doi.org/10.1021/ja404391q
199. Deganutti G et al (2017) Impact of protein–ligand solvation and desolvation on transition state thermodynamic properties of adenosine A2A ligand binding kinetics. Silico Pharmacol 5(1): 16. https://doi.org/10.1007/s40203-017-0037-x
200. Zwanzig RW (1954) High-temperature equation of state by a perturbation method. I. Nonpolar gases. J Chem Phys 22(8):1420–1426. https://doi.org/10.1063/1.1740409
201. Fratev F, Sirimulla S (2019) An improved free energy perturbation FEP+ sampling protocol for flexible ligand-binding domains. Sci Rep 9(1):16829. https://doi.org/10.1038/s41598-019-53133-1
202. Wu D et al (2022) Free energy perturbation (FEP)-guided scaffold hopping. Acta Pharm Sin B 12(3):1351–1362. https://doi.org/10.1016/j.apsb.2021.09.027
203. Wang L et al (2015) Accurate and reliable prediction of relative ligand binding potency in prospective drug discovery by way of a modern free-energy calculation protocol and force field. J Am Chem Soc 137(7):2695–2703. https://doi.org/10.1021/ja512751q

204. Matricon P et al (2017) Fragment optimization for GPCRs by molecular dynamics free energy calculations: probing druggable subpockets of the a 2A adenosine receptor binding site. Sci Rep 7(1):6398. https://doi.org/10.1038/s41598-017-04905-0
205. Matricon P, Suresh RR, Gao Z-G, Panel N, Jacobson KA, Carlsson J (2021) Ligand design by targeting a binding site water. Chem Sci 12(3):960–968. https://doi.org/10.1039/D0SC04938G
206. Jespers W et al (2021) Deciphering conformational selectivity in the A2A adenosine G protein-coupled receptor by free energy simulations. PLoS Comput Biol 17(11):e1009152. https://doi.org/10.1371/journal.pcbi.1009152
207. Jespers W et al (2020) X-ray crystallography and free energy calculations reveal the binding mechanism of a 2A adenosine receptor antagonists. Angew Chem Int Ed 59(38): 16536–16543. https://doi.org/10.1002/anie.202003788
208. Wang X et al (2021) Identification of V6.51L as a selectivity hotspot in stereoselective A2B adenosine receptor antagonist recognition. Sci Rep 11(1):14171. https://doi.org/10.1038/s41598-021-93419-x
209. Deflorian F et al (2020) Accurate prediction of GPCR ligand binding affinity with free energy perturbation. J Chem Inf Model 60(11):5563–5579. https://doi.org/10.1021/acs.jcim.0c00449
210. Cherkasov A et al (2014) QSAR modeling: where have you been? Where are you going to? J Med Chem 57(12):4977–5010. https://doi.org/10.1021/jm4004285
211. Voet A et al (2014) Pharmacophore modeling: advances, limitations, and current utility in drug discovery. J Receptor Ligand Channel Res:81. https://doi.org/10.2147/JRLCR.S46843
212. Laurens van der Maaten GH (2008) Visualizing Data using t-SNE. J Mach Learn Res 9:2579–2605
213. Stojanović L, Popović M, Tijanić N, Rakočević G, Kalinić M (2020) Improved scaffold hopping in ligand-based virtual screening using neural representation learning. J Chem Inf Model 60(10):4629–4639. https://doi.org/10.1021/acs.jcim.0c00622
214. Floris M, Sabbadin D, Medda R, Bulfone A, Moro S (2012) Adenosiland: walking through adenosine receptors landscape. Eur J Med Chem 58:248–257. https://doi.org/10.1016/j.ejmech.2012.10.022
215. Xu Z, Cheng F, Da C, Liu G, Tang Y (2010) Pharmacophore modeling of human adenosine receptor A2A antagonists. J Mol Model 16(12):1867–1876. https://doi.org/10.1007/s00894-010-0690-z
216. Bacilieri M et al (2013) Revisiting a receptor-based pharmacophore hypothesis for human a 2A adenosine receptor antagonists. J Chem Inf Model 53(7):1620–1637. https://doi.org/10.1021/ci300615u
217. Tafi A et al (2006) Pharmacophore based receptor modeling: the case of adenosine A3 receptor antagonists. An approach to the optimization of protein models. J Med Chem 49(14): 4085–4097. https://doi.org/10.1021/jm051112+
218. Samuel AL (1959) Some studies in machine learning using the game of checkers. IBM J Res Dev 3(3):210–229. https://doi.org/10.1147/rd.33.0210
219. Alzubaidi L et al (2021) Review of deep learning: concepts, CNN architectures, challenges, applications, future directions. J Big Data 8(1):53. https://doi.org/10.1186/s40537-021-00444-8
220. James G, Witten D, Hastie T, Tibshirani R (2013) An introduction to statistical learning, vol 103. Springer, New York. https://doi.org/10.1007/978-1-4614-7138-7
221. Chicco D, Jurman G (2020) The advantages of the Matthews correlation coefficient (MCC) over F1 score and accuracy in binary classification evaluation. BMC Genomics 21(1):6. https://doi.org/10.1186/s12864-019-6413-7
222. Hicks SA et al (2022) On evaluation metrics for medical applications of artificial intelligence. Sci Rep 12(1):5979. https://doi.org/10.1038/s41598-022-09954-8
223. Abiodun OI et al (2019) Comprehensive review of artificial neural network applications to pattern recognition. IEEE Access 7:158820–158846. https://doi.org/10.1109/ACCESS.2019.2945545

224. Qi G-J, Luo J (2022) Small data challenges in big data era: a survey of recent progress on unsupervised and semi-supervised methods. IEEE Trans Pattern Anal Mach Intell 44(4): 2168–2187. https://doi.org/10.1109/TPAMI.2020.3031898
225. Chu X, Ilyas IF, Krishnan S, Wang J (2016) Data cleaning. In: Proceedings of the 2016 international conference on management of data, pp 2201–2206. https://doi.org/10.1145/2882903.2912574
226. Chen J, Si Y-W, Un C-W, Siu SWI (2021) Chemical toxicity prediction based on semi-supervised learning and graph convolutional neural network. J Cheminform 13(1):93. https://doi.org/10.1186/s13321-021-00570-8
227. Venkatraman V (2021) FP-ADMET: a compendium of fingerprint-based ADMET prediction models. J Cheminform 13(1):75. https://doi.org/10.1186/s13321-021-00557-5
228. Blaschke T et al (2020) REINVENT 2.0: an AI tool for de novo drug design. J Chem Inf Model 60(12):5918–5922. https://doi.org/10.1021/acs.jcim.0c00915
229. Gupta A, Müller AT, Huisman BJH, Fuchs JA, Schneider P, Schneider G (2018) Generative recurrent networks for de novo drug design. Mol Inform 37(1–2):1700111. https://doi.org/10.1002/minf.201700111
230. Schneider G (2018) Automating drug discovery. Nat Rev Drug Discov 17(2):97–113. https://doi.org/10.1038/nrd.2017.232
231. Meli R, Morris GM, Biggin PC (2022) Scoring functions for protein-ligand binding affinity prediction using structure-based deep learning: a review. Front Bioinforma 2. https://doi.org/10.3389/fbinf.2022.885983
232. Noé F, Tkatchenko A, Müller K-R, Clementi C (2020) Machine learning for molecular simulation. Annu Rev Phys Chem 71(1):361–390. https://doi.org/10.1146/annurev-physchem-042018-052331
233. Carpenter KA, Huang X (2018) Machine learning-based virtual screening and its applications to Alzheimer's drug discovery: a review. Curr Pharm Des 24(28):3347–3358. https://doi.org/10.2174/1381612824666180607124038
234. Liu X, Ye K, van Vlijmen HWT, IJzerman AP, van Westen GJP (2019) An exploration strategy improves the diversity of de novo ligands using deep reinforcement learning: a case for the adenosine A2A receptor. J Cheminform 11(1):35. https://doi.org/10.1186/s13321-019-0355-6
235. Goßen J et al (2023) Machine learning-aided discovery of novel chemotype antagonists for G protein-coupled receptors: the case of the adenosine A_{2A} receptor. bioRxiv. https://doi.org/10.1101/2023.03.31.535043
236. Tang M, Wen C, Lin J, Chen H, Ran T (2023) Discovery of novel A2AR antagonists through deep learning-based virtual screening. Artif Intell Life Sci 3:100058. https://doi.org/10.1016/j.ailsci.2023.100058
237. Puhl AC, Gao Z-G, Jacobson KA, Ekins S (2022) Machine learning for discovery of new ADORA modulators. Front Pharmacol 13. https://doi.org/10.3389/fphar.2022.920643
238. Böselt L, Thürlemann M, Riniker S (2021) Machine learning in QM/MM molecular dynamics simulations of condensed-phase systems. J Chem Theory Comput 17(5):2641–2658. https://doi.org/10.1021/acs.jctc.0c01112
239. Pozzan A (2020) QM calculations in ADMET prediction. Quantum Mech Drug Discov:285–305. https://doi.org/10.1007/978-1-0716-0282-9_18

Top Med Chem (2023) 41: 317–330
https://doi.org/10.1007/7355_2023_160

Published online: 23 June 2023

P2X4 Receptors in Immunity and Inflammation

Luca Antonioli, Matteo Fornai, Carolina Pellegrini, Laura Benvenuti, Clelia Di Salvo, Vanessa D'Antongiovanni, and György Haskó

Contents

Abstract The $P2X_4$ receptor is an ATP-gated cation channel that is widely expressed in cells of the immune system. It is able to form heterotrimers by combining with $P2X_2$, $P2X_5$, and/or $P2X_6$. Signal transduction is accompanied by fast influx of Ca^{2+} and Na^+ and efflux of K^+. This receptor differs from other members of the P2X family since it is one of the most sensitive purinergic receptors (at nanomolar ATP concentrations), about 1,000 times more than the archetypal $P2X_7$. In the immune system, this receptor has been implicated in the processing and release of cytokines as well as in the modulation of several immune cell activities.

Keywords ATP, Immune cells, Inflammation, P2X receptors, Purine

L. Antonioli (✉), M. Fornai, C. Pellegrini, L. Benvenuti, C. Di Salvo, and V. D'Antongiovanni
Unit of Pharmacology and Pharmacovigilance, Department of Clinical and Experimental Medicine, University of Pisa, Pisa, Italy
e-mail: luca.antonioli@unipi.it

G. Haskó
Department of Anesthesiology, Columbia University, New York, NY, USA

1 Introduction

In 1972, Professor Geoffrey Burnstock postulated the "purinergic hypothesis" recognizing the nucleotides, ATP, ADP, UTP, UDP, and UDP-glucose, previously considered merely as ubiquitous biochemical sources of energy, as extracellular transmitters at the intestinal level [1]. Since these initial observations, a number of evidence highlighted a relevant contribution of such nucleotides also in the modulation of several cell populations, including the immune system [2].

In this regard, it has been observed that ATP is important for cell-to-cell communication in the immune system. In particular, ATP released by T cells is a key autocrine player in T-cell receptor (TCR) signaling, driving the phenotype differentiation of T helper cells [3, 4]. Further investigation highlighted that ATP as signaling molecule at sites of inflammation/cell injury, holding a critical role in immune cell migration, chemotaxis, and cytokine release [5, 6]. Indeed, an increased extracellular ATP level represents a "danger signal" able to trigger pro-inflammatory responses, particularly of the innate immune system [5–7].

Burnstock and Kennedy suggested pharmacological criteria for discerning two types of ATP P2 receptors (P2X and P2Y) [8]. Successively, set of studies on transduction mechanisms and cloning of nucleotide receptors allowed to identify P2X as ligand-gated ion channel receptors, and P2Y as G-protein-coupled receptor family [8]. Both P2X and P2Y receptors play important roles in leukocyte biology and modulation. Both receptor subtypes fuel several signaling pathways, aimed at regulating cell volume, eicosanoid release, phosphatidylserine exposure, hemolysis, impaired ATP release, reactive oxygen species formation, and apoptosis [9].

P2X receptors in particular are distributed widely throughout the body in neural and other tissues, and they mediate several physiological and pathophysiological responses mainly in microglia [10]. P2X receptor stimulation triggers multiple key responses fundamental for initiation, propagation, and resolution of inflammation [11].

2 Purinergic Signaling: Receptors, Enzymes, Transporters

Purinergic signaling is originated by the release of nucleotides and nucleosides into the extracellular milieu extruded via anion channels, maxi-anion channels, transporters, connexins, and pannexins [12], as well as through exocytotic pathways and membrane damage [13].

Once released in the extracellular environment, the nucleotides and nucleosides elicit a number of cellular responses via the engagement of the ionotropic P2X ($P2X_{1–7}$) and metabotropic P2Y ($P2Y_{1,2,4,6,11–14}$) receptors [13]. The P2X receptors are characterized by a trimeric structure, with two putative transmembrane domains, gating primarily Na^+, K^+, and Ca^{2+}, and, in some cases, Cl^- [14]. The engagement of the $G_{q/11}$-coupled $P2Y_{1,2,4,6}$ and $P2Y_{11}$ receptors induced the stimulation of phospholipase C, thus leading to an increased production of inositol-(1,4,5)-trisphosphate and diacylglycerol (DAG) [15]. In particular, the inositol-(1,4,5)-trisphosphate rises the intracellular levels of Ca^{2+} and DAG activates protein kinase C (PKC) [15]. In addition, the stimulation of $P2Y_{11}$ receptor can activate, whereas $P2Y_{12,13}$ receptor activation can inhibit the adenylate cyclase [15].

Once released into the extracellular space, ATP and/or ADP are quickly converted by CD39 (ecto-nucleoside triphosphate diphosphohydrolase 1, E-NTPDase1) into AMP, and then, via CD73 (ecto-5′-nucleotidase) AMP is dephosphorylated into adenosine [16].

Adenosine exerts homeostatic functions through the engagement of four specific G protein-coupled receptors [17–19]. In particular, the adenosine receptors A_1 and A_3 receptors are coupled to G_i, G_q, and G_o proteins, whereas the A_{2A} and A_{2B} receptor subtypes activate adenylyl cyclase via G_s or G_{olf} [17]. The stimulation of A_{2B} receptors can also trigger the phospholipase C via G_q [17, 18].

The extracellular concentration of adenosine is closely modulated by the activity of nucleoside transporters as well as by the adenosine-metabolizing enzyme adenosine deaminase [20, 21]. The nucleoside transporters, distinguished into equilibrative nucleoside transporters (ENTs; ENT1, ENT2, ENT3, and ENT4) or concentrative nucleoside transporters (CNTs; CNT1, CNT2, and CNT3), carry the purines across cell membranes from the intra to the extracellular space and vice versa [22, 23]. Adenosine deaminase is a critical checkpoint to reduce extracellular adenosine levels converting it into inosine [24]. Finally, inosine is converted into the stable end product uric acid by xanthine oxidase [24].

3 Structure and Molecular Biology of $P2X_4$ Receptor

The P2X receptors are a distinct family of non-selective trimeric ligand-gated channels, allowing Na^+, K^+, and Ca^{2+}ion fluxes, upon the activation via extracellular ATP [25].

Nowadays, seven subtypes of P2X receptors have been cloned and classified as $P2X_1$ to $P2X_7$, with functional channels assembled by homo- or heterotrimers [25]. Their activity is molded by several factors such as the pH, the extracellular concentration of ions (Zn, Cu, Hg, Ni, Cd), protons, lipids, steroids, and ethanol [10].

The $P2X_4$ is a ligand-gated ion channel receptor composed by three subunits [26]. Each subunit is characterized by two transmembrane domains linked by a large extracellular loop and intracellularly located N- and C-termini [26]. This receptor subtype displays a high affinity profile susceptible to nanomolar concentrations of extracellular ATP. Once stimulated, the opening of this channel receptor permits ion fluxes (North, 2002). Of note, it has been observed that the Ca^{2+} permeability of $P2X_4$ is the highest among the P2X family [27].

Immunoprecipitation investigations pointed out that the $P2X_4$ receptors can heteromerically assemble with other P2X subtypes, such as $P2X_1$, $P2X_2$, and $P2X_6$

[28, 29]. In this regard, Schneider et al. [30] demonstrated that $P2X_4$ and $P2X_7$ subunits can form heterotrimeric $P2X_4/P2X_7$ receptors, widely co-expressed particularly on the immune/inflammatory cells [30].

The $P2X_4$ has a wide distribution throughout the body [31]. Indeed, it is widely expressed in central and peripheral neurons, in microglia, and also in various glandular tissues such as pancreatic acinar cells and salivary glands as well as in endothelial cells [32]. $P2X_4$ is also expressed throughout the gastrointestinal tract, liver, lung, kidney, and reproductive system [23, 24]. At the intracellular level, $P2X_4$ is expressed not only on the plasma membrane, but also in intracellular compartments such as lysosomes, vesicles, vacuoles, and lamellar bodies [33].

An increasing attention has been paid to the $P2X_4$ receptors on the basis of the growing evidence obtained through pharmacological and genetic experiments that outlined a role of this receptor subtype in several disorders, such as neuropathic pain, neurodegenerative disorders, metabolic syndrome [33]. In addition, a deep involvement of $P2X_4$ receptors in the pathophysiology of several immuno-mediated disorders (i.e., rheumatoid arthritis, multiple sclerosis, brain post-ischemic inflammation) [10, 34, 35] has been also demonstrated.

4 $P2X_4$ Receptor and in Innate and Adaptive Immunity

All immune cell populations, whether of the myeloid or lymphoid lineage, are endowed at least one P2X or P2Y receptor subtype (see Table 1) [36]. Several evidence highlighted a critical role played by both receptor subtypes in immune cell biology and activation [37–40]. In addition to the autocrine mechanisms modulating the function of healthy immune cells, it has been observed that purinergic receptors allow immune cells to recognize ATP, that acting as a "find-me signal" drives the phagocytes to inflammatory sites, promoting the clearance of damaged and apoptotic cells [37]. Despite several evidences demonstrated an involvement of $P2X_4$ receptors in the regulation of immune system, only recently it has begun to consider the actual involvement of this receptor subtype in the immunopathological mechanisms underlying inflammation.

4.1 P2X4 and Innate Immune Cells

4.1.1 Monocytes and Macrophages

Monocytes and macrophages are relevant for phagocytosis and killing of microorganisms, release of cytokines, and presentation of pathogens to T cells, eliciting both cellular and humoral immune responses [41].

All monocyte and macrophage express both P2X and P2Y receptors. Monocytes showed an abundant $P2X_4$ receptor transcripts, followed by transcripts for $P2X_7$ and

Table 1 Expression of P2 receptors on immune cells

Immune cell population	Purinergic receptors	References
Innate immunity		
Monocyte/ macrophage	$P2Y_1$, $P2Y_2$, $P2Y_4$, $P2Y_6$, $P2Y_{11}$, $P2Y_{12}$, $P2Y_{13}$, $P2Y_{14}$ $P2X_1$, $P2X_4$, $P2X_7$	[38, 72]
Dendritic cells	$P2Y_2$, $P2Y_4$, $P2Y_6$, $P2Y_{11}$, $P2Y_{12}$, $P2Y_{14}$ $P2X_1$, $P2X_4$, $P2X_7$	[60, 73]
Neutrophils	$P2Y_1$, $P2Y_4$, $P2Y_6$, $P2Y_{11}$ $P2X_1$, $P2X_4$, $P2X_7$	[39]
Eosinophils	$P2Y_1$, $P2Y_4$, $P2Y_6$, $P2Y_{11}$, $P2Y_{14}$ $P2X_1$, $P2X_4$, $P2X_7$	[74, 75]
Mast cells	$P2Y_1$, $P2Y_{12}$, $P2Y_{13}$, $P2Y_{14}$ $P2X_1$, $P2X_3$, $P2X_4$, $P2X_7$,	[76, 77]
Natural killer cells	$P2Y_{11}$ $P2X_1$, $P2X_4$, $P2X_7$	[78]
Adaptive immunity		
T lymphocytes	$P2Y_1$, $P2Y_4$, $P2Y_6$, $P2Y_{11}$, $P2Y_{12}$, $P2Y_{13}$, $P2Y_{14}$ $P2X_1$, $P2X_4$, $P2X_7$	[79]
B lymphocytes	$P2Y_1$, $P2Y_4$, $P2Y_6$, $P2Y_{11}$, $P2Y_{12}$, $P2Y_{13}$, $P2Y_{14}$ $P2X_1$, $P2X_4$, $P2X_7$	[40]

Reproduced with permission from Antonioli L, Blandizzi C, Fornai M, Pacher P, Lee HT, Haskó G. P2X4 receptors, immunity, and sepsis. Curr Opin Pharmacol. 2019;47:65–74

$P2X_1$, indicating a critical role for these receptors in monocyte chemotaxis and activation [42].

Murine peritoneal macrophages revealed the presence of functionally active $P2X_4$ and $P2X_7$ receptors [43, 44]. In particular, low concentrations of ATP evoked a small $P2X_4$-driven ion current, whereas marked ATP concentrations induced a large $P2X_7$-driven ion current [44]. When considering the role of such receptor subtype in macrophage activity, $P2X_7$ displayed to engage pro-inflammatory signaling pathways in immune cells, mainly via the activation of caspase-1 containing inflammasome [44]. Unfortunately, despite a role in the activation of macrophages might be suggested also for $P2X_4$, a demonstration about this effect was not possible in this study [44]. The slight and transient increase in intracellular Ca^{2+} elicited by the stimulation of $P2X_4$ might be overshadowed by the Ca^{2+} release induced by stimulation of $P2Y_2$ receptors in macrophages characterized by a comparable ATP sensitivity [44].

It has been observed a role of $P2X_4$ in shaping the macrophage activity [45]. In particular, the authors demonstrated a dynamic regulation for this receptor subtype during distinct phases of macrophage activation [45]. Indeed, resting macrophages showed very low levels of functional $P2X_4$ receptors, whereas there was a quick up-regulation in response to phagocytosis [45]. By contrast, a classical activation of macrophage reduced the surface and functional $P2X_4$ receptor expression not affecting the total $P2X_4$ receptor protein levels [45]. The alternative activation of these cells via incubation with IL-4 or IL-13 did not modify the expression or the activity of this receptors [45]. Based on these premises, it has been delineated a model in which an initial bacterial or inflammatory stimulus determines a rapid traffic of $P2X_4$ receptors to the macrophage cell surface followed by an amplified Ca^{2+} influx, thus promoting their activation [45]. Subsequent to the macrophage activation, a feedback mechanism modulates $P2X_4$ receptor trafficking and function, facilitating the resolution inflammation [45].

It has been observed the existence of a structural and functional interplay between $P2X_7$ and $P2X_4$ receptors on macrophages [46–48]. In particular, a protein–protein association between $P2X_4$ and $P2X_7$ receptors occurs via the C-terminus of $P2X_7$ [46] and that this physical interaction detected in unstimulated cells is functionally active only in the presence of ATP [46]. Of note, the $P2X_4$ receptors, allowing Ca^{2+}influx, modulate numerous mechanisms underlying the $P2X_7$ receptor-mediated inflammation, including the release of HMGB1 and IL-1β [47, 48].

Set of in vitro experiments allowed to demonstrate a role of ATP, via $P2X_4$ receptors, in regulating bacterial killing in macrophages [49]. The selective $P2X_4$ receptor antagonist 5-BDBD as well as the shRNA-mediated receptor silencing prevented the stimulatory effect of ATP on *E. coli* killing [49]. In line with this evidence, in macrophages isolated from $P2X_4$ receptor knockout mice the ATP failed to stimulate bacterial killing, whereas in cells obtained from wild-type animals, ATP maintained its ability to kill bacteria [49]. The $P2X_4$ receptor activation specifically stimulates mitochondrial, but not cellular, ROS production, even in the absence of bacterial stimulus [49].

In parallel, Haskó's research group demonstrated that macrophage $P2X_4$ receptors are involved in controlling bacterial spread and inflammation in a murine model of sepsis [49]. An increased bacterial load and a decreased survival in comparison with wild-type animals have been reported in $P2X_4$ receptor knockout mice [49], indicating that endogenous ATP released during sepsis exerts a protective effect through $P2X_4$ receptor engagement [49].

4.1.2 Microglia

Microglia represent a specialized population of macrophages-like cells in the central nervous system (CNS) considered immune sentinels that are capable of orchestrating a potent inflammatory response [50]. Microglia are also involved in synaptic organization, trophic neuronal support during development, phagocytosis of apoptotic cells in the developing brain, myelin turnover, control of neuronal excitability, phagocytic debris removal as well as brain protection and repair [50]. Set of immunohistochemical studies revealed that expression of $P2X_4$ receptors in the spinal cord was upregulated entirely in microglia, suggesting a pivotal role of these receptor subtype in neuropathic pain transmission [51]. In this regard, Tsuda et al. reported that the activation of $P2X_4$ receptors in hyperactive microglia is essential for tactile allodynia after nerve injury and is appropriate to induce tactile allodynia in normal animals [52].

$P2X_4$ receptors evoke a rise in intracellular Ca^{2+} levels that leads to activation of intracellular Ca^{2+}-sensitive signaling cascades, such as that which activates p38 MAPK [53]. The downstream intracellular signaling could in turn lead to activation of contact factors or to release of bioactive diffusible factors from microglia, which are known to be a source neuroactive mediators in the CNS [53].

Interestingly, it was demonstrated that activation of microglial $P2X_4$ receptors stimulated the synthesis and release of brain-derived neurotrophic factor (BDNF), which determined an alteration of transmembrane anion gradient in a subpopulation of dorsal horn lamina I neurons [54]. In particular, it has been proposed that $P2X_4$ receptors through the downregulation of the neuronal chloride transporter KCC2 render GABA and glycine effects depolarizing, rather than hyperpolarizing, in these neurons [55]. This event causes an abnormal nociceptive output that determines the neuropathic pain [56]. It has been observed that pharmacological blockade or antisense-oligonucleotide knockdown of $P2X_4$ receptors suppressed the pain hypersensitivity [53], thus indicating this receptor as an innovative way to manage pain.

4.1.3 Dendritic Cells

Dendritic cells (DCs) are professional antigen presenting cells, uniquely able to induce naïve T-cell activation and effector differentiation [57]. DCs represent a promising target for extracellular nucleotides, since several data reported the expression of P2 receptors and the functional responses associated with their stimulation [58]. The DCs are endowed with P2Y receptors ($P2Y_1$, $P2Y_2$, $P2Y_4$, $P2Y_6$, $P2Y_{11}$, and $P2Y_{14}$) and P2X receptors ($P2X_1$, $P2X_4$, $P2X_7$) [59, 60].

Sakaki et al. [61] described the role of $P2X_4$ in modulating the activity of DCs, reporting the involvement of such receptor subtype in facilitating the $P2X_7$-mediated inflammatory functions [61]. The $P2X_4$ receptor stimulation, eliciting a quick Ca^{2+} influx, positively sustains $P2X_7$-dependent IL-1β and IL-18 release from DCs [61]. The co-expression of $P2X_4$ and $P2X_7$ receptors in this immune cell population participating in the regulation of these pro-inflammatory cytokines, critically involved in supporting inflammation, is a critical step in shaping the inflammatory responses associated with immunopathological disorders.

At present, limited attention has been focused to the role of $P2X_4$ receptors in modulating DC activity. In particular, particular attention could be addressed to investigate the effect of a pharmacological modulation of this receptor subtype on DC survival and function.

4.1.4 Neutrophils

Neutrophils, the most abundant leukocytes in the circulation, have been regarded as first line of defense in the innate arm of the immune system [62]. These cells capture and destroy invading microorganisms, via phagocytosis and intracellular degradation, release of granules, and formation of neutrophil extracellular traps after detecting pathogens [62].

It is emerging that ATP closely modulates a number of neutrophil functions, including oxidative burst, phagocytosis, chemotaxis, rolling adhesion and transmigration, extracellular traps formation, degranulation, and apoptosis [39]. de Rivero Vaccari et al. [63] investigating the role of $P2X_4$ receptor in the modulation of neutrophil activity highlighted that $P2X_4$ knockout mice were characterized by a reduced infiltration of the number of neutrophils at the injury site [63].

The lack of evidence about the involvement of $P2X_4$ receptors on neutrophil activity is a point that needs to be carefully investigated.

4.2 P2X4 and Adaptive Immune Cells

4.2.1 T Cells

T cells coordinate multiple aspects of adaptive immunity throughout life, including responses to pathogens, allergens, and tumors [64].

Despite the mechanisms underlying this condition of T-cell suppression are not well defined, several lines of evidence indicate that purinergic signaling controls T cells in many different ways [5, 65]. Indeed, the release of ATP via pannexin-1 pores during lymphocyte activation enhances IL-2 release, in an autocrine manner, through the engagement of $P2X_1$, $P2X_4$, and $P2X_7$ receptors [66, 67]. In particular, ATP, eliciting a Ca^{2+} influx through these receptors, stimulates the nuclear factor of activated T cells, the mitogen-activated protein kinase signaling as well as the IL-2 expression [3, 66, 68]. By contrast, the pharmacological blockade of purinergic receptors as well as the removal of extracellular ATP by apyrase or alkaline phosphatase treatment, blunting the T-cell Ca^{2+} influx, leads to T-cell anergy, thus confirming the pivotal role of ATP in shaping T-cell functions [3].

The relevance of purinergic signaling in regulating the events underlying the T-cell activation has been substantiated by evidences describing the accumulation of several purinergic components, such as pannexin 1 channels, $P2X_1$ and $P2X_4$ receptors, at the immune synapse a membrane domain facilitates T cell–accessory cell communication and antigen recognition [68]. Interestingly, mitochondria appear to hold a critical role in this context. Indeed, they translocate to the immune synapse where, providing the ATP useful to engage $P2X_1$ and $P2X_4$ receptors, fuel the autocrine purinergic signaling in the synaptic cleft [68]. The $P2X_4$ receptors colocalize with mitochondria in clusters at the front of cells, facilitating the Ca^{2+} influx to maintain local mitochondrial ATP production at the levels needed for cell migration [69].

In patients with sepsis, the purinergic signaling appears defective [70]. Indeed, resting cells from such patients displayed a lack of the autocrine feedback loops that maintain basal cytosolic Ca^{2+} levels and mitochondrial function, leading to a state of cellular paralysis that impairs vigilance and precludes proper functional responses to

T-cell stimulation [70]. However, the suppression of purinergic signaling is not due to a reduction of P2X receptor expression, since the mRNA levels of these receptors appear unchanged ($P2X_4$ and $P2X_7$) or elevated ($P2X_1$) in patients with sepsis [70]. For this reason, it is conceivable that the impaired basal mitochondrial function and purinergic signaling undermine the T-cell vigilance and their ability to mount appropriate responses to infections.

4.2.2 B Cell

B cells are at the center of the adaptive humoral immune system and are responsible for mediating the production of antigen-specific immunoglobulin (Ig) directed against invasive pathogens (typically known as antibodies) [71]. At present, a role for $P2X_4$ receptors in B cells has not yet been determined.

Compliance with Ethical Standards Conflict of Interest

The authors declare that they have no conflict of interest.
Funding No funds to declare.

Ethical Approval This chapter does not contain any studies with human participants or animals performed by the authors.

References

1. Burnstock G (1972) Purinergic nerves. Pharmacol Rev 24:509–581
2. Antonioli L, Blandizzi C, Pacher P, Hasko G (2019) The purinergic system as a pharmacological target for the treatment of immune-mediated inflammatory diseases. Pharmacol Rev 71: 345–382
3. Schenk U, Westendorf AM, Radaelli E, Casati A, Ferro M, Fumagalli M, Verderio C, Buer J, Scanziani E, Grassi F (2008) Purinergic control of T cell activation by ATP released through pannexin-1 hemichannels. Sci Signal 1:ra6
4. Atarashi K, Nishimura J, Shima T, Umesaki Y, Yamamoto M, Onoue M, Yagita H, Ishii N, Evans R, Honda K et al (2008) ATP drives lamina propria T(H)17 cell differentiation. Nature 455:808–812
5. Ledderose C, Bao Y, Kondo Y, Fakhari M, Slubowski C, Zhang J, Junger WG (2016) Purinergic signaling and the immune response in sepsis: a review. Clin Ther 38:1054–1065
6. Bours MJ, Swennen EL, Di Virgilio F, Cronstein BN, Dagnelie PC (2006) Adenosine 5′--triphosphate and adenosine as endogenous signaling molecules in immunity and inflammation. Pharmacol Ther 112:358–404
7. Csoka B, Nemeth ZH, Toro G, Idzko M, Zech A, Koscso B, Spolarics Z, Antonioli L, Cseri K, Erdelyi K et al (2015) Extracellular ATP protects against sepsis through macrophage P2X7 purinergic receptors by enhancing intracellular bacterial killing. FASEB J 29:3626–3637
8. Kennedy C (2021) The P2Y/P2X divide: how it began. Biochem Pharmacol 187:114408
9. Rayah A, Kanellopoulos JM, Di Virgilio F (2012) P2 receptors and immunity. Microbes Infect 14:1254–1262
10. Suurvali J, Boudinot P, Kanellopoulos J, Ruutel Boudinot S (2017) P2X4: a fast and sensitive purinergic receptor. Biomed J 40:245–256

11. Di Virgilio F, Sarti AC, Grassi F (2018) Modulation of innate and adaptive immunity by P2X ion channels. Curr Opin Immunol 52:51–59
12. Taruno A (2018) ATP release channels. Int J Mol Sci 19
13. Antonioli L, Colucci R, Pellegrini C, Giustarini G, Tuccori M, Blandizzi C, Fornai M (2013) The role of purinergic pathways in the pathophysiology of gut diseases: pharmacological modulation and potential therapeutic applications. Pharmacol Ther 139:157–188
14. Pawson AJ, Sharman JL, Benson HE, Faccenda E, Alexander SP, Buneman OP, Davenport AP, McGrath JC, Peters JA, Southan C et al (2014) The IUPHAR/BPS guide to PHARMACOLOGY: an expert-driven knowledgebase of drug targets and their ligands. Nucleic Acids Res 42: D1098–D1106
15. Franke H, Krugel U, Illes P (2006) P2 receptors and neuronal injury. Pflugers Arch 452:622–644
16. Antonioli L, Pacher P, Vizi ES, Hasko G (2013) CD39 and CD73 in immunity and inflammation. Trends Mol Med 19:355–367
17. Antonioli L, Fornai M, Blandizzi C, Pacher P, Hasko G (2018) Adenosine signaling and the immune system: when a lot could be too much. Immunol Lett
18. Antonioli L, Blandizzi C, Pacher P, Hasko G (2013) Immunity, inflammation and cancer: a leading role for adenosine. Nat Rev Cancer 13:842–857
19. Hasko G, Antonioli L, Cronstein BN (2018) Adenosine metabolism, immunity and joint health. Biochem Pharmacol 151:307–313
20. Antonioli L, Fornai M, Colucci R, Awwad O, Ghisu N, Tuccori M, Del Tacca M, Blandizzi C (2011) Differential recruitment of high affinity A1 and A2A adenosine receptors in the control of colonic neuromuscular function in experimental colitis. Eur J Pharmacol 650:639–649
21. Antonioli L, Fornai M, Colucci R, Ghisu N, Blandizzi C, Del Tacca M (2006) A2a receptors mediate inhibitory effects of adenosine on colonic motility in the presence of experimental colitis. Inflamm Bowel Dis 12:117–122
22. Boswell-Casteel RC, Hays FA (2017) Equilibrative nucleoside transporters-A review. Nucleosides Nucleotides Nucleic Acids 36:7–30
23. Young JD (2016) The SLC28 (CNT) and SLC29 (ENT) nucleoside transporter families: a 30-year collaborative odyssey. Biochem Soc Trans 44:869–876
24. Antonioli L, Colucci R, La Motta C, Tuccori M, Awwad O, Da Settimo F, Blandizzi C, Fornai M (2012) Adenosine deaminase in the modulation of immune system and its potential as a novel target for treatment of inflammatory disorders. Curr Drug Targets 13:842–862
25. Wang J, Yu Y (2016) Insights into the channel gating of P2X receptors from structures, dynamics and small molecules. Acta Pharmacol Sin 37:44–55
26. Stojilkovic SS, Yan Z, Obsil T, Zemkova H (2010) Structural insights into the function of P2X4: an ATP-gated cation channel of neuroendocrine cells. Cell Mol Neurobiol 30:1251–1258
27. Egan TM, Khakh BS (2004) Contribution of calcium ions to P2X channel responses. J Neurosci 24(13):3413–3420.
28. Antonio LS, Stewart AP, Xu XJ, Varanda WA, Murrell-Lagnado RD, Edwardson JM (2011) P2X4 receptors interact with both P2X2 and P2X7 receptors in the form of homotrimers. Br J Pharmacol 163:1069–1077
29. Antonio LS, Stewart AP, Varanda WA, Edwardson JM (2014) Identification of P2X2/P2X4/P2X6 heterotrimeric receptors using atomic force microscopy (AFM) imaging. FEBS Lett 588: 2125–2128
30. Schneider M, Prudic K, Pippel A, Klapperstuck M, Braam U, Muller CE, Schmalzing G, Markwardt F (2017) Interaction of purinergic P2X4 and P2X7 receptor subunits. Front Pharmacol 8:860
31. Bo X, Kim M, Nori SL, Schoepfer R, Burnstock G, North RA (2003) Tissue distribution of P2X4 receptors studied with an ectodomain antibody. Cell Tissue Res 313:159–165
32. Burnstock G, Kennedy C (2011) P2X receptors in health and disease. Adv Pharmacol 61:333–372

33. Xu J, Chai H, Ehinger K, Egan TM, Srinivasan R, Frick M, Khakh BS (2014) Imaging P2X4 receptor subcellular distribution, trafficking, and regulation using P2X4-pHluorin. J Gen Physiol 144:81–104
34. Li F, Guo N, Ma Y, Ning B, Wang Y, Kou L (2014) Inhibition of P2X4 suppresses joint inflammation and damage in collagen-induced arthritis. Inflammation 37:146–153
35. Sadovnick AD, Gu BJ, Traboulsee AL, Bernales CQ, Encarnacion M, Yee IM, Criscuoli MG, Huang X, Ou A, Milligan CJ et al (2017) Purinergic receptors P2RX4 and P2RX7 in familial multiple sclerosis. Hum Mutat 38:736–744
36. Burnstock G (2016) P2X ion channel receptors and inflammation. Purinergic Signal 12:59–67
37. Junger WG (2011) Immune cell regulation by autocrine purinergic signalling. Nat Rev Immunol 11:201–212
38. Di Virgilio F, Vuerich M (2015) Purinergic signaling in the immune system. Auton Neurosci 191:117–123
39. Wang X, Chen D (2018) Purinergic regulation of neutrophil function. Front Immunol 9:399
40. Przybyla T, Sakowicz-Burkiewicz M, Pawelczyk T (2018) Purinergic signaling in B cells. Acta Biochim Pol 65:1–7
41. Haveman JW, Muller Kobold AC, Tervaert JW, van den Berg AP, Tulleken JE, Kallenberg CG, The TH (1999) The central role of monocytes in the pathogenesis of sepsis: consequences for immunomonitoring and treatment. Neth J Med 55:132–141
42. Cekic C, Linden J (2016) Purinergic regulation of the immune system. Nat Rev Immunol 16: 177–192
43. Sim JA, Park CK, Oh SB, Evans RJ, North RA (2007) P2X1 and P2X4 receptor currents in mouse macrophages. Br J Pharmacol 152:1283–1290
44. Brone B, Moechars D, Marrannes R, Mercken M, Meert T (2007) P2X currents in peritoneal macrophages of wild type and P2X4 –/– mice. Immunol Lett 113:83–89
45. Stokes L, Surprenant A (2009) Dynamic regulation of the P2X4 receptor in alveolar macrophages by phagocytosis and classical activation. Eur J Immunol 39:986–995
46. Perez-Flores G, Levesque SA, Pacheco J, Vaca L, Lacroix S, Perez-Cornejo P, Arreola J (2015) The P2X7/P2X4 interaction shapes the purinergic response in murine macrophages. Biochem Biophys Res Commun 467:484–490
47. Kawano A, Tsukimoto M, Mori D, Noguchi T, Harada H, Takenouchi T, Kitani H, Kojima S (2012) Regulation of P2X7-dependent inflammatory functions by P2X4 receptor in mouse macrophages. Biochem Biophys Res Commun 420:102–107
48. Kawano A, Tsukimoto M, Noguchi T, Hotta N, Harada H, Takenouchi T, Kitani H, Kojima S (2012) Involvement of P2X4 receptor in P2X7 receptor-dependent cell death of mouse macrophages. Biochem Biophys Res Commun 419:374–380
49. Csoka B, Nemeth ZH, Szabo I, Davies DL, Varga ZV, Paloczi J, Falzoni S, Di Virgilio F, Muramatsu R, Yamashita T et al (2018) Macrophage P2X4 receptors augment bacterial killing and protect against sepsis. JCI Insight:3
50. Bachiller S, Jimenez-Ferrer I, Paulus A, Yang Y, Swanberg M, Deierborg T, Boza-Serrano A (2018) Microglia in neurological diseases: a road map to brain-disease dependent-inflammatory response. Front Cell Neurosci 12:488
51. Inoue K, Tsuda M (2012) P2X4 receptors of microglia in neuropathic pain. CNS Neurol Disord Drug Targets 11:699–704
52. Tsuda M, Shigemoto-Mogami Y, Koizumi S, Mizokoshi A, Kohsaka S, Salter MW, Inoue K (2003) P2X4 receptors induced in spinal microglia gate tactile allodynia after nerve injury. Nature 424:778–783
53. Tsuda M (2016) Microglia in the spinal cord and neuropathic pain. J Diabetes Investig 7:17–26
54. Trang T, Beggs S, Wan X, Salter MW (2009) P2X4-receptor-mediated synthesis and release of brain-derived neurotrophic factor in microglia is dependent on calcium and p38-mitogen-activated protein kinase activation. J Neurosci 29:3518–3528

55. Coull JA, Beggs S, Boudreau D, Boivin D, Tsuda M, Inoue K, Gravel C, Salter MW, De Koninck Y (2005) BDNF from microglia causes the shift in neuronal anion gradient underlying neuropathic pain. Nature 438:1017–1021
56. Beggs S, Liu XJ, Kwan C, Salter MW (2010) Peripheral nerve injury and TRPV1-expressing primary afferent C-fibers cause opening of the blood-brain barrier. Mol Pain 6:74
57. Rossi M, Young JW (2005) Human dendritic cells: potent antigen-presenting cells at the crossroads of innate and adaptive immunity. J Immunol 175:1373–1381
58. Di Virgilio F (2005) Purinergic mechanism in the immune system: a signal of danger for dendritic cells. Purinergic Signal 1:205–209
59. Berchtold S, Ogilvie AL, Bogdan C, Muhl-Zurbes P, Ogilvie A, Schuler G, Steinkasserer A (1999) Human monocyte derived dendritic cells express functional P2X and P2Y receptors as well as ecto-nucleotidases. FEBS Lett 458:424–428
60. Ferrari D, La Sala A, Chiozzi P, Morelli A, Falzoni S, Girolomoni G, Idzko M, Dichmann S, Norgauer J, Di Virgilio F (2000) The P2 purinergic receptors of human dendritic cells: identification and coupling to cytokine release. FASEB J 14:2466–2476
61. Sakaki H, Fujiwaki T, Tsukimoto M, Kawano A, Harada H, Kojima S (2013) P2X4 receptor regulates P2X7 receptor-dependent IL-1beta and IL-18 release in mouse bone marrow-derived dendritic cells. Biochem Biophys Res Commun 432:406–411
62. Rosales C (2018) Neutrophil: a cell with many roles in inflammation or several cell types? Front Physiol 9:113
63. de Rivero Vaccari JP, Bastien D, Yurcisin G, Pineau I, Dietrich WD, De Koninck Y, Keane RW, Lacroix S (2012) P2X4 receptors influence inflammasome activation after spinal cord injury. J Neurosci 32:3058–3066
64. Kumar BV, Connors TJ, Farber DL (2018) Human T cell development, localization, and function throughout life. Immunity 48:202–213
65. Burnstock G, Boeynaems JM (2014) Purinergic signalling and immune cells. Purinergic Signal 10:529–564
66. Yip L, Woehrle T, Corriden R, Hirsh M, Chen Y, Inoue Y, Ferrari V, Insel PA, Junger WG (2009) Autocrine regulation of T-cell activation by ATP release and P2X7 receptors. FASEB J 23:1685–1693
67. Woehrle T, Yip L, Manohar M, Sumi Y, Yao Y, Chen Y, Junger WG (2010) Hypertonic stress regulates T cell function via pannexin-1 hemichannels and P2X receptors. J Leukoc Biol 88: 1181–1189
68. Woehrle T, Yip L, Elkhal A, Sumi Y, Chen Y, Yao Y, Insel PA, Junger WG (2010) Pannexin-1 hemichannel-mediated ATP release together with P2X1 and P2X4 receptors regulate T-cell activation at the immune synapse. Blood 116:3475–3484
69. Ledderose C, Liu K, Kondo Y, Slubowski CJ, Dertnig T, Denicolo S, Arbab M, Hubner J, Konrad K, Fakhari M et al (2018) Purinergic P2X4 receptors and mitochondrial ATP production regulate T cell migration. J Clin Invest 128:3583–3594
70. Ledderose C, Bao Y, Ledderose S, Woehrle T, Heinisch M, Yip L, Zhang J, Robson SC, Shapiro NI, Junger WG (2016) Mitochondrial dysfunction, depleted purinergic signaling, and defective T cell vigilance and immune defense. J Infect Dis 213:456–464
71. Ollila J, Vihinen M (2005) B cells. Int J Biochem Cell Biol 37:518–523
72. Desai BN, Leitinger N (2014) Purinergic and calcium signaling in macrophage function and plasticity. Front Immunol 5:580
73. Ferrari D, Gorini S, Callegari G, la Sala A (2007) Shaping immune responses through the activation of dendritic cells' P2 receptors. Purinergic Signal 3:99–107
74. Ferrari D, Idzko M, Dichmann S, Purlis D, Virchow C, Norgauer J, Chiozzi P, Di Virgilio F, Luttmann W (2000) P2 purinergic receptors of human eosinophils: characterization and coupling to oxygen radical production. FEBS Lett 486:217–224
75. Kobayashi T, Soma T, Noguchi T, Nakagome K, Nakamoto H, Kita H, Nagata M (2015) ATP drives eosinophil effector responses through P2 purinergic receptors. Allergol Int 64(Suppl): S30–S36

76. Bulanova E, Bulfone-Paus S (2010) P2 receptor-mediated signaling in mast cell biology. Purinergic Signal 6:3–17
77. Gao ZG, Jacobson KA (2017) Purinergic signaling in mast cell degranulation and asthma. Front Pharmacol 8:947
78. Gorini S, Callegari G, Romagnoli G, Mammi C, Mavilio D, Rosano G, Fini M, Di Virgilio F, Gulinelli S, Falzoni S et al (2010) ATP secreted by endothelial cells blocks CX(3)CL 1-elicited natural killer cell chemotaxis and cytotoxicity via P2Y(1)(1) receptor activation. Blood 116: 4492–4500
79. Wang L, Jacobsen SE, Bengtsson A, Erlinge D (2004) P2 receptor mRNA expression profiles in human lymphocytes, monocytes and CD34+ stem and progenitor cells. BMC Immunol 5:16

Top Med Chem (2023) 41: 331–344
https://doi.org/10.1007/7355_2022_157

Published online: 11 February 2023

CD73 Inhibitors as Antitumor Agents

Clemente Capasso, Alessio Nocentini, and Claudiu T. Supuran

Contents

Abstract 5′-Ectonucleotidase (e5NT), a metalloenzyme incorporating two zinc ions in its structure, also known as CD73, hydrolyzes AMP to phosphate and adenosine. It is involved in several physiological and pathological processes, among which immune homeostasis, inflammation, tumor progression. The last decade saw the development of small molecule and monoclonal antibodies (MAbs) acting as CD73 inhibitors in the search of novel immunotherapeutic agents for the management of many tumor types. CD73 X-ray crystal structure, alone as well as in complex with a non-hydrolyzable adenosine diphosphate (ADP) analog, adenosine (α,β)-methylene diphosphate (AMPCP), but also with other types of inhibitors, has been reported. This has been useful for the design of effective nucleotide/nucleoside-based CD73 inhibitors. Medicinal chemistry campaigns based on the identified leads thereafter afforded many other such inhibitors. The main strategies for obtaining them included the modifications of the nucleobase,

C. Capasso (✉)
Department of Biology, Agriculture and Food Sciences, CNR, Institute of Biosciences and Bioresources, Naples, Italy
e-mail: clemente.capasso@ibbr.cnr.it

A. Nocentini and C. T. Supuran
NEUROFARBA Department, Pharmaceutical and Nutraceutical Section, University of Florence, Florence, Italy

changes at the sugar moiety as well as to zinc-binding groups of the lead compounds. Non-nucleotide CD73 inhibitors were also reported, belonging to the following chemical classes: flavones, anthraquinone sulfonates, primary sulfonamides, etc. One of the most effective in vitro small molecule CD73 inhibitors, known as AB680, is nowadays in clinical trials as an immunotherapeutic, antitumor agent with a new mechanism of action.

Keywords 5′-Ectonucleotidase, Adenosine, AMP, Baicalin, CD73, Nucleotide, Phosphonate

1 Introduction

CD73 (E.C. 3.1.3.5), also known as ecto-5′-nucleotidase (e5NT), is an enzyme which catalyzes the hydrolysis of phosphoric acid esters incorporating nucleobases to the corresponding nucleoside and inorganic phosphate, as shown in the below equation [1, 2]:

$$\text{Nucleoside} - 5' - \text{phosphate} + H_2O \rightarrow \text{nucleoside} + \text{inorganic phosphate (Pi)}.$$

This enzyme also hydrolyzes other substrates, among which 5′-dinucleotide-, 5′-trinucleotide-, and complex, mixed nucleotides such as uridine diphosphate glucose (UDP-glucose), flavin adenine dinucleotide (FAD), and others [2]. CD73 seems to represent an important pharmacological target, due to its role in regulating adenosine (A) concentrations outside the cell, by hydrolyzing AMP to A and Pi [2–4]. Many cellular responses are regulated through the changes in extracellular nucleotide concentrations, which in turn act on their specific receptors, most of which are G-protein coupled receptors (GPCR). In the case of A, there are at least 4 types of such GPCR receptors, the A_1, A_{2A}, A_{2B}, and A_3 adenosine receptors [1, 2].

Balanced concentrations of ATP, ADP, AMP, and adenosine are essential for controlling inflammatory reactions and as thus tissue damage, and for this reason the regulation of enzymes which produce and/or degrade these molecules is a tightly controlled process [5, 6]. For example, the two isotypes of P_2 purinergic receptors P_{2X}Rs and P_{2Y}rs mediate inflammatory responses via regulation of ATP levels. Indeed, ATP undergoes enzymatic hydrolysis mediated by two ecto-5′--nucleotidases: CD39 converts ATP to ADP, whereas as mentioned above CD73 converts AMP to A (and inorganic phosphate, see Scheme 1). The adenosine produced in this way binds to immune cells rich in adenosine receptor isotypes A_{2A} and A_{2B}, which stimulates potent anti-inflammatory and immunosuppressive responses [7–9].

Considering these premises related to the involvement of A and its precursors in inflammation and cancer, most CD73 inhibitor research has been concentrated on

Scheme 1 Hydrolysis of AMP to adenosine and inorganic phosphate, catalyzed by the ecto-5′-nucleotidase CD73

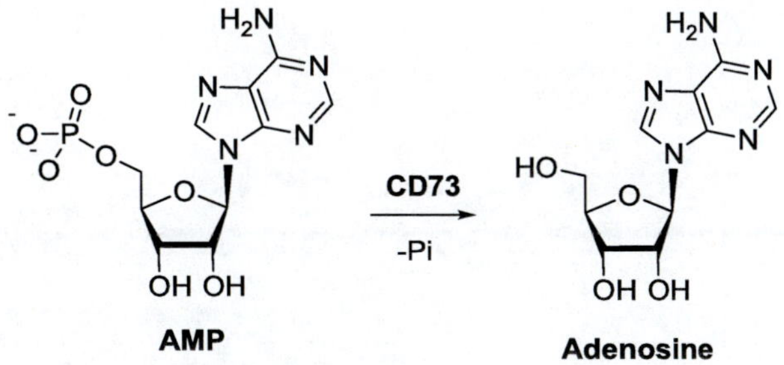

these two fields. It is now ascertained that in hypoxic environment, pH dysregulation, metabolic changes, overexpression of many proteins normally not found in normal cells induce inflammation and other pathological/biochemical events that facilitate tumor progression. Indeed, overexpression and/or upregulation of CD73 levels was often associated with an increased expression of A_{2B} receptors as well as adenosine levels on the surface of many types of cancer cells [4, 8, 10], which leads to an immunosuppressive tumor microenvironment. Thus, the suppression/disruption of CD73 enzymatic activity is thought to mediate resistance to cancer and eventually inhibit tumor cell growth, with profound antitumor effects, by a mechanism of action poorly exploited to date to obtain anticancer agents [11–14].

2 CD73 Structure and Functions

e5NTs are widely expressed in prokaryotes and eukaryotes, with a large number of representatives present in bacteria, plants, and animals. The metal ion(s) present in their active sites are Mg^{2+} or Zn^{2+} and they represent cofactors used in the catalytic process. Often, a dinuclear metal center is required for the hydrolysis of phosphate esters mediated by these enzymes (Fig. 1). The animals e5NTs are classified into two classes: the cytosolic and extracellular membrane-bound forms, which are structurally not related [8]. The cytosolic forms are crucial for the nucleotides metabolism [1, 2], whereas the membrane-bound enzymes are primarily involved in the synthesis of adenosine through the extracellular hydrolysis of AMP, but they are also implicated in transmembrane signaling, cell–matrix as well as cell–cell interactions [2].

The human CD73 (hCD73) is a transmembrane, membrane-anchored protein via glycosylphosphatidylinositol (GPI) tails to the cell membrane and is primarily expressed on lymphocytes and endothelial/epithelial cells [5, 6]. The catalytic domain with the metallic center is found within the lipid bilayer region and faces the extracellular environment [2]. Intriguingly, the extracellular portion of this enzyme was also characterized in soluble form after cleavage of the GPI tail [2]. The enzyme preferentially cleaves nucleoside 5′-monophosphates as substrates, with AMP being its best substrate (K_M in the low micromolar range), whereas ATP

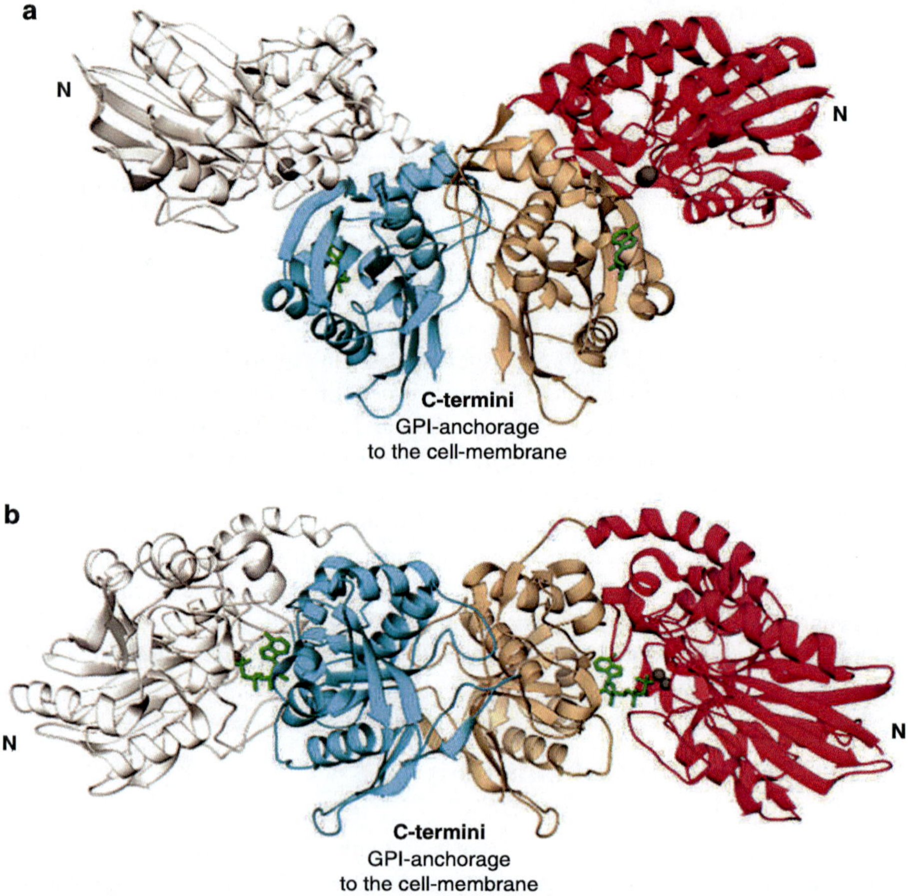

Fig. 1 Crystal structure of e5NT of human origin, in: (**a**) an open conformation (PDB 4H2G) butterfly-like, and (**b**) closed conformation (PDB 4H2I) form. The *N*- and *C*-terminus domains of one subunit of the dimer are shown in gray and sky blue, respectively [15]. The *N*- and *C*-terminus domains of the other subunit are shown in magenta and brown, respectively. The ligands adenosine and adenosine (α,β)-methylene diphosphate (AMPCP, 1) in the open and closed forms, respectively, are shown as green sticks while the metal ions are shown as gray spheres

and ADP are not good substrates, but act as rather effective competitive inhibitors with AMP as substrate [7, 8].

2.1 CD73 Structure

The X-ray crystal structure of the soluble form of hCD73 has been reported, with a resolution of 2.2 Å. The protein is a homodimer, being composed of two identical 70 kDa polypeptide chains. They are connected by a flexible stretch of oligopeptides, which is important for allowing the enzyme to assume two very distinct conformations: a close one and an open one, making the overall structure to look like a butterfly, which is crucial for catalysis, but also for the binding of substrates and inhibitors [12]. The monomers composing hCD73 are each formed by two distinct domains: the *N*-terminal domain, involved in the binding of the two divalent cations essential for catalysis, and the *C*-terminal domain, responsible for the interaction with the nucleotide(s) substrate [12]. hCD73 seems to be a zinc-dependent enzyme, and the two cations are involved in the catalytic mechanism of this enzyme. It has been demonstrated that by adding Zn^{2+} salts to a solution of apo-hCD73 leads to reconstitution of the enzyme activity as ecto-5′-nucleotidase [12].

A disulfide bridge between two cysteine residues, Cys353 and Cys358, which are located in a loop area is involved in the dimerization and formation of the quaternary structure of hCD73 [12]. Mutations of these two Cys residues lead to the loss of structural integrity and result in an enzyme without catalytic activity as ecto-5′--nucleotidase [12, 13]. The enzyme hydrolytic activity only occurs when the protein adopts its close conformation, which places the nucleotide substrate within the middle of the *N*- and *C*-terminal domains constituting the active site. In this way, the phosphate moiety which will be hydrolyzed is placed nearby the catalytic zinc ions and the nucleophiles associated with them. Thus, the ecto-5′-nucleotidase active site is formed at the intersection of these two domains. One should mention that hCD73 is a highly glycosylated enzyme, which incorporates various oligosaccharide moieties on diverse amino acid residues of the protein [12].

2.2 *CD73 Catalytic Mechanism*

CD73 catalyzes the hydrolysis of monophosphate nucleoside substrates. The catalytic mechanism of this process has been proposed by Knöfel and Sträter [14]. The first step consists of a nucleophilic attack of a hydroxyl moiety bound to one of the two Zn^{2+} metal ions present in the active site, on the trigonal bipyramidal phosphorus of the substrate, which releases the nucleoside and leads to the formation of Pi [12] (Fig. 2). hCD73 contains two zinc ions, but the first crystallographic studies were performed on an ortholog e5NT enzyme from *Escherichia coli*. Those experiments were performed in the presence of Mn(II) salts, which in fact have been observed bound to the protein [14, 15]. Thus, it is probable that the two metal ions (M1 and M2) present in the active site are coordinated in either a trigonal bipyramidal geometry (as for Zn^{2+}) or in an octahedral geometry (as for Mn^{2+}), as illustrated in Fig. 2. In the non-inhibited enzyme, M1 is coordinated by residues Asp36, His38, Asp85 (which acts as bridging ligand between two metal ions), a hydroxide anion (W1, also acting as a bridging ligand), and two water molecules (W2 and W3) [15]. M2 was observed to be coordinated by Asn117, His220, His243, as well as the bridging ligands mentioned above (W1 and Asp85). In e5NT adducts with the non-hydrolyzable ADP analog adenosine (α,β)-methylene diphosphate

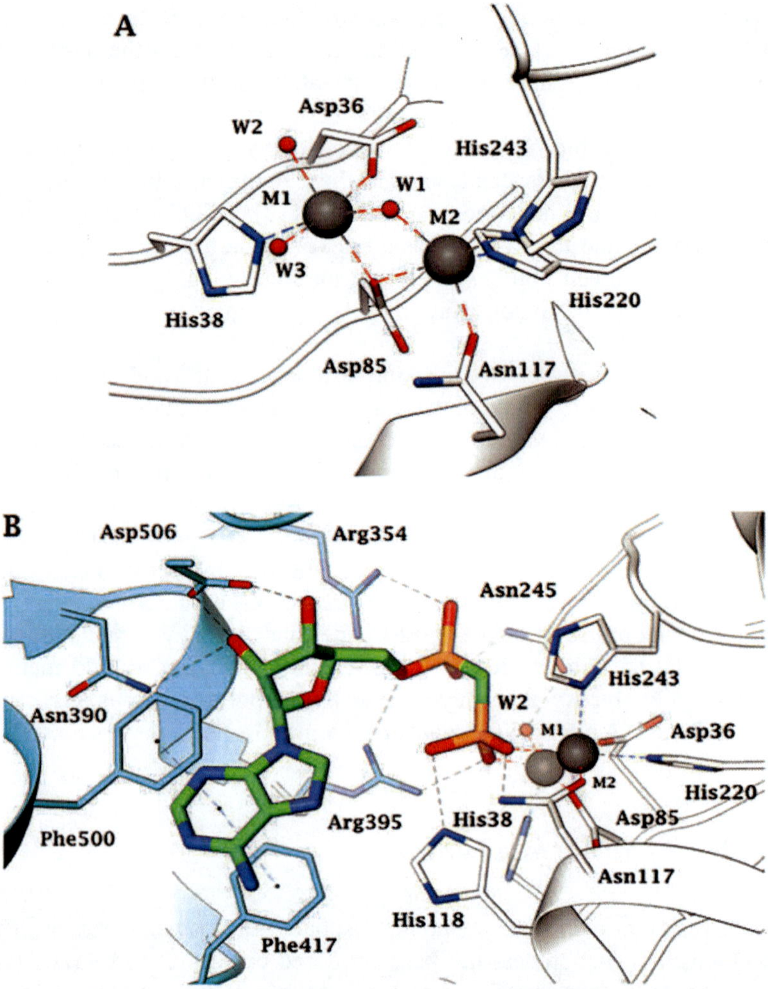

Fig. 2 (**a**) Active site view of apo-e5NT open conformation. (**b**) View of the ADP analog AMPCP (**1**) bound to the active site of e5NT in the closed conformation [15]. The two zinc ions are shown as gray spheres. The protein residues of the *C*-terminal domain forming the substrate binding site are shown in blue whereas the amino acids of the *N*-terminal domain are depicted in gray. Water molecules (small red spheres), unless coordinating metal ions, are omitted for clarity. H-bonds and π-π interactions are depicted as black and blue dashed lines, respectively

(AMPCP), investigated by means of crystallographic experiments, the terminal phosphonate group of the ligand was seen bidentately bridged to the two metal ions, thus replacing W1 [14, 15].

A catalytic dyad conserved in all e5NTs investigated so far, composed of residues His118 and Asp121, is probably involved in catalysis too, as it stabilizes the P (V) intermediate in its trigonal bipyramidal geometry, during the hydrolysis. What the attacking nucleophile is, it is also a matter of debate, but most probably it is the water molecule/hydroxide ion coordinated to one of the metal ions, M1, as it is placed in a favorable geometry for attacking the scissile phosphate bond [14, 15]. Positively charged residues such as Arg354 and Arg395 seem to also take part in the catalytic process, and they clearly stabilize the transition state during the hydrolysis of the phosphate [14, 15].

3 Nucleoside/Nucleotide and Non-nucleotide CD73 Inhibitors

3.1 *Nucleoside/Nucleotide CD73 Inhibitors*

Muller's group reported most of the drug design studies of nucleotide- and nucleoside-hCD73 inhibitors [15–17]. ADP may serve as a substrate for other enzymes involved in purinergic signaling, and for this reason, a non-hydrolyzable derivative of it, designated as AMPCP (compound **1**, Fig. 3) constituted the starting point for designing hCD73 inhibitors, as this compound has an interesting inhibition constant, $K_I = 0.87$ μM [7, 15–19]. In this lead compound, the bridging oxygen present in the diphosphate group of ADP has been replaced by a methylene moiety, which represents a classical isosteric replacement on medicinal chemistry [15–19]. X-ray crystal structures of AMPCP bound to hCD73 afforded crucial structural information for the design of hCD73 inhibitors [15]. All fragments of this compound were observed to be important for inhibition of the enzyme: (1) the terminal phosphonate moiety is anchored to both metal ions from the active site. As thus, they act as zinc-binding groups (ZBGs), a case which is frequently observed in many other zinc-enzyme inhibitors [20]; (2) the sugar moiety and the phosphate group near the sugar participate in hydrophobic interactions and hydrogen bonds with the following residues: Arg354, Arg395, and Asp506 (Fig. 2); (3) the purine ring of the inhibitor was observed to participate in π-stacking interactions which involve two phenyl aromatic rings from the enzyme active site of Phe417 and Phe500. They participate into a very effective π-stacking with the heterocyclic aromatic ring of the inhibitor. Thus, many drug design studies were focused on analogs of compound 1 (Fig. 3), in which all these three structural elements mentioned above were changed: the ZBG group, the sugar moiety, and the heterocyclic moiety [7]. The X-ray crystal structures of the close form of hCD73 with two different derivatives (**2** and **3**) bound within the active site were again reported by Muller's group [7]. Compound **2** incorporates a hydrazinyl functionality at the purine ring and acts as a low nM inhibitor (K_I of 15.5 nM), whereas derivative **3**, with a bulkier substitution pattern (piperazine instead of the hydrazine group), was 10 times a less efficient inhibitor (K_I of 184 nM) compared to **2**, presumably due to steric clash involving the piperazine functionality **2**. Other interactions between different inhibitors and residues from hCD73 active site were observed to be similar to those

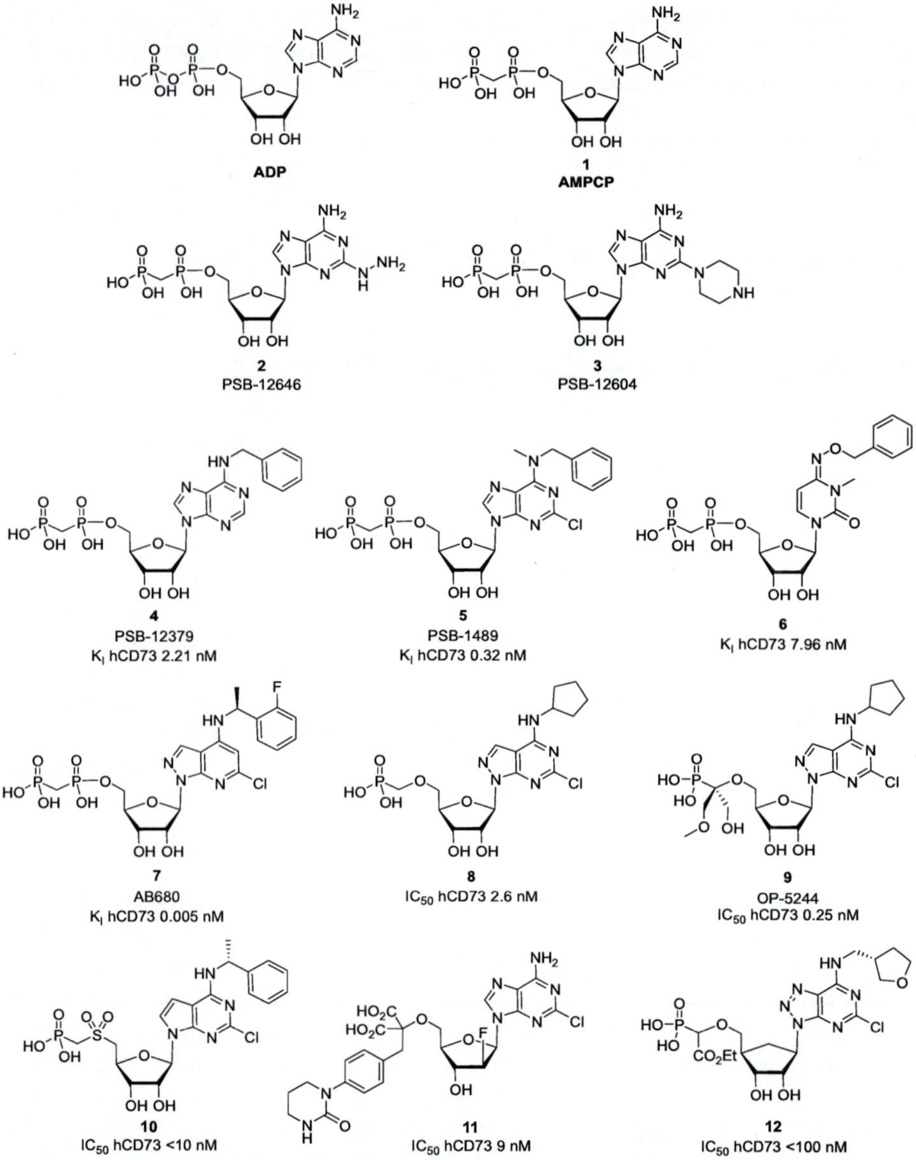

Fig. 3 Examples of compounds designed as hCD73 inhibitors considering ADP AMPCP, **1** and PSB-12646 **2**/PSB-12604 **3** as leads. The CD73 inhibitors **4-12** incorporate nucleoside/nucleotide scaffolds

mentioned for the AMPCP **1** adduct, providing thus an explanation for the favorable inhibitory profile reported for derivatives of this class.

In previous works [7, 18, 19], Muller's group demonstrated that AMPCP structure may be used as lead and a large number of derivatives, among which compounds **4-12** (Fig. 3) show significant inhibitory activity against the enzyme. Many of these hCD73 inhibitors were highly efficient, with inhibition constants in the subnanomolar–low nanomolar range [18, 19].

3.2 Non-nucleotide Inhibitors

X-ray crystallography has been extremely useful also in the process of identifying non-nucleotide hCD73 inhibitors, with Muller's group accomplishing the most significant achievements [15]. As for the previously mentioned hCD73 inhibitors, the effectiveness of the binding is governed by the π-stacking interactions between aromatic ring(s) of the inhibitor and the phenyl moieties of Phe500 and Phe417 from the enzyme active site. Furthermore, the endocyclic oxygen and the OH of the ribose moiety were observed to participate in H-bonds with residues Arg395, Arg354, Asn390, and Asp506, as illustrated in Fig. 4. However, most of these inhibitors seem to lack a well-defined ZBG, which leads to weaker inhibition compared to the nucleoside/nucleotide derivatives, which possess the phosphate as a ZBG.

The natural product baicalin **13** [21] (Fig. 4), a flavone derivative acting as hCD73 inhibitor, was crystallized when bound to the open conformation of hCD73 (Fig. 5) [15]. Baicalin was observed bound in the same active site region as adenosine, with its benzopyrone ring interacting with Phe417 and Phe500 by means of π stacking, similar to the purine ring present in adenosine as discussed above. The glucuronic acid moiety from baicalin is oriented toward the same part of the active site as the ribose of adenosine, although the two moieties are rather different. Indeed, in the case of baicalin the sugar moiety is bulkier than the ribose from adenosine, and thus, it is protruding toward the metal ions, without interacting with any of the two zinc ions from the enzyme active site.

Another interesting class of hCD73 inhibitors is constituted by the anthraquinone sulfonates [16]. Compounds such as 1-amino-4-[4-fluoro-2-carboxyphenylamino]-9,10-dioxo-9,10-dihydroanthracene-2-sulfonate (**16a**, PSB-0952, K_I of 260 nM) and

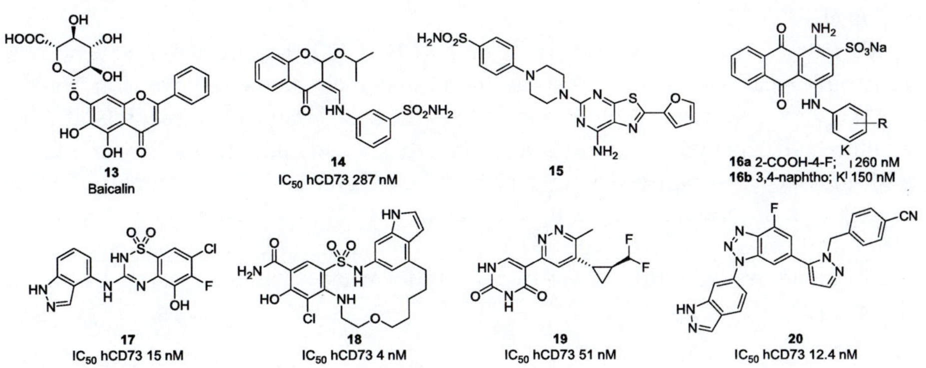

Fig. 4 Non-nucleoside hCD73 inhibitors **13-20**

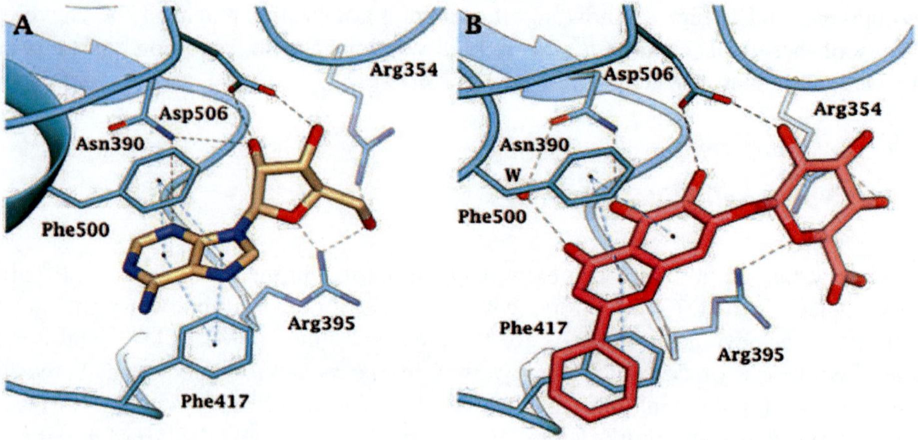

Fig. 5 The open conformation e5NT, its substrate binding site view in adduct with (**a**) adenosine (PDB 4H2G) and (**b**) baicalin (PDB 4H2B) [15]. The H-bonds and π-π interactions involving the inhibitor and active site amino acid residues are shown as black and blue dashed lines, respectively

1-amino-4-[2-anthracenylamino]-9,10-dioxo-9,10-dihydroanthracene-2-sulfonate (**16b**, PSB-0963, K_I of 150 nM) show highly effective hCD73 inhibition among all non-nucleotide inhibitors described so far [16].

3.2.1 Monoclonal Antibodies (MAbs) as hCD73 Inhibitors

Antibodies bind to antigens on a target cell triggering cell-membrane destruction, blocking cell growth, preventing blood vessel growth, blocking the immune response, or directly attacking abnormal cancer cells [22]. Antibodies used to treat cancer recognize surface-bound antigens, including enzymes, such as CD73. They may also neutralize enzymes by inhibiting their catalytic activity or interfering with their signaling role in tumor cells. CD73 inhibition by MAbs was demonstrated to induce apoptosis of the tumor cells, which may constitute an anticancer treatment approach.

Various MAbs were developed against hCD73 [23] and, recently, several new such biological agents, e.g., Ab001/Ab002 and their humanized version Hu001/Hu002 were reported [24]. These MAbs showed high CD73 binding affinity, inhibited the enzyme activity, and in vivo, also inhibited the growth of RAS-mutant non-small cell lung cancer (NSCLC) tumors, in an animal model of cancer [24]. Several antibody-drug conjugates (ADCs) of the two humanized MAbs mentioned above were also reported and showed enhanced efficacy compared to the agents alone, providing the proof of concept for using such therapeutics in clinical settings [24].

4 Conclusions

hCD73 inhibitor research is focused on inflammation and cancer. Several clinical trials using anti-CD73 monoclonal antibodies, such as Oleclumab (MEDI9447), are currently being conducted for the treatment of solid tumors (pancreatic, breast, prostate, NSLCC, etc.), either alone or in combination with other anticancer agents [25].

However, mAbs have low penetrability in solid tumors and their pharmacokinetics are rather complex, which prompted research for the identification of small molecule compounds acting as hCD73 inhibitors. Such compounds might be more resilient for anticancer drug development for several reasons, including the ease of administration, lack of allergic reactions, etc. Indeed, the small molecule CD73 inhibitor AB680 [26] (compound **7**, $K_I = 0.005$ nM Fig. 3) recently entered Phase I clinical trials as an antitumor agent [27]. This potent, competitive h CD73 inhibitor is possibly the first of many derivatives that will be studied in clinical contexts in the next years. It should be noted that CD73 is a hypoxia-inducible factor (HIF) regulated enzyme, and targeting it may be particularly relevant for a range of malignancies associated with hypoxia and extracellular acidification [28, 29]. Thus, a combination of CD73 inhibitors with carbonic anhydrase (CA) IX/XII inhibitors may significantly improve cancer immunotherapy [30–33]. In general, the most recent breakthroughs in the CD73 inhibitors are particularly relevant for the development of cutting-edge cancer treatments and therapies.

Compliance with Ethical Standards **Conflict of Interest**: The authors declare that they have no conflict of interest.

Funding: Original research of our team is funded by the MIUR (Italian Ministry for University and Research), projects FISR2019_04819 (BacCAD) and by Ente Cassa di Risparmio di Firenze (ECRF), grant CRF2020.1395.

Ethical Approval: This chapter does not contain any studies with human participants or animals performed by the authors.

References

1. Bogan KL, Brenner C (2010) 5'-Nucleotidases and their new roles in NAD(+) and phosphate metabolism. New J Chem 34:845–853
2. Strater N (2006) Ecto-5′-nucleotidase: structure function relationships. Purinergic Signal 2: 343–350
3. Zimmermann H (1992) 5'-Nucleotidase: molecular structure and functional aspects. Biochem J 285(Pt 2):345–365
4. Zhang B (2010) CD73: a novel target for cancer immunotherapy. Cancer Res 70:6407–6411
5. Zimmermann H, Zebisch M, Strater N (2012) Cellular function and molecular structure of ecto-nucleotidases. Purinergic Signal 8:437–502
6. Burnstock G (2007) Purine and pyrimidine receptors. Cell Mol Life Sci 64:1471–1483
7. Bhattarai S, Pippel J, Scaletti E, Idris R, Freundlieb M, Rolshoven G, Renn C, Lee SY, Abdelrahman A, Zimmermann H, El-Tayeb A, Muller CE, Strater N (2020) 2-Substituted alpha,beta-methylene-ADP derivatives: potent competitive ecto-5′-nucleotidase (CD73) inhibitors with variable binding modes. J Med Chem 63:2941–2957

8. Burnstock G, Verkhratsky A (2010) Long-term (trophic) purinergic signalling: purinoceptors control cell proliferation, differentiation and death. Cell Death Dis 1:e9
9. Antonioli L, Fornai M, Blandizzi C, Pacher P, Hasko G (2019) Adenosine signaling and the immune system: when a lot could be too much. Immunol Lett 205:9–15
10. Stagg J, Divisekera U, McLaughlin N, Sharkey J, Pommey S, Denoyer D, Dwyer KM, Smyth MJ (2010) Anti-CD73 antibody therapy inhibits breast tumor growth and metastasis. Proc Natl Acad Sci U S A 107:1547–1552
11. Vijayan D, Young A, Teng MWL, Smyth MJ (2017) Targeting immunosuppressive adenosine in cancer. Nat Rev Cancer 17:765
12. Heuts DP, Weissenborn MJ, Olkhov RV, Shaw AM, Gummadova J, Levy C, Scrutton NS (2012) Crystal structure of a soluble form of human CD73 with ecto-5′-nucleotidase activity. Chembiochem 13:2384–2391
13. St Hilaire C, Ziegler SG, Markello TC, Brusco A, Groden C, Gill F, Carlson-Donohoe H, Lederman RJ, Chen MY, Yang D, Siegenthaler MP, Arduino C, Mancini C, Freudenthal B, Stanescu HC, Zdebik AA, Chaganti RK, Nussbaum RL, Kleta R, Gahl WA, Boehm M (2011) NT5E mutations and arterial calcifications. New Engl J Med 364:432–442
14. Knofel T, Strater N (2001) Mechanism of hydrolysis of phosphate esters by the dimetal center of 5′-nucleotidase based on crystal structures. J Mol Biol 309:239–254
15. Knapp K, Zebisch M, Pippel J, El-Tayeb A, Muller CE, Strater N (2012) Crystal structure of the human Ecto-5 '-Nucleotidase (CD73): insights into the regulation of purinergic signaling. Structure 20:2161–2173
16. Baqi Y, Lee SY, Iqbal J, Ripphausen P, Lehr A, Scheiff AB, Zimmermann H, Bajorath J, Muller CE (2010) Development of potent and selective inhibitors of ecto-5'-nucleotidase based on an anthraquinone scaffold. J Med Chem 53:2076–2086
17. Nocentini A, Capasso C, Supuran CT (2021) Small-molecule CD73 inhibitors for the immunotherapy of cancer: a patent and literature review (2017-present). Expert Opin Ther Pat 31: 867–876
18. Bhattarai S, Freundlieb M, Pippel J, Meyer A, Abdelrahman A, Fiene A, Lee SY, Zimmermann H, Yegutkin GG, Strater N, El-Tayeb A, Muller CE (2015) α,β-Methylene-ADP (AOPCP) derivatives and analogues: development of potent and selective ecto-5'-nucleotidase (CD73) inhibitors. J Med Chem 58:6248–6263
19. Junker A, Renn C, Dobelmann C, Namasivayam V, Jain S, Losenkova K, Irjala H, Duca S, Balasubramanian R, Chakraborty S, Borgel F, Zimmermann H, Yegutkin GG, Muller CE, Jacobson KA (2019) Structure-activity relationship of purine and pyrimidine nucleotides as ecto-5'-nucleotidase (CD73) inhibitors. J Med Chem 62:3677–3695
20. Nocentini A, Angeli A, Carta F, Winum JY, Zalubovskis R, Carradori S, Capasso C, Donald WA, Supuran CT (2021) Reconsidering anion inhibitors in the general context of drug design studies of modulators of activity of the classical enzyme carbonic anhydrase. J Enzyme Inhib Med Chem 36:561–580
21. Atanasov AG, Zotchev SB, Dirsch VM, Supuran CT, Taskforce INPS (2021) Natural products in drug discovery: advances and opportunities. Nat Rev Drug Discov 20:200–216
22. Zahavi D, Weiner L (2020) Monoclonal antibodies in cancer therapy. Antibodies:9
23. Al-Rashida M, Qazi SU, Batool N, Hameed A, Iqbal J (2017) Ectonucleotidase inhibitors: a patent review (2011-2016). Expert Opin Ther Pat 27:1291–1304
24. Jin R, Liu L, Xing Y, Meng T, Ma LP, Pei JP, Cong Y, Zhang XS, Ren ZQ, Wang X, Shen JK, Yu K (2020) Dual mechanisms of novel CD73-targeted antibody and antibody-drug conjugate in inhibiting lung tumor growth and promoting antitumor immune-effector function. Mol Cancer Ther 19:2340–2352
25. Geoghegan JC, Diedrich G, Lu X, Rosenthal K, Sachsenmeier KF, Wu H, Dall'Acqua WF, Damschroder MM (2016) Inhibition of CD73 AMP hydrolysis by a therapeutic antibody with a dual, non-competitive mechanism of action. MAbs 8:454–467
26. Jeffrey JL, Lawson KV, Powers JP (2020) Targeting metabolism of extracellular nucleotides via inhibition of ectonucleotidases CD73 and CD39. J Med Chem 63:13444–13465

27. Bowman CE, da Silva RG, Pham A, Young SW (2019) An exceptionally potent inhibitor of human CD73. Biochemistry 58:3331–3334
28. Giatromanolaki A, Kouroupi M, Pouliliou S, Mitrakas A, Hasan F, Pappa A, Koukourakis MI (2020) Ectonucleotidase CD73 and CD39 expression in non-small cell lung cancer relates to hypoxia and immunosuppressive pathways. Life Sci 259
29. Petruk N, Tuominen S, Akerfelt M, Mattsson J, Sandholm J, Nees M, Yegutkin GG, Jukkola A, Tuomela J, Selander KS (2021) CD73 facilitates EMT progression and promotes lung metastases in triple-negative breast cancer. Sci Rep:11
30. McDonald PC, Chia S, Bedard PL, Chu Q, Lyle M, Tang LR, Singh M, Zhang ZH, Supuran CT, Renouf DJ, Dedhar S (2020) A phase 1 study of SLC-0111, a novel inhibitor of carbonic anhydrase IX, in patients with advanced solid Tumors. Am J Clin Oncol-Canc 43:484–490
31. Angeli A, Carta F, Nocentini A, Winum JY, Zalubovskis R, Akdemir A, Onnis V, Eldehna WM, Capasso C, De Simone G, Monti SM, Carradori S, Donald WA, Dedhar S, Supuran CT (2020) Carbonic anhydrase inhibitors targeting metabolism and tumor microenvironment. Metabolites 10:412
32. Supuran CT (2021) Carbonic anhydrase inhibitors: an update on experimental agents for the treatment and imaging of hypoxic tumors. Expert Opin Investig Drugs 30:1197–1208
33. McDonald PC, Chafe SC, Supuran CT, Dedhar S (2022) Cancer therapeutic targeting of hypoxia induced carbonic anhydrase IX: from bench to bedside. Cancers (Basel) 14:3297

Top Med Chem (2023) 41: 345–354
https://doi.org/10.1007/7355_2023_159

Published online: 17 March 2023

Bacterial Ectonucleotidases: Underexplored Antibacterial Drug Targets

Alessio Nocentini, Clemente Capasso, and Claudiu T. Supuran

Contents

Abstract The bacterial ecto-5′-nucleotidases (5′-NTs) are metalloenzymes that hydrolyze 5′-nucleoside-monophosphates (NMPs) to generate nucleosides and phosphate. These metallophosphoesterases are found in the periplasmic space and cytoplasm or are connected to the bacterial outer membrane. Membrane-bound and periplasmic 5′-NTs break down extracellular nucleotides for microbe feeding requirements, whereas cytosolic 5′-NTs control the bacterial ribo- and deoxyribonucleoside monophosphates and nucleotides levels. Both membrane-associated and periplasmic 5′-NTs have been linked to enhanced microbial virulence and pathogenicity for various pathogenic microorganisms, including *Staphylococcus aureus*, *Vibrio cholerae*, *Streptococcus agalactiae*, *Pseudomonas aeruginosa*, and *Legionella pneumophila*. Although these enzymes are considered as potential targets for developing inhibitors with pharmacological action as antibacterials, few drug design studies have been described in the literature. At the moment, only

A. Nocentini and C. T. Supuran (✉)
Dipartimento Neurofarba, Sezione di Scienze Farmaceutiche e Nutraceutiche, Università Degli Studi di Firenze, Florence, Italy
e-mail: claudiu.supuran@unifi.it

C. Capasso
Department of Biology, Agriculture and Food Sciences, CNR, Institute of Biosciences and Bioresources, Naples, Italy

protein-based products, natural polyphenolic compounds, and 1-amino-4-ar(alk)-ylamino-2-sulfoanthraquinones with low-micromolar IC_{50} values were described as bacterial 5′-NT inhibitors. Thus, this chapter may serve as a warning for the scientific community to find inhibitors of bacterial 5-NTs, which may represent a potential solution to the troublesome issue of antibiotic resistance to presently used anti-infective agents.

Keywords Ecto-5′-nucleotidases, Inhibitors, Metallophosphoesterases, Nucleosides

1 Introduction

Several enzymes are involved in the biosynthesis and breakdown of nucleic acids as well as in the homeostasis of nucleotides (composed of a nucleobase, a five-carbon sugar, and one or more phosphate groups) and nucleosides (composed of a nucleobase and a five-carbon sugar) pool [1, 2]. Among these enzymes, the ecto-5′-nucleotidases (5′-NTs) hydrolyze 5′-nucleoside-monophosphates (NMPs) to generate nucleosides and inorganic phosphate [2] (Scheme 1).

There are two nucleosidases expressed in mammals: the ectonucleoside triphosphate diphosphohydrolase 1, also known as CD39 (EC 3.6.1.5), and the ecto-5′--nucleotidase indicated with the acronym CD73 (EC 3.1.3.5) [1, 2]. CD39 generates adenosine monophosphate (AMP) from the extracellular adenosine triphosphate (ATP) and/or adenosine diphosphate (ADP) by hydrolyzing the terminal phosphate moiety. CD73 produces adenosine (Ado) from the extracellular AMP, also generating inorganic phosphate during the hydrolytic process [3]. Purinergic signaling, which employs the released ATP, ADP, and AMP, as well as other nucleotides into the extracellular space, is triggered by tissue injury, cellular stress, or apoptosis [1, 4]. It plays significant physiological roles in the mammalian central nervous system, including neurotransmission and neuromodulation. Furthermore, purinergic signaling is also correlated to the homeostasis of the immune system [1, 4]. Mammalian 5′-nucleotidases have been studied extensively and thoroughly characterized, whereas the field of bacterial 5′-nucleotidases is still largely unexplored.

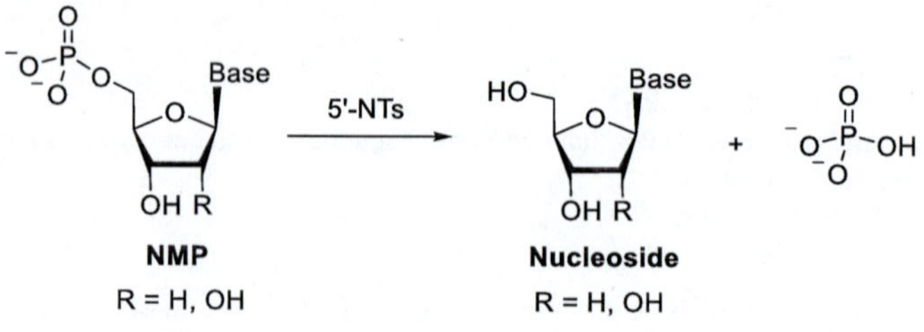

Scheme 1 Reactions catalyzed by 5′-NTs (NMP = nucleoside monophosphate)

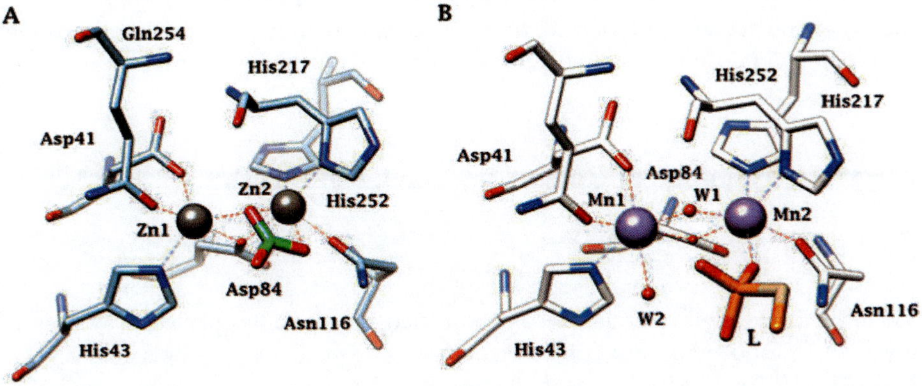

Fig. 1 (**a**) The di-zinc(II) metal ion center reported for *E. coli* e5NT in 1999 [5]. A (bi)carbonate ion (in green) from the dissolved carbon dioxide was observed as a metal ligand; (**b**) the di-manganese(II) metal ion center reported for *E. coli* e5NT in 2001 [6]. The bridging water molecule (W1) acting as nucleophile in the hydrolytic process and the manganese ion ligands are shown in detail, as determined by X-ray crystallography [6]. Only the β-methylene-phosphate group and the α-phosphorus atom of a ligand/inhibitor (L) are shown for clarity (in orange)

Based on similarities in the primary sequence, it is readily apparent that the animal ecto-5′-NTs are related to the bacterial 5′-NTs, which are part of a group of proteins called metallophosphoesterases characterized by the presence of a dinuclear metal-binding center in their active sites (Fig. 1). The metal ion nature present in these enzymes is still rather controversial. The human CD73 enzyme contains two zinc ions, but the *E. coli* ortholog e5NT structure has been obtained with manganese(II) ions bound within the active site [5]. However, the enzyme comprises domains linked by a flexible stretch of amino acids, allowing it to take on a closed or open butterfly-like shape [5, 6]. The N-terminal domain (residues 25–342) binds the two metal ions and contains the catalytic Asp-His dyad. The C-terminal domain (residues 362–550) has a unique structure and binds the substrate [5]. In the closed conformation, the active site is located between the two domains, and the substrate, which is located nearby the catalytically active metal ions (Zn (II) or Mn(II)), is hydrolyzed through the nucleophilic attack of a bridging water molecule activated for catalysis by the metal ions [2, 5, 6]. The amino acid residues coordinating the bimetallic catalytic center in the e5NT from *Escherichia coli* are shown in Fig. 1 and are His, Glu, Gln, and Asn residues [5, 6]. Interestingly, Xf50-Nt, which is the recombinant 5′-NT encoded by the *Xylella fastidiosa* genome, exhibited the highest affinity for Mg(II), followed by Mn(II) > Co(II) > Cu (II) > Ni(II) > Ca(II) > Zn(II) [7].

The 5′-nucleotidases encoded by the bacterial genome are polypeptides anchored to the bacterial outer membrane [8], located in the periplasmic space [9] or in the cellular cytoplasm [10]. Depending on the type of acceptor of the phosphoryl group, which can be an activated water molecule or a nucleophilic amino acid residue, the bacterial 5′NTs are catalytically categorized as a type I or type II, respectively

[11]. *E. coli* 5′-NT hydrolyzes AMP, ADP, ATP, and other 5′-ribo- and 5′- -deoxyribonucleotides [6]. Indeed, the enzyme can hydrolyze the artificial substrate *p*-nitrophenyl phosphate, which is not a substrate for ecto-5′-NT [6].

2 The Physiological Role of the Bacterial 5′NTs

2.1 *Intracellular 5′-NTs*

The bacterial level of ribo- and deoxyribonucleoside monophosphates, as well as all the pool of ribo- and deoxyribonucleotides, is regulated by cytoplasmic 5′-NTs [12]. The 5′-NT activity has been identified in the cytoplasmic soluble fraction of many Gram-positive and Gram-negative bacteria. Only a few of these enzymes have been defined as having a functional role. Among the bacterial species, one well-studied microbe for its 5′-NTs is represented by the Gram-negative opportunistic pathogen, *Escherichia coli*. The cytoplasm of *E. coli* contains four 5′-NTs (UmpH, SurE, YfbR, and YjjG) [13]. UmpH allows the degradation of UMP and is essential for bacterial pyrimidine homeostasis [14]; SurE is involved in the regulation of dNTP and NTP pools, playing a significant physiological role in the stress response and is essential for the survival of cells in the stationary growth phase; YfbR plays a critical role in the deoxycytidine pathway and de novo synthesis of thymidylate; YjjG can dephosphorylate a wide range of non-canonical pyrimidine derivatives, preventing the incorporation of these potentially mutagenic compounds into DNA and RNA.

Moreover, it has been hypothesized that. *E. coli* SuhB, a polypeptide chain that is homologous to mammalian inositol monophosphatase, could also function as 5′- -nucleotidase since it can hydrolyze adenosine 2′-monophosphate, also considering that *E. coli* does not present the phosphatidylinositol [15], which is essential for eukaryotic membrane function, serving as organelle markers and signaling cascades [13]. Recently, five genes encoding for 5′-NTs have been identified in the genome of *Bacillus subtilis* and heterologously expressed in *E. coli.* Surprisingly, the recombinant proteins displayed a 5′-nucleosidase activity with different intensity and substrate specificities [10]. For example, only one of them (YcsE) resulted in a powerful 5′-nucleotidase, having a broader specificity (IMP, inosine monophosphate; AMP, adenosine monophosphate; GMP, guanosine monophosphate; UMP, uridine monophosphate; CMP, cytidine monophosphate; G6P, glucose 6-phosphate; and β-glycerophosphate), while the others had a feeble 5′-nucleotidase activity [10]. Again, some of these enzymes have an essential role in the ability of the organism to withstand oxidative stress, while others are crucial for development and proliferation of the microorganisms on solid media [10].

2.2 *Membrane-Bound and Periplasmic 5′-NTs*

Generally, the membrane-bound and periplasmic 5′-NTs have the physiological function of breaking down the extracellular nucleotides to meet the nutritional needs of the microorganism. DNA degradation provides phosphate, carbon, and nitrogen to the microbes. Thus, membrane-bound and periplasmic 5′-NT activities allow bacterial survival when phosphate is limited, but extracellular DNA is available [16]. As described in the literature, *E. coli* periplasmic 5′NTs are involved in the assimilation of exogenous nucleotides, CDP-alcohols, UDP-sugars, DNA, and other phosphorylated substances that cannot cross the cell membrane [17].

3 Bacterial 5′-NTs as Pharmacological Targets

3.1 *5′-NTs and Bacterial Virulence*

Membrane-associated 5′-NTs are linked to virulence in several pathogenic bacteria since pathogenic microorganisms and their hosts frequently interact via surface-exposed proteins. For example, the bacterial membrane-associated 5′-NT aided bacteria in their ability to infect the host by increasing extracellular quantities of potent immunosuppressive chemical compounds, i.e., decreasing the concentration of the pro-inflammatory ATP and/or increased concentrations of Ado. This observation has been reported for many microbes, such as *Staphylococcus aureus*, *Vibrio cholerae*, *Streptococcus agalactiae*, *Pseudomonas aeruginosa*, and others [18–21].

Periplasmic 5′-nucleotidases can alter the pathogenicity of these bacteria, enhancing the damage they cause when invading host cells [11]. For example, it has been reported that the periplasmic 5′-NT activity disrupts critical intermediates in the host's phospholipid biosynthetic pathways, such as CDP-ethanolamine and CDP-choline. As a result, the synthesis of cell wall components of the host is impaired [22]. It has been shown that *Burkholderia cepacia* and *P. aeruginosa* produce secreted ATP-hydrolyzing enzymes, which via activating purinergic nucleotide receptors interfere with the ATP levels of macrophages and mast cells, causing their demise [23, 24]. In 2013, the recombinant *Xylella fastidiosa* 5′-NT (Xf50 -Nt) was considered an essential enzyme in the biofilm formation of this microorganism. *X. fastidiosa* is a Gram-negative plant microbe responsible for illnesses that affect significant crops all over the world [25]. Its biofilm formation is responsible for the occlusion of the plant xylem vessels, provoking devastating plant disease [7]. In 2015, a novel extracellular nucleotidase (S5nA) has been identified in *Streptococcus pyogenes*, an opportunistic pathogen that causes pharyngitis, impetigo, necrotizing fasciitis, and toxic shock syndrome [26]. S5nA hydrolyzes AMP and ADP, but not ATP, to generate Ado, demonstrating how the conversion of dAMP into dAdo resulted in a potent inhibitor of macrophages and monocytes. Thus, S5nA is being considered as a novel *S. pyogenes* virulence factor able to deregulate the host immune response [26]. However, in 2016, Fiedler's group demonstrated that S5nA was unessential for the growth of S. *pyogenes* in the human blood, as well as for the evasion of phagocytosis by neutrophils, bacterial biofilms formation, and virulence [27].

Intracellular 5′-NTs, as described above, are mainly involved in controlling intracellular nucleotide pools for DNA/RNA synthesis. It has been reported that the deletion of the nagD gene encoding for an intracellular 5′-NT in *Staphylococcus aureus* drastically reduces the virulence of this pathogen [28].

3.2 Bacterial 5′-NT Inhibitors

Already in 1966 and 1967, the first 5′-NT inhibitors were discovered in the cytoplasm of several bacteria [29]. It was demonstrated that in *E. coli* the activity of the 5′-NT-UshA (a surface-located metallophosphoesterase/5′-nucleotidase) was blocked by proteinaceous substances which protect cytoplasmic nucleotides before 5′-NT-UshA secretion [29]. However, the complete characterization of such proteinaceous bacterial 5′-NT inhibitors is a poorly explored field and no new convincing results have been obtained in the last few decades. On the other hand, it has been reported that polyphenolic natural compounds isolated from betel nuts (*Areca catechu*) or wine grapes also possess inhibitory effects against bacterial 5′-NTs [30–32].

Although these enzymes seem to show a relevant potential for the development of inhibitors with pharmacological activity as antibacterials, there are few drug design studies in the literature that considered bacterial 5′-NTs as drug targets. The most interesting one has been reported by Muller's group and targets a 5′-NT from *Legionella pneumophila*, the etiological agent of Legionnaire's disease [33, 34]. The nucleoside triphosphate diphosphohydrolase (NTPDase) from *L. pneumophila*, denominated Lp1NTPDase, was found to be a structural and functional homolog of mammalian such enzymes and to catalyze the hydrolysis of ATP to ADP and ADP to AMP, and its activity contributes to the virulence of this pathogen [33, 34]. Thus, the capillary electrophoresis (CE)-based enzyme assay for studying the Lp1NTPDase activity and inhibition was established by Muller's group, which led to the identification of several compounds belonging to 1-amino-4-ar(alk)ylamino-2-sulfoanthraquinones, possessing IC_{50} values in the low-micromolar range (Fig. 2).

Starting from the structures of two anthraquinone dyes, Reactive Blue **1** and Cibacron Blue 3GA **2**, which were observed to induce a 45–75% inhibition of *L. pneumophila* Lp1NTPDase at 30 μM concentration of compound in the assay system [34], a series of simpler 1-amino-4-ar(alk)ylamino-2-sulfoanthraquinones **3–6** (Fig. 2) were synthesized, which incorporated smaller moieties at the 4-amino functionality, such as alkyls, cycloalkyls, phenyl and substituted phenyls, 2- and 1-naphthyl moieties, or even the phenanthryl one (as in **6**) [34]. The most effective inhibitors of Lp1NTPDase were derivatives **5** and **6**, which showed an inhibition of

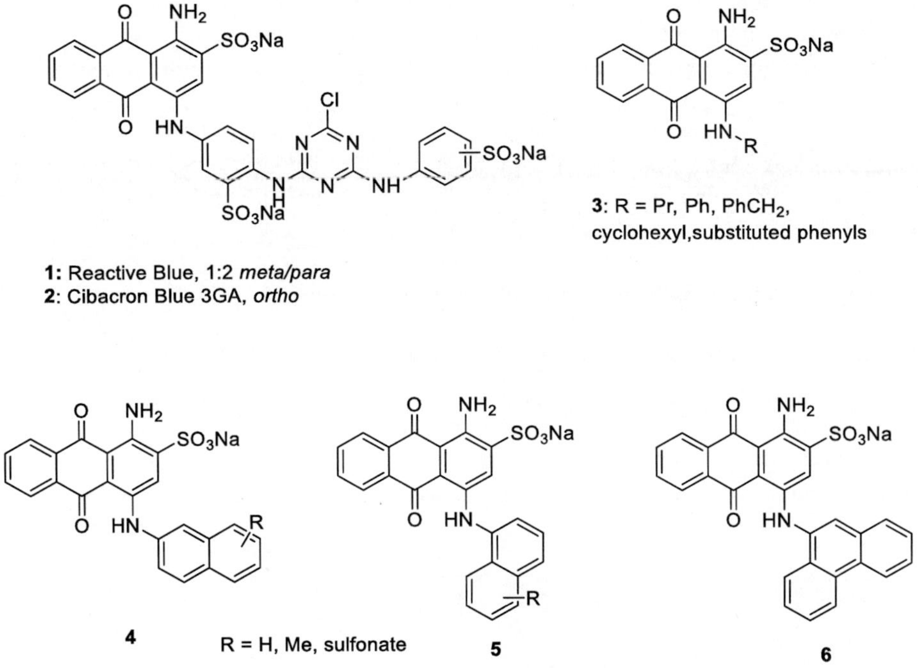

Fig. 2 1-Amino-4-ar(alk)ylamino-2-sulfoanthraquinones of types **1–6** investigated as *L. pneumophila* Lp1NTPDase inhibitors

62–90% at 30 μM concentration of compound in the assay system [34]. Although these derivatives are not exceedingly potent bacterial enzyme inhibitors, they may represent an interesting starting point for the drug design of more effective such inhibitors.

4 Conclusions

The field of bacterial 5′-NTs is still wholly unknown compared to mammalian 5′-nucleotidases, which have been the subject of intensive research and have been characterized in great detail. AMP, ADP, ATP, and other 5′-ribo- and 5′--deoxyribonucleotides are hydrolyzed by bacterial 5′-NTs. Intriguingly, numerous harmful bacterial pathogens seem to have connected such enzymes to their infectiousness and virulence. For example, *Staphylococcus aureus*, *Vibrio cholerae*, *Streptococcus agalactiae*, and *Pseudomonas aeruginos*a membrane-associated 5′-NTs have been connected to microbial virulence since the surface-exposed proteins are typically the medium through which pathogenic microbes and hosts communicate. It has been also demonstrated that the host invasion by *Burkholderia* *cepacia, Pseudomonas aeruginosa, Xylella fastidiosa*, and *Streptococcus pyogenes* is dependent on the activity of the periplasmic 5′-nucleotidases.

Unfortunately, very few inhibitors of the bacterial 5-NTs were reported so far. Among them, it is possible to list the following ones: proteinaceous compounds, natural polyphenolic compounds, and 1-amino-4-ar(alk)ylamino-2-sulfoanthraquinones, showing low-micromolar IC_{50} values. Thus, although these enzymes show potential as antibacterial therapeutic targets, still few drug design studies concerning the bacterial 5′-NTs are available.

Compliance with Ethical Standards Conflict of Interest

The author declares that they have no conflict of interest.

Funding Original research of our team is funded by the MIUR (Italian Ministry for University and Research) projects FISR2019_04819 (BacCAD) and by Ente Cassa di Risparmio di Firenze (ECRF), grant CRF2020.1395.

Ethical Approval This chapter does not contain any studies with human participants or animals performed by the authors.

References

1. Illes P, Xu GY, Tang Y (2020) Purinergic signaling in the central nervous system in health and disease. Neurosci Bull 36:1239–1241
2. Nocentini A, Capasso C, Supuran CT (2021) Small-molecule CD73 inhibitors for the immunotherapy of cancer: a patent and literature review (2017-present). Expert Opin Ther Pat 31: 867–876
3. Camici M, Garcia-Gil M, Tozzi MG (2018) The inside story of adenosine. Int J Mol Sci 19
4. Cekic C, Linden J (2016) Purinergic regulation of the immune system. Nat Rev Immunol 16: 177–192
5. Knöfel T, Sträter N (1999) X-ray structure of the Escherichia coli periplasmic 5′-nucleotidase containing a dimetal catalytic site. Nat Struct Biol 6:448–453
6. Knofel T, Strater N (2001) Mechanism of hydrolysis of phosphate esters by the dimetal center of 5'-nucleotidase based on crystal structures. J Mol Biol 309:239–254
7. Santos CA, Saraiva AM, Toledo MA, Beloti LL, Crucello A, Favaro MT, Horta MA, Santiago AS, Mendes JS, Souza AA, Souza AP (2013) Initial biochemical and functional characterization of a 5′-nucleotidase from Xylella fastidiosa related to the human cytosolic 5′-nucleotidase I. Microb Pathog 59-60:1–6
8. Tamao Y, Noguchi K, Sakai-Tomita Y, Hama H, Shimamoto T, Kanazawa H, Tsuda M, Tsuchiya T (1991) Sequence analysis of nutA gene encoding membrane-bound Cl(−)-dependent 5′-nucleotidase of vibrio parahaemolyticus. J Biochem 109:24–29
9. Burns DM, Beacham IR (1986) Nucleotide sequence and transcriptional analysis of the E. coli ushA gene, encoding periplasmic UDP-sugar hydrolase (5′-nucleotidase): regulation of the ushA gene, and the signal sequence of its encoded protein product. Nucleic Acids Res 14:4325–4342
10. Terakawa A, Natsume A, Okada A, Nishihata S, Kuse J, Tanaka K, Takenaka S, Ishikawa S, Yoshida K (2016) Bacillus subtilis 5'-nucleotidases with various functions and substrate specificities. BMC Microbiol 16
11. Zakataeva NP (2021) Microbial 5'-nucleotidases: their characteristics, roles in cellular metabolism, and possible practical applications. Appl Microbiol Biotechnol 105:7661–7681

12. Proudfoot M, Kuznetsova E, Brown G, Rao NN, Kitagawa M, Mori H, Savchenko A, Yakunin AF (2004) General enzymatic screens identify three new nucleotidases in Escherichia coli – biochemical characterization of SurE, YfbR, and YjjG. J Biol Chem 279:54687–54694
13. Botero S, Chiaroni-Clarke R, Simon SM (2019) Escherichia coli as a platform for the study of phosphoinositide biology. Sci Adv:5
14. Reaves ML, Young BD, Hosios AM, Xu YF, Rabinowitz JD (2013) Pyrimidine homeostasis is accomplished by directed overflow metabolism. Nature 500:237
15. Kates M (1980) Citation classic – bacterial lipids. Life Sci 12-12
16. Pinchuk GE, Ammons C, Culley DE, Li SMW, McLean JS, Romine MF, Nealson KH, Fredrickson JK, Beliaev AS (2008) Utilization of DNA as a sole source of phosphorus, carbon, and energy by Shewanella spp.: ecological and physiological implications for dissimilatory metal reduction. Appl Environ Microbiol 74:1198–1208
17. Kakehi M, Usuda Y, Tabira Y, Sugimoto S (2007) Complete deficiency of 5'-nucleotidase activity in Escherichia coli leads to loss of growth on purine nucleotides but not of their excretion. J Mol Microbiol Biotechnol 13:96–104
18. Thammavongsa V, Kern JW, Missiakas DM, Schneewind O (2009) Staphylococcus aureus synthesizes adenosine to escape host immune responses. J Exp Med 206:2417–2427
19. Punj V, Zaborina O, Dhiman N, Falzari K, Bagdasarian M, Chakrabarty AM (2000) Phagocytic cell killing mediated by secreted cytotoxic factors of vibrio cholerae. Infect Immun 68:4930–4937
20. Firon A, Dinis M, Raynal B, Poyart C, Trieu-Cuot P, Kaminski PA (2014) Extracellular nucleotide catabolism by the group B streptococcus ectonucleotidase NudP increases bacterial survival in blood. J Biol Chem 289:5479–5489
21. Zaborina O, Dhiman N, Ling Chen M, Kostal J, Holder IA, Chakrabarty AM (2000) Secreted products of a nonmucoid Pseudomonas aeruginosa strain induce two modes of macrophage killing: external-ATP-dependent, P2Z-receptor-mediated necrosis and ATP-independent, caspase-mediated apoptosis. Microbiology (Reading) 146(Pt 10):2521–2530
22. Alves-Pereira I, Canales J, Cabezas A, Cordero PM, Costas MJ, Cameselle JC (2008) CDP-alcohol hydrolase, a very efficient activity of the 5'-nucleotidase/UDP-sugar hydrolase encoded by the ushA gene of yersinia intermedia and Escherichia coli. J Bacteriol 190:6153–6161
23. Melnikov A, Zaborina O, Dhiman N, Prabhakar BS, Chakrabarty AM, Hendrickson W (2000) Clinical and environmental isolates of Burkholderia cepacia exhibit differential cytotoxicity towards macrophages and mast cells. Mol Microbiol 36:1481–1493
24. Zaborina O, Misra N, Kostal J, Kamath S, Kapatral V, El-Idrissi ME, Prabhakar BS, Chakrabarty AM (1999) P2Z-independent and P2Z receptor-mediated macrophage killing by Pseudomonas aeruginosa isolated from cystic fibrosis patients. Infect Immun 67:5231–5242
25. Mansfield J, Genin S, Magori S, Citovsky V, Sriariyanum M, Ronald P, Dow M, Verdier V, Beer SV, Machado MA, Toth I, Salmond G, Foster GD (2012) Top 10 plant pathogenic bacteria in molecular plant pathology. Mol Plant Pathol 13:614–629
26. Zheng LS, Khemlani A, Lorenz N, Loh JMS, Langley RJ, Proft T (2015) Streptococcal 5'-Nucleotidase A (S5nA), a novel streptococcus pyogenes virulence factor that facilitates immune evasion. J Biol Chem 290:31126–31137
27. Dangel ML, Dettmann JC, Hasselbarth S, Krogull M, Schakat M, Kreikemeyer B, Fiedler T (2019) The 5'-nucleotidase S5nA is dispensable for evasion of phagocytosis and biofilm formation in streptococcus pyogenes. PloS One:14
28. Begun J, Sifri CD, Goldman S, Calderwood SB, Ausubel FM (2005) Staphylococcus aureus virulence factors identified by using a high-throughput Caenorhabditis elegans-killing model. Infect Immun 73:872–877
29. Zimmermann H (1992) 5'-Nucleotidase – molecular-structure and functional-aspects. Biochem J 285:345–365

30. Iwamoto M, Matsuo T, Uchino K, Tonosaki Y, Fukuchi A (1988) New 5′-nucleotidase inhibitors, NPF-86IA, NPF-86IB, NPF-86IIA, and NPF-86IIB from Areca catechu; Part II anti-tumor effects. Planta Med 54(5):422–425
31. Uchino K, Matsuo T, Iwamoto M, Tonosaki Y, Fukuchi A (1988) New 5′-nucleotidase inhibitors, NPF-86IA, NPF-86IB, NPF-86IIA, and NPF-86IIB from Areca catechu; part I isolation and biological properties. Planta Med 54(5):419–422
32. Toukairin T, Uchino K, Iwamoto M, Murakami S, Tatebayashi T, Ogawara H, Tonosaki Y (1991) New polyphenolic 5′-nucleotidase inhibitors isolated from the wine grape "Koshu" and their biological effects. Chem Pharm Bull(Tokyo) 39(6):1480–1483
33. Sansom FM, Riedmaier P, Newton HJ, Dunstone MA, Müller CE, Stephan H, Byres E, Beddoe T, Rossjohn J, Cowan PJ, d'Apice AJ, Robson SC, Hartland EL (2008) Enzymatic properties of an ecto-nucleoside triphosphate diphosphohydrolase from legionella pneumophila: substrate specificity and requirement for virulence. J Biol Chem 283(19): 12909–12918
34. Fiene A, Baqi Y, Malik EM, Newton P, Li W, Lee SY, Hartland EL, Müller CE (2016) Inhibitors for the bacterial ectonucleotidase Lp1NTPDase from *Legionella pneumophila.* Bioorg Med Chem 24(18):4363–4371